Progress in Physics
Volume 18

Picture on the cover is a simplified version of a figure created by Oliver Conradt (Department of Physics and Astronomy, University of Basel, Klingelbergstrasse 82, CH-4056 Basel, Switzerland). It depicts projective plane coordinates of a pencil of planes. For more information we refer to Conradt's paper *The Principle of Duality in Clifford Algebra and Projective Geometry* in this volume.

Clifford Algebras and their Applications in Mathematical Physics

Volume 1: Algebra and Physics

Rafał Abłamowicz
Bertfried Fauser
Editors

Springer Science+Business Media, LLC

Rafał Abłamowicz
Department of Mathematics
Box 5054
Tennessee Technological University
Cookeville, TN 38505
U.S.A.

Bertfried Fauser
Fachbereich Physik
Fach M678
Universität Konstanz
78457 Konstanz
Germany

Library of Congress Cataloging-in-Publication Data

Clifford algebras and their applications in mathematical physics.
p. cm. – (Progress in physics ; v. 18-19)
Includes bibliographical references and indexes.
Contents: v. 1. Algebra and physics / Rafał Abłamowicz, Bertfried Fauser, editors–v. 2. Clifford analysis / John Ryan, Wolfgang Sprößig, editors.
ISBN 978-1-4612-7116-1 ISBN 978-1-4612-1368-0 (eBook)
DOI 10.1007/978-1-4612-1368-0
1. Clifford algebras. 2. Mathematical physics. I. Abłamowicz, Rafał. II. Progress in physics (Boston, Mass.); v. 18-19.
QC20.7.C55 C55 2000
530.15'257–dc21 00-034310

AMS Subject Classifications: 03G12, 03G25, 08A99, 14A22, 14L35, 14N10, 15A03, 15A18, 15A23, 15A33, 15A63, 15A66, 16W30, 16W50, 17A35, 17B10, 17B37, 17C90, 30C30, 22E60, 32L25, 34A26, 51A25, 51B10, 51B20, 51P05, 53A50, 53B25, 53B30, 53B99, 53C20, 53C50, 53Z05, 54H15, 58A99, 68T15, 70G35, 70H40, 70G99, 78A35, 81E10, 81Q05, 81R25, 81R50, 81R99, 81S05, 81T15, 81T15, 81T75, 81U05, 81U20, 83A05, 83C10, 83E99

Printed on acid-free paper.

ISBN 978-1-4612-7116-1

Typeset by the editors in LATEX.

9 8 7 6 5 4 3 2 1

Contents

Preface to Volume 1

The last conference on "Clifford Algebras and Their Applications in Mathematical Physics," the 5th of this well-known series, took place in Ixtapa-Zihuatanejo, Mexico, from June 27–July 4, 1999, in the beautiful surroundings of the Pacific Coast. The first conference of this series was organized in 1985 in Canterbury, United Kingdom, and was initiated by J.S.R. Chisholm at a time when Clifford algebras were just becoming recognized tools. Under the leadership of D. Hestenes, among others, Clifford algebras had not only entered various fields by providing an elegant and powerful tool for solving geometric problems, but, more importantly, Clifford algebras have provided a unique approach to reasoning in mathematics and physics. As a natural consequence of this development, the conferences of this series have had a large impact and have managed to form a "Clifford community."

The topics covered by the recent conference can be divided into two major parts: Clifford analysis and mathematical physics. This structure is reflected by the division of the presented contributions into two volumes: *Algebra and Physics* (Volume 1), and *Clifford Analysis* (Volume 2). The majority of papers are an outgrowth and further development of the talks given by the contributors at the conference. All papers in these two volumes have been refereed and were further developed after the conference.

There will also be a special issue of the *International Journal of Theoretical Physics,* edited by D. R. Finkelstein and Z. Oziewicz and containing invited papers. This too demonstrates the innovative and flourishing ideas in Clifford algebras.

During the main conference, two special sessions were organized. One was Applied Clifford Algebra in Cybernetics, Robotics, Image Processing, and Engineering (ACACSE), organized by E. J. Bayro-Corrochano and G. Sobczyk; the other was Global and Local Problems for Dirac Operators, organized by E. R. de Arellano, J. Ryan, and W. Sprößig. The goal was to gather people with interests in applications of Clifford algebras in engineering, robotics, computer vision, and symbolic computer algebra, or in the mathematics of Dirac Operators. The ACACSE activities will be presented elsewhere, while the Dirac operator contributions belong to *Clifford Analysis* (Volume 2).

The increasing interest in Clifford algebras has some deep foundations. Geometrical methods have seen a revival of making use of Clifford algebras which have been exactly designed to serve as geometrical algebras, combining the power of geometric intuition with the power of algebraic manipulations – which had been a dream of Leibniz followed by Graßmann, Peano, Hamilton, Cayley, Clifford, Boole, Rota, Hestenes, and others. It is remarkable to note the wealth of contributions to these volumes using conformal, projective, and hyperbolic geometries, not only in algebraic settings but also in Clifford analysis. The advent of these types of geometries during the 19th century gave birth to quaternions, Graßmann and Clifford algebras, and finally, to modern algebraic geometry.

Clifford algebras are used almost everywhere in mathematics and physics. Most problems can be encoded via a pair of a linear space and a quadratic form – some people even assign a value to such objects as line segments, areas, etc. However, this is already sufficient to construct a unique Clifford algebra. Multiplication makes the considered objects behave just like "numbers" and makes them easier to manipulate.

Clifford algebras have produced valuable applications. Besides their fundamental aspects, Clifford algebra, or quaternion, methods have produced well-recognized applications – even if sometimes disguised by matrix representations. Computer vision, robotics, navigation, space flight, and other areas also use these techniques.

Clifford algebras and their accompanying Graßmann-Cayley algebras broaden fields of thought. Automatic theorem proving, which might be important for autonomous robot systems, is based on this connection to Clifford algebras. Pioneered by Gian-Carlo Rota, the idea of connecting Hopf algebras with Clifford structures was also developed. Taking Clifford numbers as entities leads to new physical principles. Deformed Clifford algebras are used to solve problems in quantum field theory. Clifford algebraic computations provide a challenge for computer algebraic systems and open the new and fascinating area of experimental mathematics. Clifford algebras are bound to play a major role in quantum computing and the design of quantum computers.

Most of these currents in the Clifford community have found their way into these volumes. In this way, these books will contribute to the development of the field. The following outline will provide a subjective guide to the contributions (avoiding technical terms as much as possible) and will highlight their main features.

Dedication to Gian-Carlo Rota

Gian-Carlo Rota was invited to be a plenary speaker at this 5th International Conference on Clifford Algebras and Their Application in Mathematical Physics. His talk would have been a highlight of the conference. But he died suddenly, shortly before the conference. For this reason, the organizers of this conference arranged a special session dedicated to the memory of Professor Rota.

Several lecturers including David Ritz Finkelstein and Bernd Schmeikal, who knew Rota personally, took a chance to present some warm remarks about Professor Rota, showing him as a person, scientist, philosopher and a beloved man.

Physics – Applications and Models

Baylis: The linear space underlying a Clifford algebra is coming conventionally equipped with multi-vector gradings. The common use is that physically different entities are mapped onto the particular gradings. The advantage of this convention is that physical transformations become inner automorphisms while the disadvantage is the introduction of possibly nonphysical or unnecessary variables. The present paper studies paravectors, which are sums of scalars and vectors and constitute themselves a graded space of a coarser grading than that of the multi-vector grading. Some mathematical outcomes are explained, and the usefulness of the paravector picture is demonstrated for plane waves and wave packets in electrodynamics. A paravector arises from a spacetime split, as discussed by Conradt.

Dray & Manogue: Two component Weyl spinors, also extensively used by Penrose, and four component Dirac spinors are widely known in mathematics and high energy physics. Usually the latter are constructed by the former in adding two unequivalent representations of Weyl spinors, differing by the outer automorphism of complex conjugation, in a direct sum. However, the authors present a generalization of Weyl spinors by extending the base field of complex numbers $\mathbb{C}$ to the quaternions $\mathbb{H}$, a skew field and a non-commutative division ring. As an advantage, one is able to describe massless and massive Dirac equations on the same footing by two-component quaternionic spinors. As a logical step, the two-component octonionic Dirac equation, which turns out to be 10-dimensional, is examined. Dimensional reduction takes place by singling out a unit – i.e., invertible element – in the octonions, or equivalently, a unique complex subalgebra. This allows one to find the variety of particles of a generation of leptons. The three distinct possibilities for singling out a sub-quaternion algebra are conjectured to carry the three families or generations. This mechanism should be compared with Schmeikal's finding several copies of $SU(3)$ as presented in his paper.

Just & Thevenot: As is well known, Dirac's theory contains the mass-term which comes with the time-like γ-matrix. In some phenomenological models, potentials are used which are not only vectorial in nature – having a γ-matrix – but involve scalar, vector, tensor, axial or pseudo vector, and pseudo scalar contributions. Especially the anomalous magnetic moment bears a tensorial character. Such terms are called Pauli terms. This paper considers the question of whether or not such general potential is possible and whether it is compatible with Dirac theory. As a surprising outcome, the authors present arguments against the presence of Pauli terms. One should note, however, that these terms are not necessary in the standard model and may be rendered superfluous.

Lewis, Lasenby, & Doran: Scattering experiments are still the common source of information in experimental high energy physics. S-matrix theory establishes the theoretical counterpart. An awkward and technical detail of this theory is the calculation of spin-sums of spinning particles. Using spacetime algebra (STA) – the Clifford algebra of Dirac theory – a spin direction can be introduced when the S-matrix is replaced by an STA operator. This operator can depend on the spin direction. In this way, a plain and straightened method is developed to work directly with the invariant spin direction avoiding spin sums and a choice of basis. However, some improvements have to be made for multi-particle spin states. Some achievements toward a proper formulation of multi-particle STA are made by the contributions of Fauser & Abłamowicz.

Physics – Structures

Bette's contribution reviews a twistor phase space picture which was developed in the past. After having introduced the twistor phase space, the first point is the introduction of shifted position coordinates which fulfill a non-trivial Poisson bracket and are, thereby, non-commuting. These positions are physically motivated by the requirement that the inner product of momentum and Pauli-Lubanski spin-part vanish. This requirement is motivated by Dirac quantization and conservation laws. In fact, since one wants to have momenta represented by derivatives, one looks for Noether currents. A classical spinning particle is given as an example in $\mathbf{Tp(2)}$, and the resulting equations of motion are shown to be different from Bargman-Michel-Telegdi equations.

Johnson: Clifford algebras have, as an interesting sub-structure, discrete groups sometimes called Dirac groups. In fact, one can construct Clifford algebras over the reals by a ring extension from the group algebras of these cyclic and dihedral discrete groups. Based on this idea, the author tensors two such structures to get a prototype of a fiber of a configuration space manifold. After defining a suitable subspace as the tangent space, he is able to incorporate the action in the complement which finally constitutes a constraint in the tensor space. A $1+1$ dimensional model is considered.

Pezzaglia criticizes that a mere reformulation of physics in new mathematical formalisms does not lead to new physics even if there might be a great achievement in straightening out the problem at hand. However, changing the perspective might be a key step in being able to generalize physical principles and to reach new physics. The author's key point is to look at a Clifford element as an entity even if it can be split into multi-vectors. Therefore, he interprets every multi-vector part of a Clifford number as a physical quantity with its own coordinate. Indeed, this is the situation in quaternion theory, where the "vector" part is given by the linear span of **i**, **j**, **k** = **ij**, and **k** is algebraically –but not linearly– dependent. As an example, special relativity is revisited, and a "polydimensional" (ungraded) formulation of physics is developed. Multi-vector valued, or "matrix," derivatives and the action principle are formulated. Papapetrou's equation and Crawford's hypergravity provide examples of the method in classical physics.

Piazzese: A real linear space can be equipped with different metrical forms, e.g., Euclidean or Minkowski. In this contribution, a relation is established for (time-like) vectors of a Minkowski space to such a Euclidean space without resorting to complex numbers. The remarkable fact is that one is able to transport transformation laws into the Euclidean picture. This allows the author to propose a quasi-classical description of a particle. Besides the classical energy, formed with use of the relativistic velocity – i.e., quasi classical – a second term arises in this description, which can be connected – via de Broglie's idea that every particle has an "internal clock," or an internal frequency – to an internal degree of freedom. This freedom might be of rotatory nature, and it might be connected to spin. The quasi-classical energy becomes frame independent.

Vargas & Torr provide a clear introduction to the unification program which had been developed by the authors in the past. The second section concentrates on notation not only for clarity but also to point out some difficulties, which are usually ignored, but which prove to be most important in the later development. A particular point presented in the paper is the distinction between the Dirac-Hestenes equation and Kähler theory of Dirac equations. Differential forms are introduced. Using the Hestenesian idea of mathematical viruses, two major diseases of common treatments, the "transmutation virus" and the "bachelor algebra virus" are described, and their prevention is discussed. A motivation to alter the Kähler approach by introducing a Kaluza-Klein type theory is given which is later on related to Finsler geometry. Connection is made with previous developments, pointing out a new way of deriving the results. Finally, the interior derivative in the particular case of the previously given Kaluza-Klein theory is introduced. The conclusion comes up with some very interesting outlooks about the further development of the theory. A full geometrical (invariant) Kähler equation for Clifford valued cliffoms is given.

Geometry and Logic

Conradt: Duality is currently a well-recognized structure in physical theories. In string theory, duality connects strong and weakly interacting models which allows a perturbative approach to the former. Duality originated in projective geometry where it appeared as the striking fact that every projective theorem has a dual theorem if one interchanges several notions as point with plane, join with meet, etc. Since Clifford geometric algebras are known to describe metric and projective geometries, the paper calls for an implementation of this rather fundamental projective concept. Starting by defining a Clifford algebra, the author uses Poincaré duality of the underlying multi-vector space and asks the question if one could use $(n-1)$-vectors – the isomorphic picture of 1-vectors – to construct a dualized Clifford algebra. It turns out that the meet of the $(n-1)$-vectors can be seen to establish the dual outer product. In the same way, a dual Clifford product can be defined, and a dual Clifford algebra can be formulated in the same vector space as the original one, but with all r-multi-vectors mapped to $(n-r)$-multi-vectors of the Poincaré dual space. Such a duality for meet and join has been investigated by Rota and others. The resulting algebra is called double or Grassmann-Cayley algebra and constitutes a substructure of the Clifford approach presented here. The interpretation of such dual Clifford algebras in terms of projective geometry is given. Projective coordinate systems for points, lines, and planes are exemplified. The linear complex is introduced and a motivation provided as to where to use this structure in physics.

Li: Non-Euclidean geometry was one of the main working fields of W. K. Clifford, which might well have influenced his algebraic ideas. Furthermore, non-Euclidean geometries have already been described by Euclidean models by Felix Klein in the 19th century. This fact provides the basis of the study of hyperbolic geometries by Clifford algebraic techniques, singling out appropriate subspaces. Geometric facts can be encoded in the Grassmann-Cayley algebra, called "double algebra" by Rota. The main achievement of such an algebraization is that it opens an analytic and invariant approach to geometric problems. Automated geometric theorem proving is a possibility explored in this contribution. Not only machine based recalculations of theorems but the quite more interesting proving of new theorems, thus, becomes possible. Beside its beauty, this method is important also for applications in robotics and visualization.

Schmeikal: A ring of idempotent elements, or (alternatively) a ring over $\mathbb{Z}_2$, is a Boolean ring. Any assertions in such a ring can be interpreted as true or false, while the operations in this algebra become inference in a certain logic. Boole, de Morgan, Frege as McCulloch, Parry, or Peirce, or Zellweger have considered pictographical notations of logical conjunctions. Schmeikal adds in this contribution a description by the subring of idempotent elements of a Clifford algebra. The set of all idempotents constitutes a lattice, and, thereby, one can define "basic reflections," i.e., involutive automorphisms. Since representation spaces can be seen to be ideals generated by idempotents, such reflections connect different spaces. Following an idea of Chisholm, one can recover the fundamental $SU(3)$ representation by fixing one primitive idempotent out of four and then by looking at its stabilizer group. The octahedral symmetry of the idempotent lattice of the Dirac algebra is examined, and six copies of $SU(3)$ are detected. However, the picture gets more complicated by the introduction of generalized logic operators, which act on logic assertions. This tool, after being developed, is applied to establish the logic of Dirac spinors. Discrete symmetries are discussed as examples. Finally, it is outlined how this model can be enlarged to $C\ell_{n,n}$. It is pointed out that using wave functions, and not $\mathbb{Z}_2$, provides an example of quantum logic where "tertium datur."

Mathematics – Deformations

Abłamowicz & Fauser: Indistinguishable particles are described by wave or partition functions invariant under the permutation group. This observation, made by physicists in thermodynamics and quantum mechanics, has influenced the theory of group representations to a large degree. Already Weyl showed that multi-particle states can be classified by a method developed by Young. Multi-particle Clifford algebras have, thus, to carry an action of the symmetric group. This article studies representations of the deformation of the group algebra of the symmetric group known as the Hecke algebra. Deformations have been proposed to serve as symmetries for composed entities. This provides the main motivation for this investigation and is in full accord with Fiore's contribution. Representations are constructed in ideals generated by q-idempotents in quantum Clifford algebras. These idempotents are Young operators. To get an intimate relation between both structures, reversion is taken to act as conjugation on Young operators, i.e., Young idempotents. Detailed computations are possible only by computer algebra. CLIFFORD, a Maple V package for (quantum) Clifford algebras – developed by one of the authors R.A. – has made algebraic computations possible. The representation theory in the 2- and 3-dimensional cases is developed in full detail.

Fiore: Any Clifford algebra naturally has a Lie algebra substructure. Therefore, it is interesting to ask if it is possible to q-deform this Lie algebra, or, more precisely, its enveloping algebra U_q. As a natural outcome of covariance under the action of such a quasi-triangular Hopf algebra, one obtains a q-deformed Clifford algebra. However, due to finite dimensionality in the orthogonal case, it is possible to express the deformed generators as polynomials in the undeformed ones. As a consequence, q-deformed creation and annihilation operators might be interpreted as creation and annihilation operators of effective, or compound, entities. A detailed analysis of the situation gives a connection between deformed and undeformed invariants.

It is shown under which conditions non-trivial deformations, having new invariants, actually occur.

Rosenbaum & Vergara point out that recently at two different places closely related Hopf algebra structures have popped up. The first one is the Connes-Moscovici Hopf algebra, which originates from non-commutative extensions of Riemannian geometry. The second is the Hopf algebra of rooted trees, which has been employed by Connes and Kreimer to produce the combinatorics of renormalization in perturbative quantum field theory. Decorated rooted trees are used to establish the forest theorems of renormalization, and the antipode action generates the counter terms in all orders. The aim of this work is to present both types of Hopf algebras in an invariant, coordinate-free language. Furthermore, it is shown that there might be a connection between the Dirac operator, spacetime at Planck scale, and the above two Hopf algebras, which may lead toward a finite quantum field theory. The connection to Schwinger-Dyson equations is discussed.

Vancliff: Neither physics nor mathematics is currently able to present a concise model of a non-commutative space as a coordinate space of a quantum group or other deformed algebras. Common methods fail to work since in such non-commutative spaces one cannot find a Poincaré-Birkhoff-Witt like basis. Following an idea of S.P. Smith, this work, being projective and geometric in nature, connects geometric data to deformed algebras. The Sklyanin algebra serves as a model to exemplify the problem at hand. The deep connection between projective point, line, etc., schemes and quantum spaces is explored. Roughly speaking, the quantum space of an algebra is a quotient category of graded modules. Certain modules play the role of points; others play the role of lines, and so forth. In the commutative setting, this idea can be traced back to J.-P. Serre. Quantum spaces are constructed via Poisson geometry; examples are provided.

Mathematics – Structures

Belinfante: Spinors have been discovered by Élie Cartan when classifying complex semi-simple Lie algebras. During further development done by Cartan, Freudenthal, Dynkin, Chevalley, and others, it became clear that one can construct spinor modules of **B** and **D** type Lie algebras over the integers. The integers play a fundamental role in the classification also. Spinors are connected, however, to Clifford algebras constituting their natural irreducible representation spaces. As a natural approach, spinor and semi-spinor modules are constructed for complex orthogonal Lie algebras – i.e., Lie algebras of types **B** and **D**. First and second Clifford algebras are introduced, and the mechanism using Dynkin diagrams to construct spinor modules is explicitely given. $\mathbf{B}_1$ and $\mathbf{B}_2$ provide examples; spin groups are discussed. MATHEMATICA code is provided which was used to check the results and invites the reader to redo the computations for gaining deeper insights into the theory.

Fauser & Abłamowicz: Clifford algebras over real and complex numbers are classified. As an outcome of this classification, every Clifford algebra can be decomposed into graded tensor products of "atomic," i.e., indecomposable, Clifford algebra factors. These factors are at most of algebraic dimension 4. This decomposition is the origin of periodicity theorems and vice versa. Albert Crumeyrolle stated that the decomposition properties of Clifford algebras provide the Mendeleïev periodic system of elementary particles. However, the situation can be much more complicated. Defining quantum Clifford algebras, i.e., Clifford algebras of an arbitrary bilinear form, it can be shown that common periodicity theorems in general fail to hold. A detailed introduction provides arguments that this is not the exception, but, instead, it should be seen as the rule. The Wick theorem of normal-ordering in quantum mechanics and quantum field theory establishes a quantum Clifford algebra structure. After the detailed and rigorous development of the theory, three examples are provided showing the theory at work.

Fernández, Moya & Rodrigues present in their paper the theory of covariant derivative operators on a Minkowski manifold. As a main tool of their investigation, the concept of multiform calculus is used. It is developed by passing directional covariant derivative operators and associated operators to covariant derivative operators which are compatible with a non-degenerate symmetric tensor in a Minkowski manifold. This seems to be one of the main goals of the paper because this result allows us to construct Riemann-Cartan geometries. As examples of applications of the presented theory, the Levi-Civita derivative and the Hestenes derivative are discussed in a detailed way, and they are interpreted in the framework of the developed theory. The paper presents an elegant way to handle covariant derivative operators, and it should initiate discussions about the different approaches.

Ławrynowicz & Suzuki: Twistors, introduced by Penrose, have been successfully used in gravity, the theory of non-linear differential equations, and representation theory of conformal groups. The "twistor program" has recently been geometrized by Ławrynowicz and Rembieliński. The geometric approach makes it possible to connect the Hurwitz problem of (de)composing quadratic forms with the so-called Hurwitz pairs, which constitute pseudo twistors. These pseudo twistors can be constructed for arbitrary signature and are not necessarily connected to conformal symmetry. This contribution deals with a special class of Hurwitz twistors, which includes signature $(3,2)$ and its dual $(1,4)$. Cohomological aspects and a generalization of Cartan's triality to a doubled triality as an atomization theorem and holomorphic embeddings are presented.

Oziewicz & Zeni: Symmetries can be used to reduce the order of differential equations via Lie's theorem. Sometimes this is not directly possible, but only after having prolonged the differential equation by a suitable integrating factor. The well-known multiplication of Newton's equations of motions by an $\dot{x}$ can be integrated yielding the energy conservation law. Lie had already shown under which conditions a "last multiplier" can be found, turning a differential equation into an exact one which can then be integrated. This article tackles the problem of finding last multiplier and the corresponding symmetries for ordinary differential equations in $n+1$ dimensions. Differential forms are shown to be the natural language for the problem at hand, and a generalized Lie theorem is proved. Moreover, this method is constructive and allows one to find the integrating multiplier by direct calculations. The method is independent of a Riemannian or symplectic structure and does not rely on coordinate methods.

Tian: Besides their invariant coordinate-free character, Clifford algebras are still used by many physicists via matrix representations. Starting from a generator and relations approach to complex Clifford algebras, it is shown that there are only two distinct types of representations. One is for simple Clifford algebras, if the number of generators is even, and one is over a double field, in the semi-simple case when the dimension is odd. Complex similarity factorizations provide a one-to-one mapping of the abstract algebras into certain matrix algebras. In this way, each element a in the $\mathbb{C}$ algebra $C\ell_n$ could be regarded as an eigenvalue of its complex representation matrix $\phi_{n\times n}(a)$. This might be useful in application as calculating exponentials of Clifford numbers.

Acknowledgment: Since editing such a book cannot be done without valuable help of other people, we would like to express our gratitude to all contributors: Ann Kostant, Tom Grasso, and Caroline Graf of Birkhäuser, Elizabeth Loew of TEXniques for her help with TEXing, Amy Knox for her monumental proofreading, and all referees for their constructive criticism.

The first editor, R.A., thanks his wife Halina for her patience during this project. The second editor, B.F., would like to thank his wife Mechthild for her support during the laborious period when the book was completed.

Rafał Abłamowicz, Cookeville, Tennessee, U.S.A.
Bertfried Fauser, Konstanz, Germany April 1, 2000

Preface to Volume 2

This volume of contributed papers arises from the section on Clifford analysis which took place as part of the "5th International Conference on Clifford Algebras and Their Applications in Mathematical Physics," held in Ixtapa, Mexico, June, 1999. Like in Volume 1, the majority of papers are an outgrowth and further development of the talks presented at the conference. All papers in this volume were refereed and were further developed after the conference. The editors are grateful to the referees for their extremely valuable assistance in creating this volume. The papers gathered here reflect some of the latest developments in the field of Clifford analysis and its applications. Topics range from the study of generalized Schwarzian derivatives to applications to boundary value problems and singular integrals. Topics covered also include links to a Möbius invariant function theory on hyperbolic space, analogues of Ahlfors-Beurling inequalities and their applications, differentiability properties of monogenic functions, links to supersymmetry, hyperbolic Dirac equations, and scattering theory. The papers appearing here can be broadly subdivided into the following categories.

Partial Differential Equations and Boundary Value Problems

As is well known, the complex Beltrami equation is of enormous importance in the general theory of elliptic equations and has many applications to other fields of analysis and geometry. Because of the many degrees of freedom in the combination of partial differential operators in higher dimensions, there is a large variety of generalized Beltrami equations. To motivate the treatment by Clifford analytic methods, U. Kähler shows some typical difficulties (integrability conditions, conditions on the Jacobian, etc.) in higher dimensions. The paper is a comprehensive survey of the study of Beltrami type equations in the three-dimensional setting. The results are based on integral operator methods and include detailed norm estimates for integral operators in different function spaces.

In her article, Xinhua Ji shows the relationship between Green's function, Möbius transformations and the Laplace-Beltrami operator acting on the one point compactification of $\mathbb{R}^n$. The author gives a complete description of all non-Euclidean translations of the corresponding Möbius group. With the aid of a geodesic distance, she obtains the fundamental solution of the degenerate Laplace-Beltrami equation. Finally, she solves the Dirichlet problem for the non-homogeneous Laplace-Beltrami equation. This paper contains a very nice example of a generalized Poisson kernel illustrating that the maximum principle is no longer valid for degenerate elliptic equations.

W. Sprößig considers a stationary problem from fluid mechanics using methods from Clifford analysis. Stationary Navier-Stokes equations are combined with field induction. Under certain conditions the solution can be obtained by an iteration procedure which converges rapidly in Sobolev spaces.

Singular Integral Operators

In their paper, Tao Qian, John Ryan and Xinhua Ji study Fourier multipliers and singular integral operators on Möbius images of Lipschitz graphs and starlike Lipschitz surfaces. All basic notation is explained. It is shown that the singular integral operators form an operator algebra. The main ideas are the application of a generalized Fourier transform and the use of Möbius transformations to pullback known results. One of the main results is the fact that, in the case of Lipschitz graphs, the operator algebra of singular integrals may be identified with the bounded holomorphic Fourier multipliers. The same problems are considered for starlike Lipschitz surfaces.

J. B. Reyes and R. A. Blaya study quaternionic Cauchy integrals on Ahlfors regular surfaces. In this very general setting Plemelj-Sokhotzkij formulae are deduced. These results can be applied to prove the solvability of a special case of Riemann's problem.

In M. Martin's article he proves some Hedberg type inequalities for the convolution operator associated with the Cauchy kernel in Euclidean space. Each of these inequalities involves a specific maximal operator and they all provide the best possible constants. Applications of Clifford analysis are also presented. Among them, one should single out a higher-dimensional generalization of a classical inequality in one-variable complex function theory due to Ahlfors and Beurling and some extensions of Alexander's inequality.

Applications in Geometry and Physics

S. Bernstein describes a new application for the Borel-Pompeiu formula in $\mathbb{C}^n$. It is obtained as a direct analogue of the Martinelli-Bochner formula. An extremely interesting feature here is her application of these results to scattering theory.

In G. Kaiser's contribution complex distance is applied to describe a useful extension of potential theory in $\mathbb{R}^n$ to $\mathbb{C}^n$. The resulting Newtonian potential is generated by an extended source distribution $\tilde{\delta}(z)$ in $\mathbb{C}^n$ whose restriction to $\mathbb{R}^n$ is the point source $\delta(x)$. This provides a possible model for extended particles in physics. In $\mathbb{C}^{n+1}$, interpreted as complex spacetime, $\tilde{\delta}$ acts as a propagator generating solutions of the wave equation from their initial values. This gives a new connection between elliptic and hyperbolic equations that does not assume analyticity of the Cauchy data. Generalized to Clifford analysis, it induces a similar connection between solutions of elliptic and hyperbolic Dirac equations, thereby extending earlier work of J. Ryan. There is a natural application to the time-dependent, inhomogeneous Dirac and Maxwell equations and the so-called electromagnetic wavelets.

The paper by J. Snygg deals with the use of Clifford algebras in differential geometry, especially involving the holonomy group. The Clifford algebra is used to express isometry operators.

In his paper F. Sommen presents an extension of Clifford analysis using both commuting as well as anti-commuting variables, thus, following the lines of thinking of supersymmetry. For abstract vector variables, the calculus remains the same. He illustrates that the radial algebra can be represented by both the use of commuting and anti-commuting variables. Another important fact is that the action of both the symplectic group and the rotation group are united in the super spin-group. This contribution suggests that the radial algebra is a natural background for supersymmetry because it is independent of dimension and invariant in the group theoretic sense.

J. Tolksdorf investigates bosonic and fermionic action in a Euclidean version of the standard model of particle physics in terms of elliptic Dirac operators on compact smooth even-dimensional manifolds without boundary. He explains the specific role of "standard Dirac operators" for fermionic action in contrast to the role of non-standard operators of Dirac type for bosonic action and he relates the Wodzicki residue (WR) to this action, thereby illuminating also an interesting interrelation with Connes' non-commutative geometry.

Möbius Transformations and Monogenic Functions

In their contribution M. Wada and O. Kobayashi define Schwarzian derivatives of transformations of $\mathbb{R}^n$ in terms of Schwarzian derivatives of regular curves and prove that a transformation of $\mathbb{R}^n$ is Möbius if and only if its Schwarzian derivatives are constantly zero. The framework then is modified to prove a result on immersions between Riemannian manifolds. The paper gives a new definition for Schwarzian derivative (cf. [1] and [3]) and shows accordingly the necessary and sufficient condition for a transformation being Möbius in terms of its Schwarzian derivatives.

In the last decades the construction of infinitesimal generators of the conformal group has played an important role in the study of monogenic functions. Several authors, including P. Lounesto, J. Ryan, F. Sommen and H. Leutwiler, have results on this topic. For instance F. Sommen [4] has obtained in his paper on monogenic operators a Taylor series formula using the conformal group. Here Y. Krasnov presents an explicit Taylor series formula by making use of the conformal group and its generators.

The paper of T. Hempfling is devoted to the hyperbolic modification of Clifford analysis. He considers the Cauchy-Riemann operator in $\mathbb{R}^{n+1}$. By splitting this operator into a radial and spherical part, he can formulate conditions for the radial part to vanish. This is especially true if the corresponding functions are hypermonogenic.

In 1978, A. Sudbery (cf. [5]) was the first to define in a quaternionic setting the derivative as differential coefficient between two forms of higher degree. Here H. Malonek succeeds in getting a similar result within the Clifford algebra $C\ell_{0,n}$ by considering differential forms of degree $n-1$ and n as well as using the Hodge star operator; see also earlier results of J. Ryan cited in H. Malonek's paper. Here H. Malonek introduces the notion of the hypercomplex derivability of a function defined for paravectors with values in the n-dimensional real Clifford algebra. For instance, a real valued function f is called (left) derivable if and only if $A_{f,\ell}(z) \in C\ell_{0,n}$ exists and

$$d(d\tau f) = d\sigma A_{f,\ell}(z),$$

where $d\sigma$ denotes a hypercomplex differential of degree n and $d\tau$ a hypercomplex differential of degree $n-1$. The main result is contained in a theorem that shows that a real differentiable function f is (left) derivable if and only if f is (left) monogenic.

H. Leutwiler shows in [2] and some other papers that the power functions are solutions of the modified Cauchy-Riemann system

$$x_n Df + (n-1)f_n = 0.$$

This equation is closely related to the hyperbolic metric. Consequently, these functions are closely connected to the Laplace-Beltrami equation. H. Leutwiler and Sirkka-Liisa Eriksson-Bique introduce in their paper hypermonogenic functions as a generalization of classical complex functions. These hypermonogenic functions behave in relation to the Laplace-Beltrami operator like monogenic functions to the Laplacian. They introduce operators P and Q which can be understood as projections onto the real and "imaginary" parts of the Clifford algebra $C\ell_{n-1}$. The main theorems describe hypermonogenic functions, their relation to solutions of the hyperbolic Dirac operator and a representation formula.

Another paper devoted to modified Clifford analysis is presented by P. Cerejeiras. On the basis of results obtained by H. Leutwiler, J. Cnops, and J. Ryan, the existence of a Poisson-Szegö kernel for the Laplace-Beltrami equation associated with an n-dimensional hyperbolic space is proven. Poisson-Szegö kernels are explicitely constructed for a large class of orientable manifolds by using the initial kernel for the spherical model of hyperbolic spaces and the usual Möbius transform. These kernels solve a generalized Dirichlet problem.

The editors express their gratitude to Amy Knox for her very thorough proofreading of all papers.

John Ryan, Fayetteville, Arkansas, U.S.A.
Wolfgang Sprößig, Freiberg, Germany April 1, 2000

References

[1] L. V. Ahlfors, Cross-ratios and Schwarzian derivatives in $\mathbb{R}^n$, *Complex Analysis,* J. Hersch and A. Huber, eds., articles dedicated to Albert Pfluger on the occasion of his 80th birthday, Birkhäuser, 1988, 1–15.

[2] H. Leutwiler, Rudiments of a function theory in $\mathbb{R}^3$, *Forum Math.* **7** (1995), 279–305.

[3] J. Ryan, Generalized Schwarzian derivatives for generalized fractional linear transformations, *Ann Polon. Math.* **LVII** (1992), 29–44.

[4] F. Sommen, N. Van Acker, Monogenic differential operators, *Results in Math.* Vol. **22** (1992), 781–798.

[5] A. Sudbery, Quaternionic analysis, *Math. Proc. Cambr. Phil. Soc.* **85** (1979), 199–225.

Dedication to Gian-Carlo Rota

Shortened version of a presentation given at the Special Session dedicated to the memory of Gian-Carlo Rota at the 5th International Conference on Clifford Algebras and their Applications in Mathematical Physics, Ixtapa, Mexico, June 27 – July 4, 1999.

Gian-Carlo Rota — philosopher and mathematician — inspired and befriended several generations of colleagues and students; indeed his presence continues to impact on many of us even after his death.

Rota the man and Rota the highly original thinker drew from his foundations in phenomenological thinking, namely, the traditions of Husserl and Heidegger whose works became Rota's "livres de chevet" throughout his life. From phenomenological traditions, Rota developed an altogether unique philosophy of mathematics, which is engagingly presented in the profound essays from his book *Indiscrete Thoughts*. *Indiscrete Thoughts* is also an autobiographical accounting of Rota's tastes, what and whom he admired or disliked, his perspectives on the golden age of 20th century science, and his intense friendship with Stan Ulam (intricately exposed in *The Lost Café*).

Ulam's death created a deep void in Rota. Although he had always realized that death is a natural part of life, the sadness that would accompany such a realization left him with a sense of "amertume." The art of dying as the art of living was important, and in Ulam's passing, Rota sought to extend Stan's influence beyond the grave. As he said, *"I am at a loss to tell where Ulam ends and where I really begin. Perhaps this is one way he chose to survive."* It is a rather phenomenological statement, taken more from a living experience than from science, and akin perhaps to the notion of existential psychology and interpersonal perception — concepts dear to Rota.

In connection with interpersonal perception, Rota describes how Ulam used to comfort him when he was feeling depressed about death. Ulam would say to Rota, *"You are not the best mathematician I ever met because von Neumann was better. You are not the best Italian I have ever met because Fermi was better. But you are the best psychologist I have ever met,"* [1] in an attempt to lighten his mood. This devotion helped ease the transition from life to death.

Ulam also influenced Rota's mathematical ideas to some extent through his phenomological *"flashes of foresight."* [2] One such idea involved the concept of surface. In the nineteenth century, mathematicians conceived of a surface by explicit equations. After some further abstraction, topologists redefined the concept in terms of extension and neighborhood. Rota and N. Metropolis [3] picked up an essential part of the problem, namely codification and enumeration of the faces of the n-cube. Their paper *On the lattice of the faces of the n-cube* makes us aware of the Dilworth partition and its symmetries.

Along with Ulam, Rota further envisioned a refinement of Maxwell's equations that would rid them of algebraic irrelevancies. Such equations can be found in *Clifford Algebras and Spinors* [4], by Pertti Lounesto. Rota believed there were differences in algebra one, consisting of algebraic geometry or algebraic number theory, and algebra two, connected with Boole's name and containing logic algebra and invariant theory. Recent developments in Clifford algebra have removed those differences that once seemed invincible. Rota himself made advances in algebra two.

With Rota's departure we are left alone to decipher the symbols of instant vision and their relations to phenomenology and what is meant by them. Following Husserl, Rota proposed to investigate connectives of the form "A is absent from B," or "A anticipates B." The meaning of "anticipation" apparently bridges living experience and mathematics.

The culmination of Rota's ideas on death, life, and God came in April 1999 with his own death. It seems that Rota had at last managed to connect the big with the little death, sleep with eternity. We live with fond memories of Gian-Carlo: his manifold generosities and kindnesses, his lifelong quest for understanding and truth, and his perceptions of mathematics and its relation to logical and clear thinking.

Note. This dedication was shortened and reformulated in part by Ann Kostant, Executive Editor of Mathematics and Physics, Birkhäuser.

References

[1] Gian-Carlo Rota, *Indiscrete Thoughts,* Birkhäuser, Boston, 1997, ch. 6: *The Lost Café*, p. 83.

[2] Ronald D. Laing, *Conversations with Children,* Harmondsworth, 1978, p. 19.

[3] N. Metropolis, Gian-Carlo Rota, On the lattice of faces of the n-cube, *Bulletin of the American Mathematical Society* Vol. **84**, No. 2. March 1978.

[4] P. Lounesto, *Clifford Algebras and Spinors,* Cambridge University Press, Cambridge, 1997.

Bernd Schmeikal
Biofield Laboratory
Kundmanngasse 26, A-1030 Vienna, Austria
Email: schmeika@isis.wu-wien.ac.at

Received October 4, 1999; Revised: April 28, 2000

1.

PHYSICS: APPLICATIONS AND MODELS

Multiparavector Subspaces of $C\ell_n$: Theorems and Applications

William E. Baylis

ABSTRACT While Clifford algebras are known to provide appropriate mathematical structures for describing many geometrical relations and physical phenomena, traditional applications use only homogeneous elements (elements of a single grade) to model physical entities such as spacetime vectors in relativity and their transformations. Lower-dimensional realizations of the structures inherent in physical systems are sometimes afforded by exploiting mixed-grade representations of such entities, for example by modeling spacetime vectors by paravectors (sums of scalars and vectors). This contribution explores the geometry of subspaces generated by paravectors of $C\ell_n$, the Clifford algebra of n-dimensional Euclidean space, and its applications to physical phenomena.
Keywords: Paravector subspaces, multiparavectors, Clifford algebra, Minkowski metric, electrodynamics, Lorentz transformations, Lorentz force, plane-wave pulse.

1 Introduction

An important feature of Clifford algebras is their rich internal structure [1] [2]. A simple but important example is the Pauli algebra $C\ell_3$ [3], the Clifford algebra of 3-dimensional Euclidean space $\mathbb{E}^3$, which operates in an eight-dimensional linear space

$$\mathcal{V}_{\{1,\mathbf{e}_1,\mathbf{e}_2,\mathbf{e}_3,\mathbf{e}_2\mathbf{e}_3,\mathbf{e}_3\mathbf{e}_1,\mathbf{e}_1\mathbf{e}_2,\mathbf{e}_1\mathbf{e}_2\mathbf{e}_3\}} \tag{1.1}$$

by which we mean the linear span of $\{1,\mathbf{e}_1,\mathbf{e}_2,\mathbf{e}_3,\mathbf{e}_2\mathbf{e}_3,\mathbf{e}_3\mathbf{e}_1,\mathbf{e}_1\mathbf{e}_2,\mathbf{e}_1\mathbf{e}_2\mathbf{e}_3\}$. It is common to classify graded elements of the algebra according to whether they are scalars, vectors, bivectors, etc., and the full linear space is a direct sum of the corresponding homogeneous linear subspaces

$$\mathcal{V}_{\{1\}} \oplus \mathcal{V}_{\{\mathbf{e}_1,\mathbf{e}_2,\mathbf{e}_3\}} \oplus \mathcal{V}_{\{\mathbf{e}_2\mathbf{e}_3,\mathbf{e}_3\mathbf{e}_1,\mathbf{e}_1\mathbf{e}_2\}} \oplus \mathcal{V}_{\{\mathbf{e}_1\mathbf{e}_2\mathbf{e}_3\}}. \tag{1.2}$$

AMS Subject Classification: 15A66, 51P05, 78A35, 83A05.

However, inhomogeneous (mixed-grade) subspaces may also play important roles. For example, the center of the algebra comprises scalars and trivectors and is isomorphic to the complex field

$$\mathbb{C} \cong \mathcal{V}_{\{1\}} \oplus \mathcal{V}_{\{\mathbf{e}_1\mathbf{e}_2\mathbf{e}_3\}} \tag{1.3}$$

and the space of even elements can be identified with quaternions

$$\mathbb{H} \cong \mathcal{V}_{\{1\}} \oplus \mathcal{V}_{\{\mathbf{e}_2\mathbf{e}_3,\mathbf{e}_3\mathbf{e}_1,\mathbf{e}_1\mathbf{e}_2\}}. \tag{1.4}$$

Another important inhomogeneous subspace of $C\ell_3$ contains all linear combinations of scalars and vectors. This is the linear space of *paravectors* [4]

$$\mathcal{V}_{\{1,\mathbf{e}_1,\mathbf{e}_2,\mathbf{e}_3\}} = \mathcal{V}_{\{1\}} \oplus \mathcal{V}_{\{\mathbf{e}_1,\mathbf{e}_2,\mathbf{e}_3\}}. \tag{1.5}$$

Since one can identify the volume element of $\mathbb{E}^3$ with the unit imaginary $\mathbf{e}_1\mathbf{e}_2\mathbf{e}_3 = i$ (a right-handed basis $\{\mathbf{e}_1, \mathbf{e}_2, \mathbf{e}_3\}$ is assumed), the eight-dimensional space of $C\ell_3$ over $\mathbb{R}$ maps to a paravector space over $\mathbb{C}$.

As we will see more generally below, the metric of the paravector space of $C\ell_3$ turns out to be the metric of Minkowski spacetime. In the observer's frame, the time component of the spacetime vector is represented by a scalar and the spatial components by a vector in the paravector space. However, not all scalars of the algebra are time components of spacetime vectors; they may instead be Lorentz invariants. The double role of scalars reflects a duality of roles common in physics: the product mc of the mass m of a particle and the speed of light c, for example, is both the time component of the spacetime momentum in the particle rest frame and also the Lorentz-invariant length of the momentum.

Of course, the split of a spacetime vector into time and space parts is not covariant; it will be different for different observers. Nevertheless, the paravector model in $C\ell_3$ does admit a covariant description of relativistic phenomena. The critical point is to view paravectors as identifiable entities, independent of the observer. The noncovariant split occurs only once the paravector is expanded in a tetrad of basis paravectors. The natural tetrad to use is one at rest relative to the observer, and in such a tetrad, the timelike paravector is $\mathbf{e}_0 = 1$. However, bases in motion relative to the observer can also be used. Indeed Lorentz transformations in the paravector model take the form of spin transformations that act on the basis elements:[1]

$$\mathbf{e}_\mu \to \mathbf{u}_\mu = L\mathbf{e}_\mu L^\dagger,\ L \in \mathbf{\$pin}_+(1,3) \simeq SL(2,\mathbb{C}). \tag{1.6}$$

[1]The group $\mathbf{\$pin}_+(1,3)$, as defined by Lounesto in ref. [2], p. 126, contains all unimodular elements of $C\ell_3$. It is isomorphic to $\mathbf{Spin}_+(1,3)$ as well as to the matrix group $SL(2,\mathbb{C})$.

pv-grade	pv-type	no.	v-grade	basis elmts
0	scalar	1	0	$1 = \mathbf{e}_0$
1	paravector	4	$0+1$	$\mathbf{e}_\mu$
2	biparavector	6	$1+2$	$\langle \mathbf{e}_\mu \bar{\mathbf{e}}_\nu \rangle_V$
3	triparavector	4	$2+3$	$\langle \mathbf{e}_\lambda \bar{\mathbf{e}}_\mu \mathbf{e}_\nu \rangle_{\mathrm{Im}}$
4	pseudoscalar	1	3	$\langle \mathbf{e}_\lambda \bar{\mathbf{e}}_\mu \mathbf{e}_\nu \bar{\mathbf{e}}_\rho \rangle_{\mathrm{Im}\, S}$

TABLE 1.1. Multiparavectors of spacetime of a given paravector (pv) grade are formed from one or two grades of multivectors in 3-D vector (v) space. The subscripted angular brackets in the last column are explained below.

In particular, $\mathbf{u}_0 = LL^\dagger$ is the proper velocity of the transformed reference frame with respect to the observer. (The dagger conjugation is explained below in the section on $C\ell_n$.)

Biparavectors[2] and triparavectors can also be introduced within $C\ell_3$. Biparavectors represent directed spacetime surfaces and are composed of vector plus bivector parts. A prime example is the electromagnetic field, the Faraday $\mathbf{F}$ [5] whose vector and bivector parts are the electric and magnetic fields, respectively. Similarly, triparavectors represent hypersurfaces in spacetime and comprise bivector and trivector parts. Finally, the spacetime 4-volume is represented by a quadriparavector, which is just a trivector of $C\ell_3$ and hence commutes with all other elements. The relations are summarized in Table 1. Each subspace has inherent symmetries that reveal corresponding symmetries in nature. Consider, for example, biparavectors that generate rotations in spacetime. They come in commuting dual pairs, such as $\mathbf{e}_0\bar{\mathbf{e}}_1$ and $\mathbf{e}_2\bar{\mathbf{e}}_3 = i\mathbf{e}_0\bar{\mathbf{e}}_1$, representing the generators of orthogonal spacetime planes (see Fig. 1; the bar conjugation is explained below). The compact symmetry group for biparavector space is a combination of (spinor) rotations in three-dimensional space and duality rotations, namely $SU(2) \times U(1)$. Paravectors of $C\ell_3$ generate with their products a multiparavector algebra that is clearly the same as $C\ell_3$ itself. It is useful to identify multiparavectors as distinguishable units since, as described above, they provide a covariant model of relativistic phenomena. For such purposes, the paravector model is thus an alternative to the larger Clifford algebras $C\ell_{3,1}$ or $C\ell_{1,3}$ which are commonly invoked for relativistic problems. Physical systems often admit many mathematical models. Different models tend to offer different insights, and it can often be difficult to choose the most appropriate model for a given problem. If one looks for a model with a minimum of excess baggage that is well adapted to our physical

[2]The term parabivector was used by the author in ref. [1] for the nonscalar part of the product of two paravectors, and this name has also been adopted by Lounesto in ref. [2]. However, the terminology biparavector, triparavector, and multiparavector seem better to describe the multivector spaces generated by paravectors. This choice, which is the terminology used in ref. [3], is also motivated by the fact that the author encountered readers who assumed a parabivector to be the sum of a scalar and a bivector.

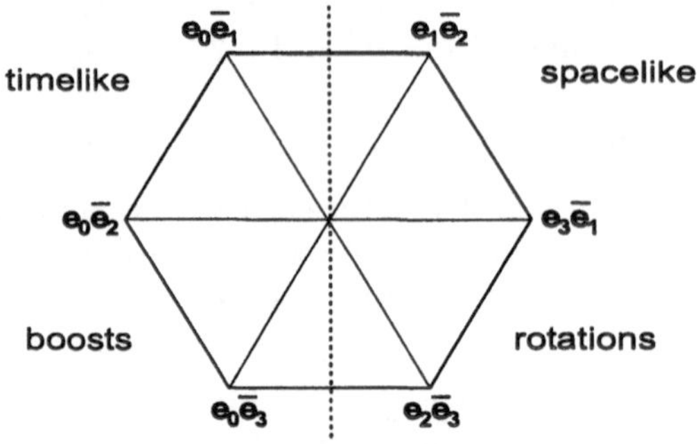

FIGURE 1. The unit biparavectors that generate restricted Lorentz transformations come in commuting dual pairs (located on opposite ends of the diagonal lines through the center) and are transformed into one another by $SU(2) \times U(1)$.

intuition, the paravector model appears to be a worthy candidate.

Many of the results in $C\ell_3$ can be extended to paravectors of $C\ell_n$. This contribution explores such extensions, gathers common results, and considers applications to physical systems. Because of numerous possible mappings between Clifford algebras of different spaces, the theorems we give for paravectors of $C\ell_n$ have counterparts in other algebras and, indeed, some are well-known in other guises. The present contribution displays them in the context of paravectors where, as indicated above, some physical applications may be more apparent. In the following section, we consider the possible motivation for adding one or more Euclidean dimensions to physical space. Next we establish several useful theorems for paravectors and their products in $C\ell_n$. Finally, we look at applications to a problem in classical electrodynamics.

2 Physical applications of $C\ell_n$

Spacetime rotations of paravectors in $C\ell_3$ form the restricted Lorentz group. The full homogeneous Lorentz group also includes discrete operations such as Clifford conjugation, $p \to \bar{p}$, which changes the sign of the vector part of p (see below). We can incorporate these "improper" transformations into the inner antiautomorphisms of the algebra by introducing $\mathbf{e}_4$ an additional spatial direction $(\mathbf{e}_4^2 = 1)$ orthogonal to $\mathbf{e}_1, \mathbf{e}_2, \mathbf{e}_3$.

Since orthogonal vectors anticommute, any paravector[3] $p = p^{\mu}\mathbf{e}_{\mu}$, $\mu = 0, 1, 2, 3$, obeys

$$p \rightarrow \bar{p} = \mathbf{e}_4 p \mathbf{e}_4. \tag{2.1}$$

The volume element $\mathbf{e}_5 = \mathbf{e}_1\mathbf{e}_2\mathbf{e}_3\mathbf{e}_4$ of $\mathbb{E}^4$ spanned by $\{\mathbf{e}_1, \mathbf{e}_2, \mathbf{e}_3, \mathbf{e}_4\}$ anticommutes with all vectors and thus acts as a fifth spatial dimension orthogonal to the first four. We are thus at $\mathbb{E}^5$. But because its volume element is

$$\mathbf{e}_1\mathbf{e}_2\mathbf{e}_3\mathbf{e}_4\mathbf{e}_5 = (\mathbf{e}_5)^2 = 1, \tag{2.2}$$

our algebra is not the universal algebra $C\ell_5$ but only half of it, namely, $C\ell_4$. Additional dimensions may be added to relate other discrete transformations to continuous ones or to extend the Euclidean space to a projective space or to a conformal space.

The point is that physicists may have interest in Euclidean spaces of $n > 3$. The equivalence

$$C\ell_{p,q} \simeq C\ell_{q+1,p-1}, \tag{2.3}$$

established by one-to-one mappings between Clifford algebras, also extends the scope of algebras based on Euclidean spaces. For example,

$$C\ell_4 \simeq C\ell_{1,3} \tag{2.4}$$

follows by the invertible mapping

$$\mathbf{e}_4 \mapsto \gamma_0, \quad \mathbf{e}_k\mathbf{e}_4 \mapsto \gamma_k, \quad k = 1, 2, 3. \tag{2.5}$$

Both algebras can model the same phenomena, but an advantage of $C\ell_4$ is that in Euclidean spaces all basis elements have Hermitian representations. Such a mapping mixes grades of elements.

Let's look next at the paravector space of $\mathbb{E}^n$.

3 $C\ell_n$ and its paravectors

For $n > 3$ one can distinguish *simple bivectors* and *compound bivectors*. A simple bivector is associated with a *single* plane and can be expressed as the wedge product, say $\mathbf{v} \wedge \mathbf{w} = \frac{1}{2}(\mathbf{vw} - \mathbf{wv})$, of two noncollinear vectors $\mathbf{v}, \mathbf{w}$, in the plane. Its square is a scalar. A compound bivector is a linear superposition of two or more commuting bivectors, representing

[3] The summation convention is adopted for indices that appear both as an upper and lower index in a term.

nonintersecting planes; its square is a real scalar-like element comprising a scalar and a 4-vector.

The concept of simple and compound elements may be extended to products of higher grade. Thus, we refer to k-vectors that square to a real scalar as *simple* and others as *compound.* Because any simple k-vector $\mathbf{K}$ in $C\ell_n$ can be factored into a scalar $K = \sqrt{|\mathbf{K}^2|}$ and a unit k-vector $\hat{\mathbf{K}}$ that squares to $\hat{\mathbf{K}}^2 = \pm 1$, calculations with simple k-vectors can often be significantly reduced. Thus, whereas an analytic function of any element can be defined by a power-series expansion, functions of a simple k-vector $\mathbf{K} = K\hat{\mathbf{K}}$ reduce to a real linear combination of a scalar and $\mathbf{K}$:

$$f(\mathbf{K}) = f_+(K) + \hat{\mathbf{K}} f_-(K), \tag{3.1}$$

where $f_\pm(x)$ are the even and odd parts of $f(x)$

$$f_\pm(x) = \frac{f(x) \pm f(-x)}{2}. \tag{3.2}$$

Wide use is made of such expansions in relativistic applications [3]. In Euclidean spaces, $\hat{\mathbf{K}}^2$ is positive if k is $(0 \text{ or } 1) \bmod 4$ and negative otherwise.

3.1 Hodge duals

A volume element in $C\ell_n$ is proportional to the unit n-vector

$$\mathbf{e}_T = \mathbf{e}_1\mathbf{e}_2 \cdots \mathbf{e}_n, \tag{3.3}$$

which commutes with all elements if n is odd and anticommutes with vectors if n is even. If $n = 1 \bmod 4$, $\mathbf{e}_T^2 = 1$; the vector space of the algebra can then be reduced to 2^{n-1}-dimensions by mapping $\mathbf{e}_T$ to ± 1. If $n = 3 \bmod 4$, $\mathbf{e}_T^2 = -1$ and the algebra has a natural complex structure in a space of 2^{n-1}-dimensions.

The volume element of $C\ell_n$ can be used to map multivectors of grade g into the space of grade $n - g$. Except for a possible minus sign depending on the grade and the convention used, the mapping is equivalent to the Hodge dual relation of elements, but in the Clifford algebra, the relation is easily extended from homogeneous elements to arbitrary ones, possibly of mixed grade. We take

$$^*x = x\mathbf{e}_T^{-1} \tag{3.4}$$

to be the (Hodge) dual of x.[4]

[4]This is the inverse of what Lounesto [2] calls the *Clifford dual.*

3.2 Conjugations

The conjugations of $C\ell_3$ can be directly extended to $C\ell_n$. In particular, *reversion* is defined to reverse the order of vector factors in any multivector. In $C\ell_n$ it is natural to take all the basis vectors of $\mathbb{E}^n$ to be Hermitian since they can be represented by Hermitian matrices; then reversion is the same as *Hermitian conjugation*

$$x \rightarrow x^{\dagger}. \tag{3.5}$$

(By contrast, elements that square to negative numbers cannot be represented by Hermitian matrices, so that in spaces of indefinite metric, one must generally distinguish reversion and Hermitian conjugation.) Any element $x \in C\ell_n$ can be split into real (Re) and imaginary (Im) parts

$$x = \langle x \rangle_{\text{Re}} + \langle x \rangle_{\text{Im}}\,, \tag{3.6}$$

where

$$\langle x \rangle_{\text{Re}} = \frac{1}{2}\left(x + x^{\dagger}\right) \tag{3.7}$$

$$\langle x \rangle_{\text{Im}} = \frac{1}{2}\left(x - x^{\dagger}\right). \tag{3.8}$$

In *Clifford conjugation* or *spatial reversal*, not only is the order of vector factors reversed, but they are also each multiplied by -1. The conjugation is denoted by a bar

$$x \rightarrow \bar{x}. \tag{3.9}$$

Elements of the algebra can be split into scalar-like and vector-like parts

$$x = \langle x \rangle_S + \langle x \rangle_V\,, \tag{3.10}$$

where

$$\langle x \rangle_S = \frac{1}{2}\left(x + \bar{x}\right) \tag{3.11}$$

$$\langle x \rangle_V = \frac{1}{2}\left(x - \bar{x}\right). \tag{3.12}$$

Reversion and Clifford conjugations are antiautomorphic involutions. Their combination gives the *grade automorphism*

$$x \rightarrow \bar{x}^{\dagger}. \tag{3.13}$$

Elements can be split into *even* and *odd* parts by the automorphism

$$x = \langle x \rangle_+ + \langle x \rangle_-\,, \tag{3.14}$$

$k\,(\mathrm{mod}\,4)$	S/V	Re / Im	+/−
0	S	Re	+
1	V	Re	−
2	V	Im	+
3	S	Im	−

TABLE 1.2. The properties of k-vectors.

where

$$\langle x\rangle_+ = \frac{1}{2}\left(x+\bar{x}^\dagger\right) \tag{3.15}$$

$$\langle x\rangle_- = \frac{1}{2}\left(x-\bar{x}^\dagger\right). \tag{3.16}$$

We can split elements into four parts by specifying both the reality and the vector-like properties, or, equivalently, both the reality and evenness properties, or both vector-like and evenness properties. For example,

$$x = \langle x\rangle_{\mathrm{Re}\,S} + \langle x\rangle_{\mathrm{Im}\,S} + \langle x\rangle_{\mathrm{Re}\,V} + \langle x\rangle_{\mathrm{Im}\,V}, \tag{3.17}$$

where

$$\begin{aligned}\langle x\rangle_{\mathrm{Re}\,S} &\equiv \langle\langle x\rangle_{\mathrm{Re}}\rangle_S = \langle\langle x\rangle_S\rangle_{\mathrm{Re}}\\ &= \frac{1}{4}\left(x+x^\dagger+\bar{x}+\bar{x}^\dagger\right)\\ &= \langle x\rangle_0 + \langle x\rangle_4 + \langle x\rangle_8 + \cdots,\end{aligned} \tag{3.18}$$

and similarly for the other terms. Note that scalars, i.e., 0-vectors, are real, even, and scalar-like; vectors (1-vectors) are real, odd, and vector like; bivectors (2-vectors) are imaginary, even, and vector-like; trivectors (3-vectors) are imaginary, odd, and scalar-like; 4-vectors have the same set of these properties as scalars, and, more generally, the sequence repeats in cycles of $\Delta k = 4$ as indicated in the accompanying table.

3.3 The paravector metric

A *paravector* in $C\ell_n$ is the sum of a scalar and a vector in $\mathbb{E}^n$: $p = p^0 + \mathbf{P}$. Unless stated explicitly otherwise, a paravector is assumed to be real. It is of mixed grade. Its square is not generally a scalar and, therefore, its inverse is not generally p/p^2. However, its Clifford conjugate $\bar{p} = p^0 - \mathbf{P}$ always forms a scalar product with p:

$$p\bar{p} = \bar{p}p = \overline{p\bar{p}} = \left(p^0\right)^2 - \mathbf{P}^2. \tag{3.19}$$

Any paravector p thus has an inverse

$$p^{-1} = \bar{p}/p\bar{p} \tag{3.20}$$

provided $p\bar{p}$ does not vanish. We will refer to $p\bar{p}$ as the *square norm* of the paravector p.

It is possible for the paravector p to be nonzero but to have a vanishing norm. If $p\bar{p} = 0$, then p is a *null paravector.* Such noninvertible elements are proportional to a *projector* P, which is defined to be a real idempotent

$$p = p^0 (1 + \hat{\mathbf{p}}) = 2p^0 P, \tag{3.21}$$

where

$$P = \frac{1}{2}(1 + \hat{\mathbf{p}}) = P^{\dagger} = P^2 \tag{3.22}$$

is a projector in the direction $\hat{\mathbf{p}} = \mathbf{p}/p^0$. The Clifford conjugate of P is the *complementary projector*

$$P + \bar{P} = 1, \quad P\bar{P} = \bar{P}P = 0. \tag{3.23}$$

An important property of paravector projectors is their ability to "gobble up" factors of the direction $\hat{\mathbf{p}}$:

$$\hat{\mathbf{p}}P = P\hat{\mathbf{p}} = P, \quad \hat{\mathbf{p}}\bar{P} = \bar{P}\hat{\mathbf{p}} = -\bar{P}. \tag{3.24}$$

This property, which has become known as the *pacwoman property* of projectors, can be applied repeatedly to analytic functions f of any paravector $p' = p'^0 + \mathbf{P}'$ whose vector part is parallel to $\hat{\mathbf{p}}$ $(\hat{\mathbf{p}}' = \hat{\mathbf{p}})$ to give the eigenvalue equations

$$f(p')P = f(p'^0 + |\mathbf{P}'|)P, \quad f(p')\bar{P} = f(p'^0 - |\mathbf{P}'|)\bar{P} \tag{3.25}$$

and a *spectral decomposition* of the function

$$f(p) = (P + \bar{P})f(p) = Pf(p^0 + |\mathbf{P}|) + \bar{P}f(p^0 - |\mathbf{P}|). \tag{3.26}$$

Paravectors are elements of the linear $(n+1)$-dimensional paravector space $\mathcal{V}_{p1}$ spanned by the paravector basis $\{\mathbf{e}_0, \mathbf{e}_1, \mathbf{e}_2, \mathbf{e}_3, \ldots, \mathbf{e}_n\}$ where $\mathbf{e}_0 = 1$. The square norm $p\bar{p}$ of a paravector p gives the quadratic form on the space. By replacing p by the sum $p + q$ we also find the scalar product of two paravectors p and q :

$$\langle p\bar{q}\rangle_S = \frac{p\bar{q} + q\bar{p}}{2}. \tag{3.27}$$

The paravectors $p, q \in C\ell_n$ are said to be *orthogonal* when $\langle p\bar{q}\rangle_S = 0$. If p and q are expanded in the basis, then

$$\langle p\bar{q}\rangle_S = p^{\mu}q^{\nu}\langle \mathbf{e}_{\mu}\bar{\mathbf{e}}_{\nu}\rangle_S, \tag{3.28}$$

and since the basis elements obey

$$\langle \mathbf{e}_{\mu}\bar{\mathbf{e}}_{\nu}\rangle_S =: \eta_{\mu\nu} = \begin{cases} 1, & \mu = \nu = 0, \\ -1, & \mu = \nu = 1, 2, \ldots, n, \\ 0, & \mu \neq \nu, \end{cases} \tag{3.29}$$

the paravector space is seen to be a Minkowski space with a $(1, n)$ metric signature.

3.4 Biparavectors

Biparavectors can be formed from products of paravectors p, q by

$$\langle p\bar{q}\rangle_V = \frac{1}{2}(p\bar{q} - q\bar{p}) . \tag{3.30}$$

More generally, a biparavector is defined to be any linear combination of such products over $\mathbb{R}$. The linear space $\mathcal{V}_{p2}$ of biparavectors is spanned by the $\frac{1}{2}n(n+1)$-dimensional basis of elements $\langle \mathbf{e}_\mu \bar{\mathbf{e}}_\nu \rangle_V = -\langle \mathbf{e}_\nu \bar{\mathbf{e}}_\mu \rangle_V$ with $\mu, \nu = 0, 1, \ldots, n$ and $\mu < \nu$.

In paravector spaces with $n \geq 3$, one distinguishes simple and compound biparavectors: a *simple biparavector* represents a single plane in paravector space and can be written as $\mathbf{B} = \langle p\bar{q}\rangle_V$. It is easy to show the following property.

Theorem 1. *Every simple biparavector squares to a real scalar.*

For the proof, note

$$\langle p\bar{q}\rangle_V^2 = \frac{1}{4}(p\bar{q} - q\bar{p})^2 = \frac{1}{4}(p\bar{q} + q\bar{p})^2 - p\bar{p}q\bar{q} = \langle p\bar{q}\rangle_S^2 - p\bar{p}q\bar{q}. \tag{3.31}$$

Both terms in the last line are seen to be real scalars.

We can use this property to extend the concept of *simplicity* to arbitrary elements of the algebra:

Definition 1. *An element is said to be* simple *if its product with its spatial reverse is a real scalar. Otherwise, it is* compound.

In particular, this definition encompasses simple k-paravectors and ensures that each simple k-paravector represents a k-dimensional linear space of paravectors. An analogous property is well-known for simple k-vectors.

Even simple biparavectors generally have both real (vector) and imaginary (bivector) parts. Since the product of a bivector and a vector contains only vector and trivector (imaginary scalar) parts, the last theorem implies the following:

Corollary 1. *If* $\mathbf{B}$ *is a simple biparavector, its vector and bivector parts anticommute:*

$$\langle \mathbf{B}\rangle_{\mathrm{Im}} \langle \mathbf{B}\rangle_{\mathrm{Re}} + \langle \mathbf{B}\rangle_{\mathrm{Re}} \langle \mathbf{B}\rangle_{\mathrm{Im}} = 0. \tag{3.32}$$

Geometrically, this implies that the vector part $\langle \mathbf{B}\rangle_{\mathrm{Re}}$ lies in the plane of the bivector $\langle \mathbf{B}\rangle_{\mathrm{Im}}$ and that the product $\langle \mathbf{B}\rangle_{\mathrm{Im}} \langle \mathbf{B}\rangle_{\mathrm{Re}}$ is an orthogonal real vector in the same plane. Furthermore, since $\mathbf{B}$ and $\mathbf{B}^\dagger$ differ only in the sign of their imaginary part, we can establish another useful equality.

Corollary 2. *If* $\mathbf{B}$ *is a simple biparavector,* $\mathbf{B}^2 = \left(\mathbf{B}^\dagger\right)^2$.

The product of a biparavector $\mathbf{F}$ and a paravector u generally has paravector and *triparavector* parts, and these can be isolated by taking real and imaginary parts:

$$\mathbf{F}u = \langle \mathbf{F}u \rangle_{\mathrm{Re}} + \langle \mathbf{F}u \rangle_{\mathrm{Im}}, \tag{3.33}$$

and the paravector part is orthogonal to u :

$$\langle \langle \mathbf{F}u \rangle_{\mathrm{Re}} \bar{u} \rangle_S = \frac{1}{2} \langle \mathbf{F}u\bar{u} + u\mathbf{F}^{\dagger}\bar{u} \rangle_S = \langle \mathbf{F} \rangle_{\mathrm{Re}\,S} = 0. \tag{3.34}$$

A similar relation is readily established for the imaginary part because the real and imaginary parts of $\mathbf{F}$ are vector-like:

$$\langle \langle \mathbf{F}u \rangle_{\mathrm{Im}} \bar{u} \rangle_S = 0. \tag{3.35}$$

Another important theorem shows that the paravector that results when a simple biparavector $\langle p\bar{q} \rangle_V$ acts on a paravector r lies in the plane of the biparavector.

Theorem 2. *Let p, q, r be paravectors. Then*

$$\langle \langle p\bar{q} \rangle_V \, r \rangle_{\mathrm{Re}} = p \langle \bar{q}r \rangle_S - q \langle \bar{p}r \rangle_S = \langle r \langle \bar{q}p \rangle_V \rangle_{\mathrm{Re}} \,.$$

This is a generalization of the triple vector product for vectors in $\mathbb{E}^3$:

$$(\mathbf{p} \times \mathbf{q}) \times \mathbf{r} = \mathbf{q}\,\mathbf{p} \cdot \mathbf{r} - \mathbf{p}\,\mathbf{q} \cdot \mathbf{r} = \mathbf{r} \times (\mathbf{q} \times \mathbf{p}) \,. \tag{3.36}$$

The proof follows by expanding $\langle \langle p\bar{q} \rangle_V \, r \rangle_{\mathrm{Re}}$, according to (3.12) and (3.7) and recombining.

By taking the scalar product of the paravector $\langle \langle p\bar{q} \rangle_V \, r \rangle_{\mathrm{Re}}$ with r we find that $\langle \langle p\bar{q} \rangle_V \, r \rangle_{\mathrm{Re}}$ and r are orthogonal:

Corollary 3. *Let p, q, r be paravectors. Then $\langle \langle \langle p\bar{q} \rangle_V \, r \rangle_{\mathrm{Re}} \, \bar{r} \rangle_S = 0$.*

3.5 Multiparavectors

Multiparavectors of higher grade can be constructed from the antisymmetric part of products of paravectors in which every second factor is bar-conjugated. Thus, the k-paravector formed from products of the paravectors $p_1, p_2, \ldots, p_k$ is

$$\frac{1}{k!} \sum_{\text{perm}} p_1 \bar{p}_2 p_3 \cdots, \tag{3.37}$$

where the last factor is p_k if k is odd and $\bar{p}_k$ if k is even, and the sum is over all $k!$ distinct permutations of the indices. More generally, a k-vector is defined to be any linear combination of such products over $\mathbb{R}$. The k-paravectors of $C\ell_n$ form an $\binom{n+1}{k}$-dimensional linear space $\mathcal{V}_{pk}$.

After biparavectors, the most common multiparavectors are the triparavectors. The triple product $p\bar{q}r$ of paravectors $p, q,$ and r contains paravectors and triparavectors. The imaginary part isolates the trivector part as shown by the following theorem.

Theorem 3. *Let p, q, r be paravectors. Then $\langle p\bar{q}r\rangle_{\mathrm{Im}}$ is fully antisymmetric, in the sense that*

$$\begin{aligned}\langle p\bar{q}r\rangle_{\mathrm{Im}} &= \langle\langle p\bar{q}\rangle_V\, r\rangle_{\mathrm{Im}} = \langle\langle q\bar{r}\rangle_V\, p\rangle_{\mathrm{Im}} = \langle\langle r\bar{p}\rangle_V\, q\rangle_{\mathrm{Im}} \\ &= \frac{1}{6}\left(p\bar{q}r + q\bar{r}p + r\bar{p}q - q\bar{p}r - p\bar{r}q - r\bar{q}p\right). \qquad (3.38)\end{aligned}$$

The proof is by expansion and the addition and subtraction of equivalent terms such as

$$r\,\langle\bar{p}q\rangle_S - \langle\bar{p}q\rangle_S\, r = 0. \qquad (3.39)$$

The triparavector represents a 3-volume in paravector space, as shown by the following:

Corollary 4. *The triparavector $\langle p\bar{q}r\rangle_{\mathrm{Im}}$ vanishes if and only if the paravectors p, q, r are coplanar in paravector space.*

Three coplanar paravectors are linearly dependent, and the triparavector must then vanish by antisymmetry. If on the other hand p, q, r are linearly independent, they constitute the basis of a 3-dimensional paravector subspace whose volume can be shown not to vanish.

The volume element of the $(n+1)$-dimensional paravector space is the product of the paravector basis elements, in which every second element is Clifford-conjugated. It is, thus, equal to the volume element $\mathbf{e}_T$ of $C\ell_n$ to within a sign.[5] The paravector dual of a k-paravector is an $(n+1-k)$-paravector, as one would expect, and it is given to within the above-mentioned sign by the Hodge dual (3.4) of a general element in $C\ell_n$.

3.6 Slices of biparavectors and paravectors dual to triparavectors

The slice of a simple biparavector $\mathbf{B}$ by the subspace normal to the paravector direction $\mathbf{e}_\nu$ is the paravector $\langle\mathbf{B}\mathbf{e}_\nu\rangle_{\mathrm{Re}}$ since, as shown above, this lies in $\mathbf{B}$ and is orthogonal to $\mathbf{e}_\nu$. The slice of a compound biparavector is the sum of the paravectors that result from slicing each of the contributing paravector planes. An example is the electromagnetic field $\mathbf{F}$, a biparavector whose slice normal to $\mathbf{e}_0$ gives the electric field

$$\mathbf{e} = \langle\mathbf{F}\mathbf{e}_0\rangle_{\mathrm{Re}}\,. \qquad (3.40)$$

[5] The sign is $+1$ if n is $(0 \text{ or } 3) \bmod 4$ and -1 otherwise.

The magnetic field, on the other hand, is (within a factor of c) the paravector dual in $C\ell_3$ to the triparavector formed from the product of $\mathbf{F}$ with $\mathbf{e}_0$:

$$c\mathbf{B} = -i \langle \mathbf{F}\mathbf{e}_0 \rangle_{\mathrm{Im}} . \tag{3.41}$$

3.7 Structure constants of $SL(2,\mathbb{C})$

Another theorem shows that the commutator of biparavectors is a biparavector. The result allows us to find the structure constants for the generators of Lie group of restricted Lorentz transformations.

Theorem 4. *Let p, q, r, s be any paravectors. Then*

$$\frac{1}{2}\left[\langle p\bar{q}\rangle_V , \langle r\bar{s}\rangle_V\right] = \langle q\bar{r}\rangle_S \langle p\bar{s}\rangle_V + \langle p\bar{s}\rangle_S \langle q\bar{r}\rangle_V - \langle p\bar{r}\rangle_S \langle q\bar{s}\rangle_V - \langle q\bar{s}\rangle_S \langle p\bar{r}\rangle_V . \tag{3.42}$$

The proof follows by expanding parts according to the definitions (3.11) and (3.12) and collecting terms. To find the structure constants for $SL(2,\mathbb{C})$, note that the generators of the group are the biparavectors

$$\mathbf{N}_{\mu\nu} = \frac{1}{2} \langle \mathbf{e}_\mu \bar{\mathbf{e}}_\nu \rangle_V , \tag{3.43}$$

where $\mathbf{e}_\mu, \mathbf{e}_\nu$, $\mu, \nu = 0, 1, \ldots, n$ are basis vectors which here do not need to be orthonormal. The paravector space has the metric $\eta_{\mu\nu} = \langle \mathbf{e}_\mu \bar{\mathbf{e}}_\nu \rangle_S$, so that the theorem gives

$$[\mathbf{N}_{\alpha\beta}, \mathbf{N}_{\mu\nu}] = \eta_{\beta\mu}\mathbf{N}_{\alpha\nu} + \eta_{\alpha\nu}\mathbf{N}_{\beta\mu} - \eta_{\alpha\mu}\mathbf{N}_{\beta\nu} - \eta_{\beta\nu}\mathbf{N}_{\alpha\mu} . \tag{3.44}$$

For $SL(2,\mathbb{C})$ we take $n = 3$.

4 Application to charge motion

4.1 Lorentz force

As an application of many of the above results, consider the motion of a charge e of mass m in an electromagnetic wave $\mathbf{F}$. The relevant equation of motion is the Lorentz-force equation, which in $C\ell_3$ can be written [3]

$$\dot{p} = e \langle \mathbf{F} u \rangle_{\mathrm{Re}} , \tag{4.1}$$

where $\dot{p}$ is the proper-time derivative of the spacetime momentum $p = mcu$ and u is the proper velocity in units of the speed of light c. From the above theorems, $\dot{p}$ is seen immediately to lie in the spacetime plane of $\mathbf{F}$, to be orthogonal to u, and to have a magnitude proportional to the projection of u in the biparavector plane $\mathbf{F}$.

The equation of motion (4.1) is simpler to solve in its spinorial form, which arises when the momentum paravector p is expressed in terms of the active Lorentz transformation $\Lambda(\tau) \in SL(2,\mathbb{C})$ that takes the charge from rest to its proper velocity and orientation at proper time τ :

$$p = \Lambda mc \Lambda^{\dagger}. \tag{4.2}$$

The element $\Lambda(\tau)$ used here is the spinorial form of the Lorentz transformation and is known as the *eigenspinor* [3] of the charge. It is related to the charge's proper velocity by

$$u = \Lambda \mathbf{e}_0 \Lambda^{\dagger} = \Lambda \Lambda^{\dagger}. \tag{4.3}$$

From the unimodular property of Λ, namely $\Lambda\bar{\Lambda} = 1$, the eigenspinor is seen to satisfy an equation of motion of the form

$$\dot{\Lambda} = \frac{1}{2}\Omega\Lambda, \tag{4.4}$$

where the biparavector $\Omega \equiv 2\dot{\Lambda}\bar{\Lambda}$ gives the spacetime rotation rate of the charge. By taking the proper time derivative of p (4.2) and substituting (4.4), we obtain

$$\dot{p} = \dot{\Lambda} mc \Lambda^{\dagger} + \Lambda mc \dot{\Lambda}^{\dagger} = mc \left\langle \Omega \Lambda \Lambda^{\dagger} \right\rangle_{\mathrm{Re}} = mc \left\langle \Omega u \right\rangle_{\mathrm{Re}}. \tag{4.5}$$

This has the same form as the Lorentz-force equation (4.1), and a comparison reveals that the spacetime rotation rate of the charge in the field $\mathbf{F}$ is $\Omega = e\mathbf{F}/mc$. It also indicates that the Lorentz-force equation (4.1) can be replaced by the simpler spinorial form

$$\dot{\Lambda} = \frac{e\mathbf{F}}{2mc}\Lambda. \tag{4.6}$$

4.2 Plane-wave pulse

Let the electromagnetic field be that of the directed plane-wave pulse

$$\mathbf{F} = \left(1 + \hat{\mathbf{k}}\right) \mathbf{e}(s) = \frac{c}{\omega} \mathbf{e}(s) \bar{k}, \tag{4.7}$$

where the electric field $\mathbf{e}(s)$ is a real function of the scalar $s = \langle k\bar{x} \rangle_S = \omega t - \mathbf{k} \cdot \mathbf{x}$, and the null paravector $k = \omega/c + \mathbf{K}$ is the nominal propagation paravector in the direction $\hat{\mathbf{k}}$, which is orthogonal to $\mathbf{E}$. The functional form of $\mathbf{e}(s)$ is arbitrary, but for a linearly polarized Gaussian pulse, it could have the form

$$\mathbf{e}(s) = \mathbf{e}(0)\, e^{-\alpha s^2} \cos s, \tag{4.8}$$

where α is a constant proportional to the inverse width of the wave packet. If $\mathbf{e}(s)$ is of finite duration, the wave is not monochromatic, but ω is taken as a representative angular frequency. The field (4.7) is seen to be a null biparavector $(\mathbf{F}\bar{\mathbf{F}} = 0)$ and, hence, simple. Its vector part is the electric field $\mathbf{e}(s)$, and its bivector part gives the plane whose normal is the magnetic field: $\hat{\mathbf{k}}\mathbf{E}(s) = ic\mathbf{B}$. The fields are seen to satisfy Corollary 1.

As simple as the spinorial equation (4.6) is, it appears at first to be impossible to solve since $\mathbf{F}$ is to be evaluated at the spacetime position of the charge, and the position is known only after the equation is solved and u is integrated. However, a remarkable symmetry permits direct solution. The key to the solution is the surprising invariance of the wave paravector k not only in the lab, but also in the instantaneous rest frame of the accelerating charge. Because $kc = \omega\left(1 + \hat{\mathbf{k}}\right)$ is a null paravector and thus proportional to a projector, the spinorial Lorentz-force equation (4.6) in the field (4.7) gives

$$\bar{k}\dot{\Lambda} = \frac{e\bar{k}\mathbf{F}}{2mc}\Lambda = 0. \tag{4.9}$$

From conjugations of this result, $\bar{\dot{\Lambda}}k = 0 = k\bar{\dot{\Lambda}}^{\dagger}$, and it follows that

$$k_{\text{rest}} = \frac{\omega_{\text{rest}}}{c}\left(1 + \hat{\mathbf{k}}\right) = \bar{\Lambda}k\bar{\Lambda}^{\dagger} = \text{const.} \tag{4.10}$$

Furthermore, the proper time-rate of change of the Lorentz scalar s is simply the fixed rest-frame frequency

$$\dot{s} = c\left\langle k\bar{u}\right\rangle_S = \omega_{\text{rest}}. \tag{4.11}$$

Consequently, the evolution (4.4) of the eigenspinor can be expressed

$$\frac{d\Lambda}{ds} = \frac{1}{\dot{s}}\dot{\Lambda} = \frac{e\mathbf{F}\Lambda}{2mc\omega_{\text{rest}}} = \frac{e\mathbf{e}(s)\bar{k}\Lambda(s)}{2m\omega\omega_{\text{rest}}} = \frac{ek\mathbf{e}(s)\Lambda(0)}{2m\omega\omega_{\text{rest}}} \tag{4.12}$$

since integration of $\bar{k}\dot{\Lambda} = 0$ implies that $\bar{k}\Lambda$ is constant, and $k\mathbf{e} = \mathbf{e}\bar{k}$. In terms of the paravector potential $A(s)$,

$$\mathbf{F} = c\left\langle\partial\bar{A}(s)\right\rangle_V = c\left\langle k\bar{A}'(s)\right\rangle_V, \tag{4.13}$$

where $A'(s) = dA(s)/ds$ and, with the Lorenz-gauge condition $\left\langle\partial\bar{A}\right\rangle_S = 0$,

$$\mathbf{F} = ck\bar{A}'(s) = -cA'(s)\bar{k} \tag{4.14}$$

so that $d\Lambda/ds$ can be integrated to give $\Delta\Lambda \equiv \Lambda(s) - \Lambda(0)$ proportional to the change ΔA in the paravector potential

$$\Delta\Lambda = \frac{ek\Delta\bar{A}}{2m\omega_{\text{rest}}}\Lambda(0). \tag{4.15}$$

This beautiful and simple result can be used to find the proper velocity $u = \Lambda\Lambda^{\dagger}$ and spacetime position $x = c\int u d\tau$ of the charge at later times. The method, which depends critically on the eigenspinor formalism and projectors in the paravector model of relativistic phenomena, can be extended to treat charge motion in plane-wave pulses superimposed on static axial electric or magnetic fields. Such a treatment has recently been applied to the motion of electrons in laser fields, elucidated ponderomotive momentum and its contributions to the dressed-particle mass, and yielded important analytic solutions for the autoresonant laser accelerator. [7]

Our ability to solve the spinorial Lorentz-force equation rests largely on the surprising invariance, noted above, of the propagation vector k in the rest frame of the charge. While Hestenes [8] proved this invariance for the case of linearly polarized monochromatic waves, its significance has evidently been largely overlooked. We showed above that it holds also for more general directed plane waves of any polarization and pulse shape. We can now use the geometry of some of the theorems of the previous section to shed light on the origin of this invariance. First, note that the

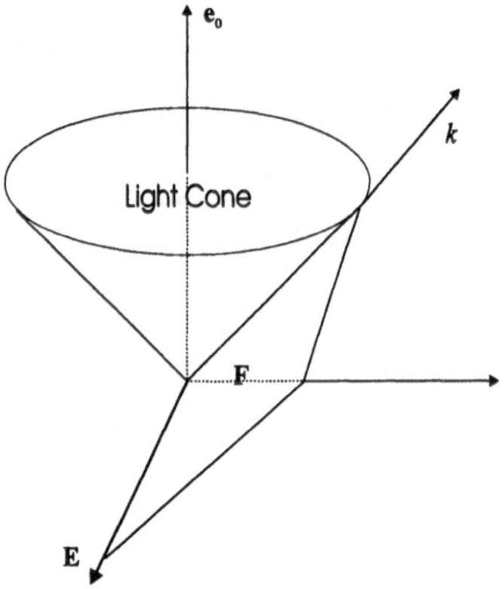

FIGURE 2. The plane-wave pulse of electromagnetic field is a null flag $\mathbf{F}$. The propagation paravector k lies in the flag but is also orthogonal to it and is consequently invariant under rotations in the flag plane.

electromagnetic field of the directed plane-wave pulse is a *null flag*, [9] formed from the product of the null propagation paravector k and the orthogonal electric field $\mathbf{e}$. The flagpole is given by k, and the flag plane is tangent to the light cone at k (see Fig. 2). Because k is null, it is orthogonal to itself as well as to $\mathbf{e}$; therefore, it is orthogonal to any linear combination of k and $\mathbf{E}$. Consequently, k, in addition to lying *in* the flag plane, is also orthogonal to it. The spinorial Lorentz-force equation (4.6) predicts a change in the eigenspinor by a spacetime rotation at the rate $\Omega = e\mathbf{F}/mc$ in the flag plane. It is a covariant equation valid for any inertial frame, including the frame instantaneously moving with the charge. The spacetime rotation has both spatial rotation and boost components and is responsible for the acceleration of the electron interaction with the field. However, since k is orthogonal to Ω, it is invariant under rotations in it. Thus, the electron accelerates and rotates in such a way that the Doppler shift of the radiation and its propagation direction in the commoving frame are constant.

5 Conclusions

Many applications of Clifford algebras, while exploiting the powerful geometrical tools of the algebras, largely overlook the rich internal structure associated with overlapping subspaces of mixed grades. This contribution has investigated the use of paravector and multiparavector subspaces of $C\ell_n$. A number of theorems have been derived as an aid to interpreting the paravector formalism, and applications to charge motion in electromagnetic plane-wave pulses illustrate the power and beauty of the covariant paravector model in $C\ell_3$ for problems in relativistic dynamics. Applications to higher-dimensional spaces are currently in progress.

Acknowledgement

The author thanks the Natural Sciences and Engineering Research Council of Canada for support of his research. Enlightening conversations with colleagues are also gratefully acknowledged, in particular with Dr. John Huschilt and with students Yuan Yao, Greg Trayling, Jacob Alexander, Shazia Hadi, and David Keselica.

REFERENCES

[1] W. E. Baylis, ed., *Clifford (Geometric) Algebra with Applications to Physics, Mathematics, and Engineering,* Birkhäuser, Boston, 1996.

[2] P. Lounesto, *Clifford Algebras and Spinors,* Cambridge University Press, Cambridge, 1997.

[3] W. E. Baylis, ed., *Electrodynamics: A Modern Geometric Approach,* Birkhäuser, Boston, 1999.

[4] J. G. Maks, PhD thesis, Technische Universiteit Delft, the Netherlands, 1989.

[5] K. S. T. Charles W. Misner and J. A. Wheeler, *Gravitation,* W. H. Freeman and Co., San Francisco, 1970.

[6] J. D. Jackson, *Classical Electrodynamics, Third Edition,* J. Wiley and Sons, New York, 1999.

[7] W. E. Baylis and Y. Yao, Relativistic dynamics of charges in electromagnetic fields: an eigenspinor approach, *Phys. Rev. A* **60** (1999), 785–795.

[8] D. Hestenes, Proper dynamics of a rigid point particle, *J. Math. Phys.* **15** (1974), 1778–1786.

[9] R. Penrose and W. Rindler, *Spinors and Space-Time Volume I: Spinors and Spacetime,* Cambridge University, Cambridge, 1984.

William E. Baylis
Department of Physics
University of Windsor
Windsor, Ontario
Canada N9B 3P4
E-mail: baylis@uwindsor.ca

Received: September 30, 1999; Revised: March 15, 2000

Quaternionic Spin

Tevian Dray and Corinne A. Manogue

ABSTRACT We rewrite the standard 4-dimensional Dirac equation in terms of quaternionic 2-component spinors, leading to a formalism which treats massive and massless particles on an equal footing. The resulting description has the correct number of spin/helicity states to be a generation of leptons. Furthermore, precisely three such generations naturally combine into an octonionic description of the 10-dimensional massless Dirac equation, as previously discussed in [1].
Keywords: Dirac equation, dimensional reduction, quaternions, octonions.

1 Introduction

We recently outlined a new dimensional reduction scheme [1]. We show here in detail that applying this mechanism to the 10-dimensional massless Dirac equation on Majorana-Weyl spinors quaternionic description of the full 4-dimensional (free) Dirac equation which treats massive and massless particles on an equal footing. Furthermore, there are naturally 3 such descriptions, each of which corresponds to a generation of leptons with the correct number of spin/helicity states.

The massive Dirac equation is usually formulated in the context of 4-component *Dirac spinors.* The 4 degrees of freedom correspond to the choice of spin (up or down) and the choice of particle or antiparticle. Similarly, 2-component *Penrose spinors*, which can be thought of as the square roots of null vectors, correspond to massless objects such as photons. In Section 2 we set the stage by reviewing these standard properties of the chiral description of the momentum-space Dirac equation.

Penrose spinors are usually thought of as Weyl projections of Dirac spinors; a Dirac spinor contains twice the information of a single Penrose spinor. As an alternative to doubling the number of (complex) components, however, we double the dimension of the underlying division algebra from the complex numbers $\mathbb{C}$ to the quaternions $\mathbb{H}$. The anticommutativity of the quaternions then enables us to package two complex representations of opposite chirality into the (now quaternionic) 2-component formalism. In Section 3 we show how to replace the usual 4-component complex Dirac description with an equivalent 2-component quaternionic Penrose description and further discuss how this puts the massive and massless Dirac equations

AMS Subject Classification: 81R99, 17A35, 17C90, 15A33, 15A18.

on an equal footing.

We then consider in Section 4 the massless Dirac equation on Majorana-Weyl spinors (in momentum space) in 10 dimensions, which can be nicely described in terms of 2-component spinors over the octonions $\mathbb{O}$, the only other normed division algebra besides $\mathbb{R}$, $\mathbb{C}$, and $\mathbb{H}$. Solutions of this equation are automatically quaternionic and thus lend themselves to the preceding quaternionic description.

The final, and most important, ingredient in our approach is the dimensional reduction scheme introduced in [1]. In Sections 5.1 and 5.2 we describe how the choice of a preferred octonionic unit, or equivalently of a preferred complex subalgebra $\mathbb{C} \subset \mathbb{O}$, naturally reduces 10 spacetime dimensions to 4, and further allows us to use the standard representation of the Lorentz group $SO(3,1)$ as $SL(2,\mathbb{C}) \subset SL(2,\mathbb{O})$. Putting this all together, we show in Section 5.3 that the quaternionic spin/helicity eigenstates correspond precisely to the particle spectrum of a generation of leptons, consisting of 1 massive and 1 massless particle and their antiparticles.

In Section 5.4, we discuss the remarkable fact that the quaternionic spin eigenstates are, in fact, simultaneous eigenstates of all 3 orthogonal projections of the spin operator although two of the eigenvalues are not real. Finally, in Section 6 we discuss our results – in particular that – in a natural sense there are precisely 3 such quaternionic subalgebras of the octonions, which we interpret as generations.

There is a long history of trying to use the quaternions in 4-dimensional quantum mechanics; see the comprehensive treatment in [2] and references therein. Our approach is different in that we use the additional degrees of freedom to repackage existing information, rather than increasing the size of the underlying space of scalars. Ultimately, this leads us to work in more than 4 spacetime dimensions.

We also note a relatively unknown paper by Dirac [3] which, much to our surprise, contains the precursors of several of the key ideas presented here.

The octonions were first introduced into quantum mechanics by Jordan [4, 5]. There has, in fact, been much recent interest in using the octonions in (higher-dimensional) field theory; excellent modern treatments can be found in [6, 7, 8].

After much of this work was completed, we became aware of the recent work of Schücking *et al.* [9, 10], who also uses a quaternionic formalism to describe a single generation of leptons. They further speculate that extending the formalism to the octonions would yield a description of a single generation of quarks as well. Although the language is strikingly similar, our approach differs fundamentally from theirs in its description of momentum. Ultimately, this hinges on our interpretation of the obvious $SU(2)$ as spin, whereas Schücking and coworkers interpret it as isospin.

2 Complex formalism

The standard Weyl representation of the gamma matrices is

$$\gamma_t = \begin{pmatrix} 0 & I \\ I & 0 \end{pmatrix}, \qquad \gamma_a = \begin{pmatrix} 0 & \sigma_a \\ -\sigma_a & 0 \end{pmatrix} \tag{2.1}$$

where σ_a for $a = x, y, z$ denote the usual Pauli matrices [1]

$$\sigma_x = \begin{pmatrix} 0 & 1 \\ 1 & 0 \end{pmatrix}, \qquad \sigma_y = \begin{pmatrix} 0 & -\ell \\ \ell & 0 \end{pmatrix}, \qquad \sigma_z = \begin{pmatrix} 1 & 0 \\ 0 & -1 \end{pmatrix}, \tag{2.2}$$

I is the 2×2 identity matrix, and our signature is $(+---)$.

The original formulation of the Dirac equation involves the even part of the Clifford algebra, historically written in terms of the matrices $\alpha_a = \gamma_t \gamma_a$ and $\beta = \gamma_t$. Explicitly, the momentum-space Dirac equation in this signature can be written as

$$(\gamma_t \gamma_\alpha \, p^\alpha - m \, \gamma_t) \, \Psi = 0 \tag{2.3}$$

where $\alpha = t, x, y, z$ and Ψ is a 4-component complex (Dirac) spinor.

Writing Ψ in terms of two 2-component complex Weyl (or Penrose) spinors θ and η as

$$\Psi = \begin{pmatrix} \theta \\ \eta \end{pmatrix} \tag{2.4}$$

and expanding (2.3) leads to

$$\begin{pmatrix} p^t I - p^a \sigma_a & -m \\ -m & p^t I + p^a \sigma_a \end{pmatrix} \begin{pmatrix} \theta \\ \eta \end{pmatrix} = 0. \tag{2.5}$$

This leads us to identify the momentum 4-vector with the 2×2 Hermitian matrix

$$\mathbf{p} = p^\alpha \sigma_\alpha = \begin{pmatrix} p^t + p^z & p^x - \ell p^y \\ p^x + \ell p^y & p^t - p^z \end{pmatrix} \tag{2.6}$$

where we have set $\sigma_t = I$, which reduces (2.5) to the two equations

$$-\widetilde{\mathbf{p}} \, \theta - m\eta = 0 \tag{2.7}$$

$$-m\theta + \mathbf{p} \, \eta = 0 \tag{2.8}$$

where the tilde denotes trace-reversal. Explicitly,

$$\widetilde{\mathbf{p}} = \mathbf{p} - \operatorname{tr}(\mathbf{p}) I \tag{2.9}$$

[1] For later compatibility with our octonion conventions we use ℓ rather than i to denote the complex unit.

which reverses the sign of p^t, so that $-\widetilde{\mathbf{p}}$ can be identified with the 1-form dual to $\mathbf{p}$. This interpretation is strengthened by noting that

$$-\widetilde{\mathbf{p}}\mathbf{p} = \det(\mathbf{p})I = p_\alpha p^\alpha I = m^2 I \tag{2.10}$$

where the identification of the norm of p^α with m is just the compatibility condition between (2.7) and (2.8).

3 Quaternionic formalism

The *quaternions* $\mathbb{H}$ are the associative, noncommutative, normed division algebra over the reals. The quaternions are spanned by the identity element 1 and three imaginary units, usually denoted $i,\ j,\ k := ij$. Quaternionic conjugation, denoted by a bar, is given by reversing the sign of each imaginary unit. Each imaginary unit squares to -1, and they anticommute with each other; the full multiplication table then follows using associativity.[2] However, in order to avoid conflict with our subsequent conventions for the octonions, we will instead label our quaternionic basis $\ell,\ k,\ \ell k$. The imaginary unit ℓ will play the role of the complex unit i, and, as we will see later, k will label this particular quaternionic subalgebra of the octonions. In terms of the Cayley-Dickson process [12, 13], we have

$$\mathbb{H} = \mathbb{C} + \mathbb{C}k = (\mathbb{R} + \mathbb{R}\ell) + (\mathbb{R} + \mathbb{R}\ell)k. \tag{3.1}$$

As vector spaces, $\mathbb{H} = \mathbb{C}^2$, which allows us to identify $\mathbb{H}^2$ with $\mathbb{C}^4$ in many different ways. We make the particular choice

$$\begin{pmatrix} A \\ B \\ C \\ D \end{pmatrix} \longleftrightarrow \begin{pmatrix} C - kB \\ D + kA \end{pmatrix} \tag{3.2}$$

with $A, B, C, D \in \mathbb{C}$. Equivalently, we can write this identification in terms of the Weyl (Penrose) spinors θ and η as

$$\Psi = \begin{pmatrix} \theta \\ \eta \end{pmatrix} \longleftrightarrow \eta + \sigma_k \theta \tag{3.3}$$

where we have introduced the generalized Pauli matrix

$$\sigma_k = \begin{pmatrix} 0 & -k \\ k & 0 \end{pmatrix}. \tag{3.4}$$

[2] The use of $\vec{\imath},\ \vec{\jmath},\ \vec{k}$ for Cartesian basis vectors originates with the quaternions, which were introduced by Hamilton as an early step towards vectors [11]. Making the obvious identification of vectors $\vec{v},\ \vec{w}$ with imaginary quaternions $v,\ w$, the real part of the quaternionic product vw is then (minus) the dot product $\vec{v} \cdot \vec{w}$, while the imaginary part is the cross product $\vec{v} \times \vec{w}$.

Since (3.3) is clearly a vector space isomorphism, it induces an isomorphism relating the linear maps on these spaces. We can use this induced isomorphism to rewrite the Dirac equation (2.3) in 2-component quaternionic language. Direct computation yields the correspondences

$$\gamma_t\gamma_a = \begin{pmatrix} -\sigma_a & 0 \\ 0 & \sigma_a \end{pmatrix} \longleftrightarrow \sigma_a \tag{3.5}$$

and

$$\gamma_t \longleftrightarrow \sigma_k \tag{3.6}$$

and, of course, also

$$\gamma_t\gamma_t \longleftrightarrow \sigma_t \tag{3.7}$$

since the left-hand side is the 4×4 identity matrix and the right-hand side is the 2×2 identity matrix. In each case, the 4×4 complex matrix on the left yields the same linear map on the components A, B, C, D as the 2×2 quaternionic matrix on the right when acting on the appropriate side of (3.2). Direct translation of (2.3) now leads to the quaternionic Dirac equation

$$(\mathbf{p} - m\sigma_k)(\eta + \sigma_k\theta) = 0. \tag{3.8}$$

Working backwards, we can separate this into an equation not involving k, which is precisely (2.8), and an equation involving k, which is

$$\mathbf{p}\,\sigma_k\theta - m\sigma_k\eta = 0. \tag{3.9}$$

Multiplying this equation on the left by σ_k and using the remarkable identity

$$\sigma_k\,\mathbf{p}\,\sigma_k = -\widetilde{\mathbf{p}} \tag{3.10}$$

reduces (3.9) to (2.7), as expected.

So far, we have done nothing than rewrite the usual momentum-space Dirac equation in 2-component quaternionic language. However, the appearance of the term $m\sigma_k$ suggests a way of putting in the mass term on the same footing as the other terms, which we now exploit. Multiplying (3.8) on the left by $-\sigma_k$ and using (3.10) brings the Dirac equation to the form

$$(\widetilde{\mathbf{p}} + m\sigma_k)\,\psi = 0 \tag{3.11}$$

where we have introduced the 2-component quaternionic spinor

$$\psi = \sigma_k(\eta + \sigma_k\theta) = \theta + \sigma_k\eta. \tag{3.12}$$

When written out in full, (3.11) takes the form

$$\begin{pmatrix} -p^t + p^z & p^x - \ell p^y - km \\ p^x + \ell p^y + km & -p^t - p^z \end{pmatrix} \psi = 0. \tag{3.13}$$

This clearly suggests viewing the mass as an additional spacelike component of a higher-dimensional vector. Furthermore, since the matrix multiplying ψ has determinant zero, this higher-dimensional vector is null. We thus appear to have reduced the massive Dirac equation in 4 dimensions to the massless Dirac, or Weyl, equation in higher dimensions, thus, putting the massive and massless cases on an equal footing. This expectation is indeed correct as we show in the next section in the more general octonionic setting.

4 Octonionic formalism

4.1 Octonionic Penrose spinors

The *octonions* $\mathbb{O}$ are the nonassociative, noncommutative, normed division algebra over the reals. The octonions are spanned by the identity element 1 and seven imaginary units, which we label as $\{i, j, k, k\ell, j\ell, i\ell, \ell\}$. Each imaginary unit squares to -1

$$i^2 = j^2 = k^2 = ... = \ell^2 = -1, \tag{4.1}$$

and the full multiplication table can be conveniently encoded in the 7-point projective plane, as shown in Figure 1; each line is to be thought of as a

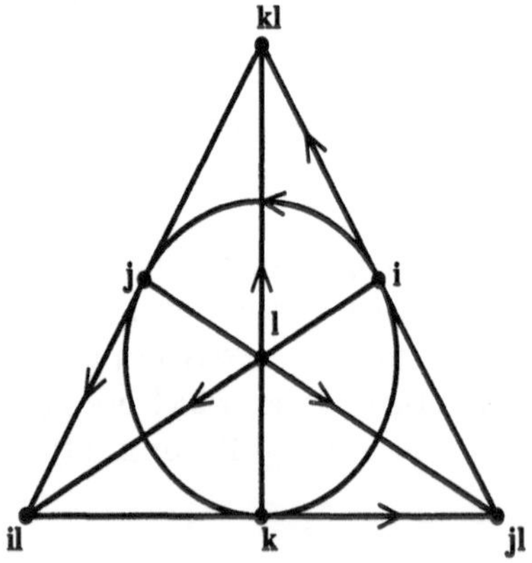

FIGURE 1. The representation of the octonionic multiplication table using the 7-point projective plane. Each of the 7 oriented lines gives a quaternionic triple.

circle. The octonionic units can be grouped into (the imaginary parts of) quaternionic subalgebras in 7 different ways, corresponding to the 7 lines in the figure; these will be referred to as quaternionic triples. Within each triple, the arrows give the orientation, so that, e.g.,

$$ij = k = -ji. \tag{4.2}$$

Any three imaginary basis units which do not lie in such a triple anti-associate. Note that any two octonions automatically lie in (at least one) quaternionic triple, so that expressions containing only two independent imaginary octonionic directions do associate. *Octonionic conjugation* is given by reversing the sign of the imaginary basis units, and the norm is just

$$|p| = \sqrt{p\overline{p}} \tag{4.3}$$

which satisfies the defining property of a normed division algebra, namely

$$|pq| = |p||q|. \tag{4.4}$$

We follow [14, 15] in representing real $(9+1)$-dimensional Minkowski space in terms of 2×2 Hermitian octonionic matrices.[3] In analogy with the complex case, a vector field q^μ with $\mu = 0, ..., 9$ can be thought of under this representation as a matrix

$$Q = \begin{pmatrix} q^+ & \overline{q} \\ q & q^- \end{pmatrix} \tag{4.5}$$

where $q^\pm = q^0 \pm q^9 \in \mathbb{R}$ are the components of q^μ in 2 null directions, and $q = q^1 + q^2 i + \ldots + q^8 \ell \in \mathbb{O}$ is an octonion representing the transverse spatial coordinates. Following [19], we define

$$\widetilde{Q} = Q - \operatorname{tr}(Q) I. \tag{4.6}$$

Furthermore, since Q satisfies its characteristic polynomial, we have

$$-Q\widetilde{Q} = -\widetilde{Q}Q = -Q^2 + \operatorname{tr}(Q)Q = \det(Q)I = g_{\mu\nu} q^\mu q^\nu I \tag{4.7}$$

where $g_{\mu\nu}$ is the Minkowski metric (with signature $(+-\ldots-)$). We can therefore identify the tilde operation with the metric dual, so that $-\widetilde{Q}$ represents the covariant vector field q_μ.

---

[3] A number of authors, such as [16], have used this approach to describe supersymmetric theories in 10 dimensions. Fairlie & Manogue [17, 18] and Manogue & Sudbery [19] described solutions of the superstring equations of motion using octonionic parameters, and Schray [20, 21] described the superparticle. A more extensive bibliography appears in [20].

Just as in the complex case (compare (2.1)), this can be thought of (up to associativity issues) as a Weyl representation of the underlying Clifford algebra $Cl(9,1)$ in terms of 4×4 gamma matrices of the form

$$q^\mu \gamma_\mu = \begin{pmatrix} 0 & Q \\ -\widetilde{Q} & 0 \end{pmatrix} \tag{4.8}$$

which are now octonionic. It is readily checked that

$$\gamma_\mu \gamma_\nu + \gamma_\nu \gamma_\mu = 2\, g_{\mu\nu} \tag{4.9}$$

as desired. In this language, a Majorana spinor $\Psi = \begin{pmatrix} \psi \\ \chi \end{pmatrix}$ is a 4-component octonionic column, whose chiral projections are the Majorana-Weyl spinors $\begin{pmatrix} \psi \\ 0 \end{pmatrix}$ and $\begin{pmatrix} 0 \\ \chi \end{pmatrix}$, which can be identified with the 2-component octonionic columns ψ and χ, which in turn can be thought of as generalized Penrose spinors. Writing

$$\gamma_\mu = \begin{pmatrix} 0 & \sigma_\mu \\ -\widetilde{\sigma}_\mu & 0 \end{pmatrix} \tag{4.10}$$

or, equivalently,

$$Q = q^\mu \sigma_\mu = q_\mu \sigma^\mu \tag{4.11}$$

defines the *octonionic Pauli matrices* σ_μ. The matrices σ_a, with $a = 1, \dots, 9$, are the natural generalization of the ordinary Pauli matrices to the octonions, and $\sigma_0 = I$. In analogy with our treatment of the complex case, we have

$$\gamma_0 \gamma_\mu = \begin{pmatrix} -\widetilde{\sigma}_\mu & 0 \\ 0 & \sigma_\mu \end{pmatrix}. \tag{4.12}$$

For completeness, we record some useful relationships. The adjoint $\overline{\Psi}$ of the Majorana spinor Ψ is given as usual by

$$\overline{\Psi} = \Psi^\dagger \gamma_0 \tag{4.13}$$

since

$$\gamma_\mu^\dagger \gamma_0^\dagger = \gamma_0 \gamma_\mu. \tag{4.14}$$

Given a Majorana spinor $\Psi = \begin{pmatrix} \psi \\ \chi \end{pmatrix}$, we can construct a real vector[4]

$$q^\mu[\Psi] = \mathrm{Re}(\Psi^\dagger \gamma^0 \gamma^\mu \Psi) \tag{4.15}$$

---

[4]We assume here that the components of our spinors are *commuting*, as we believe that the anticommuting nature of fermions may be carried by the octonionic units themselves. An analogous result for anticommuting spinors was obtained in of [18, 20].

corresponding in traditional language to $\overline{\Psi}\gamma^\mu\Psi$. We can further identify this with a 2×2 matrix $Q[\Psi]$ as in (4.8) above. Direct computation using the cyclic property of the trace, e.g., for octonionic columns Ψ_1, Ψ_2, and octonionic matrices γ

$$\mathrm{Re}(\Psi_1^\dagger\gamma\Psi_2) = \mathrm{Re}\left(\mathrm{tr}\,(\Psi_1^\dagger\gamma\Psi_2)\right) = \mathrm{Re}\left(\mathrm{tr}\,(\Psi_2\Psi_1^\dagger\gamma)\right) \tag{4.16}$$

shows that [17, 18]

$$Q[\Psi] = 2\,\psi\psi^\dagger - 2\,\widetilde{\chi\chi^\dagger}. \tag{4.17}$$

4.2 Octonionic Dirac equation

The momentum-space massless Dirac equation (Weyl equation) in 10 dimensions can be written in the form

$$\gamma_0\gamma_\mu\,p^\mu\,\Psi = 0. \tag{4.18}$$

Choosing $\Psi = \begin{pmatrix}\psi\\0\end{pmatrix}$ to be a Majorana-Weyl spinor, and using (4.12) and (4.11), (4.18) finally takes the form

$$\widetilde{P}\psi = 0 \tag{4.19}$$

which is the octonionic Weyl equation. In matrix notation, it is straightforward to show that the momentum p^μ of a solution of the Weyl equation must be null: (4.19) says that the 2×2 Hermitian matrix P has 0 as one of its eigenvalues,[5] which forces $\det(P) = 0$, which is

$$\widetilde{P}P = 0 \tag{4.20}$$

which in turn is precisely the condition that p^μ be null.

Equations (4.20) and (4.19) are algebraically the same as the octonionic versions of two of the superstring equations of motion, as discussed in [17, 18, 19], and are also the octonionic superparticle equations [20]. As implied by those references, (4.20) implies the existence of a 2-component spinor θ such that

$$P = \pm\theta\theta^\dagger \tag{4.21}$$

where the sign corresponds to the time orientation of P, and the general solution of (4.19) is

$$\psi = \theta\xi \tag{4.22}$$

[5] It is *not* true in general [22, 23] that the determinant of an $n\times n$ Hermitian octonionic matrix is the product of its (real) eigenvalues, unless $n = 2$; however, see also [24].

where $\xi \in \mathbb{O}$ is arbitrary. The components of θ lie in the complex subalgebra of $\mathbb{O}$ determined by P, so that (the components of) θ and ξ (and, hence, also P) belong to a quaternionic subalgebra of $\mathbb{O}$. Thus, for solutions (4.22), the Weyl equation (4.18) itself becomes quaternionic.

Furthermore, it follows immediately from (4.22) that

$$\psi\psi^\dagger = \pm|\xi|^2 P. \tag{4.23}$$

Comparing this with (4.17), we see that the vector constructed from ψ is proportional to P, or in more traditional language

$$\overline{\Psi}\gamma^\mu\Psi \sim p^\mu \tag{4.24}$$

which can be interpreted as the requirement that the Pauli-Lubanski spin vector be proportional to the momentum for a massless particle.

5 Dimensional reduction and spin

5.1 Choosing a preferred complex subalgebra

The description in the preceding section of 10-dimensional Minkowski space in terms of Hermitian octonionic matrices is a direct generalization of the usual description of ordinary (4-dimensional) Minkowski space in terms of complex Hermitian matrices. If we fix a complex subalgebra $\mathbb{C} \subset \mathbb{O}$, then we single out a 4-dimensional Minkowski subspace of 10-dimensional Minkowski space. The projection of a 10-dimensional null vector onto this subspace is a causal 4-dimensional vector, which is null if and only if the original vector was already contained in the subspace, and timelike otherwise. The time orientation of the projected vector is the same as that of the original, and the induced mass is given by the norm of the remaining 6 components. Furthermore, the ordinary Lorentz group $SO(3,1)$ clearly sits inside the Lorentz group $SO(9,1)$ via the identification of their double-covers, the spin groups $\mathrm{Spin}(d,1)$, namely,[6]

$$\mathrm{Spin}(3,1) = SL(2,\mathbb{C}) \subset SL(2,\mathbb{O}) = \mathrm{Spin}(9,1). \tag{5.1}$$

Therefore, all it takes to break 10 spacetime dimensions to 4 is to choose a preferred octonionic unit to play the role of the complex unit. We choose ℓ rather than i to fill this role, preferring to save i, j, k for a (distinguished) quaternionic triple. We choose the particular projection π from $\mathbb{O}$ to $\mathbb{C}$ given by

$$\pi(q) = \frac{1}{2}(q + \ell q \bar{\ell}), \tag{5.2}$$

[6]The last equality is more usually discussed at the Lie algebra level. Manogue & Schray [25] gave an explicit representation using this language of the *finite* Lorentz transformations in 10 spacetime dimensions. For further discussion of the notation $SL(2,\mathbb{O})$, see also [26].

and we thus obtain a preferred $SL(2,\mathbb{C})$ subgroup of $SL(2,\mathbb{O})$, corresponding to the "physical" Lorentz group.

5.2 Spin

Since we now have a preferred 4-d Lorentz group, we can use its rotation subgroup $SU(2) \subset SL(2,\mathbb{C})$ to define spin. However, care must be taken when constructing the Lie algebra $su(2)$, due to the lack of commutativity.

Under the usual action of $M \in SU(2)$ on a Hermitian matrix Q (thought of as a spacetime vector via (4.11)), namely,

$$Q \mapsto MQM^\dagger \tag{5.3}$$

we can identify the basis rotations as usual as

$$R_z = \begin{pmatrix} e^{\ell\frac{\phi}{2}} & 0 \\ 0 & e^{-\ell\frac{\phi}{2}} \end{pmatrix}, \quad R_y = \begin{pmatrix} \cos\frac{\phi}{2} & \sin\frac{\phi}{2} \\ -\sin\frac{\phi}{2} & \cos\frac{\phi}{2} \end{pmatrix},$$
$$R_x = \begin{pmatrix} \cos\frac{\phi}{2} & \ell\sin\frac{\phi}{2} \\ \ell\sin\frac{\phi}{2} & \cos\frac{\phi}{2} \end{pmatrix} \tag{5.4}$$

corresponding to rotations by the angle ϕ about the z, y, and x axes, respectively.

The infinitesimal generators of the Lie algebra $su(2)$ are obtained by differentiating these group elements, via

$$L_a = \left.\frac{dR_a}{d\phi}\right|_{\phi=0} \tag{5.5}$$

where as before $a = x, y, z$. For reasons which will become apparent, we have *not* multiplied these generators by $-\ell$ to obtain Hermitian matrices. We have instead

$$2L_z = \begin{pmatrix} \ell & 0 \\ 0 & -\ell \end{pmatrix}, \quad 2L_y = \begin{pmatrix} 0 & 1 \\ -1 & 0 \end{pmatrix}, \quad 2L_x = \begin{pmatrix} 0 & \ell \\ \ell & 0 \end{pmatrix}, \tag{5.6}$$

which satisfy the commutation relations

$$[L_a, L_b] = \epsilon_{abc} L_c \tag{5.7}$$

where ϵ is completely antisymmetric and

$$\epsilon_{xyz} = 1. \tag{5.8}$$

Spin eigenstates are usually obtained as eigenvectors of the Hermitian matrix $-\ell L_z$, with real eigenvalues. Here we must be careful to multiply by ℓ in the correct place. We define

$$\widehat{L}_z \psi := -L_z \psi \ell \tag{5.9}$$

which is well-defined by alternativity, so that

$$\widehat{L}_z = -\ell_R \circ L_z \tag{5.10}$$

where the operator ℓ_R denotes right multiplication by ℓ and where $\circ$ denotes composition. The operators $\widehat{L}_a$ are self-adjoint with respect to the inner product

$$\langle \psi, \chi \rangle = \pi(\psi^\dagger \chi) \, . \tag{5.11}$$

We therefore consider the eigenvalue problem

$$\widehat{L}_z \psi = \psi \lambda \tag{5.12}$$

with $\lambda \in \mathbb{R}$. It is straightforward to show that the real eigenvalues are

$$\lambda_\pm = \pm \frac{1}{2} \tag{5.13}$$

as expected. However, the form of the eigenvectors is a bit more surprising:

$$\psi_+ = \begin{pmatrix} A \\ kD \end{pmatrix}, \qquad \psi_- = \begin{pmatrix} kB \\ C \end{pmatrix} \tag{5.14}$$

where $A, B, C, D \in \mathbb{C}$ are any elements of the preferred complex subalgebra, and k is any imaginary octonionic unit orthogonal to ℓ, so that k and ℓ anticommute. Thus, the components of spin eigenstates are contained in the quaternionic subalgebra $\mathbb{H} \subset \mathbb{O}$ which is generated by ℓ and k.

Therefore, if we wish to consider spin eigenstates, ℓ must be in the quaternionic subalgebra $\mathbb{H}$ defined by the solution. We can further assume without loss of generality that $\mathbb{H}$ takes the form given in (3.1). Thus, the only possible nonzero components of p_μ are $p_t = p_0$, $p_x = p_1$, $p_k = p_4$, $p_{k\ell} = p_5$, $p_y = p_8$, and $p_z = p_9$, corresponding to the gamma matrices with components in $\mathbb{H}$. We can further assume (via a rotation in the $(k, k\ell)$-plane if necessary) that $p_5 = 0$, so that

$$P = \pi(P) + m\,\sigma^k \tag{5.15}$$

where

$$\pi(P) = p_\alpha \sigma^\alpha \equiv \mathbf{p} \tag{5.16}$$

with $\alpha = 0, 1, 8, 9$ (or equivalently $\alpha = t, x, y, z$) is complex, and corresponds to the 4-dimensional momentum of the particle, with squared mass

$$m^2 = p_\alpha p^\alpha = -\det(\pi(P)). \tag{5.17}$$

Inserting (5.15) into (4.19), we recover precisely (3.11), and we see that we have come full circle: solutions of the *octonionic* Weyl equation (4.18) are described precisely by the *quaternionic* formalism of Section 3, and the dimensional reduction scheme determines the mass term.

5.3 Particles

For each solution ψ of (4.22), the momentum is proportional to $\psi\psi^\dagger$ by (4.23). Up to an overall factor, we can therefore read off the components of the 4-dimensional momentum p_α directly from $\pi(\psi\psi^\dagger)$. We can use a Lorentz transformation to bring a massive particle to rest or to orient the momentum of a massless particle to be in the z-direction.

If $m \neq 0$, we can distinguish particles from antiparticles by the sign of the term involving m, which is the coefficient of σ_k in P. Equivalently, we have the particle/antiparticle projections (at rest)

$$\Pi_\pm = \frac{1}{2}\left(\sigma_t \pm \sigma_k\right). \tag{5.18}$$

If $m = 0$, however, we can only distinguish particles from antiparticles in momentum space by the sign of p^0, as usual; this is the same as the sign in (4.23). Similarly, in this language, the chiral projection operator is constructed from

$$\Upsilon^5 = \sigma^t\sigma^x\sigma^y\sigma^z = -\begin{pmatrix} \ell & 0 \\ 0 & \ell \end{pmatrix}. \tag{5.19}$$

However, as with spin, we must multiply by ℓ in the correct place, obtaining

$$\widehat{\Upsilon}^5 = \ell_R \circ \Upsilon^5. \tag{5.20}$$

As a result, even though Υ^5 is a multiple of the identity, $\widehat{\Upsilon}^5$ is not, and the operators $\frac{1}{2}(\sigma_t \pm \widehat{\Upsilon}^5)$ project $\mathbb{H}^2$ into the Weyl subspaces $\mathbb{C}^2 \oplus \mathbb{C}^2 k$ as desired.

Combining the spin and particle information over the quaternionic subalgebra $\mathbb{H} \subset \mathbb{O}$ determined by k and ℓ, we thus find 1 massive spin-$\frac{1}{2}$ particle at rest, with 2 spin states, namely,

$$e_\uparrow = \begin{pmatrix} 1 \\ k \end{pmatrix}, \qquad e_\downarrow = \begin{pmatrix} -k \\ 1 \end{pmatrix} \tag{5.21}$$

whose antiparticle is obtained by replacing k by $-k$ (and changing the sign in (4.23)). We also find 1 massless spin-$\frac{1}{2}$ particle involving k moving in the z-direction, with a single helicity state,

$$\nu_z = \begin{pmatrix} 0 \\ k \end{pmatrix} \tag{5.22}$$

which corresponds, as usual, to a particle and its antiparticle. It is important to note that

$$\nu_{-z} = \begin{pmatrix} k \\ 0 \end{pmatrix} \tag{5.23}$$

corresponds to a massless particle with the same helicity moving in the opposite direction, not to a different particle with the opposite helicity. Each of the above states may be multiplied (on the *right*) by an arbitrary complex number.

There is also a single *complex* massless spin-$\frac{1}{2}$ particle, with the opposite helicity, which is given in momentum space by

$$\emptyset_z = \begin{pmatrix} 0 \\ 1 \end{pmatrix}. \tag{5.24}$$

As with the other massless momentum space states, this describes a particle and an antiparticle. Alone among the particles, this one does not contain k and, hence, does not depend on the choice of identification of a particular quaternionic subalgebra $\mathbb{H}$ satisfying $\mathbb{C} \subset \mathbb{H} \subset \mathbb{O}$.

5.4 Spin operators

We saw in the previous section that the spin up particle state $e_\uparrow$ is a simultaneous eigenvector of the spin and particle projections, that is,

$$2\widehat{L}_z e_\uparrow = e_\uparrow = \Pi_+ e_\uparrow. \tag{5.25}$$

Remarkably, $e_\uparrow$ is also an eigenvector of the remaining spin projections, namely

$$2\widehat{L}_x e_\uparrow = -e_\uparrow k, \qquad 2\widehat{L}_y e_\uparrow = -e_\uparrow k\ell \tag{5.26}$$

although the eigenvalues are not real. Similar statements hold for the corresponding spin down and antiparticle states, although with different eigenvalues. We find it illuminating to consider the equivalent right eigenvalue problem

$$L_z \psi = \psi \lambda \tag{5.27}$$

for the non-Hermitian operator L_z. The operator L_z admits imaginary eigenvalues $\pm\frac{\ell}{2}$, which correspond to the usual spin eigenstates. But L_z also admits other imaginary eigenvalues! These correspond precisely to the eigenvalues of $\widehat{L}_z$ which are not real and, in fact, not in $\mathbb{C}$. We emphasize that the projected spin operators *are* self-adjoint (with respect to (5.11)). However, over the octonions it is not true that all the eigenvalues of Hermitian matrices are real [22, 27]; the case of self-adjoint operators is similar.

How does it affect the traditional interpretation of quantum mechanics to have simultaneous eigenstates of all 3 orthogonal projections of the spin operator? The essential feature which permits this is that only one of the eigenvalues is real, and only real eigenvalues correspond to observables. Thus, from this point of view, the reason that the projected spin operators fail to commute is not that they do not admit simultaneous eigenstates, but rather that their *eigenvalues* fail to commute!

Furthermore, while (right) multiplication by a (complex) phase does not change any of the real eigenvalues, the non real eigenvalues do depend on the phase since the phase doesn't commute with the eigenvalue! Does this allow one *in principle* to determine the exact spin orientation even if no corresponding measurement exists?

6 Discussion

We have shown how the massless Dirac equation in 10 dimensions reduces to the (massive and massless) Dirac equation in 4 dimensions when a preferred octonionic unit is chosen.

The quaternionic Dirac equation discussed in Section 3 describes 1 massive particle with 2 spin states, 1 massless particle with only 1 helicity, and their antiparticles. We identify this set of particles with a generation of leptons.

Furthermore, as can be seen from Figure 1, there is room in the octonions for exactly 3 such quaternionic descriptions which have only their complex part in common, corresponding to replacing k in turn by i and j. We identify these 3 quaternionic spaces as describing 3 generations of leptons.

There is, however, one additional massless particle/antiparticle pair, given by (5.24). Being purely complex, it does not belong to any generation, and it has the opposite helicity from the other massless particles. Could this be the sterile neutrino? dark matter?

Note that the mass appears in this theory as an overall scale, which can be thought of as the length scale associated with the corresponding quaternionic direction. In particular, antiparticles must have the same mass as the corresponding particles. This suggests that the only free parameters in this theory are 3 length scales, corresponding to the masses in each generation.

The theory presented here can be elegantly rewritten in terms of *Jordan matrices*, i.e., 3×3 octonionic Hermitian matrices, along the lines of the approach to the superparticle presented in [20]. This approach, which is briefly described in [24], demonstrates that the theory is invariant under a much bigger group than the Lorentz group, namely the exceptional group E_6 (actually, E_7, since only conformal transformations are involved). We therefore believe it may be possible to extend the theory so as to include quarks and color.

Finally, as noted in [1], we have worked only in momentum space and have discussed only free particles. Perhaps our most intriguing result is the observation that the introduction of position space would require a preferred complex unit in the Fourier transform. Similarly, a description of interactions based on minimal coupling would again involve a preferred complex unit. Therefore, it does not appear to be *possible* to use the formalism presented here to give a full, interacting, 10-dimensional theory in which all 10 spacetime dimensions are on an equal footing. We view this as a tantalizing hint that not only interactions, but even 4-dimensional spacetime itself, may arise as a consequence of the symmetry breaking from 10 dimensions to 4!

Acknowledgments

It is a pleasure to thank David Griffiths, Phil Siemens, Tony Sudbery, and Pat Welch for comments on earlier versions of this work, Paul Davies for moral support, and Reed College for hospitality.

References

[1] Corinne A. Manogue and Tevian Dray, Dimensional reduction, *Mod. Phys. Lett.* **A14** (1999), 93–97.

[2] Stephen L. Adler, *Quaternionic Quantum Mechanics and Quantum Fields,* Oxford University Press, New York, 1995.

[3] P. A. M. Dirac, Application of quaternions to Lorentz transformations, *Proc. Roy. Irish Acad.* Sect. **A** Vol. **L** (1945), 261–270.

[4] P. Jordan, Über die Multiplikation quantenmechanischer Größen, *Z. Phys.* **80** (1933), 285–291 .

[5] P. Jordan, J. von Neumann, and E. Wigner, On an algebraic generalization of the quantum mechanical formalism, *Ann. Math.* **35** (1934), 29–64.

[6] Feza Gürsey and Chia-Hsiung Tze, *On the Role of Division, Jordan, and Related Algebras in Particle Physics,* World Scientific, Singapore, 1996.

[7] S. Okubo, *Introduction to Octonion and Other Non-Associative Algebras in Physics,* Cambridge University Press, Cambridge, 1995.

[8] Geoffrey M. Dixon, *Division Algebras: Octonions, Quaternions, Complex Numbers and the Algebraic Design of Physics*, Kluwer Academic Publishers, Boston, 1994.

[9] E. L. Schücking, Jerome Epstein, William P. Kowalski, and Salvatore Lauro, What is space-time made of? in *Quantum Gravity,* Proceedings of the XIV Course of the International School of Cosmology and Gravitation (Erice 1995), eds. P. G. Bergmann, V. de Sabbata, and H.-J. Treder, World Scientific, Singapore, 1996, 342–365.

[10] Engelbert Schücking and Jerome Epstein, The (sub)standard theory, in preparation.

[11] Michael J. Crowe, *A History of Vector Analysis,* Dover, Mineola, 1984 (originally published 1967).

[12] L. E. Dickson, On quaternions and their generalization and the history of the eight square theorem, *Ann. Math.* **20** (1919), 155–171.

[13] Richard D. Schafer, *An Introduction to Nonassociative Algebras,* Academic Press, New York, 1966 & Dover, Mineola, 1995.

[14] A. Sudbery, Division algebras, (pseudo)orthogonal groups and spinors, *J. Phys.* **A17** (1984), 939–955.

[15] K. W. Chung and A. Sudbery, Octonions and the Lorentz and conformal groups of ten-dimensional space-time, *Phys. Lett.* **B198** (1987), 161–164.

[16] T. Kugo and P. Townsend, Supersymmetry and the division algebras, *Nucl. Phys.* **B221** (1983), 357–380.

[17] David B. Fairlie and Corinne A. Manogue, Lorentz invariance and the composite string, *Phys. Rev.* **D34** (1986), 1832–1834.

[18] David B. Fairlie and Corinne A. Manogue, A parameterization of the covariant superstring, *Phys. Rev.* **D36** (1987), 475–479.

[19] Corinne A. Manogue and Anthony Sudbery, General solutions of covariant superstring equations of motion, *Phys. Rev.* **D40** (1989), 4073–4077.

[20] Jörg Schray, The general classical solution of the superparticle, *Class. Quant. Grav.* **13** (1996), 27–38.

[21] Jörg Schray, *Octonions and Supersymmetry,* Ph.D. Thesis, Department of Physics, Oregon State University, 1994.

[22] Tevian Dray and Corinne A. Manogue, The octonionic eigenvalue problem, *Adv. Appl. Clifford Algebras* **8** (1998), 341–364.

[23] Tevian Dray and Corinne A. Manogue, Finding octonionic eigenvectors using *Mathematica*, *Comput. Phys. Comm.* **115** (1998), 536–547.

[24] Tevian Dray and Corinne A Manogue, The exceptional Jordan eigenvalue problem, *Internat. J. Theoret. Phys.* **38** (1999), 2901–2916.

[25] Corinne A. Manogue and Jörg Schray, Finite Lorentz transformations, automorphisms, and division algebras, *J. Math. Phys.* **34** (1993), 3746–3767.

[26] Corinne A. Manogue and Tevian Dray, Octonionic Möbius transformations, *Mod. Phys. Lett.* **A14** (1999), 1243–1255.

[27] Tevian Dray, Jason Janesky, and Corinne A. Manogue, Octonionic hermitian matrices with non-real eigenvalues, *Adv. Appl. Clifford Algebras* (submitted).

Tevian Dray
Department of Mathematics
Oregon State University
Corvallis, OR 97331
E-mail: tevian@math.orst.edu

Corinne A. Manogue
Department of Physics
Oregon State University
Corvallis, OR 97331
E-mail: corinne@physics.orst.edu

Received: September 30, 1999; Revised: February 2, 2000

Pauli Terms Must Be Absent in the Dirac Equation

Kurt Just and James Thevenot

ABSTRACT It should be of interest whether Dirac's equation involves all 16 basis elements of his Clifford algebra Cl_D. These include the 6 'tensorial' $\sigma^{\mu\nu}$ with which the 'Pauli terms' are formed. We find that these violate a basic axiom of any *-algebra when Dirac's Ψ is canonical. Then the Dirac operator is spanned only by the 10 elements $1, i\gamma_5, \gamma^\mu, \gamma^\mu\gamma_5$ (which don't form a basis of Cl_D because the $\sigma^{\mu\nu}$ are excluded).

Keywords: Quantum field theory, Dirac equation, Clifford algebra.

1 Motivation and conclusions

In Dirac's equation

$$i\partial\!\!\!/\Psi = \mathcal{B}\Psi \quad \text{with} \quad \partial\!\!\!/ := \gamma^\mu \frac{\partial}{\partial x^\mu}, \tag{1.1}$$

the Bose field $\mathcal{B}$ is a member of the Clifford algebra Cl_D. Hence it can be written as

$$\mathcal{B} = \mathcal{S}^+ + i\gamma_5\mathcal{S}^- + \gamma^\mu\mathcal{V}^+_\mu + \gamma^\mu\gamma_5\mathcal{V}^-_\mu + \sigma^{\mu\nu}\mathcal{T}_{\mu\nu}. \tag{1.2}$$

Here $\mathcal{S}^\pm, \mathcal{V}^\pm, \mathcal{T}_{\mu\nu}$ are matrices which act on the flavors and colors of Ψ (the Dirac field for leptons and quarks). In the excellently verified Standard Model, the *matrices*

$$\mathcal{S} := \mathcal{S}^+ + i\gamma_5\mathcal{S}^- \quad \text{and} \quad \mathcal{V}_\mu := \mathcal{V}^+_\mu + \gamma_5\mathcal{V}^-_\mu \tag{1.3}$$

contain all the fields of Higgs and Yang-Mills. Vice versa, the Standard Model requires (1.2) to contain the 10 basis elements $1, i\gamma_5, \gamma^\mu, \gamma^\mu\gamma_5 \in Cl_D$, but not the further $\sigma^{\mu\nu}$ or $\sigma^{\mu\nu}\gamma_5$ (6 of which are linearly independent). Thus we encounter the question to what extent (1.2) can involve $\mathcal{T}_{\mu\nu} = -\mathcal{T}_{\nu\mu}$. For this problem it is irrelevant whether $\mathcal{T}_{\mu\nu}$ is a separate 'tensor potential' or a multiple of Maxwell's $F_{\mu\nu} := A_{\mu,\nu} - A_{\nu,\mu}$ (as proposed by Pauli [1]) or of its dual $\varepsilon_{\mu\nu\rho\sigma}F^{\rho\sigma}$. Hence we always call $\sigma^{\mu\nu}\mathcal{T}_{\mu\nu}$ the 'Pauli

AMS Subject Classification: 81E10, 81Q05, 81S05.

term' of (1.2). In rigorous, but not trivial ways, we find that $\mathcal{T}_{\mu\nu}$ must be *absent* for a very basic reason: For the members $\mathbf{a}, \mathbf{b}, \ldots$ of any *-algebra and their conjugates $\mathbf{a}^\dagger, \mathbf{b}^\dagger, \ldots$, one postulates $(\mathbf{a}^\dagger \mathbf{b})^\dagger = \mathbf{b}^\dagger \mathbf{a}$. This would be violated by any Pauli term. Hence we must demand

$$\mathcal{T}_{\mu\nu} = 0 \qquad \text{in order to keep} \qquad (\mathbf{a}^\dagger \mathbf{b})^\dagger = \mathbf{b}^\dagger \mathbf{a}. \tag{1.4}$$

In other words, our fields will not generate a *-algebra unless (1.2) is *restricted* by (1.4). This clear result has not been found in the literature, because a familiar reciprocity [2] is generally misnamed a theorem, whereas we prove it to be a *condition*, which excludes (1.2) unless it satisfies (1.4).

Showing in Sections 2 and 3 the adopted foundations, we indicate the proof of (1.4) briefly in Section 4, and more elaborately in Appendix A. It rests on a reciprocity *condition*, which is discussed in Appendix B. Calling this a 'relation' [2], one generally suggests that it holds without restricting the Bose fields to be prescribed. That misnomer may have caused the absence of (1.4) in the literature.

2 Gauge theory from Clifford algebra

For the vector field from (1.3) we must admit the gauge transformation

$$\mathcal{V}_\mu \quad \rightarrow \quad e^{-i\omega} (\mathcal{V}_\mu - i\partial_\mu) e^{i\omega} \quad \approx \quad \mathcal{V}_\mu + \omega_{,\mu} + i [\mathcal{V}_\mu, \omega] \, . \tag{2.1}$$

In order to state that $\mathcal{V}_\mu$ and ω are *hermitian* matrices, we write them

$$\mathcal{V}_\mu(x) = \mathbf{t}_Y \mathcal{V}_\mu^Y(x) = \mathcal{V}_\mu(x)^\dagger \quad \text{and} \quad \omega(x) = \mathbf{t}_Y \omega^Y(x) = \omega(x)^\dagger. \tag{2.2}$$

While the constant matrices $\mathbf{t}_Y$ act on flavors and colors, they contain all coupling constants and the $\gamma_5 := i\gamma^0\gamma^1\gamma^2\gamma^3$ from Dirac's Clifford algebra Cl_D. We generate this Cl_D by $\gamma^{(\mu}\gamma^{\rho)} = \eta^{\mu\rho}$ from

$$\gamma_\mu^\dagger = \gamma_\mu^{-1} = \gamma^\mu = \overline{\gamma^\mu} \quad , \quad \text{where} \quad \overline{\Gamma} := \gamma_0 \Gamma^\dagger \gamma_0 \in Cl_D. \tag{2.3}$$

A transformation similar to the homogeneous part of (2.1) follows for the $\mathcal{S}$ of (1.3). In order to prove (1.4), however, we must initially use (1.2) with (1.3) in the form

$$\mathcal{B} = \mathcal{S} + \gamma^\mu \mathcal{V}_\mu + \sigma^{\mu\nu} \mathcal{T}_{\mu\nu}, \quad \text{where} \quad \mathcal{T}_{\mu\nu} \neq 0. \tag{2.4}$$

This together with $(\mathbf{a}^\dagger \mathbf{b})^\dagger = \mathbf{b}^\dagger \mathbf{a}$ will in Section 4 yield a contradiction, which then proves (1.4). That proof holds in Quantum Induction [3] where the canonical relations

$$\left[\Psi(x), \Psi(0)^\dagger\right]_+ \delta(x^0) = \delta(x) \quad \text{and} \quad \left[\Psi(x), \Psi(0)^T\right]_+ \delta(x^0) = 0 \tag{2.5}$$

together with Dirac's equation (1.1) are *fundamental.* Its possible validity under presumptions different from (1.1) through (1.3) is discussed in Appendix B.

3 Short distance representation

For the bilocal, time-ordered Dirac matrix

$$b(x,z) := (4\pi)^2\, T\, \Psi(x+z)\overline{\Psi}(x-z), \tag{3.1}$$

(1.1) and (2.5) provide the differential equation

$$\{\not\partial^x + \not\partial^z + 2i\mathcal{B}(x+z)\}\, b(x,z) = 2\pi^2\delta(z) = i\not\partial^z \not z_-^{-3}, \tag{3.2}$$

where $\not z_-^{-3} := (z^2 - i\epsilon)^{-2}\not z$ with $\epsilon \to +0$. The representation used here for $\delta(z) := \delta(z^0)\delta(z^1)\delta(z^2)\delta(z^3)$ follows directly from the familiar

$$\Box\,(z^2 - i\epsilon)^{-1} = (2\pi)^2 i\delta(z).$$

Writing (3.1) as

$$b(x,z) = i\not z_-^{-3} + (C^{-2} + C^{-1} + r^0)(x,z), \tag{3.3}$$

let us anticipate that the C^h can be made *homogeneous* in the sense that

$$C^h(x,\lambda z) = \lambda^h C^h(x,z) \quad \text{for} \quad h = -2,-1 \quad \text{and} \quad \lambda \in \mathbb{C}. \tag{3.4}$$

Also using the 'Taylor representation'

$$\mathcal{B}(x+z) = \mathcal{B}(x) + z^\mu \mathcal{B}_{,\mu}(x) + R(x,z)\not z, \quad \text{where} \quad R(x,0) = 0, \tag{3.5}$$

we can split (3.2) into

$$\not\partial^z C^{-2}(x,z) = 2\mathcal{B}(x)\not z_-^{-3}, \tag{3.6}$$

$$\not\partial^z C^{-1}(x,z) = 2z^\mu \mathcal{B}(x)_{,\mu}\not z_-^{-3} - \{2i\mathcal{B}(x) + \not\partial^x\}C^{-2}(x,z), \tag{3.7}$$

$$\not\partial^z r^0(x,z) = \ldots - 2iR(x,z)z_-^{-2}. \tag{3.8}$$

Here the dots symbolize infinitely many unknown terms; they will be irrelevant because they are not more singular than $\not z_-^{-1}$ (for $z \to 0$ at $z^2 \neq 0$). As one easily verifies, (3.6) with (2.4) can be solved by

$$z^4 C^{-2}(x,z) = 2\not z z^\mu \overline{\mathcal{V}}_\mu(x) - z^2\mathcal{S}(x)^\dagger + \not z\sigma^{\mu\nu}\not z\mathcal{T}_{\mu\nu}(x)^\dagger. \tag{3.9}$$

This clearly satisfies (3.4) and, due to Appendix A, is the *only* solution of (3.6) which does so. Since it makes the right side of (3.7) homogeneous in z of the order $h = -2$, we can choose also $C^{-1}(x,z)$ to obey (3.4). On the right side of (3.8), we have omitted terms which, due to (3.9), are as homogeneous as $\not z_-^{-1}$ (or less singular).

Thus (3.8) can be solved by an r^0 which (due to $R(x,0) = 0$) satisfies

$$z^\mu r^0(x,z) \to 0 \quad \text{for} \quad z \to 0. \tag{3.10}$$

Therefore, we can include in r^0 those terms from C^{-2} and C^{-1} which are left arbitrary by (3.6) and (3.7) because they are independent of z. In order to learn much more about $r^0(x,z)$ than (3.10) shows, one would need 'outer' boundary conditions (at large z). These could be obtained from heat kernels (HK), but here the 'inner' condition (by (2.5) giving to (3.2) its right side) has been sufficient.

No obstruction to *solving* (3.2) has been encountered in (3.9) or in Appendix A. Neither would any arise if we would extend our recursion for (3.3) or invoke HK[4]. Taking the Dirac adjoint of (3.1), however, we obtain the 'reciprocity' condition

$$\overline{z^4 b(x,z)} = z^4 b(x,-z) \quad \text{due to} \quad (\Psi_+\Psi_-^\dagger)^\dagger = \Psi_-\Psi_+^\dagger. \tag{3.11}$$

The latter exemplifies a general rule for *-algebras.

4 Reciprocity as a condition

The z^4 in (3.11) removes the denominators of the terms in (3.3), so that the hermitian conjugation affects only their *numerators*. Hence these must (for $h = -2, -1$) satisfy

$$\overline{z^4 C^h(x,z)} = z^4 C^h(x,-z) \quad \text{and} \quad \overline{z^4 r^0(x,z)} = z^4 r^0(x,-z) \tag{4.1}$$

because all three are linearly independent. In (4.1) we did not mention the leading $i\not{z}_-^{-3}$ of (3.3) because it satisfies (3.11) trivially. Since (2.4) equals its adjoint $\overline{\mathcal{B}} = \overline{\mathcal{S}} + \gamma^\mu \mathcal{V}_\mu^\dagger + \sigma^{\mu\nu}\overline{\mathcal{T}_{\mu\nu}}$, that reciprocity is also verified for (3.9).

Instead of (4.1) with $h = -1$, however, in Appendix A we find

$$\overline{z^4 C^{-1}(x,z)} \neq z^4 C^{-1}(x,-z) \qquad \text{unless} \qquad \mathcal{T}_{\mu\nu} = 0. \tag{4.2}$$

Hence it is *misleading* when one calls (3.11) a reciprocity 'theorem' [2] as if it were fulfilled for arbitrarily prescribed Bose fields (2.4). Since (3.9) satisfies (4.1) even with $\mathcal{T}_{\mu\nu} \neq 0$, we also see that (1.4) cannot be derived as long as one only examines that solution of (3.6). In other words, we do not know a way of reaching the conclusion (1.4) without noting that the solution of (3.7) provides (4.2).

By (1.4), however, the $C^{-1}(x,z)$ solving (3.7) is simplified greatly, and the further parts of (3.3) are *shortened* still more drastically. Hence superfluous work is done by those who pursue higher terms of (3.3) without inserting (1.4) quickly. They solve the differential equation (3.2) correctly but in excessive generality. Thus they miss the fact that (3.1) must also satisfy (3.11), which is a non-trivial condition not expressed by (3.2).

Acknowledgments

For comments we are thankful to S. A. Fulling, K. Kwong, Z. Oziewicz, W. Stoeger and E. Sucipto.

A Appendix: Linear differential equations

A.1 Exactly homogeneous solutions

In (3.2) we have used

$$i\not{\partial}^z \not{z}_-^{-3} = 2\pi^2 \delta(z) \quad \text{for} \quad \not{z}_-^{-3} := (z_-^{-2})^2 \not{z}, \tag{A.1}$$

where $z_-^{-2} = (z^2 - i\epsilon)^{-1}$ with $\epsilon \to +0$. These obviously provide

$$\begin{aligned} \not{\partial}^z \not{z}_-^{-3} z^\mu &= \gamma^\mu \not{z}_-^{-3}, \quad \not{\partial}^z z_-^{-2} = -2\not{z}_-^{-3}, \\ \not{\partial}^z \not{z}_-^{-3} \sigma^{\mu\nu} \not{z} &= \gamma^\rho \not{z}_-^{-3} \sigma^{\mu\nu} \gamma_\rho = 2\sigma^{\mu\nu} \not{z}_-^{-3}, \end{aligned} \tag{A.2}$$

which are derived most *easily*, when one uses (A.1) as often as possible. Thus (3.9) makes

$$\not{\partial}^z C^{-2}(x,z) = 2\gamma^\mu \not{z}_-^{-3} \overline{\mathcal{V}}_\mu(z) + 2\not{z}_-^{-3} \mathcal{S}(x)^\dagger + 2\not{z}_-^{-3} \sigma^{\mu\nu} \mathcal{T}_{\mu\nu}(x)^\dagger, \tag{A.3}$$

so that (3.6) with (2.4) is fulfilled.

All these calculations of course do not make sense on the cone $z^2 = 0$. Hence throughout this paper we assume $z^2 \neq 0$ as we must clearly do in (3.3) through (3.8). This is also true for the *limits* with $z \to 0$, as needed in (3.10) and in the proof of (A.1). Hence such limit transitions can proceed on any path which ends at $z = 0$, except that it must not touch the cone $z^2 = 0$.

While (3.6) with (2.4) is due to (A.1) satisfied by (3.9), its most general solution follows when we add any matrix H which fulfills the homogeneous Dirac equation $\not{\partial} H = 0$. For choosing this H we need the

Main Theorem (Singularity). *Every Poincaré covariant member H of Dirac's Clifford algebra Cl_D, which solves*

$$\not{\partial} H(z) = 0 \quad \textit{in a neighborhood of} \quad z = 0, \tag{A.4}$$

becomes for $z \to 0$ with $z^2 \neq 0$ either more singular than $\not{z}^{-3}$ or less than $\not{z}^{-1}$, hence yields either

$$\lim_{z \to 0} \not{z}^3 H(z) = \infty \quad \textit{or} \quad \lim_{z \to 0} z^\mu H(z) = 0. \tag{A.5}$$

One proves this easily when H depends only on z. The general proof is lengthy and scarcely of interest to physicists; hence, we shall show it whenever requested.

Obvious solutions of (A.4) and (A.5) are all H which do not depend on z. Already the solution of (3.6) by (3.9) says that (3.4) with $h=-2$ can be satisfied. The theorem (A.5) proves that (3.9) yields the only $C^{-2}(x,z)$ which does so (such that (3.6) and (3.4) make (3.9) necessary). It also shows that the leading term of (3.3) is determined uniquely. In the same way, (A.5) says that the solution of (3.7) and (3.4) will be *unique* when such a $C^{-1}(x,z)$ can be found at all.

A.2 Relevant short distance terms

Inserting (2.4) and (3.9) in (3.7), we obtain

$$\begin{aligned}\not\partial^z C^{-1}(x,z) = {} & 2z^\mu\gamma^\rho \not{z}_-^{-3}(\overline{\mathcal{V}}_{\rho,\mu} - \overline{\mathcal{V}}_{\mu,\rho} - 2i\overline{\mathcal{V}}_\rho\overline{\mathcal{V}}_\mu) \\ & + z_-^{-2}\gamma^\rho(\mathcal{S}^\dagger_{,\rho} + 2i\mathcal{V}_\rho\mathcal{S}^\dagger) + 2\not{z}_-^{-3}z^\rho(\mathcal{S}^\dagger_{,\rho} - 2i\mathcal{S}^\dagger\overline{\mathcal{V}}_\rho) \\ & + 2iz_-^{-2}\mathcal{S}\mathcal{S}^\dagger + 2\not{z}_-^{-3}\gamma^{\mu\nu}\not{z}\mathcal{S}\mathcal{T}_{\mu\nu} \\ & + 2z^\rho\gamma^{\mu\nu}\not{z}_-^{-3}(2\mathcal{T}_{\mu\nu}\overline{\mathcal{V}}_\rho + i\mathcal{T}_{\mu\nu,\rho}) - 2z_-^{-2}\gamma^{\mu\nu}\mathcal{T}_{\mu\nu}\mathcal{S}^\dagger \\ & + 2i\gamma^{\mu\nu}\not{z}_-^{-3}\gamma^{\rho\sigma}\not{z}\mathcal{T}_{\mu\nu}\mathcal{T}_{\rho\sigma} \\ & + \gamma^\rho\not{z}_-^{-3}\gamma^{\mu\nu}\not{z}(2\mathcal{V}_\rho\mathcal{T}_{\mu\nu} - i\mathcal{T}_{\mu\nu,\rho}) \end{aligned} \tag{A.6}$$

with $\gamma^{\mu\nu} := \gamma^{[\mu}\gamma^{\nu]} = -i\sigma^{\mu\nu} = \gamma^\mu\gamma^\nu - \eta^{\mu\nu}$. Here all the fields $\mathcal{S}, \mathcal{V}_\mu, \mathcal{T}_{\mu\nu}$ and their partial derivatives are localized at x; hence, this parameter has been suppressed in the notation. Although (A.6) is for C^{-1} a linear differential equation in z of the *first* order, deriving a solution was tedious; but after such a C^{-1} has been found, only differentiations are needed to verify its validity. Only a few readers, however, would perform this extremely easy but time-consuming task; thus, let us merely say that the $C^{-1}(x,z)$ solving the differential equation (A.6) is twice as lengthy as this.

It implies (4.2) and thus contradicts (4.1) *unless* all 5 local field polynomials

$$\begin{aligned} \mathcal{Z}^1_{\rho\sigma\tau\mu} &:= [\mathcal{T}_{[\rho\sigma}, \mathcal{T}_{\tau]\mu}]_+ , \\ \mathcal{Z}^2_{\rho\sigma\tau} &:= \mathcal{T}_{[\rho\sigma}\overline{\mathcal{V}}_{\tau]} + \mathcal{V}_{[\tau}\mathcal{T}_{\rho\sigma]}, \\ \mathcal{Z}^3_\rho &:= [\mathcal{V}^\mu, \mathcal{T}_{\mu\rho}] \\ \mathcal{Z}^4_{\rho\sigma} &:= \mathcal{T}_{\rho\sigma}\mathcal{S}^\dagger + \mathcal{S}\mathcal{T}_{\rho\sigma}, \\ \mathcal{Z}^5_{\rho\sigma\tau} &:= \mathcal{T}_{\rho\sigma,\tau} + i\overline{\mathcal{V}}_\tau\mathcal{T}_{\rho\sigma} - i\mathcal{T}_{\rho\sigma}\mathcal{V}_\tau \end{aligned} \tag{A.7}$$

satisfy

$$\mathcal{Z}^n_{\dots}(x) = 0. \tag{A.8}$$

These derivations have been rigorous because we did not admit any approximation (rarely possible in physics). Our solution (1.4) is clearly the only one which holds regardless of $\mathcal{S}$ and $\mathcal{V}_\mu$. If (A.8) were mathematically solvable by any $\mathcal{T}_{\mu\nu} \neq 0$ (a case we can't examine exactly), it would restrict $\mathcal{S}$ and $\mathcal{V}_\mu$ in complicated and extremely unphysical ways.

Within the permitted size of this paper, we can't show the complete proof for the necessity of (A.8); the known extension to quantum fields [3] would enlarge it enormously. A result as simple as (1.4), however, should find a much shorter proof. Hence, we would prefer to delay the publication until that simplicity is achieved. If we would not show the result (1.4) and indicate our lengthy derivation from (3.11), however, it would be hard to get anyone interested in such a problem.

B The reciprocity violation

Let us finally collect further remarks about our result (1.4) and its absence from the literature:

(a) The reciprocity condition (3.11) has been called [2] a *relation*, as if it had been proved with (1.1) containing the most general Bose field (1.2).

(b) We had to perform extensive *computer* algebra for the derivation of $C^{-1}(x, z)$ from its differential equation (A.6). Still more would be required if one were to make the compact result from HK explicit. Neither is needed any longer because it is easy to verify a known solution of any differential equation, no matter how hard its integration had been. Because of (a), however, nobody found this tedious search worthwhile.

(c) In the mathematics of HK, the boundary conditions [4] at *large* z are presently more interesting than the behavior of (3.3) at small z.

(d) It is fashionable, instead of the differential equation (3.2), to solve a related integral equation, with an 'inner' boundary condition given by the right side of (3.2) and a purely mathematical condition at some outer boundary (where z is large or infinite). For our problem, no choice of the latter makes sense because the result (A.8) depends only on (3.2) and its boundary condition at $z = 0$. Why should we use a physically irrelevant integral equation for a conclusion which is *completely* determined by a differential equation together with a single, well-justified boundary condition?

(e) Methods of HK have been initiated [5] for *classical* field theories, where the reciprocity arises from a symmetry of their Green functions.

(f) In many treatments by HK, not only the Bose field $\mathcal{B}$, but also Dirac's Ψ is non-quantized (or not even mentioned). Then (3.2) is regarded as an equation for a classical Green function $b(x,z)$, not related to any quantum field such as (3.1). In that approach, one hardly sees whether a reciprocity (3.11) should be desired.

(g) Under *infinite* renormalizations, (1.1) and (2.5) and, therefore, (3.3) break down [6]. Hence we can't prove (1.4) in familiar settings (although it might be true even there).

(h) Any significant $\mathcal{T}_{\mu\nu} \neq 0$ would damage the excellent *verification* of the Standard Model [7] by the magnetic moment of the electron. This agreement [8] had formerly been regarded as a brilliant confirmation of renormalized QED. Under the present philosophy [9] of 'effective' actions, however, it is an unimportant result of imprecise measurements.

(i) Instead of (3.1), one often uses $\beta(e,t) := (4\pi)^2 \mathrm{T}\Psi(e+t)\overline{\Psi}(e) = b(e+\frac{1}{2}t, \frac{1}{2}t)$ with the *eccentric* coordinates $e = x-z$ and $t = 2z$. These simplify (especially under gravity) the derivation of (3.2) but make the analysis of (3.11) complicated.

(j) The singularity at $z^2 = 0$ makes (3.3) dependent on the *time ordering* of (3.1). Hence (3.11) can't simply be written $\overline{b}(x,z) = b(x,-z)$ because an hermitian conjugation reverses the time order.

(k) Wherever basic 'tensor potentials' $\mathcal{T}_{\mu\nu}$ have found any attention, one has *coupled* them to each other or further Bose fields [10], leaving their interaction with Dirac's Ψ open.

(l) Whenever 'tensor couplings' are mentioned in phenomenology [11], it is unclear whether they are *fundamental* or caused by bound states or by 'radiative' corrections.

(m) The *differential* or integral equation for (3.1) can 'mathematically' be solved [4] without any concern about the reciprocity (3.11) needed in physics. However, (3.2) without (3.11) does not exhaust the contents of (3.1).

(n) The leading terms $i\not{z}_-^{-3}$ and $C^{-2}(x,z)$ of (3.3) satisfy (3.11) even when (2.4) has $\mathcal{T}_{\mu\nu} \neq 0$. Hence the *contradiction* between (4.1) and (4.2) is not recognized until one also determines $C^{-1}(x,z)$.

(o) A further reason for the usual rejection of (1.4) may be that such a clear result deserves a *simple* derivation. Instead our proof has required the lengthy (but straightforward) deduction of the differential equation (A.6) and its explicit solution. That $C^{-1}(x,z)$ is twice as long as (A.6) and, therefore, not shown here, but we hope that others can simplify our arguments.

(p) In order to examine (3.11) completely, the reciprocity (4.1) should also be checked for the 'remainder' $r^0(x,z)$. Analyzing those parts of it which in z are homogenous of the orders $h=0$ and $h=1$, we have not found any restriction beyond $\mathcal{T}_{\mu\nu}=0$ (which simplifies those parts enormously). Our *local* approach (without outer boundary conditions) cannot extend that result to all orders. This should be taken as an incentive to treat the condition (3.11) globally, but not as excuse for discarding our result (4.2). Doing so would correspond to ignoring the singular part of a Laurent series until all its orders are known.

(q) For many authors, the Higgs field does not contribute to $\mathcal{B}$ because they attach it in *isospinors* to Ψ.

(r) Many authors prefer *two-component* spinors instead of Dirac's Ψ.

(s) Some authors use other *notations* for Dirac matrices, for instance, α, β instead of γ^μ or explicit 4 x 4 squares.

(t) For Dirac's γ^μ, one sometimes uses *representations* in which γ^μ is not its own adjoint or the β in $\overline{\Psi}=\Psi^\dagger\beta$ differs from γ^0.

(u) From (2.4) we derived (1.4) by *reductio ad absurdum,* which not every mathematician appreciates.

(v) Readers may dislike (1.4) because beautiful theories are no longer expected to be simple but to offer rich mathematical structures.

(w) Instead of the Dirac equation for physics (which is of first order in Minkowski space), mathematicians prefer the elliptic equation given by its iteration in Euclidean space.

(x) One often uses Dirac's Clifford algebra without any basis, hence, not separating the scalar, vectorial, and tensorial parts of $\mathcal{B}$.

References

[1] W. Pauli, Relativistic field theories of elementary particles, *Rev. Mod. Phys.* **13** (1941), eq. (91), 203–232.

[2] B. S. DeWitt, *Dynamical Theory of Groups and Fields,* Gordon & Breach, New York, 1965, p. 44; F. G. Friedlander, *The Wave Equation in a Curved Space-Time,* Univ. Press, Cambridge, 1975, 178–181.

[3] K. Just and E. Sucipto, Basic versus practical quantum induction, in *Group 21, Vol. 2,* Doebner, Scherer, and Schulte, eds., World Sci., Singapore, 1997, p. 739.

[4] G. Esposito, *Dirac Operators and Spectral Geometry,* Univ. Press, Cambridge, 1998.

[5] J. Hadamard, *Lectures on Cauchy's Problem,* Yale Univ. Press, New Haven, 1923.

[6] R. A. Brandt, Gauge invariance in quantum electrodynamics, *Ann. Phys.* **52** (1969), 122–175; Field equations in quantum electrodynamics, *Fortschr. Physik* **18** (1970), 249–283.

[7] K. Huang, *Quantum Field Theory,* Wiley, New York, 1998, p. 231.

[8] V. W. Hughes and T. Kinoshita, Anomalous g-values of the electron and muon, *Rev. Mod. Phys.* **71** (1999), 133–139.

[9] S. Weinberg, *The Quantum Theory of Fields, Vol. I,* Univ. Press, Cambridge, 1995/96, 517–521.

[10] V. Dvoeglazov, Quantized $(1,0) \oplus (0,1)$ fields, *Int. J. Theor. Phys.* **37** (1998), 1915–1944; A. Khoudeir, Non-Abelian antisymmetric-vector coupling from self-interaction, *Mod. Phys. Lett.* **A11** (1996), 2489–2496; C. Y. Lee and D. W. Lee, preprint, hep-th/9709020.

[11] M. V. Chizov, Search for tensor interactions in Kaon decays at DAΦNE, *Physics Lett.* **B 381** (1996), 359–364; M. G. Negrão, et al., preprint, hep-th/9808174.

Kurt Just
Department of Physics
University of Arizona
Tucson AZ 85721
E-mail: just@physics.arizona.edu

James Thevenot
Department of Physics
University of Arizona
Tucson AZ 85721
E-mail: jimthev@physics.arizona.edu

Received: September 30, 1999; Revised: February 6, 2000

Electron Scattering in the Spacetime Algebra

Antony Lewis, Anthony Lasenby, and Chris Doran

ABSTRACT The Spacetime Algebra provides an elegant language for studying the Dirac equation. Cross section calculations can be performed in an intuitive way following a method suggested by Hestenes [1]. The S-matrix is replaced with an operator which rotates the initial states into the scattered states. We show how the method neatly handles spin dependence by allowing the scattering operator to become a function of the initial spin. When the operator is independent of spin, we can provide manifestly spin-independent results. Spin basis states are not needed, and we do no spin sums, instead dealing with the spin orientation directly. We perform some example calculations for single electron scattering and briefly discuss more complicated cases in QED.
Keywords: Spacetime Algebra, scattering, cross sections, spin.

1 Introduction

Methods for calculating spinor cross sections are well-known; however, these often involve complicated abstract calculations with gamma matrices. In this paper we show how to calculate cross sections in a more transparent and intuitive way. Instead of using spin basis states, summing over spins and using spin projection operators, we incorporate the spin orientation directly. This greatly streamlines the calculation of spin dependent results, and makes it clear when results are independent of spin. We first consider single electron scattering, where our method is most naturally applied, and then briefly discuss multi-particle scattering.

Spacetime Algebra (STA) is the geometric (Clifford) algebra of Minkowski spacetime, first developed by Hestenes [1, 2, 3]. The formulation of Dirac theory within the algebra replaces the matrices of the conventional theory with multivectors. We introduce the STA form of the Dirac equation and show how the theory can be developed within the STA. Using the STA formulation, Hestenes [1] has demonstrated an elegant method for performing cross section calculations. We extend and clarify this work,

---

AMS Subject Classification: 81U05, 81U20, 15A66.

handling spin-dependence in a natural way.

Throughout, we make use of the Geometric Algebra. We present a brief summary of the STA below to clarify our notation and conventions. Full details of Geometric Algebra can be found elsewhere [2, 4].

2 Spacetime algebra

We shall use the four orthogonal basis vectors of spacetime γ_μ, where $\gamma_0^2 = 1$, and $\gamma_k^2 = -1$ for $k = 1, 2, 3$. The Geometric Algebra has an associative product and the basis vectors satisfy the Dirac algebra

$$\gamma_\mu \cdot \gamma_\nu \equiv \tfrac{1}{2}(\gamma_\mu \gamma_\nu + \gamma_\nu \gamma_\mu) = \mathrm{diag}(+ - - -).$$

The antisymmetric part of the product defines the *outer product*

$$\gamma_\mu \wedge \gamma_\nu \equiv \tfrac{1}{2}(\gamma_\mu \gamma_\nu - \gamma_\nu \gamma_\mu).$$

By repeated multiplication of the basis vectors we can build up the 16 basis elements of STA multivectors:

1	$\{\gamma_\mu\}$	$\{\boldsymbol{\sigma}_k, \boldsymbol{\Sigma}_k\}$	$\{I\gamma_\mu\}$	I
scalar	vectors	bivectors	pseudovectors	pseudoscalar

The bivectors $\boldsymbol{\sigma}_k \equiv \gamma_k \gamma_0$ are isomorphic to the basis vectors for Euclidean 3-space satisfying $\boldsymbol{\sigma}_j \cdot \boldsymbol{\sigma}_k = \delta_{jk}$. Similarly, the $\boldsymbol{\Sigma}_k = I\boldsymbol{\sigma}_k$ are isomorphic to the basis bivectors of Euclidean 3-space. We define the highest grade element

$$I \equiv \sigma_1 \sigma_2 \sigma_3 = \gamma_1 \gamma_0 \gamma_2 \gamma_0 \gamma_3 \gamma_0 = \gamma_0 \gamma_1 \gamma_2 \gamma_3$$

which satisfies $I^2 = -1$ and anticommutes with all vectors.

We usually take γ_0 to be the lab frame velocity vector. The $\boldsymbol{\sigma}_k$ then represent a frame in space relative to the γ_0 vector. We can do a spacetime split of a vector into the γ_0 frame by multiplying by γ_0. Bold letters are now used for relative 3-vectors (spacetime bivectors). The split is

$$a\gamma_0 = a_0 + \boldsymbol{a}$$

where $a_0 = a \cdot \gamma_0$ and $\boldsymbol{a} = a \wedge \gamma_0$. We use natural units throughout, where $c = \epsilon_0 = \hbar = 1$, so, for example, the momentum p is split

$$p\gamma_0 = p \cdot \gamma_0 + p \wedge \gamma_0 = E + \boldsymbol{p}.$$

It is often useful to project out a particular grade from a multivector. Angled brackets $\langle A \rangle_r$ are used to denote the grade r projection of A. For the scalar part, $r = 0$, and we write it as $\langle A \rangle$. The inner and outer products for grade r and s multivectors are defined as

$$A_r \cdot B_s \equiv \langle A_r B_s \rangle_{|r-s|}, \qquad A_r \wedge B_s \equiv \langle A_r B_s \rangle_{r+s}.$$

In the case where $r = 1$, so that $a \equiv A_r$ is a vector, we have the relation

$$aB_s = a\wedge B_s + a\cdot B_s.$$

The symmetry of the inner and outer product alternate with increasing grade of B_s :

$$a\cdot B_s \equiv \langle aB_s\rangle_{s-1} = \frac{1}{2}(aB_s - (-1)^s B_s a)$$
$$a\wedge B_s \equiv \langle aB_s\rangle_{s+1} = \frac{1}{2}(aB_s + (-1)^s B_s a).$$

We adopt the useful convention that when the operands of the inner and outer product are in bold type, and are therefore spatial vectors, the inner and outer products take their three-dimensional meaning.

Another important operation is that of *reversion*. We write the reverse of A as $\tilde{A}$, which reverses all the vector products making up the multivector. This has the property that $(AB)\tilde{} = \tilde{B}\tilde{A}$.

Lorentz transformations are spacetime rotations and can be performed by use of a *rotor*, which can be written $R = \pm\exp(B/2)$. Here B is a bivector in the plane of the rotation, and $|B|$ determines the amount of rotation. The rotation of a multivector M is then given by $M \to RM\tilde{R}$.

Finally, we need the vector derivative $\nabla \equiv \gamma^\mu \partial_\mu$, the derivative with respect to position. Its definition implies that it has the spacetime split

$$\nabla\gamma_0 = \partial_t - \boldsymbol{\nabla}.$$

3 The Dirac equation

The Hestenes STA form of the Dirac equation is entirely equivalent to the usual equation [5]. However, the STA approach brings out the geometric structure, leading to more physically transparent calculations. Here we show that it is possible to arrive at the Dirac equation by quantizing a classical equation. This "derivation" has the advantage that the observables are clearly related to the classical parameters, and the geometric structure of the theory is brought out. Our classical model will consist of a small spinning symmetric top with four velocity v. We can represent v as a boosted version of the lab frame time vector γ_0 :

$$v = L\gamma_0\tilde{L},$$

where L is a boosting rotor. In this way the velocity can be represented by the rotor L. Similarly, we can use a spatial rotor U to encode the spin plane as a rotation of some fixed reference plane. We write the rest spin of the top as

$$\hat{S}^0 = U\Sigma\tilde{U},$$

where Σ is some arbitrary constant reference bivector orthogonal to γ_0 ($\Sigma = \Sigma_3$ is often chosen). Since U is a spatial rotor, it does not affect the γ_0 direction; so the momentum can be written

$$p = mR\gamma_0\widetilde{R}$$

where $R = LU$. This equation for p squares to give $p^2 = m^2$ which gives the Klein-Gordon equation on quantization. However the rotor equation contains much more information than the scalar equation given by its square.

As well as encoding the rotation of Σ into the spin plane, the spatial rotor U can also include an arbitrary unobservable rotation in the reference plane Σ. We can boost up $\hat{S}^0$ to define the relativistic spin bivector

$$\hat{S} = L\hat{S}^0\widetilde{L} = R\Sigma\widetilde{R}.$$

The rotor R now encodes everything about the four velocity and spin direction of the top, plus some arbitrary unobservable rotation in the spin plane.

In the quantum version we wish to have probability densities. In the rest frame of the top this corresponds to some probability density ρ of finding it at each point. We want this to be the $v\cdot J$ component of a four vector probability current J, with the lab frame probability density given by $\gamma_0\cdot J$. We therefore define the four vector $J = \rho v$ which can be written

$$J = \rho R\gamma_0\widetilde{R}.$$

We now wrap up ρ and R into a single even multivector $\psi = \rho^{\frac{1}{2}}R$ so that

$$J = \psi\gamma_0\widetilde{\psi},$$

and the rest frame probability density is given by $\rho = \psi\widetilde{\psi}$. We now want to put the equation for the momentum in terms of ψ. Multiplying the equation on the right by R we have

$$pR = mR\gamma_0 \implies p\psi = m\psi\gamma_0.$$

This equation now contains all the ingredients for successful quantization. The usual procedure is to make the replacement $p_\mu \rightarrow \jmath\nabla_\mu$, so we get

$$\jmath\nabla\psi = m\psi\gamma_0$$

as our form of the Dirac equation. For a plane wave $\psi(x) = \psi e^{-\jmath p\cdot x}$ this gives us back our classical equation, as expected.

There is a remaining ambiguity as to what $\jmath$ is. It could be an imaginary scalar, or could it be something more physical? Multiplication by $\jmath$ should affect the phase of the wave function; we don't want the spin or momentum to be affected. So for plane waves, writing $\psi' = \hat{\jmath}\psi$, we want

$$S' = \psi'\boldsymbol{\Sigma}\widetilde{\psi}' = \psi\boldsymbol{\Sigma}\widetilde{\psi} \quad \text{and} \quad J' = \psi'\gamma_0\widetilde{\psi}' = \psi\gamma_0\widetilde{\psi}.$$

These can be satisfied if

$$\psi' \equiv \hat{\jmath}\psi = \hat{S}\psi \quad \text{or} \quad \psi' \equiv \hat{\jmath}\psi = \psi\boldsymbol{\Sigma},$$

and indeed for plane wave states these are equivalent since

$$\psi\boldsymbol{\Sigma} = \frac{1}{\rho}\psi\boldsymbol{\Sigma}\widetilde{\psi}\psi = \hat{S}\psi.$$

So the "complex" phase factors of the form $e^{\hat{\jmath}\alpha}$ encode rotations in the spin plane – the rotations that were unobservable in the classical case.

4 Dirac theory

Having "derived" the Dirac equation we now take that equation as given and see what it implies. For positive energy plane wave states, all the classical results still hold. However we now have two sets of plane wave solutions,

$$\psi^{(+)} = u(p)e^{-\hat{\jmath}p\cdot x} \quad \text{and} \quad \psi^{(-)} = v(p)e^{\hat{\jmath}p\cdot x}$$

where

$$mu - pu\gamma_0 = 0 \quad \text{and} \quad mv + pv\gamma_0 = 0.$$

We therefore have positive and negative energy states, with energy projection operators given by

$$\Lambda_{\pm}(\psi) = \frac{1}{2m}(m\psi \pm p\psi\gamma_0).$$

Since $\psi\widetilde{\psi}$ reverses to itself, it can only contain scalar and pseudoscalar parts; we define

$$\rho e^{I\beta} \equiv \psi\widetilde{\psi}$$

where β and ρ are scalars. The general form for ψ is now

$$\psi = \rho^{\frac{1}{2}}e^{\frac{I\beta}{2}}R.$$

In addition to encoding a rotation and a dilation, the spinor also contains a "β-factor". This determines the ratio of particle to anti-particle solutions since

$$\Lambda_{\pm}(I\psi) = I\Lambda_{\mp}(\psi).$$

The transformation properties of ψ are inherited from its component rotor, so we have

$$\psi_L(x) = R\psi(\tilde{R}xR).$$

We call an element of the STA which transforms as a rotor a *spinor*.

The Dirac equation can be obtained from the Lagrangian

$$\mathcal{L} = \langle \hat{\jmath}\nabla\psi\gamma_0\tilde{\psi} - m\psi\tilde{\psi}\rangle$$

by using the multivector form of the Euler-Lagrange equations [6, 7]. The Lagrangian is invariant under

$$\psi \to \psi e^{\hat{\jmath}\theta}$$

corresponding to invariance under rotation in the spin plane. Using the multivector form of Noether's theorem [6], we find the corresponding conserved probability current

$$J = \psi\gamma_0\tilde{\psi},$$

in agreement with our classical definition. The Dirac equation ensures that $\nabla\cdot J = 0$.

5 Plane waves and basis states

Using the decomposition $R = LU$ of a rotor into a spatial rotation U and a boost L, we can write a spinor ψ as

$$\psi = \rho^{\frac{1}{2}} e^{\frac{I\beta}{2}} LU.$$

Consider a positive energy spinor $u = \Lambda_+(u)$ and a negative energy spinor $v = \Lambda_-(v)$. If the particle is at rest, we have

$$\gamma_0 u^0 \gamma_0 = u^0 \quad \text{and} \quad \gamma_0 v^0 \gamma_0 = -v^0$$

which implies that

$$u^0 = \rho_u^{\frac{1}{2}} U_u \quad \text{and} \quad v^0 = \rho_v^{\frac{1}{2}} I U_v.$$

We can find the more general form by performing a boost to momentum p. The boost transforms $m\gamma_0$ into the momentum p:

$$p = mL\gamma_0\tilde{L} \implies pL\gamma_0 - mL = 0$$

so that $\Lambda_-(L) = 0$. A solution is therefore of the form $L = \Lambda_+(X)$. Choosing X equal to a constant so that $\tilde{L}L = 1$ we have

$$L = \frac{m + p\gamma_0}{\sqrt{2m(E+m)}} = \frac{E + m + \boldsymbol{p}}{\sqrt{2m(E+m)}}. \tag{5.1}$$

Normalizing so that $\rho_u = \rho_v = 2m$ and performing the boost we get

$$u(p) = Lu^0 = \sqrt{E+m}\left(1 + \frac{p}{E+m}\right)U_u$$
$$v(p) = Lv^0 = I\sqrt{E+m}\left(1 + \frac{p}{E+m}\right)U_v.$$

In addition to the energy projection operators there are also the projection operators

$$\chi_\pm(\psi) = \tfrac{1}{2}(\psi \mp P\psi\boldsymbol{\Sigma})$$

where P is a bivector with $P^2 = -1$. For a state a ψ satisfying $\psi = \chi_\pm(\psi)$ we have

$$\psi = \mp P\psi\boldsymbol{\Sigma}.$$

Multiplying on the right by $\widetilde{\psi}$, this gives $\rho = \mp\rho P\hat{S}$ and so $\hat{S} = \pm P$. The projection operator therefore projects out parts corresponding the two spin orientations in the plane P. The spin projection operators commute with the energy projection operators

$$\Lambda_\pm(\psi) = \frac{1}{2m}(m\psi \pm p\psi\gamma_0)$$

since $P \cdot p = 0$. We can therefore split an arbitrary spinor (eight real components) into scalar and $\hat{\jmath}$ multiples of four basis states

$$u_1 = \chi_+(\Lambda_+(u_1)), \qquad u_2 = \chi_-(\Lambda_+(u_2)),$$
$$v_1 = \chi_-(\Lambda_-(v_1)), \qquad v_2 = \chi_+(\Lambda_-(v_2)).$$

With the normalization convention that $u\widetilde{u} = 2m$, the four basis states obey the orthogonality relations

$$\langle \widetilde{u}_r u_s\rangle_S = 2m\delta^r_s, \qquad \langle \widetilde{v}_r v_s\rangle_S = -2m\delta^r_s,$$
$$\langle \widetilde{u}_r v_s\rangle_S = 0, \qquad \langle \widetilde{v}_r u_s\rangle_S = 0,$$

where $\langle A\rangle_S$ represents the $\{1, \boldsymbol{\Sigma}\}$ projection of A :

$$\langle A\rangle_S \equiv \langle A\rangle - \langle A\boldsymbol{\Sigma}\rangle\boldsymbol{\Sigma}.$$

By writing ψ as a sum over basis states it is easy to see that

$$\sum_r u_r\langle \widetilde{u}_r\psi\rangle_S = p\psi\gamma_0 + m\psi \quad \text{and} \quad \sum_r v_r\langle \widetilde{v}_r\psi\rangle_S = p\psi\gamma_0 - m\psi.$$

So we see that the usual basis state results of Dirac theory can be formulated in the STA approach. However we shall now develop the scattering theory largely without resort to basis states.

6 Feynman propagators

We now consider how to handle scattering from a vector potential A. This requires solutions of the minimally coupled Dirac equation which can be written

$$\hat{\jmath}\nabla\psi\gamma_0 - m\psi = eA\psi\gamma_0,$$

where $e = -|e|$ is the electron charge. We use Greens' function for this equation satisfying

$$\hat{\jmath}\nabla_x S_F(x-x')\psi(x')\gamma_0 - mS_F(x-x')\psi(x') = \delta^4(x-x')\psi(x')$$

so that an integral solution can be found from

$$\psi(x) = \psi_i(x) + e\int d^4x' S_F(x-x')A(x')\psi(x')\gamma_0, \tag{6.1}$$

where ψ_i satisfies the free-particle equation. Taking the Fourier transform we have

$$pS_F(p)\psi\gamma_0 - mS_F(p)\psi = \psi,$$

where

$$S_F(x-x') = \int \frac{d^4p}{(2\pi)^4} S_F(p) e^{-\hat{\jmath}p\cdot(x-x')}.$$

Operating on both sides with the energy projection operator Λ_+, we can solve for the momentum space Feynman propagator:

$$(p^2-m^2)S_F(p)\psi = p\psi\gamma_0 + m\psi \;\Rightarrow\; S_F(p)\psi = \frac{p\psi\gamma_0 + m\psi}{p^2-m^2+\hat{\jmath}\epsilon}. \tag{6.2}$$

The $\hat{\jmath}\epsilon$ ensures that the contour integral is in the $\boldsymbol{\Sigma}$ plane and that it is causal—positive energy waves propagate into the future and negative energy waves into the past. Fourier transforming back and performing the integral over dE we get

$$S_F(x-x')\psi = -2m\hat{\jmath}\int \frac{d^3\boldsymbol{p}}{2E_p(2\pi)^3}[\theta(t-t')\Lambda_+(\psi)e^{-\hat{\jmath}p\cdot(x-x')} + \theta(t'-t)\Lambda_-(\psi)e^{\hat{\jmath}p\cdot(x-x')}] \tag{6.3}$$

where $E = +\sqrt{\boldsymbol{p}^2+m^2}$. For the photon propagator we use the Lorentz gauge $\nabla\cdot A = 0$, so we have $\nabla^2 A = J$ and Greens' function must satisfy

$$\nabla_x^2 D_F(x-x') = \delta^4(x-x').$$

Taking the Fourier transform we can solve for the Feynman propagator

$$D_F(p) = \frac{-1}{p^2+\hat{\jmath}\epsilon}.$$

7 Electron scattering

For scattering calculations we write the wavefunction as the sum of an incoming plane wave and a scattered beam, $\psi = \psi_i + \psi_{\text{diff}}$, where ψ_{diff} is the solution at asymptotically large times given by

$$\psi_{\text{diff}}(x) = -2m\hat{\jmath}e \int d^4x' \int \frac{d^3\boldsymbol{p}}{2E_p(2\pi)^3} \Lambda_+ \left[A(x')\psi(x')\gamma_0\right] e^{-\hat{\jmath}p\cdot(x-x')}.$$

This can be written as a sum over final states

$$\psi_{\text{diff}}(x) = \int \frac{d^3\boldsymbol{p}_f}{2E_f(2\pi)^3} \psi_f(x),$$

where the subscript on $\psi_f(x)$ labels the final momentum and the final states are plane waves of the form

$$\begin{aligned}\psi_f(x) &\equiv \psi_f e^{-\hat{\jmath}p_f\cdot x} \\ &\equiv -\hat{\jmath}e \int d^4x' [p_f A(x')\psi(x') + mA(x')\psi(x')\gamma_0] e^{-\hat{\jmath}p_f\cdot(x-x')}. \qquad (7.1)\end{aligned}$$

With this definition the number of scattered particles is given by

$$\int d^3\boldsymbol{x}\gamma_0 \cdot J_{\text{diff}} = \int \frac{d^3\boldsymbol{p}_f}{2E_f(2\pi)^3} \left[\frac{\gamma_0 \cdot J_f}{2E_f}\right] \equiv \int \frac{d^3\boldsymbol{p}_f}{2E_f(2\pi)^3} N_f,$$

where we have defined the number density per Lorentz invariant phase space interval to be

$$N_f \equiv \frac{\gamma_0 \cdot J_f}{2E_f} = \frac{\gamma_0 \cdot (\psi_f \gamma_0 \tilde{\psi}_f)}{2E_f} = \frac{\rho_f}{2m}.$$

The Born series perturbative solution is generated by iterating (6.1). In the first order Born approximation this amounts to simply replacing $\psi(x')$ by $\psi_i(x')$. For plane waves of particles we have

$$\psi(x) = \psi e^{-\hat{\jmath}p\cdot x} \quad \text{and} \quad m\psi\gamma_0 = p\psi,$$

so the final states become

$$\psi_f = -\hat{\jmath}e \int d^4x' [p_f A(x') + A(x')p_i]\psi_i e^{\hat{\jmath}q\cdot x'} = -\hat{\jmath}e[p_f A(q) + A(q)p_i]\psi_i$$

where $q \equiv p_f - p_i$.

More generally we define

$$\psi_f = S_{fi}\psi_i$$

where S_{fi} is the *scattering operator* which rotates and dilates the initial states into the final states. Here the f and i indices label the initial and final momenta and the initial spin, so in general $S_{fi} = S_{fi}(p_f, p_i, \hat{S}_i)$. However S_{fi} does not depend on the final spin—instead the final spin is determined from the initial spin by a rotation encoded in S_{fi}. Since S_{fi} consists of a rotation and dilation, it is convenient to decompose it as

$$S_{fi} = \rho_{fi}^{\frac{1}{2}} R_{fi},$$

where R_{fi} is a rotor. There is no $e^{I\beta}$ part since we have particles scattering to particles, not a mixture of particles and antiparticles. The cross section will be determined by the ρ_{fi} factor, as detailed in the next section. The rotor R_{fi} rotates states with momentum p_i into states with momentum p_f. This also relates the initial and final spins by

$$\hat{S}_f = R_{fi}\hat{S}_i\widetilde{R}_{fi}$$

so the rest spins are related by

$$\hat{S}_f^0 = \widetilde{L}_f\hat{S}_f L_f = \widetilde{L}_f R_{fi}\hat{S}_i\widetilde{R}_{fi}L_f = \widetilde{L}_f R_{fi}L_i\hat{S}_i^0\widetilde{L}_i\widetilde{R}_{fi}L_f.$$

We therefore define the *rest spin scattering operator*

$$U_{fi} \equiv \widetilde{L}_f R_{fi} L_i$$

so that

$$\hat{S}_f^0 = U_{fi}\hat{S}_i^0\widetilde{U}_{fi}.$$

The rest spin scattering operator and the cross section contain all the information about scattering of states with momentum p_i and spin $\hat{S}_i$ into states with momentum p_f.

The form of the external line Feynman propagator (6.3) ensures that S_{fi} is of the form

$$S_{fi} = -\hat{\jmath}(p_f M + M p_i) \tag{7.2}$$

where in the Born approximation example $M = eA(q)$. However in general M can have some $\hat{\jmath}$-dependence in which case we can write

$$S_{fi}\psi_i = -\hat{\jmath}(p_f[M_r + \hat{\jmath}M_j] + [M_r + \hat{\jmath}M_j]p_i)\psi_i$$

where M_j and M_r are independent of $\hat{\jmath}$. Using $\hat{\jmath}\psi_i = \psi_i\Sigma = \hat{S}_i\psi_i$ and the fact that $\hat{S}_i$ and p_i commute, this can be written

$$S_{fi} = -\hat{\jmath}(p_f M + M p_i)$$

where

$$M = M_r + M_j\hat{S}_i$$

now depends on the initial spin. We can thus replace dependence on the "imaginary" $\hat{\jmath}$ with dependence on the spin bivector.

Using $mL^2 = p\gamma_0$ we can obtain U_{fi} from

$$U_{fi} \propto L_f\gamma_0 M L_i + \widetilde{L}_f M\gamma_0\widetilde{L}_i.$$

8 Positron scattering and pair annihilation

Adapting the above results to positron scattering is straightforward. We consider a negative energy plane wave coming in from the future and scattering into the past, so $\psi_i(x) = \psi_2 e^{\hat{\jmath} p_2 x}$ and

$$S_{fi}\psi_i = -\hat{\jmath}(-p_1 M \psi_i + M\psi_i\gamma_0)$$

where p_1 is the incoming positron momentum, and p_2 is the outgoing momentum. This then gives

$$S_{fi} = \hat{\jmath}(p_1 M + M p_2),$$

amounting to the substitution $p_f \to -p_1, p_i \to -p_2$.

The other case to consider is when the incoming electron gets scattered into the past, corresponding to pair annihilation. In this case we have

$$S_{fi} = -\hat{\jmath}(-p_2 M + M p_1)$$

where p_1 and p_2 are the incoming momenta of the electron and positron, respectively. In this case, we can decompose S_{fi} as

$$S_{fi} = \rho_{fi}^{\frac{1}{2}} I R_{fi}$$

since S_{fi} must now contain a factor of I to map electrons into positrons. This also implies

$$S_{fi}\tilde{S}_{fi} = -\rho_{fi}.$$

9 Cross sections

The scattering rate into the final states per unit volume per unit time is given by

$$W_{fi} = \frac{1}{VT}N_f = \frac{1}{VT}\frac{\gamma_0 \cdot J_f}{2E_f} = \frac{\rho_f}{2mVT},$$

where ρ_f is given simply by

$$\rho_f = |S_{fi}|^2 \rho_i = \rho_{fi}\rho_i.$$

Here we have defined

$$|S_{fi}|^2 \equiv |S_{fi}\tilde{S}_{fi}| = \pm S_{fi}\tilde{S}_{fi}$$

where the plus sign corresponds to electron to electron and positron to positron scattering, the minus sign to electron-positron annihilation. The cross section is defined as

$$d\sigma = \frac{W_{fi}}{\text{Target density} \times \text{Incident flux}}.$$

When S_{fi} is of the form

$$S_{fi} = -\jmath(2\pi)^4\delta^4(P_f - P_i)T_{fi}$$

where the delta function ensures momentum conservation ($P_f = P_i$), we have

$$|S_{fi}|^2 = VT(2\pi)^4\delta^4(P_f - P_i)|T_{fi}|^2.$$

Working in the J_i frame, the target density is ρ_i, so writing the incident flux as χ we have

$$d\sigma = \frac{1}{2m\chi}(2\pi)^4\delta^4(P_f - P_i)|T_{fi}|^2.$$

Alternatively, we may have elastic scattering with energy conservation ($E_f = E_i$) and

$$S_{fi} = -\jmath 2\pi\delta(E_f - E_i)T_{fi}.$$

In this case

$$|S_{fi}|^2 = 2\pi T\delta(E_f - E_i)|T_{fi}|^2.$$

A target density of $1/V$ and an incident flux of $|\boldsymbol{J}_i| = \rho_i|\boldsymbol{p}_i|/m$ then gives

$$d\sigma = \frac{\pi}{|\boldsymbol{p}_i|}\delta(E_f - E_i)|T_{fi}|^2.$$

Above we have considered the total number of particles scattered. If we are interested in the final spin, we can find it using the spin scattering operator. However, we might also like to consider the cross section when we only observe particles with final spins in a certain plane $\hat{S}_o$ (where $\hat{S}_o{\cdot}p_f = 0$). This is particularly relevant in examples like electron-positron annihilation where ψ_f is actually an input state and we would like to calculate the cross section for arbitrary initial spins.

The spin projection operators into the $\hat{S}_o$ plane are

$$\chi_\pm(\psi) = \tfrac{1}{2}(\psi \mp \hat{S}_o\psi\boldsymbol{\Sigma}),$$

and we are interested in scattering into

$$\chi_\pm(\psi_f) = \chi_\pm(S_{fi}\psi_i).$$

Now if S_{fi} is in the form (7.2), we have

$$\begin{aligned}
\chi_\pm(S_{fi}\psi_i) &= -\tfrac{1}{2}\left[(p_f M + M p_i)\psi_i\boldsymbol{\Sigma} \pm \hat{S}_o(p_f M + M p_i)\psi_i\right] \\
&= -\tfrac{1}{2}\left[(p_f M + M p_i)\hat{S}_i \pm \hat{S}_o(p_f M + M p_i)\right]\psi_i \\
&= -\tfrac{1}{2}\left[p_f(M\hat{S}_i \pm \hat{S}_o M) + (M\hat{S}_i \pm \hat{S}_o M)p_i\right]\psi_i.
\end{aligned}$$

Defining $\chi_{\pm}(\psi_f) = S^{\pm}_{fi}\psi_i$ the scattering rate will be proportional to $\rho^{\pm}_{fi}$ given by

$$|S^{\pm}_{fi}|^2 = \left\langle (m^2 M + p_f M p_i)(\widetilde{M} \mp \hat{S}_i \widetilde{M} \hat{S}_o) \right\rangle . \tag{9.1}$$

If we sum over final spins, the $\hat{S}_o$ term cancels out and we get the expected result for the total ρ_{fi} :

$$|S_{fi}|^2 = \langle (p_f M + M p_i)(\widetilde{M} p_f + p_i \widetilde{M}) \rangle = 2\langle m^2 M\widetilde{M} + p_f M p_i \widetilde{M} \rangle \tag{9.2}$$

10 Coulomb scattering

As our first simple example, we consider the first Born approximation in electron Coulomb scattering where we have an external field given by

$$A(x) = \frac{-Ze}{4\pi|\boldsymbol{x}|}\gamma_0.$$

In the first Born approximation, M is given by $M = eA(q)$ where the Fourier transform of $A(x)$ is

$$A(q) = -\frac{2\pi Ze}{\boldsymbol{q}^2}\delta(E_f - E_i)\gamma_0$$

and $q\cdot\gamma_0 = E_f - E_i$. Writing

$$S_{fi} = -j 2\pi\delta(E_f - E_i) T_{fi}$$

and using energy conservation we have

$$T_{fi} = -\frac{Ze^2}{\boldsymbol{q}^2}(2E + \boldsymbol{q})$$

so that the formula for the cross section becomes

$$d\sigma = \left(\frac{Ze^2}{\boldsymbol{q}^2}\right)^2 \frac{\pi}{|\boldsymbol{p}_i|}\delta(E_f - E_i)(4E^2 - \boldsymbol{q}^2)\frac{d^3\boldsymbol{p}_f}{2E_f(2\pi)^3}.$$

Using $d^3\boldsymbol{p}_f = |\boldsymbol{p}_f| E_f dE_f d\Omega_f$ we recover the Mott cross section

$$\left(\frac{d\sigma}{d\Omega_f}\right)_{\text{Mott}} = \frac{Z^2\alpha^2}{\boldsymbol{q}^4}(4E^2 - \boldsymbol{q}^2) = \frac{Z^2\alpha^2}{4\boldsymbol{p}^2\beta^2\sin^4(\frac{\theta}{2})}(1 - \beta^2\sin^2(\frac{\theta}{2})),$$

where

$$\boldsymbol{q}^2 = (\boldsymbol{p}_f - \boldsymbol{p}_i)^2 = 2\boldsymbol{p}^2(1 - \cos\theta) \quad \text{and} \quad \beta = \frac{|\boldsymbol{p}|}{E}.$$

The derivation is manifestly independent of initial spin, so the cross section is spin independent. Of course, the final and initial spins will be related by the rest spin scattering operator U_{fi}, where

$$U_{fi} \propto L_f L_i + \widetilde{L}_f \widetilde{L}_i \propto (E+m)^2 + \boldsymbol{p}_f \boldsymbol{p}_i.$$

If U_{fi} rotates by an angle δ in the $\hat{B}$ plane ($\hat{B}^2 = -1$), it is given by

$$U_{fi} = \exp(\frac{\delta \hat{B}}{2}) = \cos(\frac{\delta}{2}) + \hat{B}\sin(\frac{\delta}{2}).$$

So, we see that the rotation is in the $\boldsymbol{p}_f \wedge \boldsymbol{p}_i$ plane and by an angle δ given by

$$\tan(\frac{\delta}{2}) = \frac{|\langle U_{fi}\rangle_2|}{\langle U_{fi}\rangle} = \frac{|\boldsymbol{p}_f \wedge \boldsymbol{p}_i|}{(E+m)^2 + \boldsymbol{p}_f \cdot \boldsymbol{p}_i} = \frac{\sin\theta}{(E+m)/(E-m) + \cos\theta}.$$

Similar derivations of these result using the STA approach have been given before [1, 8].

11 Compton scattering

In Compton scattering one electron interacts with two photons. We can therefore apply the above formalism in a somewhat heuristic way by using plane waves to represent the potentials of the two photons. There are two Feynman diagrams to consider which give two terms of the form

$$M_{12} =$$
$$e^2 \int d^4x' \int d^4x'' \int \frac{d^4p}{(2\pi)^4} A_1(x') \frac{pA_2(x'') + A_2(x'')p_i}{p^2 - m^2 + \hat{\jmath}\epsilon} e^{\hat{\jmath}x'\cdot(p_f - p)} e^{\hat{\jmath}x''\cdot(p - p_i)}$$

where

$$A(x) = \epsilon e^{\mp \hat{\jmath} k\cdot x}$$

is different at each vertex and $\epsilon^2 = -1$. Performing the integrations and summing the two contributions, we have

$$M =$$
$$e^2(2\pi)^4 \delta^4(p_f + k_f - p_i - k_i) \left[\epsilon_f \frac{(p_i + k_i)\epsilon_i + \epsilon_i p_i}{2k_i \cdot p_i} + \epsilon_i \frac{(p_i - k_f)\epsilon_f + \epsilon_f p_i}{-2p_i \cdot k_f} \right].$$

Choosing $p_i \cdot \epsilon_i = p_i \cdot \epsilon_f = 0$ this is simply

$$M = e^2(2\pi)^4 \delta^4(p_f + k_f - p_i - k_i) \left(\frac{\epsilon_f h_i \epsilon_i}{2k_i \cdot p_i} + \frac{\epsilon_i k_f \epsilon_f}{2p_i \cdot k_f} \right).$$

Writing

$$S_{fi} = -\hat{\jmath}(2\pi)^4\delta^4(p_f + k_f - p_i - k_i)T_{fi}$$

and using (9.2) we then have

$$|T_{fi}|^2 = e^4\left\langle \frac{m^2\epsilon_f k_i\epsilon_i\epsilon_f k_f\epsilon_i + p_f\epsilon_f k_i\epsilon_i p_i\epsilon_f k_f\epsilon_f}{k_i\cdot p_i k_f\cdot p_i} + \frac{p_f\epsilon_f k_i\epsilon_i p_i\epsilon_i k_i\epsilon_f}{2(k_i\cdot p_i)^2} + \frac{p_f\epsilon_i k_f\epsilon_f p_i\epsilon_f k_f\epsilon_i}{2(k_f\cdot p_i)^2}\right\rangle.$$

The identities we need to calculate are now the same as in the traditional approach; only now we know that the result is independent of initial spin since we haven't done a spin sum. Using momentum conservation we know

$$p_f + k_f = p_i + k_i, \qquad k_f\cdot p_i = k_i\cdot p_f, \qquad p_i\cdot k_i = p_f\cdot k_f.$$

Applying these the result becomes, after some work,

$$|T_{fi}|^2 = e^4\left[4(\epsilon_i\cdot\epsilon_f)^2 - 2 + \frac{p_i\cdot k_f}{p_i\cdot k_i} + \frac{p_i\cdot k_i}{p_i\cdot k_f}\right].$$

To calculate the cross section we work in the frame where the electron is initially at rest $(p_i = m\gamma_0)$. The incoming photon flux is $2k_i^0$, so we have

$$d\sigma = (2\pi)^4\delta^4(p_f + k_f - p_i - k_i)\frac{|T_{fi}|^2}{2m2k_i^0}\frac{d^3k_f}{2k_f^0(2\pi)^3}\frac{d^3p_f}{2E_f(2\pi)^3}.$$

Now

$$\int d^3p_f d^3k_f\delta^4(p_f + k_f - p_i - k_i) = (k_f^0)^2\frac{E_f k_f^0}{mk_i^0}d\Omega,$$

where we have done the integral over the final electron's momentum since we are primarily interested in the scattering of the photon. In the lab frame the result is, therefore,

$$\frac{d\sigma}{d\Omega} = \left(\frac{k_f}{k_i}\right)^2\frac{|T_{fi}|^2}{4m^2(4\pi)^2} = \frac{\alpha^2}{4m^2}\left(\frac{k_f}{k_i}\right)^2\left[\frac{k_f}{k_i} + \frac{k_i}{k_f} + 4(\epsilon_f\cdot\epsilon_i)^2 - 2\right]$$

in agreement with the Klein-Nishina formula. Again, the difference is that this derivation applies regardless of the initial electron spin. Of course, if we had used circularly polarized photons, we would have introduced some $\hat{\jmath}$-dependence; the result would have become spin-dependent.

12 Pair annihilation

A process closely related to Compton scattering is electron-positron annihilation. We have to take account of the fact that the "out" state is a positron, so the final states are of the form

$$\psi_f(x) \equiv \psi_f e^{\hat{\jmath} p_f \cdot x}$$
$$\equiv -\hat{\jmath} e \int d^4x' [-p_f A(x')\psi(x') + mA(x')\psi(x')\gamma_0] e^{\hat{\jmath} p_f \cdot (x-x')}.$$

Writing

$$S_{fi} = -\hat{\jmath}(-p_f M + M p_i)$$

we have two terms of the form

$$M_{12} =$$
$$e^2 \int d^4x' \int d^4x'' \int \frac{d^4p}{(2\pi)^4} A_1(x') \frac{p A_2(x'') + A_2(x'') p_i}{p^2 - m^2 + \hat{\jmath}\epsilon} e^{-\hat{\jmath} x' \cdot (p_f + p)} e^{\hat{\jmath} x'' \cdot (p - p_i)}$$

where

$$A(x) = \epsilon e^{\hat{\jmath} k \cdot x}$$

is different at each vertex and $\epsilon^2 = -1$. As for Compton scattering, we now choose $p_i \cdot \epsilon_1 = p_i \cdot \epsilon_2 = 0$ and sum the two contributions to get

$$M = e^2 (2\pi)^4 \delta^4 (p_f + p_i - k_1 - k_2) \left(\frac{\epsilon_2 k_1 \epsilon_1}{2 k_1 \cdot p_i} + \frac{\epsilon_1 k_2 \epsilon_2}{2 p_i \cdot k_2} \right).$$

For general positron and electron spins we should use (9.1) to calculate the cross section. However if either ψ_f or ψ_i are unpolarized the spin dependence will cancel out and the average introduces a factor of two into equation (9.2). In this case $|T_{fi}|^2$ is obtained from the Compton case by the substitution $p_f \to -p_f$ and $k_i \to -k_1$, and an overall sign change because $T_{fi}\tilde{T}_{fi} < 0$ in this case:

$$|T_{fi}|^2 = -\frac{e^4}{2} \left[4(\epsilon_i \cdot \epsilon_f)^2 - 2 - \frac{p_i \cdot k_2}{p_i \cdot k_1} - \frac{p_i \cdot k_1}{p_i \cdot k_2} \right].$$

To get the cross section divide by flux factors and perform the integral as usual.

13 Second order Coulomb scattering

Second order Coulomb scattering is interesting as it is spin-dependent and so provides a good testing ground for our calculation techniques. To avoid problems with divergent integrals, the potential is replaced with the screened potential

$$A(x) = -\frac{e^{-\lambda |\boldsymbol{x}|} Ze}{4\pi |\boldsymbol{x}|} \gamma_0,$$

and the Coulomb result found in the limit λ goes to zero [9, 10]. For this potential the first order analysis above can be applied with M given by

$$eA(q) = -\frac{2\pi Z e^2}{\lambda^2 + \boldsymbol{q}^2}\delta(E_f - E_i)\gamma_0.$$

To iterate to second order (7.1) is used, with the substitution

$$\psi(x') = \\ \psi_i e^{-\hat{\jmath} p_i\cdot x'} + e\int d^4x'' \int \frac{d^4k}{(2\pi)^4}\frac{kA(x'') + A(x'')p_i}{k^2 - m^2 + \hat{\jmath}\epsilon}\psi_i e^{\hat{\jmath} x''\cdot(k-p_i)} e^{-\hat{\jmath} k\cdot x'}$$

giving the extra contribution to M

$$M' \\ = e^2 \int d^4x' \int d^4x'' \int \frac{d^4k}{(2\pi)^4} A(x')\frac{kA(x'') + A(x'')p_i}{k^2 - m^2 + \hat{\jmath}\epsilon} e^{\hat{\jmath} x'\cdot(p_f - k)} e^{\hat{\jmath} x''\cdot(k-p_i)}.$$

Carrying out the x' and x'' integrations and using one of the resultant δ-functions we have

$$M = 2\pi\delta(E_f - E_i)M_T$$

where the extra contribution to M_T is

$$M_T' = e^2 \int \frac{d^3\boldsymbol{k}}{(2\pi)^3}\frac{A_0(p_f - k)A_0(k - p_i)}{k^2 - m^2 + \hat{\jmath}\epsilon}\gamma_0[k\gamma_0 + \gamma_0 p_i]$$

and

$$A_0(p) = \int d^3\boldsymbol{x} e^{-\boldsymbol{p}\cdot\boldsymbol{x}}\gamma_0\cdot A(\boldsymbol{x}) = \frac{-Ze}{\lambda^2 + \boldsymbol{p}^2}.$$

Using $k^2 - m^2 = \boldsymbol{p}_i^2 - \boldsymbol{k}^2$ and the integrals

$$I_1 + \tfrac{1}{2}(\boldsymbol{p}_i + \boldsymbol{p}_f)I_2 \\ = \int \frac{d^3\boldsymbol{k}}{(2\pi)^3}\frac{1 + \boldsymbol{k}}{[(\boldsymbol{p}_f - \boldsymbol{k})^2 + \lambda^2][(\boldsymbol{p}_i - \boldsymbol{k})^2 + \lambda^2](\boldsymbol{p}_i^2 - \boldsymbol{k}^2 + \hat{\jmath}\epsilon)}$$

we have

$$M_T' = Z^2 e^4 \left[\gamma_0 \tfrac{1}{2}(\boldsymbol{p}_i + \boldsymbol{p}_f)I_2 + (p_i + \gamma_0 E)I_1\right].$$

In the limit $\lambda \to 0$ our total M_T to second order is therefore

$$M_T = \frac{-Ze^2}{\boldsymbol{q}^2}\gamma_0 + Z^2 e^4 \left[(E\gamma_0 - \tfrac{1}{2}[p_f + p_i])I_2 + (p_i + \gamma_0 E)I_1\right]$$

where the integrals are [10]

$$I_1 = \frac{-\hat{\jmath}}{16\pi|\boldsymbol{p}|^3 \sin^2(\frac{\theta}{2})} \ln \frac{2|\boldsymbol{p}|\sin(\frac{\theta}{2})}{\lambda},$$

$$I_2 = \frac{1}{16\pi|\boldsymbol{p}|^3 \cos^2(\frac{\theta}{2})} \left\{ \frac{\pi[\sin(\frac{\theta}{2}) - 1]}{2\sin^2(\frac{\theta}{2})} - \hat{\jmath} \ln \frac{\lambda}{2|\boldsymbol{p}|} \right\} + \frac{I_1}{\cos^2(\frac{\theta}{2})}.$$

We see that M has some $\hat{\jmath}$ dependence, so writing $I_1 = (A+C)\hat{\jmath}$ and $I_2 = B + C\hat{\jmath}$ where A, B, and C are scalars and replacing the $\hat{\jmath}$-dependence with $\hat{S}_i$-dependence, this becomes

$$M_T = \gamma_0 \left[-\frac{Ze^2}{\boldsymbol{q}^2} + EZ^2e^4 \left\{ B + (2C + A)\hat{S}_i \right\} \right] + Z^2e^4 \left[p_i(A\hat{S}_i - B) - \tfrac{1}{2}q(B + C\hat{S}_i) \right].$$

The term proportional to q does not contribute to T_{fi}. Using

$$p_f p_i + m^2 = E(2E + \boldsymbol{q}) - \boldsymbol{p}^2 - \boldsymbol{p}_f \boldsymbol{p}_i$$

we have

$$T_{fi} = (2E + \boldsymbol{q}) \left[-\frac{Ze^2}{\boldsymbol{q}^2} + 2EZ^2e^4(A + C)\hat{S}_i \right] + Z^2e^4(\boldsymbol{p}^2 + \boldsymbol{p}_f \boldsymbol{p}_i)(B - A\hat{S}_i).$$

Keeping terms up to α^3 the cross section is governed by

$$|T_{fi}|^2 = (4E^2 - \boldsymbol{q}^2)\frac{Z^2e^4}{\boldsymbol{q}^4} - \frac{4Z^3e^6}{\boldsymbol{q}^2} \left[EB(\boldsymbol{p}^2 + \boldsymbol{p}_f \cdot \boldsymbol{p}_i) + mA(\boldsymbol{p}_i \wedge \boldsymbol{p}_f) \cdot \hat{S}_i^0 \right]$$

where $\hat{S}_i^0$ is the initial rest spin. As expected the divergent parts of the integrals have cancelled out, and we are left with the finite terms B and

$$A = \frac{\ln \sin(\frac{\theta}{2})}{16\pi|\boldsymbol{p}|^3 \cos^2(\frac{\theta}{2})}.$$

The cross section for unpolarized scattering is found by averaging over the initial spin. This gives the spin-independent part of the cross section since the spin dependent part averages to zero. The α^3 contribution is therefore

$$\frac{d\sigma}{d\Omega_f}' = -\frac{4Z^3e^6BE}{4(2\pi)^2\boldsymbol{q}^2}(\boldsymbol{p}^2 + \boldsymbol{p}_f \cdot \boldsymbol{p}_i) = \frac{\pi\alpha^3 Z^3 E[1 - \sin(\frac{\theta}{2})]}{4|\boldsymbol{p}|^3 \sin^3(\frac{\theta}{2})}.$$

Hence the unpolarized cross section, including the second Born approximation but ignoring radiative corrections, is

$$\frac{d\sigma}{d\Omega_f} = \left(\frac{d\sigma}{d\Omega_f}\right)_{\text{Mott}} \left\{ 1 + Z\alpha\pi \frac{\beta \sin(\frac{\theta}{2})[1 - \sin(\frac{\theta}{2})]}{1 - \beta \sin^2(\frac{\theta}{2})} \right\}$$

in agreement with the result obtained by Dalitz [9] using the conventional matrices and spin-sums approach.

14 Spin dependence and double scattering

As an example of handling spin dependence we can work out the asymmetry parameter for double scattering from a Coulomb potential. The idea is that since the second order correction to Coulomb scattering is spin dependent, the scattered beam will be partially polarized even with an unpolarized incident beam. The scattered beam can then impinge on a second target, which leads to an observable asymmetry in the scattered intensity. The asymmetry was first worked out by Mott [11, 12].

The first thing we need to know is the spin after the first scattering. This is given by

$$\hat{S}_f = R_{fi}\hat{S}_i\tilde{R}_{fi},$$

so we have

$$\hat{S}_f \propto T_{fi}\hat{S}_i\tilde{T}_{fi} = \frac{Z^2e^4}{q^4}(2E+\boldsymbol{q})\hat{S}_i(2E-\boldsymbol{q}) \\ - \frac{2Z^3e^6A}{q^2}\left\langle(\boldsymbol{p}^2+\boldsymbol{p}_f\boldsymbol{p}_i)(2E-\boldsymbol{q})\right\rangle_2$$

where we have only kept the lowest order terms in the spin dependent and spin-independent parts. We now define S^0 to be the polarization in the plane $\hat{S}^0$. This is a bivector in the plane of $\hat{S}^0$ with modulus equal to the polarization of the beam. Since the incoming beam is taken to be unpolarized, the resultant polarization plane will be given by the spin-independent part of $\hat{S}_f$ deboosted to rest. To get the polarization we then divide by the magnitude of the spin-dependent part:

$$S_f^0 = -\frac{2Ze^2q^2A}{(4E^2-\boldsymbol{q}^2)}\tilde{L}_f\left\langle(\boldsymbol{p}^2+\boldsymbol{p}_f\boldsymbol{p}_i)(2E-\boldsymbol{q})\right\rangle_2 L_f \\ = \frac{2Ze^2q^2A}{(4E^2-\boldsymbol{q}^2)}2m\boldsymbol{p}_i\wedge\boldsymbol{p}_f.$$

The spin-dependent part of the cross section for the second scattering is then given by

$$\left(\frac{d\sigma}{d\Omega_f}\right)_{\text{spin}} = -\frac{4Z^3e^6mA_2}{\boldsymbol{q}_2^2(2\pi)^2}(\boldsymbol{p}_f\wedge\boldsymbol{p}_2)\cdot S_f^0 \\ = -\frac{64(2\pi)^2Z^4\alpha^4\boldsymbol{q}_1^2m^2A_1A_2}{\boldsymbol{q}_2^2(4E^2-\boldsymbol{q}_1^2)}(\boldsymbol{p}_f\wedge\boldsymbol{p}_2)\cdot(\boldsymbol{p}_i\wedge\boldsymbol{p}_f)$$

where the 1 and 2 subscripts refer to the first are second scattering, respectively (e.g., $\boldsymbol{q}_2 = \boldsymbol{p}_2 - \boldsymbol{p}_f$). We see that the asymmetry will depend on the cosine of the angle ϕ between the $\boldsymbol{p}_f\wedge\boldsymbol{p}_2$ and $\boldsymbol{p}_i\wedge\boldsymbol{p}_f$ planes. The asymmetry parameter δ is defined so that the final intensity depends on ϕ through the factor $1+\delta\cos\phi$.

In the case where $p_i \cdot p_f = p_f \cdot p_2 = 0$ $(p_i \cdot p_2 = -p^2 \cos\phi)$, we find that the first non-zero contribution to the asymmetry factor is

$$\begin{aligned}\delta &= \frac{64(2\pi)^2 Z^4 \alpha^4 m^2 A^2}{(4E^2 - q^2)} p^4 \frac{q^4}{Z^2\alpha^2(4E^2 - q^2)} \\ &= Z^2\alpha^2(\ln 2)^2 \frac{\beta^2(1-\beta^2)}{(2-\beta^2)^2}\end{aligned}$$

in agreement with the answer quoted by Dalitz [9]. It is of course only the first approximation and for large Z nuclei higher order corrections will be far from negligible.

15 The partial spin-sum approach

The above formalism seems to work well for single particle scattering. Here we show how we can adapt a more traditional approach in more complicated cases, demonstrating the flexibility of the STA formalism. The scattering operator approach could equally well be used in the more complicated case, as we show below.

We use the two basis states u_r to write (7.1) as

$$\begin{aligned}\psi_f &= -\hat{\jmath}e \int d^4x' \sum_r u_r(p_f) \langle \tilde{u}_r(p_f) A(x') \psi_i \gamma_0 \rangle_S e^{\hat{\jmath} p_f \cdot x'} \\ &= \sum_r u_r(p_f) S^r_{fi},\end{aligned}$$

where S^r_{fi} is the traditional S-matrix. The total number density per Lorentz invariant phase space interval is then

$$N_f = \sum_r |S^r_{fi}|^2.$$

As an example we consider electron-muon scattering (following, for example, Björken and Drell [13]) in which A is given by

$$A(x) = \int d^4x' D_F(x - x') J(x')$$

and $J(x')$ is the 'complex' conserved current given by

$$J_a = e\langle \tilde{u}_s \gamma_a \psi_2 \gamma_0 \rangle_S.$$

Defining T^{rs} as usual we have

$$T^{rs} = -\frac{e^2}{q^2} \langle \tilde{u}_r \gamma_a \psi_1 \gamma_0 \rangle_S \langle \tilde{u}_s \gamma^a \psi_2 \gamma_0 \rangle_S$$

where $q = p_1' - p_1 = p_2 - p_2'$, and dashed variables correspond to the final states. Summing over r and s

$$\begin{aligned}
|T|^2 &= \frac{e^4}{q^4}\langle\gamma_0\widetilde{\psi}_1\gamma^b(p_1'\gamma^a + \gamma^a p_1)\psi_1\rangle_S\langle\gamma_0\widetilde{\psi}_2\gamma_b(p_2'\gamma_a + \gamma_a p_2)\psi_2\rangle_S \\
&= \frac{2e^4\rho_1\rho_2}{m_1m_2q^4}\Big[\langle\gamma^b(p_1'\gamma^a + \gamma^a p_1)p_1\rangle\langle\gamma_b(p_2'\gamma_a + \gamma_a p_2)p_2\rangle \\
&\qquad - \langle\gamma^b(p_1'\gamma^a + \gamma^a p_1)p_1\hat{S}_1\rangle\langle\gamma_b(p_2'\gamma_a + \gamma_a p_2)p_2\hat{S}_2\rangle\Big] \\
&= \frac{2e^4\rho_1\rho_2}{m_1m_2q^4}\Big[p_1'\cdot p_2' p_1\cdot p_2 + p_2'\cdot p_1 p_2\cdot p_1' - m_1^2 p_2'\cdot p_2 + m_2^2 p_1'\cdot p_1 + 2m_2^2m_1^2 \\
&\qquad - [q\cdot(\hat{S}_1\wedge p_1)]\cdot[q\cdot(\hat{S}_2\wedge p_2)]\Big].
\end{aligned}$$

This approach differs from the normal one in that we have only done one spin sum over the final spins. We can therefore explicitly retain information about the initial spins, and calculations that are spin-independent will be manifestly so. Spin averaging simply amounts to removing spin dependent terms in the cross section.

The same result could be obtained using the scattering operator approach using

$$M = eD_F\gamma^a J_a$$

and summing over the final spin of the other particle. One ends up with exactly the same equation. However, the scattering operator approach may be better for calculating spin effects. If we are interested in the spin dependence of a particular fermion line, the scattering operator approach works well once we have summed over the spins of the other particles. For example we can calculate the final spin and polarization in the same way as we did for Coulomb scattering. In this approach we still have to perform a spin sum, but only over the spins of the other particles. We could of course introduce spin projection operators to single out particular spins of the other particles if necessary.

16 Conclusions

We have seen how Hestenes' STA formulation of Dirac theory provides a useful and elegant method of performing cross section calculations. Spin is handled in a simple manner, and the logic of calculating cross sections is simplified considerably. We don't perform unnecessary spin sums, and spin dependence is manifest in the spin bivector dependence of the scattering operator. It's a simple matter to calculate spin precessions, polarizations, and spin dependent results, and the results are automatically expressed in terms of physical spin bivectors and the other scattering parameters. We can perform unpolarized calculations simply by averaging over spins.

In the multiparticle case, things are more complicated. We don't have a neat method for performing arbitrary spin dependent calculations and still have to resort to spin sums over terms involving complex conserved currents. However, we can still write down a scattering operator for any given fermion line, retaining the benefits of the scattering operator for calculations involving the spin of the particle.

Acknowledgement

Antony Lewis was supported by a PPARC studentship during the course of this work.

References

[1] D. Hestenes, Geometry of the Dirac theory, in *The Mathematics of Physical Spacetime,* J. Keller, ed., UNAM, Mexico, 1982, 67.

[2] D. Hestenes, *Space-Time Algebra,* Gordon and Breach, New York, 1966.

[3] D. Hestenes, Observables, operators, and complex numbers in the Dirac theory, *J. Math. Phys.* **16** (3) (1975), 556.

[4] D. Hestenes and G. Sobczyk, *Clifford Algebra to Geometric Calculus,* Reidel, Dordrecht, 1984.

[5] C. J. L. Doran, A. N. Lasenby, and S. F. Gull, States and operators in the spacetime algebra, *Found. Phys.* **23** (9) (1993), 1239.

[6] A. N. Lasenby, C. J. L. Doran, and S. F. Gull, A multivector derivative approach to Lagrangian field theory, *Found. Phys.* **23** (10) (1993), 1295.

[7] A. N. Lasenby, C. J. L. Doran, and S. F. Gull, Gravity, gauge theories, and geometric algebra, *Phil. Trans. R. Soc. Lond. A* (1998), 356: 487–582.

[8] C. J. L Doran, A. N. Lasenby, S. F. Gull, S. S. Somaroo, and A. D. Challinor, Spacetime algebra and electron physics, *Adv. Imag. & Elect. Phys.* (1996), 95: 271.

[9] R. H. Dalitz, On higher born approximation in potential scattering, *Proc. Roy. Soc.* (1951), 206: 509.

[10] C. Itzykson and J.-B. Zuber, *Quantum Field Theory,* McGraw-Hill, New York, 1980.

[11] N. F. Mott, Scattering of fast electrons by atomic nuclei, *Proc. Roy. Soc* **A124** (1929), 425.

[12] N. F. Mott, Polarization of electrons by double scattering, *Proc. Roy. Soc* **A315** (1932), 429.

[13] J. D. Björken and S. D. Drell, *Relativistic Quantum Mechanics, Vol 1,* McGraw-Hill, New York, 1964.

Antony Lewis
Astrophysics Group
Cavendish Laboratory
Madingley Road
Cambridge CB3 0HE, U.K.
E-mail: aml1005@cam.ac.uk

Anthony Lasenby
Astrophysics Group
Cavendish Laboratory
Madingley Road
Cambridge CB3 0HE, U.K.
E-mail: a.n.lasenby@mrao.cam.ac.uk

Chris Doran
Astrophysics Group
Cavendish Laboratory
Madingley Road
Cambridge CB3 0HE, U.K.
E-mail: c.doran@mrao.cam.ac.uk

Received: September 21, 1999; Revised: November 18, 1999

2.

PHYSICS: STRUCTURES

Twistor Approach to Relativistic Dynamics and to the Dirac Equation – A Review

Andreas Bette

ABSTRACT Twistors, which can be regarded as spinors of the conformal group acting on compactified Minkowski space, form a relativistic phase space **T** of a massless particle with a non-vanishing helicity [1]. A relativistic extended phase space of a massive spinning charged particle [2, 3] may be regarded as "embedded" in a two twistor phase space [4, 5, 6] **Tp(2)** which is simply **TxT** with its diagonal deleted (in the sense of a symplectic reduction). We describe a procedure by which relativistic physical variables of a charged spinning and massive particle are identified as functions on **Tp(2)**. Those representing spin of the particle and those representing its commuting and non-commuting Minkowski positions are given special attention. A non-holomorphic relativistic canonical quantization turns the so-identified relativistic physical variables on **Tp(2)** into the corresponding relativistic quantum mechanical Poincaré covariant operators. To initiate a test of our idea, a function H, corresponding to the square of the rest mass (see (5.1)) operator arising from the Dirac equation describing a relativistic massive, charged spin $\frac{1}{2}$ quantum particle in an external electromagnetic field, is used to generate a canonical flow on **Tp(2)** giving a reasonable Poincaré covariant set of classical equations of motion. The non-holomorphic (re-)quantization turns H into a new Poincaré scalar quantum operator. Its eigenvalue equation valid for any value of the quantized spin corresponds and mimics the second order formulation of the Dirac equation (with minimally coupled external electromagnetic field). The present paper is essentially a review of previous presentations on the subject [2, 3, 4, 5, 6, 7]. Some new work, at an undeveloped introductory level, appears in sections 5 and 6.

Keywords: Twistor theory, special relativity, elementary particles.

AMS Subject Classification: 81R25, 83A05, 70G35, 70H40.

1 Introduction

The plausible relativistic physical variables describing a spinning, charged and massive particle are, besides the charge itself, its Minkowski (four) position $\tilde{X}$, its relativistic linear (four) momentum P and also its so-called Lorentz (four) angular momentum $\Sigma \neq 0$, the latter forming four translation invariant part of its total angular (four) momentum M. Expressing these variables in terms of Poincaré covariant real valued functions defined on an extended relativistic phase space [2, 7] means that the mutual Poisson bracket relations among the total angular momentum functions M^{ab} and the linear momentum functions P^a have to represent the commutation relations of the Poincaré algebra.

On any such an extended relativistic phase space, as shown by Zakrzewski [2, 7], the (natural?) Poisson bracket relations

$$\{\tilde{X}^a,\ \tilde{X}^b\} = 0 \tag{1.1}$$

imply that for the splitting of the total angular momentum into its orbital and its spin part

$$M^{ab} = P^a\tilde{X}^b - P^b\tilde{X}^a + \Sigma^{ab}, \tag{1.2}$$

one necessarily obtains

$$\Sigma^{ab}P_b \neq 0. \tag{1.3}$$

On the other hand it is always possible to shift (translate) the commuting (see (1.1)) four position $\tilde{X}^a$ by a four vector ΔX^a

$$X^a = \tilde{X}^a + \Delta X^a. \tag{1.4}$$

so that the total angular four momentum splits instead into a new orbital and a new (Pauli-Lubański) spin part

$$M^{ab} = P^aX^b - P^bX^a + R^{ab}, \tag{1.5}$$

in such a way that

$$R^{ab}P_b = 0. \tag{1.6}$$

However, as proved by Zakrzewski [2, 7], the so-defined new shifted four position functions X^a must fulfill the following Poisson bracket relations:

$$\{X^a,\ X^b\} = -\frac{1}{(P_kP^k)^2}R^{ab}. \tag{1.7}$$

Consequently, the new shifted four-position functions X^a *do not commute* among themselves if the particle is spinning.

One concrete realization of (a special case of) an extended relativistic phase space arises as a symplectic reduction of the so-called two twistor phase space **Tp(2)** 16 dimensions) as shown by Zakrzewski and Bette [5, 6].

The first objective in this paper is to review the above mentioned results in the context of two twistor phase space **Tp(2)**. The second is a presentation of a non-holomorphic Poincaré invariant quantization procedure on **Tp(2)**, and, finally, the third objective is to initiate a test of these ideas in the special case dealing with a spinning, charged massive classical and/or quantum particle moving in an external electromagnetic field (Lorentz-Bargman-Michel-Telegdi (BMT)-like and Dirac-like equations).

The paper is organized as follows: in section 2 we review some already known facts about twistors, twistor phase spaces and extended phase spaces [4, 5, 6]. Some relativistic physical variables are identified as functions on the two twistor phase space **Tp(2)**. In section 3 the commuting position functions $\tilde{X}^a$ and the corresponding translation invariant (four) angular momentum functions Σ^{ab} are found on **Tp(2)** and discussed in detail. In section 4 a Poincaré invariant twistor quantization procedure is suggested. As usual such a procedure is not unique (for example there are choices to be made because of the ordering problem). In sections 5 and 6 it is shown how the presented ideas could be tested at the classical and quantum level in the case of a massive spinning and charged particle moving in an external electromagnetic field.

2 Twistor phase spaces

The fundamental object in twistor theory is the *twistor space* [1, 8, 9, 10] **T**, which is a 4-dimensional complex vector space equipped with a (pseudo-) hermitian form

$$\rho = \rho(Z, W) := Z^\alpha \bar{W}_\alpha, \tag{2.1}$$

where $Z, W \in \mathbf{T}$ and where linear transformations of **T** preserving this (pseudo-) hermitian form and having the determinant equal to one constitute a group G isomorphic to $SU(2,2)$. The important fact is that the latter contains the (universal covering group of the) Poincaré group as a subgroup.

The set of isotropic (with respect to (2.1)) two-dimensional subspaces in **T** which are transversal to one distinguished such a subspace I *defines* the usual Minkowski space. Relative to an arbitrary but fixed origin in the so-defined (affine) Minkowski space, each twistor may be represented by two spinors [1, 8]:

$$Z^\alpha = (\omega^A,\ \pi_{A'}). \tag{2.2}$$

Similarly, the corresponding complex conjugate twistor is represented by

$$\bar{Z}_{\alpha} = (\bar{\pi}_A,\ \bar{\omega}^{A'}). \tag{2.3}$$

(Note that the splitting of a twistor into two spinors is translation dependent, i.e., depends on our choice of the fixed origin in the Minkowski space. Consequently, a twistor should not to be confused with a Dirac bispinor, the latter being translation independent.)

The SU(2,2) group is isomorphic with (the universal covering of) the conformal group of the Minkowski space. Therefore, the twistor space provides a natural framework for the study of the conformal geometry of the Minkowski space. Moreover, it has a natural structure of a *phase space* of a massless particle with an arbitrary value of its helicity (spin) which appears as a conformally scalar function on the twistor space [1]. The natural phase space character of **T** is a consequence of the fact that the imaginary part of the (pseudo-) hermitian form in (2.1) defines a G-invariant constant symplectic two-form [5, 11] on **T**. This may be expressed in terms of a canonical conformally invariant Poisson bracket algebra

$$\{\bar{Z}_{\beta},\ Z^{\alpha}\} = i\delta^{\alpha}_{\beta}, \tag{2.4}$$

with all the remaining commutation relations being equal to zero.

The *two-twistor phase space* **Tp(2)** is now defined [4] as the set of pairs of twistors

$$Z^{\alpha} = (\omega^A,\ \pi_{A'}) \quad \text{and} \quad W^{\alpha} = (\lambda^A,\ \eta_{A'}) \tag{2.5}$$

such that

$$f := \pi^{A'}\eta_{A'}\ (= I_{\alpha\beta}Z^{\alpha}W^{\beta})\ \neq 0. \tag{2.6}$$

The corresponding complex conjugate twistors are

$$\bar{Z}_{\alpha} = (\bar{\pi}_A,\ \bar{\omega}^{A'}) \quad \text{and} \quad \bar{W}_{\alpha} = (\bar{\eta}_A,\ \bar{\lambda}^{A'}). \tag{2.7}$$

The conformally invariant symplectic two-form on **Tp(2)** [4, 12, 13] defines now a canonical conformally invariant Poisson bracket algebra

$$\{\bar{Z}_{\beta},\ Z^{\alpha}\} = i\delta^{\alpha}_{\beta}, \qquad \{\bar{W}_{\beta},\ W^{\alpha}\} = i\delta^{\alpha}_{\beta}, \tag{2.8}$$

with all the remaining commutation relations equal to zero. In terms of the Poincaré covariant spinor coordinates, the only non-vanishing commutations relations are

$$\{\bar{\pi}_B,\ \omega^A\} = i\delta^A_B, \qquad \{\bar{\eta}_B,\ \lambda^A\} = i\delta^A_B. \tag{2.9}$$

On **Tp(2)** there exist six independent real valued Poincaré scalar functions [4, 12, 13]. Four of these are also conformally scalar. The first two Poincaré (but not conformally) scalar functions are those defined by f in (2.6).

The four real valued conformally scalar functions are defined by means of the hermitian form introduced in (2.1)

$$s_1 := \frac{1}{2}(Z^\alpha \bar{Z}_\alpha) \quad \text{and} \quad s_2 := \frac{1}{2}(W^\alpha \bar{W}_\alpha), \tag{2.10}$$

$$\rho := Z^\alpha \bar{W}_\alpha. \tag{2.11}$$

On each of the two single twistor phase spaces, the four Lorentz covariant and four translation invariant functions representing the two linear null four momenta will be denoted by (abstract index notation [14, 15])

$$P_{1a} := \pi_{A'}\bar{\pi}_A, \quad P_{2a} := \eta_{A'}\bar{\eta}_A, \tag{2.12}$$

while Poincaré covariant functions (six for each single twistor phase space) representing the two angular null four momenta will be denoted by

$$M_{1ab} := i\bar{\omega}_{(A'}\pi_{B')}\epsilon_{AB} + c.c., \quad M_{2ab} := i\bar{\lambda}_{(A'}\eta_{B')}\epsilon_{AB} + c.c. \tag{2.13}$$

(The above functions arise as generators of the Poincaré group action, see below.)

By forming the Pauli Lubański spin four vectors on each of the two single twistor phase spaces, one discovers that the two real valued functions in (2.10) represent [1] the classical limit of the two helicity operators corresponding to each of the two massless systems described by the two pairs (P_1, M_1) and (P_2, M_2). In addition, the 12D phase space, obtained from **Tp(2)** by symplectic reduction with respect to the two helicity functions in (2.10), coincides with a relativistic phase space (REPS) previously introduced by Zakrzewski [2, 3, 5, 6, 7].

In the two twistor phase space, by linearity, the above single twistor phase space functions representing (P_1, M_1) and (P_2, M_2) define four new functions now representing a massive four momentum and six new functions now representing a massive angular four momentum [4, 16]

$$P_a := P_{1a} + P_{2a}, \tag{2.14}$$

$$M_{ab} := M_{1ab} + M_{2ab}. \tag{2.15}$$

The absolute value of f divided by the square root of two may now be recognized as the rest mass of a system having its four-momentum given by P_a.

It is a well-established fact that the pair (P_1, M_1) and the pair (P_2, M_2) each define separately a momentum mapping of the Poincaré group into the corresponding single twistor phase space. In other words, Poisson brackets (with respect to the conformally invariant constant symplectic two-form on a single twistor phase space) among the functions defining each such a pair reproduce the algebra of the Poincaré group [4, 10, 12, 16, 17]. This automatically implies that on the two twistor phase space **Tp(2)** the same is valid for the pair (P, M).

On the two twistor phase space $\mathbf{Tp(2)}$ we may also distinguish the following four mutually commuting complex valued functions representing the four vector position coordinates in the complexified Minkowski space [1]

$$z^a := \frac{i}{f}(\omega^A \eta^{A'} - \lambda^A \pi^{A'}). \tag{2.16}$$

For future use we introduce the following mutually commuting functions in $\mathbf{Tp(2)}$, representing three mutually Lorentz orthogonal space like four-vectors in the Minkowski space

$$l_a := P_{1a} - P_{2a}, \tag{2.17}$$

$$w_a := \pi_{A'} \bar{\eta}_A. \tag{2.18}$$

It is easy to see that the so-defined four-vectors are also orthogonal to the massive four momentum P_a. Their Lorentz norm is equal to minus the square of the rest mass of the system. In addition, the functions that represent these three space like four-vectors commute with the functions representing the massive four momentum vector P_a.

Forming the Pauli-Lubański spin four-vector on the two twistor phase space $\mathbf{Tp(2)}$,

$$S^a := \frac{1}{2}\epsilon^{abcd} M_{bc} P_d \tag{2.19}$$

yields [4, 12, 16, 17, 19, 20]

$$S_a = k l_a + \rho \bar{w}_a + \bar{\rho} w_a, \tag{2.20}$$

where

$$k := s_1 - s_2. \tag{2.21}$$

Re-expressing the massive angular four momentum as a sum of the spin and the orbital four angular momentum gives

$$M^{ab} = P^a X^b - P^b X^a + \frac{1}{(P^k P_k)^2} \epsilon^{abcd} P_c S_d, \tag{2.22}$$

where X is the real part of z

$$X^a := \frac{1}{2}(z^a + \bar{z}^a). \tag{2.23}$$

The imaginary part of z may now be expressed as a Lorentz four-vector given by [4, 12, 13, 17]

$$Y^a := \frac{1}{2i}(z^a - \bar{z}^a) = \frac{1}{2f\bar{f}}(\rho \bar{w}^a + \bar{\rho} w^a - 2s_1 P_2^a - 2s_2 P_1^a). \tag{2.24}$$

Note that the scalar spin function, defined in the usual way as

$$s := \frac{1}{m}\sqrt{-S_a S^a}, \tag{2.25}$$

is, as easily follows from (2.20), in terms of the twistor scalars given by

$$s = \sqrt{k^2 + |\rho|^2}, \tag{2.26}$$

which shows that the spin of a massive particle is not only a Poincaré but also a conformal scalar, the fact first time noticed by Perjès [19]. For massive particles formed by means of more than two twistors this is no longer true [4, 20].

Besides the already mentioned Poisson bracket relations reproducing the Poincaré algebra

$$\{P_a,\ P_b\} = 0, \tag{2.27}$$

$$\{M^{ab},\ P_c\} = -P^a\delta^b_c + P^b\delta^a_c, \tag{2.28}$$

$$\{M_{ab},\ M_{cd}\} = M_{ac}g_{bd} + M_{bd}g_{ac} - M_{ad}g_{bc} - M_{bc}g_{ad}, \tag{2.29}$$

the canonical commutation relations in (2.8) and/or in (2.9) also imply that the functions representing the four position X^a, as defined in (2.23), and the four momentum P_a, as defined in (2.14), obey, as they should, the following commutation relations

$$\{P_a,\ X^b\} = \delta^b_a. \tag{2.30}$$

3 Non-commuting and commuting position

However, the position functions in (2.23), allowing the splitting of the four angular momentum as in (1.5) or/and in (2.22) into its orbital part and its pure spin part, do not commute. One obtains instead [4, 5]

$$\{X^a,\ X^b\} = -\frac{1}{(P_k P^k)^2} R^{ab} \tag{3.1}$$

where we put

$$R^{ab} := \frac{1}{(P_k P^k)^2}\epsilon^{abcd}P_c S_d. \tag{3.2}$$

According to the remarks in the introduction, this implies that the *non-commuting* position X defined in (2.23) describes the shifted Minkowski position. From Zakrzewski's theory [2, 3] it follows that there must also exist functions on the two twistor space representing the unshifted and *commuting* position $\tilde{X}$. An explicit formula, which on the two twistor phase space defines the shift ΔX, has been found jointly by the author and the late Professor Zakrzewski in 1996 [5, 6]

$$\Delta X^a := \frac{i}{m^2}(\rho\bar{w}^a - \bar{\rho}w^a). \tag{3.3}$$

Two unshifted commuting Minkowski positions may now be distinguished

$$\tilde{X}^a_\pm := X^a \pm \Delta X^a. \tag{3.4}$$

On the two twistor space they fulfill (1.1)

$$\{\tilde{X}^a_+,\ \tilde{X}^b_+\} = 0, \tag{3.5}$$

$$\{\tilde{X}^a_-,\ \tilde{X}^b_-\} = 0. \tag{3.6}$$

Note, however, that

$$\{\tilde{X}^a_-,\ \tilde{X}^b_+\} = -\frac{2}{(P_k P^k)^2} R^{ab} \neq 0. \tag{3.7}$$

The total four angular momentum may now be expressed as

$$M^{ab} = P^a X^b - P^b X^a + R^{ab} = P^a \tilde{X}^b_\pm - P^b \tilde{X}^a_\pm + \Sigma^{ab}_\pm, \tag{3.8}$$

where the Lorentz angular momentum is defined by

$$\Sigma^{ab}_\pm := R^{ab} \mp (P^a \Delta X^b - P^b \Delta X^a). \tag{3.9}$$

Σ_{ab} may be (loosely) identified with the classical limit of the relativistic spin operator $\frac{i}{2}[\gamma_a,\ \gamma_b]$ (γ_a represent Dirac matrices) arising in the Dirac theory of the electron. However, in the Dirac theory the spin is quantized and equals to $\frac{1}{2}$ whereas at this classical level, Σ_{ab} are continuous real valued functions on the two twistor phase space. After quantization, however, (see below) they become (differential) operators. The eigenvalues of the spin become quantized according to the usual scheme but may now assume as well integral as half integral values, in contrast to the Dirac theory in which the spin is always equal to $\frac{1}{2}$.

As discussed in our work [6], the plus sign in $\tilde{X}_+$ corresponds to the charge identified as $e_+ := 2s_1$, while the minus sign in $\tilde{X}_-$ corresponds to the charge identified as $e_- := 2s_2$. So, in the first case the charge appears as (twice) the helicity function of the first massless constituent, while in the second case, it appears as (twice) the helicity function of the second massless constituent. In other words, interchanging the two twistor constituents ($Z \longrightarrow W$) corresponds to a "charge conjugation" ($e_+ \longrightarrow e_-$) with a *simultaneous change of the commuting Minkowski position* ascribed to the particle $\tilde{X}_+ \longrightarrow \tilde{X}_-$.

From [2] one knows that Σ_+ commutes with $\tilde{X}_+$ (and P, of course,) while Σ_- commutes with $\tilde{X}_-$ (and also with P, of course). This may also be checked by direct (tedious) computation using (2.8) or (2.9). Consequently, this amounts to the important insight that

$$\{\tilde{X}_+^a,\ \Sigma_+^{bc}\} = 0, \quad \{\tilde{X}_-^a,\ \Sigma_-^{bc}\} = 0. \tag{3.10}$$

Note also that, due to the fact that

$$\Delta X^a P_a = 0, \tag{3.11}$$

we obtain

$$P_a \Sigma_\pm^{ab} = \mp m^2 \Delta X^b \neq 0. \tag{3.12}$$

In the case of two twistors, it may easily be calculated that

$$I_1 := \Sigma_+^{ab}\Sigma_{ab}^+ = \Sigma_-^{ab}\Sigma_{ab}^- = \bar{K}^2 - \bar{S}^2 = k^2 \tag{3.13}$$

$$I_2 := \Sigma_+^{ab}\Sigma_{ab}^{+*} = \Sigma_-^{ab}\Sigma_{ab}^{-*} = \bar{S}\cdot\bar{K} = 0, \tag{3.14}$$

where k is the difference between the helicities (of the two massless constituents) as defined in (2.21). $\bar{S}$ and $\bar{K}$ are the three vector components of $\Sigma_\pm$ in an inertial frame while $\Sigma_\pm^*$ denotes the dual of $\Sigma_\pm$.

At the quantum level eigenvalues of the differential operator corresponding to k can assume half integral and integral values. In effect, the representations of the Lorentz group obtained in this way are only of the type $(k^2, 0)$, $k = 0, \pm\frac{1}{2}, \pm 1, \ldots$. One can speculate that massive particle constructions using three or more twistors can give the remaining representations of the Lorentz group.

Note also that

$$s^2 := \frac{1}{m^2}(-S_a S^a) = \frac{1}{2}R^{ab}R_{ab} = k^2 + \rho\bar{\rho} = I_1 - m^2(\Delta X^i \Delta X_i). \tag{3.15}$$

4 Relativistic canonical twistor quantization

In quantum mechanics the relativistic Poisson bracket relations in (2.27), (2.28), (2.29), (2.30), and so on should be replaced by the quantum mechanical commutators. For example, in order to incorporate the Heisenberg uncertainty relations, one should have

$$[\hat{P}_a,\ \hat{X}^b] = [\hat{P}_a,\ \hat{\tilde{X}}{}^b_\pm] = [\hat{P}_a,\ \hat{z}^b] = [\hat{P}_a,\ \hat{\bar{z}}^b] = i\delta_a^b \tag{4.1}$$

where the square brackets denote appropriate commutators.

A consistent relativistic canonical quantization scheme (although *not* unique), which turns all the relativistic Poisson bracket relations from the previous sections into commutators, may be achieved by the replacement of the canonical twistor variables ω and λ in (2.5) and in (2.7) by the differential operators

$$\hat{\omega}^A := \frac{\partial}{\partial \bar{\pi}_A}, \quad \hat{\bar{\omega}}^{A'} := -\frac{\partial}{\partial \pi_{A'}}, \quad \hat{\lambda}^A := \frac{\partial}{\partial \bar{\eta}_A}, \quad \hat{\bar{\lambda}}^{A'} := -\frac{\partial}{\partial \eta_{A'}}, \tag{4.2}$$

and π and η in (2.5) and in (2.7) by the multiplicative operators

$$\hat{\bar{\pi}}_A := \bar{\pi}_A, \quad \hat{\pi}_{A'} := \pi_{A'}, \quad \hat{\bar{\eta}}_A := \bar{\eta}_A, \quad \hat{\eta}_{A'} := \eta_{A'}. \tag{4.3}$$

The "wave" functions representing states of the relativistic charged massive and spinning quantum two twistor particle are thus defined as complex valued (non-holomorphic) functions on the product of two (Weyl) spinor spaces with its diagonal deleted (the spinors are not allowed to coincide). In other words the two spinors π and η define the configuration space of a massive spinning two twistor quantum particle provided the condition $\pi^{A'}\eta_{A'} \neq 0$ is fulfilled. Such a representation of relativistic states of a massive, spinning and charged particle may be called the "square root" of the linear (four) momentum representation. Let us denote the space of such states ("wave" functions) by Γ.

The massive states represented by such non-holomorphic functions are composed of a mixture of positive and negative frequency states.

The holomorphic quantization suggested by Penrose in [1], on the other hand, gives a natural connection between the positive frequency and holomorphic properties (at least for the massless non-interacting case). The second non-holomorphic possibility presented above takes away this beautiful feature and corresponds to the so-called real polarization (invariant with respect to the full Poincaré group) of the (reduced) twistor phase space as discussed by Woodhouse [21] p. 265 and [22] p. 126. One could speculate here that the coherent states should be analyzed using the holomorphic representation, while the bound states are better described by the non-holomorphic representation.

In order to turn Γ into a Hilbert space there is also the issue of defining an appropriate scalar product.

Operators acting on Γ and representing physical observables must be hermitian with respect to the chosen scalar product. In a moment, such a scalar product will be defined. Let us first list some of the operators arising by means of the canonical relativistic quantization assumed in (4.2) and (4.3).

The multiplicative canonical linear (four) momentum operator is, thus, given by

$$\hat{P}_a := \bar{\pi}_A \pi_{A'} + \bar{\eta}_A \eta_{A'}. \tag{4.4}$$

The "complexified" position differential operators corresponding to the complex position (and its complex conjugate) on the two twistor phase space **Tp(2)** represented by functions in (2.16) may be defined as (due to the usual ordering problems this is not a unique way to do it but the simplest and natural in some way)

$$\hat{z}^a := \frac{i}{f}(\eta^{A'}\frac{\partial}{\partial\bar{\pi}_A} - \pi^{A'}\frac{\partial}{\partial\bar{\eta}_A}), \quad \hat{\bar{z}}^a := \frac{i}{\bar{f}}(\bar{\eta}^A\frac{\partial}{\partial\pi_{A'}} - \bar{\pi}^A\frac{\partial}{\partial\eta_{A'}}). \tag{4.5}$$

The differential operators representing the *non-commuting* position (the shifted position; see (1.7), (3.1)) are, therefore, defined by

$$\hat{X}^a := \frac{\hat{z}^a + \hat{\bar{z}}^a}{2}. \tag{4.6}$$

The shift differential operators corresponding to the functions in (3.3) may be defined by

$$\widehat{\Delta X}^a := \frac{i}{2f\bar{f}}(\bar{\pi}^A\eta^{A'}[\bar{\eta}_B\frac{\partial}{\partial\bar{\pi}_B} - \pi_{B'}\frac{\partial}{\partial\eta_{B'}}] + \pi^{A'}\bar{\eta}^A[\eta_{B'}\frac{\partial}{\partial\pi_{B'}} - \bar{\pi}_B\frac{\partial}{\partial\bar{\eta}_B}]). \tag{4.7}$$

Therefore, the commuting differential operators corresponding to the *commuting* position functions $\tilde{X}^a_+$ in (3.4) may be defined as

$$\hat{x}^a := \hat{X}^a + \widehat{\Delta X}^a. \tag{4.8}$$

To prove commutation of the operators in (4.8) directly would be a tedious and time consuming task, but – due to the already proven results [5, 6] at the level of Poisson brackets – one already knows that they really are commuting.

We note also that the operators corresponding to (2.11) (and its complex conjugate) are consequently defined by

$$\hat{\rho} := [\bar{\eta}_B\frac{\partial}{\partial\bar{\pi}_B} - \pi_{B'}\frac{\partial}{\partial\eta_{B'}}], \qquad \hat{\bar{\rho}} := [\eta_{B'}\frac{\partial}{\partial\pi_{B'}} - \bar{\pi}_B\frac{\partial}{\partial\bar{\eta}_B}]. \tag{4.9}$$

The total (four) angular momentum operator is now defined by:

$$\hat{M}^{ab} := i\epsilon^{A'B'}\bar{\pi}^{(A}\frac{\partial}{\partial\bar{\pi}_{B)}} + i\epsilon^{AB}\pi^{(A'}\frac{\partial}{\partial\pi_{B')}} + i\epsilon^{A'B'}\bar{\eta}^{(A}\frac{\partial}{\partial\bar{\eta}_{B)}} + i\epsilon^{AB}\eta^{(A'}\frac{\partial}{\partial\eta_{B')}}, \tag{4.10}$$

while the two helicity operators are given by

$$\hat{s}_1 := \frac{1}{2}(\bar{\pi}_A\frac{\partial}{\partial\bar{\pi}_A} - \pi_{A'}\frac{\partial}{\partial\pi_{A'}}), \quad \hat{s}_2 := \frac{1}{2}(\bar{\eta}_A\frac{\partial}{\partial\bar{\eta}_A} - \eta_{A'}\frac{\partial}{\partial\eta_{A'}}). \tag{4.11}$$

The charge operator is, therefore, given by [5, 6] as

$$\hat{e} := 2\hat{s}_1, \tag{4.12}$$

and the operator corresponding to the invariant in (3.13) is given by [5, 6] as

$$\hat{k}^2 := (\hat{s}_1 - \hat{s}_2)(\hat{s}_1 - \hat{s}_2). \tag{4.13}$$

The homogeneity operators in (4.11) split Γ into subspaces labeled by their eigenvalues. The latter, in turn, are simply the homogeneity degrees of their homogeneous eigenfunctions. In such a way, the integral eigenvalues of the charge operator as defined in (4.12) and half integral values of the "spinor operator" as defined in (4.13) label the different subspaces $\Gamma_{e,k}$. On each $\Gamma_{e,k}$ one can define a scalar product, e.g., in the following way:

$$< g_1|g_2 >:= \int [\bar{g}_1(\bar{\pi}_B,\ \pi_{B'},\ \bar{\eta}_B,\ \eta_{B'})g_2(\pi_{B'},\ \bar{\pi}_B,\ \eta_{B'},\ \bar{\eta}_B)]d\mu, \tag{4.14}$$

where the measure $d\mu$ (on the product of the two (Weyl) spinor spaces with its diagonal deleted) is the usual one:

$$d\mu := d\pi^{A'} \wedge d\pi_{A'} \wedge d\bar{\pi}^A \wedge d\bar{\pi}_A \wedge d\eta^{B'} \wedge d\eta_{B'} \wedge d\bar{\eta}^B \wedge d\bar{\eta}_B. \tag{4.15}$$

The subspace $\aleph_{e,k}$ of $\Gamma_{e,k}$ consisting of functions having finite norm (with respect to the scalar product in (4.14)) defines a Hilbert space of quantum states of a relativistic massive, spinning and charged particle.

5 Example: Classical relativistic dynamics on **Tp(2)**

From now on, the commuting position $\tilde{X}_+$ as it appears above will be denoted by x, and the corresponding electric charge will be denoted by e and represented by the function $e = 2s_1$ [5, 6]. Σ_+ will be denoted by Σ.

The function on the two twistor phase space **Tp(2)** representing the square of the mass of the charged relativistic spinning particle in interaction with an external electromagnetic field $F = F(x)$, the latter defined by the four potential $A = A(x)$, will be assumed to be given by

$$H = (P_i - eA_i)(P^i - eA^i) + \frac{1}{2}e\Sigma^{kl}F_{kl} = H(Z,\ W,\ \bar{Z},\ \bar{W}). \tag{5.1}$$

Such a choice corresponds to the square of the Dirac operator describing spin one half relativistic quantum particle in interaction with an external electromagnetic field. See, e.g., Schiff [23] equation 52.25 and/or Martin and Glauber [24] (thanks to the unknown referee for the last reference). $P_i - eA_i$ represents particle's mechanical (four) linear momentum in the usual way (minimal coupling) while $\frac{1}{2}e\Sigma^{kl}F_{kl}$ corresponds to the classical limit of the so-called (*not* anomalous) Pauli term.

The canonical flow, generated on the two twistor phase space $\mathbf{Tp(2)}$ by the so-defined function H, yields

$$\frac{dx^j}{d\lambda} := \{H,\ x^j\} = 2(P^j - eA^j), \tag{5.2}$$

$$\frac{dP_j}{d\lambda} := \{H,\ P_j\} = 2e(P_i - eA_i)\frac{\partial A^i}{\partial x^j} - \frac{1}{2}e\Sigma_{kl}\frac{\partial F^{kl}}{\partial x^j}, \tag{5.3}$$

$$\begin{aligned}\frac{d\Sigma^{ij}}{d\lambda} &:= \{H,\ \Sigma^{ij}\} = \frac{1}{2}eF_{kl}\{\Sigma^{kl},\ \Sigma^{ij}\}\\ &= \frac{1}{2}eF_{kl}(g^{ik}\Sigma^{lj} - g^{il}\Sigma^{kj} + g^{jl}\Sigma^{ki} - g^{jk}\Sigma^{li})\\ &= \frac{1}{2}e(F^i{}_l\Sigma^{lj} - F_k{}^i\Sigma^{kj} + F_k{}^j\Sigma^{ki} - F^j{}_l\Sigma^{li})\end{aligned} \tag{5.4}$$

where λ labels points along the lines of the flow. To obtain (5.4) we used the result stated in (3.10). Putting

$$\Sigma F := \frac{1}{2}\Sigma^{kl}F_{kl}, \tag{5.5}$$

and using (5.1) and (5.2), one obtains

$$\left(\frac{d\tau}{d\lambda}\right)^2 := \frac{dx^j}{d\lambda}\frac{dx_j}{d\lambda} = 4(P^j - eA^j)(P_j - eA_j) = 4(H - e\Sigma F)^2, \tag{5.6}$$

where τ denotes particle's proper time. Using τ instead of λ as parameter labeling points along the lines of the canonical flow generated on the two twistor space $\mathbf{Tp(2)}$ by H in (5.1) yields (we use only the positive root of the equation in (5.6))

$$\dot{x}^j := \frac{dx^j}{d\tau} = \frac{P^j - eA^j}{\sqrt{H - e\Sigma F}} \tag{5.7}$$

so that

$$\dot{x}^j\dot{x}_j = 1. \tag{5.8}$$

Differentiating (5.7) with respect to τ, using (5.3), the positive root of (5.7) and eliminating the canonical linear momentum P, yields the following set of relativistic equations of motion

$$\begin{aligned}\frac{d^2x^j}{d\tau^2} &= \frac{e}{\sqrt{H - e\Sigma F}}F_i{}^j\frac{dx^i}{d\tau}\\ &+ \frac{e}{4(H - e\Sigma F)}\left[\frac{\partial F_{kl}}{\partial x^m}\Sigma^{kl}\left(\frac{dx^m}{d\tau}\frac{dx^j}{d\tau} - g^{mj}\right)\right].\end{aligned} \tag{5.9}$$

$$\frac{d\Sigma^{ij}}{d\tau} = \frac{e}{4\sqrt{H - e\Sigma F}}(F^i{}_l\Sigma^{lj} - F_k{}^i\Sigma^{kj} + F_k{}^j\Sigma^{ki} - F^j{}_l\Sigma^{li}). \tag{5.10}$$

Note that the invariants I_1 and I_2 in (3.13), (3.14) are constants of the motion. At the quantum level this would mean that, during the "motion", the particle stays (as it should) in the same representation of the Lorentz group labeled by the eigenvalues of k^2 (and 0). See also our discussion at the end of section 3.

The four Poincaré covariant functions on **Tp(2)** representing x are the commuting ones. Its proper time derivative, in the free particle case, equals ($\pm$) the canonical four momentum P_a divided by the mass m which – in the presence of interaction – is no longer true, as may be seen from (5.7). However, we have to accept that

$$\dot{x}_a\Sigma^{ab} \neq 0. \tag{5.11}$$

If the external electromagnetic field is switched off (5.11) tends to $-m\Delta X^b$. Therefore it seems natural to *define* a four position shift in the presence of an external electromagnetic field as

$$\Delta x^i := \frac{1}{\sqrt{H}}\Sigma^{ik}\dot{x}_k. \tag{5.12}$$

This automatically modifies the non-commuting position in the presence of an external electromagnetic field in the following way:

$$X^a := x^a - \Delta x^a \tag{5.13}$$

and renders the four angular momentum relative to the non-commuting position X as

$$r^{ij} := \Sigma^{ij} - \dot{x}_k\Sigma^{kj}\dot{x}^i + \dot{x}_k\Sigma^{ki}\dot{x}^j. \tag{5.14}$$

In the absence of the external electromagnetic field, r^{ij} reduces itself to R^{ij} in (3.2). Note that the (Poincaré) scalar function (the Pauli-Lubański spin)

$$r := \sqrt{\frac{1}{2}r_{ij}r^{ij}}. \tag{5.15}$$

is not, in general, a constant of the motion.

At the quantum level it would suggest that the eigenvalues of k rather than the eigenvalues of the Pauli-Lubański spin operator r (which, in general, do not commute with the mass squared operator corresponding to H in (5.1)) should be identified with those measured in experiments.

We note also that

$$\frac{dr^2}{d\tau} = -H\frac{d(\Delta x^i\Delta x_i)}{d\tau} \tag{5.16}$$

which, as mentioned above, does not vanish for a generic external electromagnetic field.

This shows that the price we have to pay for associating two distinct positions to a charged massive and spinning particle (commuting and noncommuting, call them tentatively "center of charge" and "center of mass") is that the length of the Pauli-Lubański spin vector is not a constant of the motion in the presence of an external electromagnetic field. This seems to correspond to the so-called "Zitterbewegung". We stress again that at the quantum level, in the presence of an external electromagnetic field, it indicates that the operator corresponding to the length of the Pauli-Lubański spin vector cannot represent the quantized fixed spin of the particle. As mentioned before, the eigenvalues of k should play this part. Of course for certain special choices of the external electromagnetic field it may so happen that r^2 is also conserved.

The above-obtained equations in (5.10) describing the behavior of the spin tensor Σ in the presence of an external electromagnetic field are *not* equivalent to the so-called BMT (Bargman-Michel-Telegdi) equations. However, if the space like distance between the "center of mass" and the "center of charge" in (5.12) is ignored, the condition $e\Sigma F << M^2$ fulfilled, then the equations in (5.10) reduce themselves to the BMT equations in the special case when the gyromagnetic ratio is equal to two [17].

6 Tentative twistor quantization of the above example

Relative to any inertial frame and for any fixed value of H, the canonical energy function $P_0 = P^0$ may be solved for from the equation in (5.1), which yields

$$\begin{aligned} E &= P_0 = P^0 \\ &= \pm\sqrt{H}\sqrt{(1 + \frac{(\mathbf{p} - e\mathbf{A})\cdot(\mathbf{p} - e\mathbf{A}) - e(\mathbf{S}\cdot\mathbf{B} - \mathbf{K}\cdot\mathbf{E})}{H})} + e\,A_0, \end{aligned} \tag{6.1}$$

wherein the chosen inertial frame, the four potential, and the corresponding electromagnetic field split into its electric and its magnetic part in the usual way while Σ splits into a "boost" and a spin vector parts (see also (3.13), (3.14)):

$$A_i = (A_0,\ \mathbf{A}), \quad F_{ij} = (\mathbf{E},\ \mathbf{B}), \quad \Sigma_{ij} = (\mathbf{K},\ \mathbf{S}). \tag{6.2}$$

The energy function in (6.1) may now be turned into (however, not uniquely and only formally) an hermitian differential operator whose eigenvalues, in the usual way, define the energy spectrum of a charged spinning massive quantum particle "moving" in an external electromagnetic field. Formally, the eigenvalue equation reads

$$\hat{E}\; g(\pi^{A'},\bar{\pi}^{A},\eta^{A'},\bar{\eta}^{A}) = E\; g(\pi^{A'},\bar{\pi}^{A},\eta^{A'},\bar{\eta}^{A}). \tag{6.3}$$

The operator $\hat{E}$ is however not well-defined. This is true not only because of the ordering ambiguities, but also because of the fact that the operators, into which the above-introduced quantization procedure turns the functions representing the electromagnetic potential and the corresponding electromagnetic field, become, in general, non-local ones. Each specific choice of the external field then requires a separate treatment. Let us therefore first have a closer look at the Coulomb field case.

In the inertial frame in which the source of the Coulomb field is at rest, one obtains formally from (6.1) and from (6.3) (with the hats omitted for convenience):

$$\{\sqrt{H + (\mathbf{p}\cdot\mathbf{p}) + \frac{e\,q\,(\mathbf{K}\cdot\mathbf{r})}{r^3}} - \frac{e\,q}{r}\}\; g = E\,g. \tag{6.4}$$

Moving the last term on the right-hand side in (6.4) to the left-hand side and squaring the operators on both sides yields

$$\{(E + \frac{e\,q}{r})^2\}\; g = \{(H + (\mathbf{p}\cdot\mathbf{p}) + \frac{e\,q\,(\mathbf{K}\cdot\mathbf{r})}{r^3})\}\; g. \tag{6.5}$$

Further manipulations (removing the distance operator r from the denominator) in the same spirit (and reintroducing hats into terms and factors containing the "spinorial" differential operators) give

$$\begin{aligned}\{\hat{r}^2[(E^2 - H)\hat{r}^2 - \hat{r}^2(\mathbf{p}\cdot\mathbf{p}) + (eq)^2]^2\}\; g \\ = \{(eq)^2[(\hat{\mathbf{K}}\cdot\hat{\mathbf{r}}) - 2E\hat{r}^2]^2\}\; g,\end{aligned} \tag{6.6}$$

which is a linear "spinorial" differential equation of sixth order. The set of eigenvalues of E for which finite (with respect to the above-defined scalar product) solutions of (6.6) exist must contain the energy eigenvalues obtained within the frame of the conventional Dirac equation approach [25]. However, the study of the well-behaved solutions of (6.6) is a difficult technical task and will not be pursued any further in this paper. We hope to be able to return to this problem in our future work.

Acknowledgment

Partial financial support from KTH – Swedish Royal Institute of Technology "fakir" fund and from Stockholm University Ivar Bendixon fund is gratefully acknowledged.

REFERENCES

[1] R. Penrose, M. A. H. MacCallum, Twistor theory: an approach to the quantization of fields and space-time, *Physics Reports* **6**:4 (1972), 241–316.

[2] S. Zakrzewski, Extended phase space for a spinning particle, *J. Phys. A: Math. Gen* **28** (1995), 7347–7357.

[3] S. Zakrzewski, Noncommutative space-time implied by spin, unpublished.

[4] A. Bette, Directly interacting massless particles – a twistor approach, *J. Math. Phys.* **37**:4 (1996), 1724–1734.

[5] A. Bette and S. Zakrzewski, Extended phase spaces and twistor theory, and references therein, *J. Phys. A: Math. Gen* **30** (1997), 195–209.

[6] A. Bette and S. Zakrzewski, Massive relativistic systems with spin and the two twistor phase space, in the *Proceedings of the XIIth workshop on "soft" physics, Hadrons-96: confinement,* Novy Svet, Crimea, June 9-16, 1996, issued by National Academy of Sciences of Ukraine, Bogoliubov Institute for Theoretical Physics (Kiev), Simferopol State University (Crimea), Université Claude Bernard de Lyon, Kiev 1996, 336–346 .

[7] S. Zakrzewski, Localization of relativistic systems, *J. Phys. A: Math. Gen* **30** (1997), 8317–8323.

[8] R. Penrose, Twistor algebra, *J. Math. Phys.* **8**:2 (1967), 345–366.

[9] R. Penrose, Twistor quantization and curved space-time, *Int. J. Theor. Phys.* **1** (1968), 61–99.

[10] R. Penrose, Twistor theory, its aims and achievements, in *Quantum Gravity: an Oxford Symposium,* eds. C. J. Isham, R. Penrose, and D. W. Sciama, Clarendon Press, Oxford, 1975.

[11] R. Penrose, in *Magic Without Magic: J. A. Wheeler a Collection of Essays in Honor of His 60th Birthday,* ed. J. R. Klauder, W. H. Freeman and Co., San Francisco, 1972.

[12] K. P. Tod, *Massive Spinning Particles and Twistor Theory,* Doctoral Dissertation at Mathematical Institute, University of Oxford, 1975.

[13] K. P. Tod, Some symplectic forms arising in twistor theory, *Rep. Math. Phys.* **11** (1979), 339–346.

[14] R. Penrose, *Batelle Rencontres,* eds. C. M. de Witt, J. A. Wheeler, Princeton University, W. A. Benjamin, Inc., New York, Amsterdam, (1968), 135–149.

[15] R. Penrose, W. Rindler, *Spinors and Space-time,* Cambridge Monographs on Mathematical Physics, Vols. 1 and 2, Cambridge University Press, 1984.

[16] L. P. Hughston, *Twistors and particles,* Lectures Notes in Physics, Springer Verlag **97** (1979).

[17] A. Bette, Twistor phase space dynamics and the Lorentz force equation, *J. Math. Phys.* **34**:10 (1993), 4617–4627.

[18] L. P. Hughston, The relativistic oscillator, *Proc. R. Soc, Lond. A* **382** (1982), 459–466.

[19] Z. Perjès, Twistor variables of relativistic mechanics, *Phys. Rev. D,* **11**:8 (1975), 2031–2035.

[20] G. A. J. Sparling, Theory of massive particles, *Proc. Roy. Soc.* **301 A**:1458 (1981), 27–74.

[21] N. M. J. Woodhouse *Geometric Quantization,* Clarendon Press, Oxford, 1980.

[22] N. M. J. Woodhouse *Geometric Quantization, Second Edition,* Clarendon Press, Oxford, 1992.

[23] L. I. Schiff, *Quantum Mechanics, Third Edition,* McGraw-Hill Book Company, 1968.

[24] P. C. Martin and R. J. Glauber, *Phys. Rev D* **109** (1958), 1306–1315.

[25] J. J. Sakurai, *Advanced Quantum Mechanics,* Addison-Wesley Publishing Company, November 1990, Chapter 3.

Andreas Bette
Royal Institute of Technology
KTH-Södertälje
Campus Telge
S-151 81 Södertälje, Sweden
E-mail: andreas.bette@telge.kth.se

Received: August 25, 1999; Revised: October 20, 1999

Fiber with Intrinsic Action on a $1+1$ Dimensional Spacetime

Robert W. Johnson

ABSTRACT I construct an algebraic model for a typical fiber on a $1+1$ dimensional spacetime. The vector space comprising the fiber is composed of elements $x \otimes x$ formed from the direct product of two copies of an element x in the $D_2 = C_2 \otimes C_2$ finite group algebra over the real numbers. The fiber contains subspaces whose elements can be associated with the tangent and momentum vectors of trajectories in the manifold. The fiber also contains a subspace whose elements are associated with the local flow of action of each trajectory. The condition of minimum action translates into a constraint on the original vector x in the direct product structure.
Keywords: Tangent space, manifold, fiber, differential geometry, $1+1$ spacetime.

1 Introduction

In a recent paper, I described a construction for a vector space with metric. In this construction, one forms elements $x \otimes x$ in a direct product algebra where x is an element in an underlying finite group algebra [1]. One uses a particular decomposition of the direct product algebra to obtain a direct sum of subspaces. One then observes that vectors in the various subspaces are interrelated. In particular examples under consideration there included a 1-dimensional subspace whose measure was determined by the components of a second vector in a higher dimensional subspace. I proposed that the measure in the 1-dimensional subspace be identified with the norm of the vector in the higher dimensional subspace. In this way both a vector space and its norm are viewed as residing in particular subspaces of a given direct product algebra $x \otimes x$. In this construction the signature of the metric arises intrinsically from the particular underlying finite group algebra.

The realization of Clifford algebras in terms of underlying finite group algebras is described by Salingaros [2]. The present work differs from the approaches of e.g., Hestenes, Lounesto, and Greider [3] by considering vector spaces obtained through the decomposition of a direct product algebra (having elements of form $x \otimes x$) rather than through a decomposition of a

AMS Subject Classification: 83C10, 15A66, 53B30.

Clifford algebra. In particular cases that I consider, however, the underlying finite group algebra (containing elements x) does correspond to a Clifford algebra.

This work is motivated by the analogous construction in quantum mechanics where one forms observable vector spaces in terms of bilinear functions of an underlying state vector. The quantum mechanical 2-state problem provides one instructive example [1]. For this problem the underlying finite group is $C_4 \otimes H$, the direct product of the cyclic group of order 4 and the quaternion group. The quantum wave function ψ is an element in a left ideal of the $C_4 \otimes H$ group algebra over the real numbers[1]. The polarization vector and its norm, the total probability, reside in particular subspaces of the direct product algebra whose elements $\psi \otimes \psi$ are formed from the product of two copies of the quantum mechanical state ψ.

Vector spaces with metric that are constructed using this procedure are also of interest as models for typical fiber vector spaces residing at each location of a configurational manifold such as arises in the context of classical Lagrangian mechanics. In this paper I develop an algebraic model for a typical fiber at a location P of a configurational manifold M having 1 space and 1 time dimension. The fiber contains both the tangent and momentum vector spaces as subspaces. In addition, a subspace that can be associated with the flow of action at P is included and arises in an intrinsic way. The construction for a configuration space with 2 space dimensions follows in a completely analogous way. The details of these two 2-dimensional cases are transparent. The extension of this construction to higher dimensional cases is also straightforward; however, the multiplicity of subspaces in the corresponding direct product algebras makes the interpretation of the overall structure more challenging, and it has not been fully addressed by this author.

In section 2, I review the construction of a vector space with metric signature $(p, q) = (1, 1)$ that corresponds to the tangent space at a point of a $1 + 1$ dimensional configurational manifold [1]. This construction uses the very simple C_2 group algebra. In section 3, I use the $D_2 = C_2 \otimes C_2$ group algebra to obtain the vector space that is the primary focus of this paper. In section 4, I summarize these results and indicate work that still remains to be done.

---

[1]This algebra is also termed the complex quaternion algebra. Elements in this algebra are linear combinations of the $C_4 \otimes H$ group elements with real coefficients. The $C_4 \otimes H$ group multiplication is used along with the distributive law to induce the algebra product rule.

2 Algebraic model for tangent space at a point of a 1 + 1 dimensional spacetime

As a starting point for this paper, let us review my earlier construction [1] of an algebraic model for the vector space with signature $(p, q) = (1, 1)$. This vector space corresponds to the tangent space at each location of a $1+1$ dimensional spacetime M. The tangent space at an arbitrary location $(t_0, q_0) \in M$ consists of tangent vectors to curves passing through (t_0, q_0). Here t_0 and q_0 denote, respectively, time and spatial location. For a curve $\gamma = \gamma(\lambda)$, $\lambda \in \mathbb{R}$, the tangent vector is defined by

$$\frac{d\gamma}{d\lambda} = \lim_{\lambda\to 0} \frac{\gamma(\lambda) - \gamma(0)}{\lambda} = (\frac{dt}{d\lambda}, \frac{dq}{d\lambda})$$

where $\gamma(0) = (t_0, q_0)$ and $\gamma(\lambda) \in M$.

We begin the construction with the C_2 group which contains two elements $C_2 = \{\mathbf{1}, \mathbf{e}\}$ with $\mathbf{e}^2 = \mathbf{1}$, and then form the vector space V_{C_2} whose elements consist of formal sums of the elements of C_2 with real coefficients. An arbitrary element $x \in V_{C_2}$ can be written $x = x_0\mathbf{1} + x_1\mathbf{e}$. The product rule that is induced by the group multiplication is used to form the group algebra. We then consider the direct product of two copies of a vector x in the algebra $(\mathbf{1} \cong \mathbf{1} \otimes \mathbf{1})$:

$$\begin{aligned} x \otimes x &= (x_0\mathbf{1} + x_1\mathbf{e}) \otimes (x_0\mathbf{1} + x_1\mathbf{e}) \\ &= x_0^2(\mathbf{1} \otimes \mathbf{1}) + x_0x_1(\mathbf{1} \otimes \mathbf{e}) + x_0x_1(\mathbf{e} \otimes \mathbf{1}) + x_1^2(\mathbf{e} \otimes \mathbf{e}) \\ &= (x_0\mathbf{1} + x_1\mathbf{E}\mathbf{e})(x_0\mathbf{1} + x_1\mathbf{e}), \end{aligned}$$

where the second line follows from bilinearity of the direct product, and in the third line we introduce the notation $\mathbf{E} = \mathbf{e} \otimes \mathbf{e}$ and $\mathbf{e} = \mathbf{1} \otimes \mathbf{e}$. $\mathbf{Ee} = \mathbf{e} \otimes \mathbf{1}$ follows from the product rule in the direct product algebra. Acting on $x \otimes x$ with the projection operators $P_\pm = \frac{1}{2}(\mathbf{1} \otimes \mathbf{1} \pm \mathbf{E})$, we obtain

$$x \otimes x = [P_+(\frac{dt}{d\lambda}\mathbf{1} + \frac{dq}{d\lambda}\mathbf{e}) + P_-\frac{ds}{d\lambda}](\mathbf{1} \otimes \mathbf{1}),$$

where

$$\begin{aligned} \frac{dt}{d\lambda} &= x_0^2 + x_1^2 \\ \frac{dq}{d\lambda} &= 2x_0x_1 \\ \frac{ds}{d\lambda} &= x_0^2 - x_1^2 = (\frac{dt}{d\lambda}^2 - \frac{dq}{d\lambda}^2)^{\frac{1}{2}}. \end{aligned}$$

The measure $\frac{ds}{d\lambda}$ of the 1-dimensional $P_-x \otimes x$ subspace is determined up to sign by the two components $(\frac{dt}{d\lambda}, \frac{dq}{d\lambda})$ of the 2-dimensional $P_+x \otimes x$

subspace and can be interpreted as their norm. We note that the measure $x_0^2 + x_1^2$ associated with the $\frac{dt}{d\lambda}$ increment is positive definite and so has an intrinsic directionality. Also, since $(2x_0x_1)^2 \le (x_0^2 + x_1^2)^2$, we have $|\frac{dq}{dt}| \le 1$ so that there is a maximum speed.

In this way the product $x \otimes x$ provides a model for an element in the tangent space of a 1+1 dimensional spacetime. The set of all such elements $x \otimes x$ can be identified with the tangent space itself.

Continuing further we find that rotations of the 2-dimensional vector space $P_+ x \otimes x$ are induced by acting on x with an element $u = u_0\mathbf{1} + u_1\mathbf{e}$ and forming the product $(x \otimes x)(u \otimes u) = xu \otimes xu$. For u such that $u_0^2 - u_1^2 = 1$, $P_- x \otimes x$ is unchanged, while the $P_+ x \otimes x$ vector undergoes a proper orthochronous rotation.

A completely analogous treatment for the 2-dimensional Euclidean case is obtained by substituting the C_4 group algebra for the C_2 group algebra. This approach also extends to higher dimensional vector spaces, although in these cases, we encounter a multiplicity of subspaces in the direct product algebra which make the interpretation of the overall structure more involved [1].

In the following section we extend this treatment to the case of a typical fiber at a location P of a configurational manifold M that contains both the tangent and momentum vector spaces.

3 Algebraic model for fiber on $1+1$ spacetime with an intrinsic action

Let us briefly review the classical mechanics that motivate this construction. The action function

$$S_{t_0,q_0}(t,q) = \int_\gamma L dt$$

is the integral of the Lagrangian $L = L(\dot{q}, q, t)$ along an extremal path $\gamma(\lambda)$, $\lambda \in$ real numbers, connecting an initial point (t_0, q_0) with (t, q) [4]. The action function for a free particle with mass m in a locally Minkowski coordinate system can be written as

$$S_{t_0,q_0}(t,q) = \int_\gamma -m\frac{ds}{dt}dt$$

where ds is the proper time with measure $ds = (dt^2 - dq^2)^{\frac{1}{2}}$ and $\gamma(\lambda)$ is a locally straight line [5]. This action corresponds to the Lagrangian $L = -m\frac{ds}{dt}$. The rate of change of the action along the path γ for a fixed initial point is

$$\frac{dS}{d\lambda} = p\frac{dq}{d\lambda} - H\frac{dt}{d\lambda} \tag{3.1}$$

where

$$p = \frac{\partial L}{\partial \dot{q}} = \frac{m\dot{q}}{(1-\dot{q}^2)^{\frac{1}{2}}}$$

and

$$H = p\dot{q} - L = \frac{m}{(1-\dot{q}^2)^{\frac{1}{2}}}.$$

We now construct a model for the local tangent and momentum vectors of a trajectory $\gamma(\lambda)$ on $1+1$ dimensional manifold for which such an action function arises intrinsically. For this construction, we use the Abelian group ($\mathbf{e_{12}} := \mathbf{e_1 e_2}$)

$$D_2 = C_2 \otimes C_2 = \{\mathbf{1}, \mathbf{e_1}, \mathbf{e_2}, \mathbf{e_{12}}\}$$

where $\mathbf{e}_1^2 = \mathbf{e}_2^2 = \mathbf{e}_{12}^2 = \mathbf{1}$. A general element of the D_2 group algebra can be written

$$x = x_0\mathbf{1} + x_1\mathbf{e}_1 + x_2\mathbf{e}_2 + x_3\mathbf{e}_{12}.$$

We decompose x into a sum of two left ideals obtained by acting on the right with the projection operators $P_{\pm 2} = \frac{1}{2}(\mathbf{1} \pm \mathbf{e}_2)$. We have

$$x = x(P_{+2} + P_{-2})$$

where

$$\begin{aligned} xP_{+2} &= [(x_0 + x_2)\mathbf{1} + (x_1 + x_3)\mathbf{e}_1]P_{+2} \\ xP_{-2} &= [(x_0 - x_2)\mathbf{1} + (x_1 - x_3)\mathbf{e}_1]P_{-2}. \end{aligned}$$

We now form the tensor product of two copies of x,

$$x \otimes x = x(P_{+2} + P_{-2}) \otimes x(P_{+2} + P_{-2}).$$

Expanding this expression and acting on the left side with the projection operator $P_{\pm 1} = \frac{1}{2}(\mathbf{1} \otimes \mathbf{1} \pm \mathbf{E}_1)$, where $\mathbf{E}_1 = \mathbf{e_1} \otimes \mathbf{e_1}$, we obtain, after a change of variables,

$$\begin{aligned} x \otimes x =& [P_{+1}(\frac{dt}{d\lambda}\mathbf{1} + \frac{dq}{d\lambda}\mathbf{e}_1) + P_{-1}\frac{ds}{d\lambda}](P_{+2} \otimes P_{+2}) \\ &+ [P_{+1}([\frac{1}{2}(H\frac{dt}{d\lambda} + p\frac{dq}{d\lambda} + m\frac{ds}{d\lambda})]^{\frac{1}{2}}\mathbf{1} \\ &+ [\frac{1}{2}(H\frac{dt}{d\lambda} + p\frac{dq}{d\lambda} - m\frac{ds}{d\lambda})]^{\frac{1}{2}}\mathbf{e}_1) \\ &+ P_{-1}([\frac{1}{2}(H\frac{dt}{d\lambda} - p\frac{dq}{d\lambda} + m\frac{ds}{d\lambda})]^{\frac{1}{2}}\mathbf{1} \\ &- [\frac{1}{2}(H\frac{dt}{d\lambda} - p\frac{dq}{d\lambda} - m\frac{ds}{d\lambda})]^{\frac{1}{2}}\mathbf{e}_{12})](P_{+2} \otimes P_{-2} + P_{-2} \otimes P_{+2}) \\ &+ [P_{+1}(H\mathbf{1} + p\mathbf{e}_1) + P_{-1}m](P_{-2} \otimes P_{-2}) \end{aligned}$$

where $\mathbf{e}_1 = \mathbf{1} \otimes \mathbf{e}_1$ and $\mathbf{e}_{12} = \mathbf{1} \otimes \mathbf{e}_{12}$. In this expression

$$\frac{dt}{d\lambda} = (x_0 + x_2)^2 + (x_1 + x_3)^2$$
$$\frac{dq}{d\lambda} = 2(x_0 + x_2)(x_1 + x_3)$$
$$\frac{ds}{d\lambda} = (x_0 + x_2)^2 - (x_1 + x_3)^2 = [\frac{dt}{d\lambda}^2 - \frac{dq}{d\lambda}^2]^{\frac{1}{2}},$$

while for the momenta variables we have

$$H = (x_0 - x_2)^2 + (x_1 - x_3)^2$$
$$p = 2(x_0 - x_2)(x_1 - x_3)$$
$$m = (x_0 - x_2)^2 - (x_1 - x_3)^2 = [H^2 - p^2]^{\frac{1}{2}}.$$

In analogy to the C_2 case discussed above, we associate the $xP_{+2} \otimes xP_{+2}$ subspace with the tangent to a curve $\gamma(\lambda)$ on the configurational manifold. The $P_{+1}(xP_{+2} \otimes xP_{+2})$ portion is identified with the tangent vector $(\frac{dt}{d\lambda}, \frac{dq}{d\lambda})$ while the $P_{-1}(xP_{+2} \otimes xP_{+2})$ portion with measure $\frac{ds}{d\lambda} = [\frac{dt}{d\lambda}^2 - \frac{dq}{d\lambda}^2]^{\frac{1}{2}}$ is associated with the norm of the tangent vector.

Similarly, we identify the $xP_{-2} \otimes xP_{-2}$ subspace with the momentum of the trajectory $\gamma(\lambda)$ at λ. The 2-dimensional $P_{+1}(xP_{-2} \otimes xP_{-2})$ projection is identified with the momentum vector (H, p) while the 1-dimensional $P_{-1}(xP_{-2} \otimes xP_{-2})$ projection with measure $m = [H^2 - p^2]^{\frac{1}{2}}$ is associated with the norm of the momentum vector.

So far in this development the velocity tangent vector $(\frac{dt}{d\lambda}, \frac{dq}{d\lambda})$ is completely independent of the momentum vector (H, p). These two vectors contain the 4 degrees of freedom of the original vector x in the D_2 algebra.

Let us now consider the $(x \otimes x)(P_{+2} \otimes P_{-2} + P_{-2} \otimes P_{+2})$ subspace of this algebra. Both the P_{+1} and P_{-1} projections on this subspace are 2-dimensional. Taking the difference of the squares of the component measures for each of these two 2-vectors, we find the resultant $m\frac{ds}{d\lambda}$. The measure squared of the $\mathbf{1}$ component of the P_{-1} subspace

$$\frac{1}{2}(H\frac{dt}{d\lambda} - p\frac{dq}{d\lambda} + m\frac{ds}{d\lambda})$$

motivates its association with $\frac{dS}{d\lambda}$ of Eq. (1). Minimizing this measure while keeping m and $\frac{ds}{d\lambda}$ constant and their product $m\frac{ds}{d\lambda}$ greater than zero requires setting the $\mathbf{e}_{12}$ component of the P_{-1} subspace to zero since the difference of the squares of the two component measures is fixed. In this way we obtain the condition

$$-m\frac{ds}{d\lambda} = p\frac{dq}{d\lambda} - H\frac{dt}{d\lambda}$$

that corresponds to eq. 1. This equation links the tangent vector $(\frac{dt}{d\lambda}, \frac{dq}{d\lambda})$ and the momentum vector (H, p). It is equivalent to the condition $\frac{dq}{dt} = \frac{p}{E}$ that holds for the trajectory of a free particle. Interestingly, this condition translates to the requirement

$$x = \frac{1}{x_0}(x_0 + x_1\mathbf{e}_1)(x_0 + x_2\mathbf{e}_2) \text{ for } x_0 \neq 0$$

for the original vector x in the D_2 group algebra; or, stated differently, it ensures that x can be written as the tensor product of two vectors in the C_2 group algebra.

Transformations that preserve the norms of the tangent and momentum vectors and the action differential $\frac{dS}{d\lambda} = -m\frac{ds}{d\lambda}$ can be induced by acting on x with an element $u = u_0 + u_1\mathbf{e}_1 + u_2\mathbf{e}_2 + u_3\mathbf{e}_3$ in the D_2 algebra and forming the product $(x \otimes x)(u \otimes u) = xu \otimes xu$. For u such that $u = u_0 + u_1\mathbf{e}_1$ and $u_0^2 - u_1^2 = 1$, the $P_{-}x \otimes x$ subspace that contains the three norms is unchanged. The tangent and momentum vectors undergo a proper orthochronous transformation under this rule.

Finally, we also note that if – instead of using $D_2 = C_2 \otimes C_2$ in the construction above – we use

$$C_2 \otimes C_4 = \{\mathbf{1}, \mathbf{e}_2\} \otimes \{\mathbf{1}, \mathbf{e}_1\} = \{\mathbf{1}, \mathbf{e}_1, \mathbf{e}_2, \mathbf{e}_{12}\}$$

where $\mathbf{e}_{12} = \mathbf{e}_1\mathbf{e}_2$, $\mathbf{e}_{21} = \mathbf{e}_{12}$, $\mathbf{e}_2^2 = +1$ and $\mathbf{e}_1^2 = \mathbf{e}_{12}^2 = -1$, then we obtain the 2-dimensional Euclidean counterpart to the above $1+1$ spacetime case.

4 Summary

We associate the vector $x \otimes x$ in the $C_2 \otimes C_2$ group algebra with an element in a typical fiber residing at a point in a $1+1$ dimensional configurational manifold:

$$\begin{aligned} x \otimes x =& \{P_{+1}(\frac{dt}{d\lambda}\mathbf{1} + \frac{dq}{d\lambda}\mathbf{e}_1) + P_{-1}\frac{ds}{d\lambda}\}(P_{+2} \otimes P_{+2}) \\ &+ \Big\{P_{+1}\Big((H\frac{dt}{d\lambda})^{\frac{1}{2}}\mathbf{1} + (p\frac{dq}{d\lambda})^{\frac{1}{2}}\mathbf{e}_1\Big) \\ &\quad + P_{-1}(m\frac{ds}{d\lambda})^{\frac{1}{2}}\Big\}(P_{+2} \otimes P_{-2} + P_{-2} \otimes P_{+2}) \\ &+ \{P_{+1}(H\mathbf{1} + p\mathbf{e}_1) + P_{-1}m\}(P_{-2} \otimes P_{-2}). \end{aligned}$$

The collection of all such vectors $x \otimes x$ comprise the typical fiber. The $(x \otimes x)(P_{+2} \otimes P_{+2})$ portion of $x \otimes x$ is identified with the tangent vector and its norm. The $(x \otimes x)(P_{-2} \otimes P_{-2})$ portion is identified with the momentum vector and its norm. The

$$(x \otimes x)(P_{+2} \otimes P_{-2} + P_{-2} \otimes P_{+2})$$

subspace is associated with the flow of action and its norm. The condition of minimum action translates into the condition that x has the form

$$x = \frac{1}{x_0}(x_0 + x_1\mathbf{e}_1)(x_0 + x_2\mathbf{e}_2) \text{ for } x_0 \neq 0.$$

It remains to determine how the fibers at different locations are connected. The description of this connection should also lead to a description of extended motions on this manifold.

REFERENCES

[1] R. W. Johnson, Direct product and decomposition of certain physically important algebras, *Found. Phys.* **26** (2) (1996), 197.

[2] N. Salingaros, The relationship between finite groups and Clifford algebras, *J. Math. Phys.* **25** (4) (1984), 738.

[3] D. Hestenes, *Spacetime Algebra,* Gordon & Breach, New York, 1966; P. Lounesto, *Found. Phys.* **23** (9) (1993), 1203; K. R. Greider, *Found. Phys.* **14** (6) (1984), 467.

[4] V. I. Arnold, *Mathematical Methods of Classical Mechanics, Second Edition,* Springer-Verlag, New York, 1989, 253.

[5] P. J. E. Peebles, *Principles of Physical Cosmology,* Princeton University Press, Princeton, 1993, 244.

Robert W. Johnson
878 Sunnyhills Road
Oakland, CA 94610
E-mail: RJohnson@hoyaoptics.com

Received: August 2, 1999; Revised: January 21, 2000

Dimensionally Democratic Calculus and Principles of Polydimensional Physics

William M. Pezzaglia Jr.

ABSTRACT A solution to the 50-year-old problem of a spinning particle in curved space has been recently derived using an extension of Clifford calculus in which each geometric element has its own coordinate. This leads us to propose that all the laws of physics should obey new polydimensional metaprinciples, for which Clifford algebra is the natural language of expression, just as tensors were for general relativity. Specifically, phenomena and physical laws should be invariant under local automorphism transformations which reshuffle the physical geometry. This leads to a new generalized unified basis for classical mechanics, which includes string theory, membrane theory and the hypergravity formulation of Crawford [4]. Most important is that the broad themes presented can be exploited by nearly everyone in the field as a framework to generalize both the Clifford calculus and multivector physics.
Keywords: Clifford algebra, automorphism invariance, Papapetrou equations, anholonomic, metamorphic covariance.

1 Introduction

Taking an existing equation and generalizing it over quaternions or Clifford numbers is certainly a way of doing new mathematics. It is important however to understand that this seldom leads to new physics (for example complexifying Newton's law of gravitation is meaningless). Reformulating existing physical laws with a new mathematical language will *not* lead to new principles or to new physics. Only by generalizing principles can we hope to do something new. However, because Clifford algebra [11] encodes the structure of the underlying geometric space, we see possible bigger patterns emerge. Specifically in the description of a spinning particle the equations of motion are invariant under a non-dimensional preserving *poly-*

Financial Support for attending the 5th International Conference on Clifford Algebras and their Applications in Mathematical Physics, Ixtapa, Mexico, June 27 – July 4, 1999, was provided by the Santa Clara University.
AMS Subject Classification: 83E99, 15A66, 53C99, 70G99

dimensional transformation which rotates between vector momentum and bivector spin. We therefore propose that "what is a vector" is *dimensionally relative* to the observer's frame, and that the universe is fully *polydimensionally isotropic* in that there is no absolute "direction" which we can assign "vector" geometry over bivector, trivector, etc.

This forces us to propose a fully *Dimensionally Democratic* Clifford calculus, in which each geometric element has its own coordinate in a *Clifford manifold* [3]. In particular we show the utility of this concept in treating the classical spinning particle in several scenarios. A new action principle is proposed in which particles take paths which minimize the sum of the linear distance traveled combined with the bivector area swept out by the spin. In curved space, the velocity of the variation is not the variation of the velocity, leading to a new derivation of the Papapetrou equations [14] as autoparallels in the Clifford manifold. This leads us to propose that the physical laws might be *metamorphic covariant* under general automorphism transformations which reshuffle the geometry.

2 Relative dimensionalism

Most physicists tend to be absolute in their association of physical quantities to geometric entities. For example, mass is a scalar while force a vector. The introduction of Einstein's relativity however caused a shift in dimensional interpretation in that the world is not three-dimensional but a four-dimensional spacetime continuum. In the old-fashion three dimensional viewpoint, energy was a scalar, but in the new 4D paradigm it is the fourth component of the momentum vector. Certainly many physicists will share the opinion that the new 4D viewpoint is right, and the old 3D view is incorrect. Yet let us consider for example the recent book by Baylis [1] in which he has a complete treatment of electrodynamics and special relativity using *paravectors* (defined as the Clifford aggregate of a scalar plus three-vector). Is he wrong to call time the scalar part of a paravector instead of calling it the fourth component of a vector?

Consider also that the even subalgebra of the 16 basis element Clifford algebra associated with 4D spacetime can be interpreted as the Clifford algebra of a 3D space. Three of the planes of 4D are reinterpreted as basis vectors in the 3D space, while the four-volume is reinterpreted as a three-volume. So if you grab a particular geometric element, is it a vector or a plane? We suggest that there is no absolute right or wrong answer. We postulate a new principle of **relative dimensionalism**: *that the geometric rank an observer assigns to an object is a function of the observer's frame of reference* (or perhaps state of consciousness). There is no absolute dimension that one can assign to a geometric object. Further we consider transformations which reshuffle the basis geometry (e.g., vector line replaced by bivector plane), yet leave sets of physical laws invariant. One application provides a new treatment of the classical spinning particle, showing that the mechanical mass is enhanced by the spin motion.

2.1 Review of Special Relativity

It is useful to see how paradigm shifts, in the concept of the dimensional nature of space, have impacted the formulation of physical laws in the past for clues as how to proceed with newer ideas. The *prima facie* example is how things changed with the introduction of special relativity. Quantities such as time and energy, that were previously defined as scalars (in a 3D formulation) now are identified as fourth components of vectors in four-dimensional Minkowski spacetime. Let us consider what is gained by using the higher dimensional concept.

Unification of Phenomena

The most obvious advantage of using vectors is that one can replace a set of physical equations by a single vector equation. Let us consider the application of four-vectors in electrodynamics. The 3D scalar work-energy law and 3D vector force law,

$$\dot{\mathcal{E}} = e\vec{\mathbf{E}} \cdot \vec{\mathbf{v}}\,, \qquad \dot{\vec{\mathbf{P}}} = e\left(\vec{\mathbf{E}} + \vec{\mathbf{v}} \times \vec{\mathbf{B}}\right), \tag{1ab}$$

can be combined into one single equation,

$$\dot{p}^{\mu} = \frac{e}{mc}\, p_{\nu}\, F^{\mu\nu}, \tag{2}$$

using 4D vectors and tensors (c is the speed of light, and the dot represents differentiation with respect to proper time). Certainly the adoption of the four-dimensional viewpoint has notational economy and provides insight that the work-energy theorem (1a) is simply the fourth aspect of the vector force law (2). However, philosophically one can ask if the 4D viewpoint is any more correct than the 3D equations since they describe the same phenomena. Since special relativity was originally formulated without the concept of Minkowski spacetime, it is convenient, but apparently not necessary, to adopt the paradigm shift from 3D to 4D. Hence we are being purposely dialectic in raising the question whether one can make an absolute statement about the dimensional nature of a physical quantity such as time. Can we state (measure) that time is a part of a four-vector (as opposed to a 3D scalar), or is this relative to whether one adopts a 3D or 4D world view, hence, relative to the observer's dimensional frame of reference?

Lorentz transformations

In classical physics the fundamental laws must be invariant under rotational displacements because it is postulated that the universe is *isotropic* (has no preferred direction). When one formulates laws with vectors (which are inherently coordinate system independent), the principle of isotropy is "built in" without needing to separately impose the condition. Hence (Gibbs) vectors are a natural language to express classical (3D) physical laws because they naturally encode isotropy.

Einstein further postulated the metaprinciple that motion was relative, that there is no absolute preferred rest frame to the universe. This coupled with the postulate that the speed of light is the same for all observers leads to the principle that the laws of physics must be invariant under Lorentz transformations (which connect inertial frames of reference). A more geometric interpretation is to see that Lorentz transformations are just rotations in 4D spacetime; hence, the principle of relative motion is an extension of the metaprinciple of isotropy to four-space.

Lorentz transformations, expressed in four-space, preserve the rank of geometry (rotates a four-vector into another four-vector). In contrast, Baylis [1] would write the Lorentz boost (in the z direction with velocity v) of the *momentum paravector* in 3D Clifford algebra as

$$\mathbf{1}\frac{\mathcal{E}'}{c} + \vec{\mathbf{P}}' = \mathcal{R}\left(\mathbf{1}\frac{\mathcal{E}}{c} + \vec{\mathbf{P}}\right)\mathcal{R}^{\dagger}, \tag{3}$$

where c is the speed of light, and the transformation operator: $\mathcal{R} = \exp(-\hat{\mathbf{e}}_3\beta/2)$, where β the rapidity related to the velocity: $\tanh(\beta) = v/c$ and $\hat{\mathbf{e}}_3$ is the unit vector in the z direction. As a consequence, in this 3D perspective, what is pure scalar (e.g., $\mathcal{E}$ = energy) to one observer is part scalar, part vector to another observer. Lorentz transformations, in this 3D viewpoint, are NOT dimensional preserving. We choose to classify such transformations as *geometamorphic* or "polydimensional".

Invariant moduli

In 3D space the length (magnitude) of a vector (e.g., electric field or momentum) is invariant under rotations. Under Lorentz transformations (4D rotations), the modulus of the four-vector is invariant,

$$\| \mathbf{p} \|^2 \equiv p_\mu p^\mu = \frac{\mathcal{E}^2}{c^2} - \| \vec{\mathbf{P}} \|^2 . \tag{4}$$

Reinterpreted with a 3D viewpoint, the invariant quantity of the Lorentz transformation (3) is the difference between the square of the scalar energy minus the magnitude of the 3D momentum vector. Neither the modulus of the 3D scalar nor 3D vector is independently invariant under these transformations.

The modulus of the momentum four-vector (4) is defined to be the *rest mass* of the particle: $m_0 \equiv c^{-1} \parallel \mathbf{p} \parallel$. When in motion, the mechanical mass of the particle (e.g., in definition of momentum: $p = mv$) increases by its kinetic energy content. This is described by the *Lorentz dilation factor* γ,

$$m \equiv \gamma\, m_0, \tag{5a}$$

$$\gamma \equiv \cosh\beta = \left(1 - \frac{v^2}{c^2}\right)^{-\frac{1}{2}} = \sqrt{1 + \left(\frac{\parallel \vec{\mathbf{P}} \parallel}{m_0 c}\right)^2}\,. \tag{5b}$$

2.2 Automorphism invariance

Transformations that preserve some physical symmetry often lead to a conservation law of physics. For example, displacements of the origin leave Newton's laws unchanged, leading to a derivation of the conservation of linear momentum. Central force problems (e.g., gravitational field around a spherical star) have rotational invariance, leading to conservation of angular momentum for orbits. When formulating physics with Clifford algebra, we should ask just what new symmetries are inherently encoded in the structure of the algebra and what (if any) new physical laws they may imply.

Matrix representation invariance

Physicists usually first encounter Clifford algebras in quantum mechanics in the form of Pauli, Majorana and Dirac "spin" matrices. The matrix representation, for example, of the four generators γ^μ of the Majorana algebra is arbitrary. Hence it is obvious that the physics must be invariant under a change of the matrix representation of the algebra. A change in representation can be reinterpreted as a global rotation of the spin space basis spinors. Requiring "spin space isotropy" (no preferred direction in spin space) leads to the physical principle of conservation of quantum spin.

Algebra automorphisms

It is possible however to avoid talking about the matrix representation entirely. The more general concept is an algebra automorphism, which is a transformation of the basis generators γ_μ of the algebra which preserves the Clifford structure

$$\{\gamma_\mu, \gamma_\nu\} = 2\, g_{\mu\nu}, \tag{6}$$

where $g_{\mu\nu}$ is the spacetime metric. For example, consider the following orthogonal transformation on any element $\mathcal{Q}$ of the Clifford algebra

$$\mathcal{Q}' = \mathcal{R}\, \mathcal{Q}\, \mathcal{R}^{-1}, \tag{7}$$

$$\mathcal{R}(\phi^\mu) \equiv \exp(-\frac{1}{2}\gamma_\mu \phi^\mu) , \quad \mu = 1,2,3,4. \tag{8}$$

Proposing local covariance of the Dirac equation under such automorphism transformations is one path to gauge gravity and grand unified theory [4].

Polydimensional isotropy

If the elements γ_μ are interpreted geometrically as basis vectors, then (8) reshuffles the dimensional ranks of the geometric objects. For example, when $\phi^4 = \frac{\pi}{2}$, eq. (7) causes the permutation

$$\gamma_j \Longleftrightarrow \gamma_4\gamma_j , \quad j = 1,2,3, \tag{9a}$$

$$\gamma_1\gamma_2\gamma_3 \Longleftrightarrow \gamma_4\gamma_1\gamma_2\gamma_3, \tag{9b}$$

which exchanges three of the vectors with their associated timelike bivectors. What is a 1D vector in one "reference frame" is hence a 2D bivector in another. The transformation (8) thus "rotates" vectors into planes. Another example would be the transformation generated by (7) for: $\mathcal{R} = \exp(\hat{\epsilon}\,\phi/2)$, where $\hat{\epsilon} \equiv \gamma_1\gamma_2\gamma_3\gamma_4$. When $\phi = \frac{\pi}{2}$, this causes the duality transformation,

$$\gamma_\mu \Longleftrightarrow \hat{\epsilon}\,\gamma_j , \quad \mu = 1,2,3,4, \tag{10}$$

exchanging vectors for dual trivectors (the rest of the algebra is unchanged).

If we feel that Clifford algebra is the natural description of a geometric space, then we must ask whether these algebra automorphisms have physical interpretation. We suggest that there is a "higher octave" to the metaprinciple of the isotropy of space. We propose the **principle of polydimensional isotropy**: that *there is no absolute or preferred direction in the universe to which one can assign the geometry of a vector*. Just as which direction you choose to call the z-axis is arbitrary, it is also arbitrary just which geometric element you call the basis vector in the z-direction. Another observer may make an entirely different selection. This suggests perhaps that we should require the laws of physics to be invariant under such automorphic transformations (which will be discussed in more detail in Section 4).

2.3 Polydimensional formulation of physics

If we embrace the new principle that space is polydimensionally isotropic, then we should consider if transformations that reshuffle the basis geometry leave certain sets of physical laws invariant. Our development will attempt to parallel that which happened in the transition from classical 3D physics to 4D special relativity.

Unification with Clifford algebra

Just as four-vectors allowed us to unify two equations into one, the language of Clifford algebra allows for further notational economy. Consider that a classically spinning point charged particle obeys the torque equation of motion [17],

$$\dot{S}^{\mu\beta} = \frac{e}{mc}\left(F^{\mu}{}_{\nu}\, S^{\nu\beta} - F^{\beta}{}_{\nu}\, S^{\nu\mu}\right) = \frac{e}{mc}\left(\mathbf{F} \otimes \mathbf{S}\right)^{\mu\beta}. \tag{11}$$

This and eq. (2) can be written in the single statement,

$$\dot{\mathcal{M}} = \frac{e}{2mc}\left[\mathbf{F}, \mathcal{M}\right], \tag{12a}$$

where $\mathbf{F} \equiv \frac{1}{2}F^{\mu\nu}\,\hat{\mathbf{e}}_\mu \wedge \hat{\mathbf{e}}_\nu$ is the electromagnetic field bivector and $\hat{\mathbf{e}}_\mu$ are the basis vectors of the geometric space which obey the same Clifford algebra rules eq. (6). The *momentum polyvector* (alias the "polymomentum") is defined as the multivector sum of the vector linear momentum and the bivector spin momentum

$$\mathcal{M} \equiv p^\mu \hat{\mathbf{e}}_\mu + \frac{1}{2\lambda} S^{\mu\nu} \hat{\mathbf{e}}_\mu \wedge \hat{\mathbf{e}}_\nu, \tag{12b}$$

where λ is some fundamental length scale constant (to be interpreted in the next section) that allows us to add the quantities with different units in analogy to the use of c in eq. (3). The ability to add different ranked (dimensional) quantities is the notational advantage of Clifford geometric algebra over standard tensors. Mathematically, (12a) allows one to *simultaneously* obtain solutions to both equations (2) and (11): $\mathcal{M}(\tau) = \mathcal{R}\mathcal{M}(0)\mathcal{R}^{-1}$, where the rotation operator

$$\mathcal{R}(\tau) = \exp\left(\frac{e}{4mc}\,\hat{\mathbf{e}}_\mu \wedge \hat{\mathbf{e}}_\nu \int^{\tau} d\tau' F^{\mu\nu}\left[x(\tau')\right]\right), \tag{13}$$

involves a path (history) dependent integral; hence, the solution is formal.

Polydimensional invariance

Equation (12a) is manifestly covariant under automorphism transformations. Specifically, the set of equations (2) and (11) are invariant under the automorphism transformations generated by (8). For example, $\phi^4 = \frac{\pi}{2}$ in (8) causes a trading between momentum and mass moment of the spin tensor

$$\lambda\, p_j \Longleftrightarrow S_{4j}. \tag{14}$$

It is not at all clear what physical interpretation to ascribe to the two frames of reference. A radical assertion of the *principle of relative dimensionalism* would be to propose that what is a vector to one observer is a bivector to another and that they would partition the polymomentum (12b) into momentum and spin portions differently. *What is spin to one would be momentum to the other.*

Just as rotational invariance led to conservation of angular momentum, we might ask just what is the conserved quantity associated with this new symmetry transformation. This will be addressed in Section 3.2.

The quadratic form of a polyvector

We define the modulus of the polyvector eq. (12b) to be the square root of the scalar part of the square of the polyvector

$$\| \mathcal{M} \|^2 \equiv p_\mu p^\mu - \frac{1}{2\lambda^2} S_{\mu\nu} S^{\mu\nu}. \tag{15}$$

This quadratic form is invariant under the rotation of vectors into bivectors generated by (8). In the $(---+)$ metric signature, we define the modulus of the momentum polyvector to be the *bare mass*: $m_0 \equiv c^{-1} \| \mathcal{M} \|$. This implies that the mechanical mass (modulus of the momentum) is NOT invariant under these transformations but has been enhanced by the spin energy content

$$m \equiv c^{-1} \| \mathbf{p} \| = m_0 \sqrt{1 + \frac{S^{\mu\nu} S_{\mu\nu}}{2\,(m_0 c \lambda)^2}}, \tag{16a}$$

in analogy to (5ab). What we have described in (16a), by simple geometric construction, is a familiar result, laboriously obtained by Dixon [6] in the mechanical analysis of extended spinning bodies. Expanding (16a) non-relativistically,

$$mc^2 \simeq m_0 c^2 + \frac{\vec{\mathbf{S}}^2}{2m_0\lambda^2} + \dots, \tag{16b}$$

where $\vec{\mathbf{S}}^2 \equiv (S_{12})^2 + (S_{23})^2 + (S_{31})^2$, one sees that λ is consistent with the *radius of gyration* of a classical extended particle such that its moment of inertia is $\simeq m_0 \lambda^2$. Hence the correction to the mass is due to the rotational kinetic energy.

3 Dimensional democracy

In quaternionic analysis, a coordinate is given to each of the four elements. Reinterpreted as a Clifford algebra, it would be as if one has given a coordinate to each of the two vector directions, one to the plane and one to the scalar. We now propose that each geometric element of the Clifford algebra democratically has its own conjugate coordinate. Further *the physical laws should be multivectorial, with each geometric component meaningful.* For example, our polymomentum (12) gives the (vector) linear momentum and (bivector) spin momentum equal importance, both contributing the modulus of the vector. This suggests a generalized action principle that particles take the paths which minimize the sum of the linear distance traveled combined with the bivector area swept out. This simple geometric idea gives a new derivation of the spin enhanced mass described by the Dixon equation (15), as well as proposals for new quantum equations.

3.1 Review of Classical Relativistic Mechanics

In ancient times, Heron of Alexandria showed that light reflecting off a mirror would take the path of least distance between two endpoints. The generalized concept is that classical particles will follow paths of least spacetime distance between endpoints even when the space is curved by gravity.

Time contributes to distance

The measure of distance between two points in flat spacetime is

$$c^2 d\tau^2 \equiv c^2 dt^2 - \left(dx^2 + dy^2 + dz^2\right) = dx^\alpha dx^\beta \, g_{\alpha\beta} \tag{17a}$$

where affine parameter τ is commonly called the *proper time.* If we adopt the 3D viewpoint, we are combining (in quadrature) the "scalar" time displacement with the "vector" path displacement, utilizing a fundamental constant c (the speed of light) to combine the quantities which have different units. The metric tensor $g_{\alpha\beta}$ in flat space is diagonal with elements $(-1, -1, -1, +1)$ such that the fourth time component has the opposite signature of the spatial parts. To generalize for curved space, $g_{\alpha\beta}(x^\sigma)$ becomes a function of spacetime position.

Dividing (17a) by dt^2 recovers the *Lorentz dilation factor* eq. (5b) in terms of the nonrelativistic velocity,

$$\gamma \equiv \frac{dt}{d\tau} = \left(1 - \frac{v^2}{c^2}\right)^{-\frac{1}{2}} . \tag{17b}$$

Euler-Lagrange equations of motion

In simplest form, to obtain the equations of motion, one chooses the special path between fixed endpoints for which the *action integral* [which is based upon the quadratic form eq. (17a)],

$$\mathcal{A} \equiv \int m_0 c \, d\tau = \int \mathcal{L} d\tau = \int m_0 c \sqrt{u^\alpha u^\beta g_{\alpha\beta}(x^\sigma)} \, d\tau, \tag{18}$$

is an *extremum.* The integrand $\mathcal{L}(\tau, x^\alpha, \dot{x}^\alpha)$ is called the *Lagrangian,* which is generally a function of the coordinates and their velocities relative to the proper time: $u^\alpha \equiv \dot{x}^\alpha = dx^\alpha/d\tau$, where $x^4 \equiv ct$, hence $u^4 = \dot{x}^4 = c\gamma$.

Each coordinate x^α has a canonically conjugate momentum p_α defined as

$$p_\mu \equiv \frac{\delta \mathcal{L}}{\delta u^\mu} = m_0 u_\mu = m_0 \dot{x}_\mu . \tag{19a}$$

For our relativistic Lagrangian (18) these obey eq. (4). When reparameterized in terms of the more familiar observer's time $t = x^4/c$,

$$\texttt{Momentum:} \quad P_j \equiv \frac{\delta \mathcal{L}}{\delta \dot{x}^j} = m_0 \dot{x}_j = m\, v_j \; , \quad j = 1, 2, 3, \tag{19b}$$

$$\texttt{Energy:} \quad \mathcal{E} \equiv c \frac{\delta \mathcal{L}}{\delta \dot{x}^4} = c\, m_0\, \dot{x}_4 = m\, c^2, \tag{19c}$$

it is easy to show that the 3D part of the momentum $P_j = m v_j$ has mass m which is enhanced by the energy content according to (5ab).

Applying Hamilton's *Principle of Least Action*, one considers the total variation of the Lagrangian with respect to a variation in path (and velocity),

$$\delta \mathcal{L} = \frac{\delta \mathcal{L}}{\delta x^\alpha}\, \delta x^\alpha + \frac{\delta \mathcal{L}}{\delta \dot{x}^\alpha}\, \delta \dot{x}^\alpha . \tag{20a}$$

To get the equations of motion as that part proportional to a variation in the path only, the last term involving the variation of the velocity is integrated by parts,

$$\frac{\delta \mathcal{L}}{\delta \dot{x}^\alpha} \delta \dot{x}^\alpha \;\equiv\; p_\alpha \delta \dot{x}^\alpha \;=\; \frac{d}{d\tau}\left(p_\alpha \delta x^\alpha\right) - \dot{p}_\alpha \delta x^\alpha + p_\alpha \left(\delta \dot{x}^\alpha - \frac{d}{d\tau} \delta x^\alpha \right) . \tag{20b}$$

The total derivative term does not contribute if the variation in path has fixed endpoints. Substituting into (20a) and setting equal to zero, one obtains the generalized Euler-Lagrange equations of motion [8]

$$\left(\dot{p}_\alpha - \frac{\delta \mathcal{L}}{\delta x^\alpha} \right) \delta x^\alpha = p_\beta \left(\delta \dot{x}^\beta - \frac{d}{d\tau} \delta x^\beta \right) . \tag{20c}$$

In most elementary mechanics texts the terms on the right are argued to vanish because it is assumed the variation of the velocity is the same as the velocity of the variation. However when coordinates are path dependent (non-holonomic), $\delta d \neq d\delta$ because derivatives will not commute [8, 13]. For example, in a rotating coordinate system

$$\delta \dot{x}^\beta - \frac{d}{d\tau} \delta x^\beta = \delta x^\alpha\, \omega_\alpha{}^\beta . \tag{20d}$$

This would introduce an $\vec{\omega} \times \vec{\mathbf{P}}$ pseudoforce term on the right side of the equation of motion eq. (20c).

Rotating coordinate systems

The principle of isotropy states that there is no preferred direction in spacetime. Hence the laws of physics must be invariant under local Lorentz transformations (which include spatial rotations). Hence we can invent a new "body frame" coordinate system which has time dependent basis vectors $\mathbf{e}_\mu(\tau)$ that are related to the fixed "lab frame" basis $\hat{\mathbf{e}}_\mu$ by a time dependent orthogonal transformation,

$$\mathbf{e}_\mu(\tau) = \mathcal{R}\,\hat{\mathbf{e}}_\mu\,\mathcal{R}^{-1}, \tag{21a}$$

$$\mathcal{R}(\tau) = \exp\left(-\frac{1}{4}\hat{\mathbf{e}}_{\mu\nu}\,\Theta^{\mu\nu}(\tau)\right). \tag{21b}$$

The Cartesian angular displacement coordinates $\Theta^{\mu\nu}$ uniquely describe the orientation state of the body frame at the particular time. However, the angular velocity bivector $\underline{\omega}$ of the frame is NOT given by the time derivatives of these coordinates, rather it is defined [10] as

$$\underline{\omega} \equiv \frac{1}{2}\omega^{\mu\nu}\,\hat{\mathbf{e}}_{\mu\nu} \equiv -2\,\mathcal{R}^{-1}\dot{\mathcal{R}} \neq \frac{1}{2}\dot{\Theta}^{\mu\nu}\,\hat{\mathbf{e}}_{\mu\nu}. \tag{22}$$

The difficulty is that rotations (Lorentz transformations) do not commute; hence, the final state of the body frame is a function of path history. This was also the case in the electrodynamic problem presented earlier in eq. (13).

One can invent some new *quasi-coordinates*: $\theta^{\mu\nu}$, for which the angular velocity IS given by their time derivative, $\omega^{\mu\nu} \equiv \dot{\theta}^{\mu\nu}$ (see Greenwood [9]). Unfortunately, these new coordinates are non-integrable (path dependent) and hence *non-holonomic* such that $(\delta d - d\delta)\theta^{\mu\nu} \neq 0$. The advantage of resorting to this complexity is that the Lagrangian and (generalized) Euler-Lagrange equations have the same form in both the body frame and lab frame [8]. Further, we can show by the chain rule

$$\mathcal{R}\underline{\omega} = -2\dot{\mathcal{R}} = -\dot{\theta}^{\mu\nu}\frac{\partial\mathcal{R}}{\partial\theta^{\mu\nu}}, \tag{23a}$$

that the tangent bivectors are given by the derivatives of the rotation operator,

$$2\frac{\partial\mathcal{R}}{\partial\theta^{\mu\nu}} = -\mathbf{e}_{\mu\nu}\mathcal{R} = -\mathcal{R}\hat{\mathbf{e}}_{\mu\nu}. \tag{23b}$$

The differential and variation of the rotation operator are hence,

$$d\mathcal{R} = -\frac{1}{2}\mathcal{R}\,d\theta^{\mu\nu}\,\hat{\mathbf{e}}_{\mu\nu}\,, \qquad \delta\mathcal{R} = -\frac{1}{2}\mathcal{R}\,\delta\theta^{\mu\nu}\,\hat{\mathbf{e}}_{\mu\nu}. \tag{24ab}$$

We assume that since $\mathcal{R}$ defines the state of the body *independent of the particular coordinate parametrization*, that $\delta(d\mathcal{R}) = d(\delta\mathcal{R})$. Explicitly taking the variation δ of eq. (24a) and setting it equal to the differential d of eq. (24b) we obtain,

$$(\delta d - d\delta)\,\frac{1}{2}\theta^{\mu\nu}\hat{\mathbf{e}}_{\mu\nu} = \mathcal{R}^{-1}\,(d\mathcal{R}\,\delta\underline{\theta} - \delta\mathcal{R}\,d\underline{\theta}) = \frac{1}{2}\,[d\underline{\theta},\delta\underline{\theta}]\,. \tag{25a}$$

In component form, we see that for rotations, the variation of the angular velocity bivector is *not* the velocity of the angular variation bivector,

$$\delta\omega^{\mu\nu} - \frac{d\delta\theta^{\mu\nu}}{d\tau} = (\underline{\omega}\otimes\underline{\delta\theta})^{\mu\nu} = \delta^{\mu\nu}_{\alpha\beta}\,\omega^{\alpha}{}_{\sigma}\,\delta\theta^{\sigma\beta}. \tag{25b}$$

3.2 Polydimensional mechanics

If we fully embrace the concept of *relative dimensionalism*, then we must recognize that what one observer labels as a "point' in spacetime with vector coordinates (x, y, z, t) may be seen as an entirely different geometric object by another. This suggests that perhaps we should formulate physics in a way which is completely *dimensionally democratic* in that all ranks of quantities are equally represented.

The Clifford manifold

We propose therefore that the world is not the usual four-dimensional manifold, but instead a fully *polydimensional continuum* made of points, lines, planes, etc. Each event Σ is a geometric point in a *Clifford manifold* [3], which has a coordinate q^A associated with each basis element $\mathbf{E}_A$ (vector, bivector, trivector, etc.). Our definition of the Clifford Manifold is hence broader than the original proposal by Chisholm and Farwell [3] in that we have been "dimensionally democratic" in giving a coordinate to each geometric degree of freedom. The *pandimensional differential* in the manifold would be

$$d\Sigma \equiv \mathbf{E}_A dq^A = \mathbf{e}_\mu dx^\mu + \frac{1}{2\lambda}\mathbf{e}_\alpha\wedge\mathbf{e}_\beta da^{\alpha\beta} + \frac{1}{6\lambda^2}\mathbf{e}_\alpha\wedge\mathbf{e}_\beta\wedge\mathbf{e}_\sigma dV^{\alpha\beta\sigma} + \ldots, \tag{26a}$$

where in Clifford algebra it is perfectly valid to add vectors to bivectors and trivectors (parameterized by the antisymmetric tensor coordinates $dx^\mu, da^{\alpha\beta}, dV^{\alpha\beta\sigma}$, respectively).

In analogy to (15), we propose that the quadratic form of the Clifford manifold would be the scalar part of the square of (26a),

$$d\kappa^2 = \parallel d\Sigma \parallel^2 \equiv dx^\mu dx_\mu + \frac{1}{2\lambda^2}da^{\alpha\beta}da_{\beta\alpha} + \frac{1}{6\lambda^4}dV^{\alpha\beta\sigma}dV_{\sigma\beta\alpha} + \ldots. \tag{26b}$$

The fundamental length constant λ must be introduced in eq. (26ab) in order to add the bivector "area" coordinate contribution to the vector "linear" one. In analogy to (17a) this new quadratic form suggests we define a new affine parameter $d\kappa = \parallel d\Sigma \parallel$ which we will use to parameterize our polydimensional equations of motion.

New classical action principle

Classical mechanics assumes points which trace out linear paths. The equations of motion are based upon minimizing the distance of the path. String theory introduces one-dimensional objects which trace out areas, and the equations of motion are analogously based upon minimizing the total area. Membrane theory proposes two-dimensional objects which trace out (three-dimensional) volumes to be minimized. Our new action principle is that we should add all of these contributions together and treat particles as polygeometric objects which trace out polydimensional paths with (26b) the quantity to be minimized.

Using only the vector and bivector contributions of (26b) the Lagrangian that is analogous to (18) would be

$$\mathcal{L}(x^{\mu}, \mathring{x}^{\mu}, a^{\alpha\beta}, \mathring{a}^{\alpha\beta}) = m_0 c \sqrt{\mathring{x}^{\mu}\, \mathring{x}^{\nu}\, g_{\mu\nu} - \frac{1}{2\lambda^2}\, \mathring{a}^{\alpha\beta}\, \mathring{a}^{\mu\nu}\, g_{\alpha\mu}\, g_{\beta\nu}}, \tag{27}$$

where the open dot denotes differentiation with respect to the new affine parameter $d\kappa$ (whereas the small dot with respect to the proper time $d\tau$),

$$\mathring{Q} \equiv \frac{dQ}{d\kappa} = \dot{Q}\frac{d\tau}{d\kappa}. \tag{28}$$

The relationship of the new affine parameter to the proper time is a new *spin dilation factor* analogous to the Lorentz dilation factor (17b). Dividing (26b) by $d\tau$ or $d\kappa$, and noting $d\tau^2 \equiv dx^{\mu}\, dx_{\mu}$, this spin dilation factor is

$$\frac{d\tau}{d\kappa} \equiv \left(1 - \frac{\dot{a}^{\mu\nu}\, \dot{a}_{\mu\nu}}{2c^2\lambda^2}\right)^{-\frac{1}{2}} = \sqrt{1 + \frac{\mathring{a}^{\mu\nu}\, \mathring{a}_{\mu\nu}}{2c^2\lambda^2}}. \tag{29}$$

Note this implies the magnitude of the bivector velocity with respect to the proper time (proportional to spin angular velocity) is bounded by λc, just as linear velocity cannot exceed c.

Canonical Momenta

In analogy to eq. (19a) we interpret the spin to be the canonical momenta conjugate to the bivector coordinate,

$$\texttt{Spin:} \quad S_{\mu\nu} \equiv \lambda^2 \frac{\delta\mathcal{L}}{\delta \mathring{a}^{\mu\nu}} = m_0\, \mathring{a}_{\mu\nu} = m\, \dot{a}_{\mu\nu}, \tag{30a}$$

$$\texttt{Momentum:} \quad p_{\mu} \equiv \frac{\delta\mathcal{L}}{\delta \mathring{x}^{\mu}} = m_0\, \mathring{x}_{\mu} = m\, \dot{x}_{\mu}, \tag{30b}$$

$$\texttt{Dynamic Mass:} \quad m \equiv m_0\, \mathring{\tau} = m_0\, \frac{d\tau}{d\kappa}. \tag{30c}$$

For our Lagrangian (27), these satisfy the Dixon equation (15). When these momenta are reparameterized in terms of the more familiar proper time, they have spin enhanced mass defined by (30c), which is equivalent to eq. (16a).

Our Lagrangian (27) is invariant under the polydimensional coordinate rotation (between vectors and bivectors), generated by four the arbitrary parameters $\delta\phi^{\alpha}$ of the automorphism transformation analogous to eq. (8),

$$\delta x^{\alpha} = \frac{1}{\lambda}\, a^{\alpha}{}_{\mu}\, \delta\phi^{\mu}, \tag{31a}$$

$$\delta a^{\mu\nu} = x^{\mu}\, \delta\phi^{\nu} - x^{\nu}\, \delta\phi^{\mu}. \tag{31b}$$

Noether's theorem associates with this symmetry transformation a new set of constants of motion,

$$Q_{\mu} = \frac{\delta\mathcal{L}}{\delta \overset{\circ}{x}{}^{\alpha}} \frac{\delta x^{\alpha}}{\delta\phi^{\mu}} + \frac{1}{2} \frac{\delta\mathcal{L}}{\delta \overset{\circ}{a}{}^{\alpha\beta}} \frac{\delta a^{\alpha\beta}}{\delta\phi^{\mu}} = a_{\mu}{}^{\alpha}\, p_{\alpha} + S_{\mu\beta}\, x^{\beta}. \tag{32}$$

Taking the derivative of (32) with respect to the affine parameter yields the familiar Weysenhoff condition, that the spin is pure spacelike in the rest frame of the particle,

$$p_{\mu}\, S^{\mu\nu} = 0. \tag{33}$$

This is quite significant, because usually (33) is imposed at the onset by fiat, while we have provided an actual derivation based on the new automorphism symmetry of the Lagrangian!

3.3 Polyrotational quasi-coordinates and quantization

We propose the body frame coordinates may be rotated by a generalized geometamorphic transformation, a combination of (8) and (21b) which leaves the Lagrangian (27) invariant. This provides clues as to the nature of the derivative with respect to the bivector coordinate. With this we can define a new "spin operator" and generalized quantum wave equations based on (15).

Polydimensionally rotating coordinate frames

We propose a polydimensional generalization of Section 3.1, invoking our principles of polydimensional isotropy and relative dimensionalism discussed in Section 2 above. The poly-rotation operator is

$$\mathcal{R}(\kappa) \equiv \exp\left[-\frac{1}{4}\,\hat{\mathbf{e}}_{\mu\nu}\Theta^{\mu\nu}(\kappa) - \frac{1}{2}\,\hat{\mathbf{e}}_{\alpha}\Phi^{\alpha}(\kappa)\right]. \tag{34a}$$

We propose that the *polyvelocity* is defined in analogy to eq. (22),

$$\overset{\circ}{\mathcal{Q}} \equiv \mathcal{M}/m_0 \equiv -2\lambda\mathcal{R}^{-1}\,\overset{\circ}{\mathcal{R}} \equiv \overset{\circ}{x}{}^{\mu}\hat{\mathbf{e}}_{\mu} + \frac{1}{2\lambda}\,\overset{\circ}{a}{}^{\mu\nu}\hat{\mathbf{e}}_{\mu\nu}, \tag{34b}$$

where $\{x^\mu, a^{\alpha\beta}\}$ must therefore be anholonomic quasi-coordinates (as opposed to the holonomic coordinates $X^\mu \equiv \lambda\Phi^\mu$ and $A^{\alpha\beta} \equiv \lambda^2\Theta^{\alpha\beta}$, respectively) such that their derivatives with respect to $d\kappa$ yield the vector and bivector velocities. In analogy to the rotational development in Section 3.1, the linear velocity is NOT the (total) derivative of the usual Cartesian coordinate $X^\mu(\kappa)$ for a spinning body. For example, when moving in the y direction, an angular acceleration of the spin along the z axis will introduce an additional "effective" velocity in the x direction due to the shift in apparent relativistic mass center. All of these interdependent effects are accounted for in the history-dependent quasi-coordinate x^μ.

It follows that the tangent basis vectors are given in analogy to eq. (23b),

$$2\lambda\frac{\partial\mathcal{R}}{\partial x^\mu} = -\mathbf{e}_\mu\mathcal{R} = -\mathcal{R}\hat{\mathbf{e}}_\mu\ , \tag{35a}$$

$$2\lambda^2\frac{\partial\mathcal{R}}{\partial a^{\alpha\beta}} = -\mathbf{e}_{\alpha\beta}\mathcal{R} = -\mathcal{R}\hat{\mathbf{e}}_{\alpha\beta} = 2\lambda^2\left[\frac{\partial}{\partial x^\alpha},\frac{\partial}{\partial x^\beta}\right]\mathcal{R}. \tag{35b}$$

This implies that the bivector derivative is equivalent to the commutator derivative, an idea developed further by Erler [7] and utilized in Section 4. Two other commutators we shall find useful are

$$\left[\frac{\partial}{\partial x^\mu},\frac{\partial}{\partial a^{\alpha\beta}}\right]\mathcal{R} = -\frac{g_{\mu\sigma}}{\lambda^2}\delta^{\omega\sigma}_{\alpha\beta}\frac{\partial}{\partial x^\omega}\mathcal{R}, \tag{35c}$$

$$\left[\frac{\partial}{\partial a^{\alpha\beta}},\frac{\partial}{\partial a^{\mu\nu}}\right]\mathcal{R} = -\frac{g^{\phi\kappa}}{\lambda^2}g_{\sigma\mu}\,g_{\omega\nu}\,\delta^{\sigma\omega\theta}_{\alpha\beta\kappa}\frac{\partial}{\partial a^{\theta\phi}}\mathcal{R}. \tag{35d}$$

We note in passing that the 4 basis vectors and 6 bivectors of a 4D space can be reinterpreted as the 10 bivectors of an enveloping 5D space, with eq. (34a) as the rotation operator. In this context our calculus may be related to that proposed by Blake [2] over the multivector manifold $spin^+(4,1)$.

By parallel argument to eq. (25ab) we obtain the non-commutativity of the variation and derivative for the polyvelocity,

$$(\delta d - d\delta)\,\mathcal{Q} = \mathcal{R}^{-1}\left(d\mathcal{R}\,\delta\mathcal{Q} - \delta\mathcal{R}\,d\mathcal{Q}\right) = \frac{[d\mathcal{Q},\delta\mathcal{Q}]}{2\lambda^2}. \tag{36}$$

Extracting the vector and bivector portions,

$$(\delta d - d\delta)\,x^\alpha = \frac{\delta x^\mu\, da_\mu{}^\alpha - dx^\mu\,\delta a_\mu{}^\alpha}{\lambda^2}, \tag{37a}$$

$$(\delta d - d\delta)\,a^{\mu\nu} = \frac{da^{\mu\sigma}\delta a_\sigma{}^\nu - \delta a^{\mu\sigma}da_\sigma{}^\nu}{\lambda^2} + dx^\mu\delta x^\nu - \delta x^\mu dx^\nu. \tag{37b}$$

Comparing to (20d) and (25b) we see that $\mathring{a} \simeq \omega\lambda^2$. However, we have additional terms which validates that variations of the linear and rotational paths are not independent, hence, eq. (20c) is no longer complete.

Quantization

In classical Hamilton mechanics, functions of motion (n.b., the Hamiltonian on which quantum mechanics is based) are parameterized in terms of the coordinates and their canonical momenta. The obvious generalization of the Poisson bracket for two functions of polydimensional coordinates would be

$$\{F,G\} \equiv \frac{\delta F}{\delta x^\alpha}\frac{\delta G}{\delta p_\alpha} - \frac{\delta G}{\delta x^\alpha}\frac{\delta F}{\delta p_\alpha} + \frac{\lambda^2}{2}\left(\frac{\delta F}{\delta a^{\alpha\beta}}\frac{\delta G}{\delta S_{\alpha\beta}} - \frac{\delta G}{\delta a^{\alpha\beta}}\frac{\delta F}{\delta S_{\alpha\beta}}\right). \quad (38)$$

There are some potential complications. From eq. (37ab) it is not at all clear that δp^σ is completely independent of δx^μ, $\delta a^{\alpha\beta}$ and especially $\delta S^{\alpha\beta}$. For today we will sidestep the issues and assume for brevity that at least the canonical pairs obey the relations $\{x^\alpha, p_\beta\} = \delta^\alpha_\beta$, and $\{a^{\mu\nu}, S_{\alpha\beta}\} = \lambda^2\delta^{\mu\nu}_{\alpha\beta}$.

The *Heisenberg quantization rule* is that the commutator of quantum operators maps to the Poisson bracket of the corresponding classical quantities,

$$[\hat{F},\hat{G}] \mapsto i\hbar\{F,G\}. \quad (39)$$

It follows that $[\hat{x}^\nu, \hat{p}_\mu] = i\hbar\delta^\nu{}_\mu$ and $[\hat{a}^{\mu\nu}\hat{S}_{\alpha\beta}] = i\hbar\lambda^2\delta^{\mu\nu}_{\alpha\beta}$. In the coordinate representation the momenta operators must be

$$\hat{p}_\mu \equiv -i\hbar\frac{\partial}{\partial x^\mu}, \qquad \hat{S}_{\mu\nu} \equiv -i\hbar\lambda^2\frac{\partial}{\partial a^{\mu\nu}}. \quad (40ab)$$

This would imply that one could define a spin angular coordinate $\theta^{\mu\nu} = \lambda^{-2}a^{\mu\nu}$. We should note that other authors see potential difficulties with the definition of angular operators in quantum mechanics [18].

(Hand) Wave equations

The polydimensional analogy of the Klein-Gordon wave equation based on the Dixon equation (15) would hence be

$$\begin{aligned}\left[\hat{p}^\mu\hat{p}_\mu - \tfrac{1}{2\lambda^2}\hat{S}^{\mu\nu}\hat{S}_{\mu\nu}\right]\psi &= -\hbar^2\left[\frac{\partial}{\partial x^\mu}\frac{\partial}{\partial x_\mu} - \frac{\lambda^2}{2}\frac{\partial}{\partial a^{\mu\nu}}\frac{\partial}{\partial a_{\mu\nu}}\right]\psi, \\ &= (m_0 c)^2\psi\end{aligned} \quad (41a)$$

where the wavefunction $\psi(x^\mu, a^{\alpha\beta})$ depends upon the vector position *and* bivector spin coordinates. If the system is in an eigenstate of total spin, then eq. (41a) simply reduces to the standard Klein-Gordon equation with spin-enhanced mass given by eq. (16a).

One might expect that the generalization of the Dirac equation would simply involve factoring (15) with the Poly momenta (12b) into the linear form $\hat{\mathcal{M}}\Psi = m_0 c\Psi$. It is not quite that simple because we know the components of the standard spin operator $\hat{\mathbf{S}} = \frac{1}{2}\hat{S}^{\alpha\beta}\mathbf{e}_{\alpha\beta}$ do not commute. Consistent with (35d), we have the standard relations [12],

$$\hat{\mathbf{S}}\hat{\mathbf{S}} = -\frac{1}{2}\hat{S}^{\mu\nu}\hat{S}_{\mu\nu} + \mathbf{e}_{\mu\nu}\mathbf{e}_{\alpha\beta}[\hat{S}^{\mu\nu}, \hat{S}^{\alpha\beta}] = -\frac{1}{2}\hat{S}_{\mu\nu}\hat{S}^{\mu\nu} - 2i\hbar\mathbf{S}. \quad (42a)$$

Equation (35b) implies that the components of the momenta no longer commute: $[\hat{p}_\mu, \hat{p}_\nu] = -i\hbar\hat{S}_{\mu\nu}/\lambda^2$, such that the square of the momentum vector $\hat{\mathbf{p}} = \hat{p}^\mu \mathbf{e}_\mu$,

$$\hat{\mathbf{p}}\hat{\mathbf{p}} = \hat{p}^\mu \hat{p}_\mu + \frac{1}{2}\mathbf{e}_{\mu\nu}\,[\hat{p}^\mu, \hat{p}^\mu] \;=\; \hat{p}^\mu \hat{p}_\mu - \frac{i\hbar}{\lambda^2}\hat{\mathbf{S}}. \tag{42b}$$

Equation (35c) implies that the spin and momenta operators do not commute,

$$[\hat{p}^\mu, \hat{S}_{\alpha\beta}] = -i\hbar\delta^{\mu\sigma}_{\alpha\beta}\,\hat{p}_\sigma, \tag{42c}$$

$$\{\hat{\mathbf{p}}, \hat{\mathbf{S}}\} = 2\hat{\mathbf{p}} \wedge \hat{\mathbf{S}} - \frac{3i\hbar}{\lambda}\hat{\mathbf{p}}. \tag{42d}$$

Putting eq. (42abd) together, the poly momenta operator (12b) obeys

$$\widehat{\mathcal{M}}\widehat{\mathcal{M}} = \hat{p}^\mu\,\hat{p}_\mu - \frac{1}{2\lambda^2}\hat{S}^{\mu\nu}\,\hat{S}_{\mu\nu} - \frac{3i\hbar}{\lambda}\widehat{\mathcal{M}} + \frac{2}{\lambda}\hat{\mathbf{p}} \wedge \hat{\mathbf{S}}. \tag{43}$$

Substituting, we can rewrite eq. (41a) as

$$\left[\widehat{\mathcal{M}}\left(\widehat{\mathcal{M}} + \frac{3i\hbar}{\lambda}\right) - \frac{2}{\lambda}\hat{\mathbf{p}} \wedge \hat{\mathbf{S}} - (m_0 c)^2\right]\psi = 0. \tag{41b}$$

If we presume an idempotent structure on the wavefunction, the trivector term can be replaced by an eigenvalue

$$\frac{2}{\lambda}\hat{\mathbf{p}} \wedge \hat{\mathbf{S}}\left[\left(1 \pm \frac{i\hat{\mathbf{p}} \wedge \hat{\mathbf{S}}}{mcS}\right)\psi\right] \;\Rightarrow\; \pm\frac{2imS}{\lambda}\psi, \tag{44}$$

where $S^2 \equiv \parallel \mathbf{S} \parallel^2 = \frac{1}{2}S^{\mu\nu}S_{\mu\nu}$. Thus eq. (41b) can now be factored into a polydimensional monogenic Dirac equation with complex mass roots $N_\pm$,

$$\left[\widehat{\mathcal{M}} + N_\pm\right]\Psi = \left[-i\hbar\left(\mathbf{e}^\mu\frac{\partial}{\partial x^\mu} + \frac{\lambda}{2}\mathbf{e}^{\mu\nu}\frac{\partial}{\partial a^{\mu\nu}}\right) + N_\pm\right]\Psi = 0. \tag{45a}$$

$$\Psi \equiv \left[\widehat{\mathcal{M}} - N_\mp\right]\psi. \tag{45b}$$

Solving the quadratic equation, we can get one of the roots to be the bare mass if we impose a constraint on the magnitude of spin,

$$N_+ = m_0 c\,, \qquad N_- = m_0 c - \frac{3i\hbar}{\lambda}, \tag{45cd}$$

$$S^2 \equiv \frac{1}{2}S^{\mu\nu}\,S_{\mu\nu} \;=\; \frac{3}{2}\hbar^2\frac{m_0}{m}. \tag{45e}$$

Invoking eq. (16a), we can thus get a relationship between the fundamental constants m_0, λ and the magnitude of the spin S. In the limit of $m_0 c\lambda >> \hbar$ one recovers the standard "half integer spin" magnitude equation: $S^2 = \frac{3}{2}\hbar^2$.

4 General poly-covariance

In curved space, particles will deviate from standard geodesics due to contributions from derivatives of the basis vectors with respect to the new bivector coordinate. Further, there are additional contributions to the non-commutativity of the variation and derivative due to torsion and curvature. This leads to a new derivation of the Papapetrou equations [14] describing the motion of spinning particles in curved space. Finally we propose a principle of **Metamorphic Covariance:** *that the laws of physics should be form invariant under local automorphism transformations which reshuffle the geometry.*

4.1 Covariant derivatives in the Clifford manifold

The total derivative of a basis vector with respect to the new affine parameter $d\kappa$ must by the chain rule contain a derivative with respect to the bivector coordinate

$$\overset{\circ}{\mathbf{e}}_\mu \equiv \frac{d\mathbf{e}_\mu}{d\kappa} = \overset{\circ}{x}{}^\sigma \frac{\partial \mathbf{e}_\mu}{\partial x^\sigma} + \frac{1}{2}\,\overset{\circ}{a}{}^{\alpha\beta}\frac{\partial \mathbf{e}_\mu}{\partial a^{\alpha\beta}}. \tag{46a}$$

Our ansätze, consistent with (35b), is that the bivector derivative obeys [7, 15],

$$\frac{\partial \mathbf{e}_\mu}{\partial a^{\alpha\beta}} \equiv \left([\partial_\alpha, \partial_\beta] - \tau^\sigma_{\alpha\beta}\partial_\sigma\right)\mathbf{e}_\mu = \left(R_{\alpha\beta\mu}{}^\nu - \tau^\sigma_{\alpha\beta}\Gamma^\nu_{\sigma\mu}\right)\mathbf{e}_\nu, \tag{46b}$$

where $\tau^\sigma_{\alpha\beta}$ is the torsion, $\Gamma^\nu_{\sigma\mu}$ is the Cartan connection, and $R_{\alpha\beta\mu}{}^\nu$ is the Cartan curvature.

We can factor out the basis vectors by defining the covariant derivative

$$\frac{\partial}{\partial x^\mu}\left(p^\nu \mathbf{e}_\nu\right) = \mathbf{e}_\nu \nabla_\mu p^\nu \equiv \mathbf{e}_\nu\left(\partial_\mu p^\nu + p^\sigma \Gamma^\nu_{\mu\sigma}\right), \tag{47a}$$

$$\frac{\partial}{\partial a^{\alpha\beta}}\left(p^\nu \mathbf{e}_\nu\right) = \mathbf{e}_\nu\left[\nabla_\alpha, \nabla_\beta\right] p^\nu \equiv \mathbf{e}_\nu\left(R_{\alpha\beta\mu}{}^\nu p^\mu - \tau^\sigma_{\alpha\beta}\nabla_\sigma p^\nu\right). \tag{47b}$$

From these definitions it is clear then the covariant derivatives of the basis vectors $\mathbf{e}^\mu$ and $\mathbf{e}_\mu$ vanish as usual.

The parallel transport of the conserved canonical momenta generates new *poly-autoparallels* in the Clifford manifold

$$0 = \frac{d}{d\kappa}\left(\mathbf{e}_\mu p^\mu\right) = \mathbf{e}_\mu\left(\overset{\circ}{x}{}^\sigma \nabla_\sigma + \frac{1}{2}\,\overset{\circ}{a}{}^{\alpha\beta}\left[\nabla_\alpha, \nabla_\beta\right]\right) p^\mu, \tag{48a}$$

$$0 = \frac{d}{d\kappa}\left(\mathbf{e}_{\mu\nu} S^{\mu\nu}\right) = \mathbf{e}_{\mu\nu}\left(\overset{\circ}{x}{}^\sigma \nabla_\sigma + \frac{1}{2}\,\overset{\circ}{a}{}^{\alpha\beta}\left[\nabla_\alpha, \nabla_\beta\right]\right) S^{\mu\nu}. \tag{48b}$$

Substituting (47ab) provides a new derivation of the Papapetrou equations of motion for spinning particles [14] in contravariant form. Ours however are more general as they include torsion and all the higher order terms. In covariant form,

$$0 = \overset{\circ}{p}_\sigma - \left(\overset{\circ}{x}^\alpha \Gamma^{\ \mu}_{\alpha\sigma} + \frac{1}{2} \overset{\circ}{a}^{\alpha\beta} R'^{\ \ \ \mu}_{\alpha\beta\sigma} \right) p_\mu, \tag{49a}$$

$$0 = \overset{\circ}{S}_{\rho\omega} - \delta^{\sigma\nu}_{\rho\omega} \left(\overset{\circ}{x}^\alpha \Gamma^{\ \mu}_{\alpha\sigma} + \frac{1}{2} \overset{\circ}{a}^{\alpha\beta} R'^{\ \ \ \mu}_{\alpha\beta\sigma} \right) S_{\mu\nu}, \tag{49b}$$

$$R'^{\ \ \ \mu}_{\alpha\beta\nu} \equiv R^{\ \ \ \mu}_{\alpha\beta\nu} - \tau^\sigma_{\alpha\beta} \Gamma^{\ \mu}_{\sigma\nu}. \tag{49c}$$

4.2 An-holonomic mechanics

It has been a long-standing unsolved problem to derive the Papapetrou equations from a simple Lagrangian. We succeed where so many others have failed because of our definition of the new affine parameter, the form of the Lagrangian (27), and by noting that the introduction of the bivector coordinate has made the system an-holonomic. Consider the variation of the Lagrangian,

$$\delta\mathcal{L} = \frac{\delta\mathcal{L}}{\delta x^\alpha} \delta x^\alpha + \frac{\delta\mathcal{L}}{\delta \overset{\circ}{x}{}^\alpha} \delta \overset{\circ}{x}{}^\alpha + \frac{1}{2} \frac{\delta\mathcal{L}}{\delta a^{\alpha\beta}} \delta a^{\alpha\beta} + \frac{1}{2} \frac{\delta\mathcal{L}}{\delta \overset{\circ}{a}{}^{\alpha\beta}} \delta \overset{\circ}{a}{}^{\alpha\beta}. \tag{50}$$

As in (20b), we must integrate the spin-velocity term by parts,

$$\begin{aligned} \tfrac{\delta\mathcal{L}}{\delta \overset{\circ}{a}{}^{\alpha\beta}} \delta \overset{\circ}{a}{}^{\alpha\beta} &= S_{\alpha\beta}\, \delta \overset{\circ}{a}{}^{\alpha\beta} \\ &= \tfrac{d}{d\kappa} \left(S_{\alpha\beta}\, \delta a^{\alpha\beta} \right) - \overset{\circ}{S}_{\alpha\beta}\, \delta a^{\alpha\beta} + \\ &\quad S_{\alpha\beta} \left(\delta \overset{\circ}{a}{}^{\alpha\beta} - \tfrac{d}{d\kappa} \delta a^{\alpha\beta} \right). \end{aligned} \tag{51}$$

However, the generalized equation of motion eq. (20c) is incomplete because in general there will be an interdependence between the vector and bivector variations in curved space. Certainly we saw this feature appear before in eq. (37ab) for the poly rotating coordinate system. The difficulty is how to derive the new contributions due to torsion and curvature.

Note that while (25b) states that the variation of the angular velocity is not the velocity of the angular variation, *for the components*, one can easily show from (24ab) and (25ab) that $d(\delta\omega^{\mu\nu}\mathbf{e}_{\mu\nu}) = \delta(d\omega^{\mu\nu}\mathbf{e}_{\mu\nu})$. We therefore argue that in curved space the same idea holds, that is,

$$\delta \left(\overset{\circ}{x}{}^\mu\, \mathbf{e}_\mu \right) = \frac{d}{d\kappa} \left(\delta x^\mu\, \mathbf{e}_\mu \right), \tag{52a}$$

$$\delta \left(\overset{\circ}{a}{}^{\alpha\beta}\, \mathbf{e}_\alpha \wedge \mathbf{e}_\beta \right) = \frac{d}{d\kappa} \left(\delta a^{\alpha\beta}\, \mathbf{e}_\alpha \wedge \mathbf{e}_\beta \right). \tag{52b}$$

Performing the variations and derivatives in the above equations and rearranging terms [and ignoring the contribution of eq. (37ab)],

$$\left(\delta \overset{\circ}{x}{}^\mu - \frac{d\delta x^\mu}{d\kappa} \right) = \delta x^\alpha\, \overset{\circ}{x}{}^\beta \tau^\sigma_{\alpha\beta} + \frac{1}{2} \left(\delta x^\alpha\, \overset{\circ}{a}{}^{\mu\nu} - \overset{\circ}{x}{}^\alpha \delta a^{\mu\nu} \right) R'^{\ \ \ \sigma}_{\mu\nu\alpha}, \tag{53a}$$

$$\left(\delta \overset{\circ}{a}^{\mu\nu} - \frac{d\delta a^{\mu\nu}}{d\kappa}\right) = \delta^{\lambda\sigma}_{\omega\nu}\left[\Gamma^{\omega}_{\alpha\mu}\left(\overset{\circ}{x}^{\alpha}\delta a^{\mu\nu} - \overset{\circ}{a}^{\mu\nu}\delta x^{\alpha}\right) + \frac{1}{4}R'_{\alpha\beta\mu}{}^{\omega}\left(\overset{\circ}{a}^{\mu\nu}\delta a^{\alpha\beta} - \overset{\circ}{a}^{\alpha\beta}\delta a^{\mu\nu}\right)\right]. \tag{53b}$$

The first term on the right of (53a) involving the torsion follows Kleinert [13], the rest are new. Substituting (53a) into (20b) and (52b) into (51), and finally into (50), separating out terms proportional to δx^{μ} and $\delta a^{\mu\nu}$ respectively, we obtain polydimensionally generalized Euler-Lagrange equations

$$\frac{\delta\mathcal{L}}{\delta x^{\mu}} - \overset{\circ}{p}_{\mu} + p_{\lambda}\,\overset{\circ}{x}^{\alpha}\,\tau^{\lambda}_{\alpha\mu} + \left(\frac{1}{2}p_{\lambda}\,R'_{\alpha\beta\mu}{}^{\lambda} + S_{\omega\beta}\,\Gamma^{\omega}_{\mu\alpha}\right)\overset{\circ}{a}^{\alpha\beta} = 0, \tag{54a}$$

$$\frac{\delta\mathcal{L}}{\delta a^{\mu\nu}} - \overset{\circ}{S}_{\mu\nu} + p_{\lambda}\,\overset{\circ}{x}^{\sigma}R'_{\mu\nu\sigma}{}^{\lambda} + \frac{1}{2}\left(S_{\omega\beta}R'_{\mu\nu\alpha}{}^{\omega} - S_{\omega\nu}R'_{\alpha\beta\mu}{}^{\omega}\right)\overset{\circ}{a}^{\alpha\beta} = 0. \tag{54b}$$

The first two terms of eq. (54a) are standard; the third term appears in Kleinert [13]; the rest of (54a) and all of (54b) are new.

Explicitly performing the derivative on the Lagrangian in (54a) we recover the Papapetrou equation (49a). To get the spin equation (49b) from (54b) we must introduce a generalization of eq. (46b) for bivector variations, that is,

$$\frac{\delta\mathcal{L}}{\delta a^{\mu\nu}} \equiv \left[\frac{\delta}{\delta x^{\mu}}, \frac{\delta}{\delta x^{\nu}}\right]\mathcal{L} - \tau^{\sigma}_{\mu\nu}\frac{\delta\mathcal{L}}{\delta x^{\sigma}} = \frac{\delta\mathcal{L}}{\delta g_{\alpha\beta}}\left(R'_{\mu\nu\alpha\beta} + R'_{\mu\nu\beta\alpha}\right). \tag{54c}$$

If there is no torsion, the R' reduces to the Riemann curvature, which is antisymmetric in the last two indices, hence eq. (54c) vanishes.

4.3 Metamorphic covariance

Our Lagrangian (27) is invariant under *local* automorphism transformations, where in general the Φ^{μ} of (34) can be position dependent upon a path-dependent (history dependent) integral of a gauge field $B^{\nu}{}_{\mu}$,

$$\Phi^{\nu}(x^{\alpha}) = \int^{x^{\alpha}} B^{\nu}{}_{\mu}(y^{\sigma})\,dy^{\mu}. \tag{55}$$

This would imply that the connection of a basis vector would become *geometamorphic* [16], e.g., under parallel transport *a vector will metamorph into a plane.* We have previously proposed [15] such a "metamorphic Clifford connection" of the form

$$\mathcal{D}\mathbf{e}_{\mu} \equiv dx^{\alpha}\left(\Gamma^{\nu}_{\alpha\mu} + \tfrac{1}{2}\Xi_{\alpha\mu}{}^{\nu\sigma}\mathbf{e}_{\nu\sigma}\right) + \tfrac{1}{2}da^{\alpha\beta}\left(R_{\alpha\beta\mu}{}^{\nu}\mathbf{e}_{\nu} + \tfrac{1}{2}\Omega_{\alpha\beta\mu}{}^{\nu\sigma}\mathbf{e}_{\nu\sigma}\right), \tag{56a}$$

where $\Xi_{\alpha\mu}{}^{\nu\sigma} \simeq B^{\omega}{}_{\alpha}\delta^{\nu\sigma}_{\omega\mu}$, and the curvature $R_{\alpha\beta\mu}{}^{\nu}$ now has contributions from derivatives on both $\Gamma^{\nu}_{\alpha\mu}$ and $\Xi_{\alpha\mu}{}^{\nu\sigma}$. This means that equations (48ab) are no longer valid because each only contains a *single* dimensional piece. We are forced to implement *dimensional democracy* and write our equations only with *polyvectors*. Further one finds that the Leibniz rule does not hold over the wedge (or dot) product although it is valid for the Clifford (direct) product [15]. Hence the metamorphic connection on the bivector would be computed as

$$\mathcal{D}(\mathbf{e}_\mu \wedge \mathbf{e}_\nu) = \frac{1}{2}[(\mathcal{D}\mathbf{e}_\mu), \mathbf{e}_\nu] + \frac{1}{2}[\mathbf{e}_\mu, (\mathcal{D}\mathbf{e}_\nu)] \neq (\mathcal{D}\mathbf{e}_\mu)\wedge \mathbf{e}_\nu + \mathbf{e}_\mu \wedge (\mathcal{D}\mathbf{e}_\nu). \quad (56b)$$

With these generalizations, reworking Section 4.1, one can get a polycovariant generalization [15] of the Papapetrou equation (49a),

$$\begin{aligned} &\overset{\circ}{p}{}^\mu + p^\nu \left(\overset{\circ}{x}{}^\beta \Gamma^\mu_{\beta\nu} + \tfrac{1}{2}\, \overset{\circ}{a}{}^{\alpha\beta} R_{\alpha\beta\nu}{}^\mu \right) \\ &\quad + \tfrac{1}{2} S^\omega{}_\sigma \left(\overset{\circ}{x}{}^\alpha \Xi_{\alpha\omega}{}^{\mu\sigma} + \tfrac{1}{2}\, \overset{\circ}{a}{}^{\alpha\beta} \Omega_{\alpha\beta\omega}{}^{\mu\sigma} \right) = 0. \end{aligned} \quad (57)$$

To derive eq. (57) from a Lagrangian requires us to make the theory fully covariant under *general* polydimensional coordinate transformations. This will cause the quadratic form (26b) to acquire cross terms such that the Lagrangian would generalize to

$$\mathcal{L} = m_0 c \sqrt{\overset{\circ}{x}{}^\alpha g_{\alpha\beta}\, \overset{\circ}{x}{}^\beta + \frac{1}{2}\, \overset{\circ}{x}{}^\alpha h_{\alpha\mu\nu}\, \overset{\circ}{a}{}^{\mu\nu} + \frac{1}{4m_0^2\lambda^4}\, \overset{\circ}{a}{}^{\alpha\beta} \mathcal{I}_{\alpha\beta\mu\nu}\, \overset{\circ}{a}{}^{\mu\nu}}, \quad (58)$$

where $\mathcal{I}$ plays the role of the relativistic moment of inertia tensor. This and the interdimensional metric $h_{\alpha\mu\nu}$ will cause the linear momenta not to be parallel to the velocity and spin momenta not parallel to bivector (angular) velocity.

Equation (56a) is the classical analog to the spin-covariant derivative for the Dirac equation derived from generalized automorphism transformations of the Dirac algebra by Crawford [4],

$$\nabla_\mu = \partial_\mu + i\left(eA_\mu + \gamma^5 a_\mu\right) + \gamma_\nu \left(\frac{1}{2} B^\nu{}_\mu + \gamma^5\, i\, b^\nu{}_\mu \right) + \frac{1}{2}\gamma_{\alpha\beta}\, C^{\alpha\beta}_\mu, \quad (59a)$$

$$(-i\hbar\gamma^\mu \nabla_\mu - mc)\,\psi = 0. \quad (59b)$$

The gauge field $B^\mu{}_\sigma$ is the same as in eq. (55). In the Dirac equation (59b), the usual momentum operator eq. (40a) has been replaced by the gauge-covariant derivative $p_\mu \to -i\hbar\nabla_\mu$. To get the interacting form of the polydimensional Dirac equation (45a) one need only additionally replace the spin operator (40b) with the form suggested by (47b), the commutator covariant derivative: $S_{\mu\nu} \to -i\hbar\lambda^2 \left[\nabla_\mu, \nabla_\nu\right]$,

$$\left(-i\hbar\gamma^\mu \nabla_\mu - i\hbar \frac{\lambda}{2} \gamma^{\alpha\beta} \left[\nabla_\alpha, \nabla_\beta\right] - m_0 c \right) \Psi = 0. \quad (60)$$

Certainly one could include higher order triple commutator derivatives. In flat space, with all but the electromagnetic gauge field A_μ suppressed in (59a), the bivector (commutator) derivative will introduce an anomalous magnetic moment interaction which provides a possible interpretation of the constant λ. It remains to be shown that an application of Ehrenfest's theorem to eq. (60) can recover the equation of motion (57), in analogy with the derivation of the Papapetrou equation (49a) from (59b) by Crawford [5].

5 Summary

In introducing *dimensional democracy* we have given the bivector a coordinate and shown its utility in the treatment of the classical spinning particle problem. This system is invariant under "polydimensional" transformations which reshuffle geometry such that "what is a vector" is *dimensionally relative* to the observer's frame. A fundamentally new action principle has been introduced which is *polydimensionally isotropic.* Generalized *metamorphic covariant* equations of motion and quantum wave equations have been derived which include curvature, torsion and spin. Most important, the principles proposed have potential broad math and physics applications beyond the examples in this paper.

References

[1] W. Baylis, *Electrodynamics, A Modern Geometric Approach,* Birkhäuser, 1999.

[2] S. Blake, Calculus on the multivector manifold $spin^+(p,q)$, *J. Math. Phys.* **39** (1998), 6106–6117.

[3] J. S. R. Chisholm and R. S. Farwell, Clifford approach to metric manifolds, in *Proceedings on the Winter School on Geometry and Physics, Srni, 6-13 January, 1990, Supplemento di Reconditi del Circulo Matematico di Palermo,* Serie **2, no. 26** (1991), 123–133.

[4] J. P. Crawford, Local automorphism invariance: Gauge boson mass without a Higgs particle, *J. Math. Phys.* **35** (1994), 2701–2718.

[5] J. P. Crawford, Spinor matter in a gravitational field: covariant equations a la Heisenberg, *Found. Phys.* **28** (1998), 457–470.

[6] W. G. Dixon, Dynamics of extended bodies in general relativity, I. Momentum and angular momentum, *Proc. Roy. Soc. Lond.* **A314** (1970), 499–527.

[7] T. G. Erler, The mathematics of the spinning particle problem, preprint, gr-qc/9912024.

[8] P. Fiziev and H. Kleinert, New action principle for classical particle trajectories in spaces with torsion, *Europhys. Lett.* **35** (1996), 241.

[9] D. T. Greenwood, *Classical Dynamics,* Dover, 1997, 113, 174–175.

[10] D. Hestenes, *New Foundations for Classical Mechanics,* Kluwer, 1990, 307.

[11] D. Hestenes, *Spacetime Algebra,* Gordon and Breach Publ., 1965.

[12] I. B. Khriplovich, Particle with internal angular momentum in a gravitational field, *Sov. Phys. JETP* **69** (1989), 217–219 .

[13] H. Kleinert, Nonabelian Bosonization as a nonholonomic transformation from flat to curved field space, *Annals. Phys.* **253** (1997), 121–176.

[14] A. Papapetrou, Spinning test-particles in general relativity, I, *Proc. Roy. Soc. London* **A209** (1951), 248–258; Equations of motion in general relativity, *Proc. Phys. Soc.* **64** (1951), 57.

[15] W. Pezzaglia, Physical applications of a generalized Clifford calculus, in *Dirac Operators in Analysis,* Pitman Research Notes in Mathematics, Number 394, J. Ryan and D. Struppa, eds., Longman Science & Technology, 1998, 191–202.

[16] W. Pezzaglia, Polydimensional relativity, a classical generalization of the automorphism invariance principle, in *Clifford Algebras and Their Applications in Mathematical Physics, Proceedings of Fourth Conference, Aachen 1996,* Habetha and Jank, eds., Dietrich, Kluwer, 1998, 305-317.

[17] F. Rohrlich, *Classical Charged Particles,* Addison-Wesley, 1965, 204–205.

[18] A. C. de la Torre and J. L. Iguain, Angle states in quantum mechanics, *Am. J. Phys.* **66** (1998), 1115–1122.

William M. Pezzaglia Jr.
Department of Physics
Santa Clara University
Santa Clara, California 95053
E-mail: wpezzag@clifford.org

Received: October 11, 1999; Revised: March 7, 2000

A Pythagorean Metric in Relativity

Franco Israel Piazzese

ABSTRACT A one-to-one mapping of the class $\hat{\mathbb{R}}^{1,3}$ of the timelike vectors of $\mathbb{R}^{1,3}$ onto $\mathbb{R}^{4,0}$ is introduced in which i) the pseudo-Pythagorean norm of each element $w \in \hat{\mathbb{R}}^{1,3}$ and the Pythagorean norm of the corresponding element $v \in \mathbb{R}^{4,0}$ are equal, ii) the "space" parts of w and v are proportional, and iii) the transformation properties of v are induced by those of w. With the aid of such a mapping, a new interpretation of the rest energy of a particle of special relativity is proposed. Half of this quantity is the sum of two terms, one of which formally coincides with the classical kinetic energy of a point like particle although it involves the relativistic velocity instead of the classical one, so it is called the *quasi-classical kinetic energy*. The other one is interpreted as describing an internal degree of freedom of the particle. As a result, the rest energy, which is *invariant* (i.e., independent of frame), is regarded as twice the amount of the *total kinetic energy* of the particle. This accounts for the acronym **itke**.
Keywords: Vector spaces, Pythagorean and pseudo-Pythagorean metrics, metric signature, timelike vectors, relativistic energy and momentum, invariant total kinetic energy (itke), quasi-classical description.

1 Introduction

The signature of the metric on a real vector space $\mathbb{R}^n$ is often considered as one of the basic features of the space. Special symbols have been introduced to denote a vector space endowed with a particular metric (e.g., $\mathbb{R}^{1,3}$, $\mathbb{R}^{1,3}$, etc.). However, the metric is a mapping $\mathbb{R}^n \times \mathbb{R}^n \to \mathbb{R}$, and different metrics, also with different signatures, can be defined on the same space.

This paper deals with space $\mathbb{R}^4$ endowed with both Pythagorean and pseudo-Pythagorean metrics, which make it $\mathbb{R}^{1,3}$ and $\mathbb{R}^{4,0}$, respectively. A one-to-one mapping of the class $\hat{\mathbb{R}}^{1,3}$ of the timelike vectors of $\mathbb{R}^{1,3}$ onto $\mathbb{R}^{4,0}$ is introduced, in which

- the pseudo-Pythagorean norm of each element $w \in \hat{\mathbb{R}}^{1,3}$ and the Pythagorean norm of the corresponding element $v \in \mathbb{R}^{4,0}$ are equal,
- the "space" parts of w and v are proportional,

AMS Subject Classification: 15A03, 15A63, 83A05.

- the transformation properties of v are induced by those of w (cf. Sects. 3 and 4).

The above mapping allows a new "quasi-classical" description of the energy-momentum of a free particle, which is very similar to the classical description of a system endowed with an internal degree of freedom in which, however, the relativistic velocity replaces the classical one. This makes the energy of a free particle independent of frame (cf. Sects. 5 and 6).

2 Mathematical preliminaries

Consider the real 4-dimensional vector space $\mathbb{R}^4$, and let

$$v = (v^0, v^1, v^2, v^3) \tag{2.1}$$

be an element of this space. The elements

$$e_0 = (1,0,0,0),\ e_1 = (0,1,0,0),\ e_2 = (0,0,1,0),\ e_3 = (0,0,0,1) \tag{2.2}$$

make the *canonical* basis. With the aid of (2.2), (2.1) assumes the form

$$v = v^\alpha e_\alpha. \tag{2.3}$$

The Greek suffixes are understood to range from 0 to 3, and the Latin ones from 1 to 3. The sum on the repeated indices is implied.

Introduce in $\mathbb{R}^4$ the standard inner product (cf. [1, Sect. 7.2])

$$(v, z) = \delta_{\mu\nu} v^\mu z^\nu \tag{2.4}$$

where $v, z \in \mathbb{R}^4$, and $\delta_{\mu\nu}$ is the Kronecker symbol. Denote by $\mathbb{R}^{4,0}$ space $\mathbb{R}^4$ endowed swith inner product (2.4). The *Pythagorean* norm $|v|$ of vector $v \in \mathbb{R}^{4,0}$ is the positive square root of the following quantity

$$|v|^2 = \delta_{\mu\nu} v^\mu v^\nu = (v^0)^2 + (v^1)^2 + (v^2)^2 + (v^3)^2 \tag{2.5}$$

which is non-negative, only vanishing when v equals the zero vector.

As it follows from (2.4), basis (2.2) is orthonormal. Thus, from (2.2) and (2.4) one gets

$$(e_\alpha, e_\beta) = \delta_{\alpha\beta}. \tag{2.6}$$

Space $\mathbb{R}^4$ is the direct sum of spaces $\mathbb{R}$ and $\mathbb{R}^3$ (cf., e.g., [1, Sect. 2.4]). Thus, (2.1) can be rewritten

$$v = (v^0, \mathbf{v}) \tag{2.7}$$

where

$$\mathbf{v} = (v^1, v^2, v^3) = v^k \mathbf{e}_k \tag{2.8}$$

in which the elements

$$\mathbf{e}_1 = (1,0,0),\ \mathbf{e}_2 = (0,1,0),\ \mathbf{e}_3 = (0,0,1) \tag{2.9}$$

make the canonical basis of $\mathbb{R}^3$.

The standard inner product can be introduced also in $\mathbb{R}^3$, and the resulting space is denoted by $\mathbb{R}^{3,0}$. With respect to such an inner product, basis (2.9) is, obviously, orthonormal.

The norm $|\mathbf{v}|$ of $\mathbf{v} \in \mathbb{R}^{3,0}$ is the positive square root of the following quantity

$$|\mathbf{v}|^2 = \delta_{ik} v^i v^k = (v^1)^2 + (v^2)^2 + (v^3)^2. \tag{2.10}$$

Based on (2.10), (2.5) can be rewritten as

$$|v|^2 = (v^0)^2 + |\mathbf{v}|^2. \tag{2.11}$$

On the other hand, introducing in $\mathbb{R}^4$ a non-degenerate symmetrical indefinite bilinear function (cf. [1, Sect. 9.4]), one can define another inner product as follows

$$< v, z > = g_{\mu\nu} v^\mu z^\nu \tag{2.12}$$

where the $g_{\alpha\beta}$'s are the components of the *pseudo-Pythagorean* (or *pseudo-Euclidean*) metric tensor. It has signature equal to -2, when the only non-zero components with respect to basis (2.2) are

$$g_{00} = -g_{11} = -g_{22} = -g_{33} = 1. \tag{2.13}$$

Space $\mathbb{R}^4$ endowed with inner product (2.12) is denoted by $\mathbb{R}^{1,3}$. The pseudo-Pythagorean square norm

$$\|w\|^2 = g_{\alpha\beta} w^\alpha w^\beta = (w^0)^2 - (w^1)^2 - (w^2)^2 - (w^3)^2 \tag{2.14}$$

of a non-zero element $w \in \mathbb{R}^{1,3}$ can be positive, negative, or zero. As a result, the non-zero vectors of $\mathbb{R}^{1,3}$ can be divided into three classes. A vector u is said to be timelike, null, or spacelike according to whether quantity (2.14) is positive, zero, or negative. In particular, one has for (2.2) the following formula

$$< e_\alpha, e_\beta >= g_{\alpha\beta} \tag{2.15}$$

Vector $w \in \mathbb{R}^{1,3}$ can be written in a form similar to (2.7)

$$w = (w^0, \mathbf{w}) \tag{2.16}$$

where

$$\mathbf{w} = (w^1, w^2, w^3) = w^k \mathbf{e}_k \tag{2.17}$$

(cf. also (2.9)). Thus, based on (2.10), (2.14) can be rewritten in the following form

$$\|w\|^2 = (w^0)^2 - |\mathbf{w}|^2 \tag{2.18}$$

(cf. (2.11)). The Pythagorean norm of an element of space $\mathbb{R}^{3,0}$ is a term in both the expressions of the Pythagorean norm of an element of $\mathbb{R}^{4,0}$ and the pseudo-Pythagorean norm of an element of $\mathbb{R}^{1,3}$.

3 A one-to-one mapping

To each timelike w (i.e., $w \in \hat{\mathbb{R}}^{1,3}$) associate $v \in \mathbb{R}^{4,0}$ defined as the following

$$v^0 = \frac{\|w\|^2}{w^0}, \quad \mathbf{v} = \frac{\|w\|}{w^0}\mathbf{w} \tag{3.1}$$

(cf. (2.16) and (2.7)). (3.1) can be trivially solved for w^0 and $\mathbf{w}$:

$$w^0 = \frac{\|w\|^2}{v^0}, \quad \mathbf{w} = \frac{w^0}{\|w\|}\mathbf{v}. \tag{3.2}$$

The following property

$$|v| = \|w\| \tag{3.3}$$

can be easily proved. In fact, by introducing (3.1) one gets from (2.11)

$$|v|^2 = \frac{\|w\|^4}{(w^0)^2} + \frac{\|w\|^2}{(w^0)^2}|\mathbf{w}|^2 = \frac{\|w\|^2}{(w^0)^2}\left(\|w\|^2 + |\mathbf{w}|^2\right) = \|w\|^2 \tag{3.4}$$

(cf. (2.18)) since both $|v|$ and $\|w\|$ are positive quantities, (3.3) follows. Using (3.3), (3.2) can be rewritten as the following

$$w^0 = \frac{|v|^2}{v^0}, \quad \mathbf{w} = \frac{|v|}{v_0}\mathbf{v} \tag{3.5}$$

where the right-hand sides only depend on v. (3.5) *and* (3.1) *define a one-to-one mapping of the class* $\hat{\mathbb{R}}^{1,3}$ *of timelike elements of* $\mathbb{R}^{1,3}$ *onto* $\mathbb{R}^{4,0}$ (and the inverse one).

Introducing (3.1) and (3.5), one gets from (2.7) and (2.16)

$$v = \frac{\|w\|}{w^0}\left(\|w\|, \mathbf{w}\right) \tag{3.6}$$

and

$$w = \frac{|v|}{v^0} \left(|v|, \mathbf{v} \right), \tag{3.7}$$

respectively. Setting

$$\alpha = \frac{\|w\|}{w^0} = \frac{v^0}{|v|} = \left(1 - \frac{|\mathbf{w}|^2}{(w^0)^2} \right)^{\frac{1}{2}} \tag{3.8}$$

(cf. the former of (3.5) and (2.18)), (3.6) and (3.7) may be rewritten as the following

$$v = \alpha \left(\|w\|, \mathbf{w} \right) = \left(\alpha^2 w^0, \alpha \mathbf{w} \right) \tag{3.9}$$

and

$$w = \alpha^{-1} \left(|v|, \mathbf{v} \right) = \left(\alpha^{-2} v^0, \alpha^{-1} \mathbf{v} \right). \tag{3.10}$$

Equations (3.6) and (3.7) or (3.9) and (3.10), with (3.8), express the above mappings as well.

4 Transformation properties

A Lorentz transformation is a rotation of Minkowski's space-time (in particular $\mathbb{R}^{1,3}$). In it, the norm of any vector of such a space is preserved although—employing the "active" point of view (cf. [2, Sect. 4.2, p. 137])—the vector is, generally, transformed to another one.

This holds, in particular, for any timelike vector $w \in \hat{\mathbb{R}}^{1,3}$ which is transformed into $w' = Lw \in \hat{\mathbb{R}}^{1,3}$, where L denotes the Lorentz operator. Assume that such a transformation induces another transformation on the elements of $\mathbb{R}^{4,0}$ so that $v \in \mathbb{R}^{4,0}$ be transformed to $v' = M(v) \in \mathbb{R}^{4,0}$, where M denotes the transformation operator.

The one-to-one mapping of $\hat{\mathbb{R}}^{1,3}$ onto $\mathbb{R}^{4,0}$ introduced in the above section—here denoted by ϕ—is preserved in the above transformation (or "covariant") and the following diagram

$$\begin{array}{ccc} & \phi & \\ w & \longrightarrow & v \\ \downarrow & & \downarrow \\ w' = Lw & \longrightarrow & v' = M(v) \\ & \phi & \end{array} \tag{4.1}$$

commutes. To this end, operator M must be defined in such a way that

$$M(v) = \phi L \phi^{-1}(v) \tag{4.2}$$

(i.e., the transformation law of v is induced from that of w.)

5 Energy-momentum 4-vectors

As it is well known, the relativistic energy of a particle

$$E = c^2 m_0 \gamma \tag{5.1}$$

and its relativistic 3-momentum

$$\mathbf{P} = m_0 \mathbf{u} \gamma, \tag{5.2}$$

where m_0 and $\mathbf{u}$ are the *proper* mass and the *relativistic* 3-velocity of a particle, and

$$\gamma(u) = \left(1 - \frac{u^2}{c^2}\right)^{-\frac{1}{2}} \tag{5.3}$$

is the well-known Lorentz factor, make the 4-momentum

$$S = (\frac{E}{c}, \mathbf{P}) \tag{5.4}$$

(cf., e.g., [3, Sect. 29, eq. (5.8)]). From a geometrical point of view, S is an element of space $\mathbb{R}^{1,3}$ (cf. (2.16)). Its square norm is

$$\|S\|^2 = \left(\frac{E}{c}\right)^2 - c^2 \mathbf{P}^2 \tag{5.5}$$

(cf. (2.18)). With the aid of (5.1) and (5.2), one gets the following value

$$\|S\|^2 = \left(\frac{E_0}{c}\right)^2 \tag{5.6}$$

where

$$E_0 = c^2 m_0 \tag{5.7}$$

is the rest energy of the particle. Setting S for w and introducing (5.4) one gets from (3.8)

$$\alpha = \left(1 - \frac{|\mathbf{P}|^2}{(E/c)^2}\right)^{\frac{1}{2}} = \left(1 - \frac{u^2}{c^2}\right)^{\frac{1}{2}}. \tag{5.8}$$

Thus, α is the inverse of the Lorentz factor (cf. (5.3)). From (5.4) and (5.8), one gets, employing (3.9) and writing p^0 and $\mathbf{p}$ for v^0 and $\mathbf{v}$,

$$p = (p^0, \mathbf{p}) \tag{5.9}$$

where

$$p^0 = \frac{E_0}{c}\gamma^{-1} \equiv cm_0\gamma^{-1} \tag{5.10}$$

and

$$\mathbf{p} = \mathbf{P}\gamma^{-1} \equiv m_0\mathbf{u}, \tag{5.11}$$

respectively. It appears that (5.11) formally coincides with the classical momentum of the particle. However, it involves the *relativistic* velocity instead of the classical one; thus, contrary to the classical counterpart, it is limited. It will be called the *quasi-classical* momentum.

In accordance with (2.11), the square norm of (5.9) is

$$|p|^2 = \left[\left(\frac{E_0}{c}\right)^2 + \mathbf{P}^2\right]\gamma^{-2} \tag{5.12}$$

(cf. (5.10) and (5.11)). Using (5.2), (5.7), and (5.3), (5.12) may be rewritten as the following

$$|p|^2 = \left(\frac{E_0}{c}\right)^2. \tag{5.13}$$

6 The invariant total kinetic energy (itke)

By eliminating the left side in (5.12) and (5.13) and dividing by $2m_0$, one gets with the aid of (5.11) the following splitting of the rest energy

$$\frac{1}{2}E_0 = \frac{\mathbf{p}^2}{2m_0} + \frac{1}{2}E_0\gamma^{-2}. \tag{6.1}$$

For convenience, let's rewrite (6.1) in the following form:

$$T = T_{(1)} + T_{(2)} \tag{6.2}$$

where

$$T = \frac{1}{2}E_0, \quad T_{(1)} = \frac{\mathbf{p}^2}{2m_0} \equiv \frac{1}{2}m_0\mathbf{u}^2, \quad T_{(2)} = \frac{1}{2}E_0\gamma^{-2}. \tag{6.3}$$

Term $T_{(1)}$ formally coincides with the classical kinetic energy of a point-like particle, or the center of mass of a particle system, but involves the *relativistic* velocity instead of the classical one. Thus, this quantity is called the *quasi-classical* kinetic energy which is obviously limited as well as the quasi-classical momentum (5.11).

To interpret term $T_{(2)}$, recall that in De Broglie's matter wave theory, the following frequency is associated with any particle

$$\nu = \nu_0 \gamma \, , \qquad \nu_0 = \frac{c^2 m_0}{h} \tag{6.4}$$

where h is the well known Planck constant (cf. [3, Sect.52]). Although this frequency does not fulfill the relativistic transformation law (cf. [4]), another frequency σ, defined as

$$\sigma = \sigma_0 \gamma^{-1}, \quad \sigma_0 = \frac{c^2 m_0}{h} \tag{6.5}$$

and fulfilling the correct transformation law, can be introduced (cf. [5]). Introducing (6.5) and taking into account (5.7), (6.3) may be rewritten in the following form

$$T_{(2)} = \frac{1}{2} h \sigma \gamma^{-1} \tag{6.6}$$

or

$$T_{(2)} = \frac{1}{2} J \sigma^2 = \frac{L^2}{2J} \tag{6.7}$$

where

$$J = \frac{h}{\sigma_0} \equiv \frac{h^2}{c^2 m_0}, \qquad L = J\sigma = h\gamma^{-1}. \tag{6.8}$$

As term $T_{(2)}$ is a finite non-negative quantity, only vanishing for $\sigma = 0$ (i.e., $m_0 = 0$, cf. (6.5)), it can be interpreted as a contribution to the kinetic energy of the particle due to an internal degree of freedom. A tempting hypothesis is that this term describes some "rotation" motion; in this case L could be interpreted as describing the spin angular momentum.

Finally, quantity T, being the sum of terms $T_{(1)}$ and $T_{(2)}$ (cf. (6.2)), can be interpreted as the total kinetic energy of the particle. It is a finite, positive quantity independent of the frame. Thus, it can be called *invariant, total kinetic energy* (or **itke**).

7 Concluding remarks

In spite of the simple mathematics employed, possibly some new insight has been offered to the study of the energy aspects in the dynamics of a particle or a particle system in the framework of relativity.

In particular, the topics discussed in the above sections are the starting points of a new *quasi-classical description* of the dynamics of the particle of special relativity (cf. [6]), in which the internal degree of freedom accounts for the wave behavior. In that reference, a wave different from that introduced by De Broglie is associated with the particle.

Acknowledgments

This work has been produced under the auspices of the Italian Council for Research, C. N. R. (G. N. F. M.), with the support of the Italian Ministry of Research.

REFERENCES

[1] W. H. Greub, *Linear Algebra,* Springer-Verlag, Berlin, New York, 1967.

[2] H. Goldstein, *Classical Mechanics,* Addison-Wesley, Reading, 1980.

[3] W. Rindler, *Special Relativity,* Oliver and Boyd, Edinburgh, 1960.

[4] L. De Broglie, L'interprétation de la mécanique ondulatorie par la théorie de la double solution, in *Foundations of Quantum Mechanics,* B. D'Espagnat, ed., Academic Press, New York, 1971, 345–367.

[5] F. Piazzese, Energy and frequencies of a particle in special relativity, *Journal of Natural Geometry* **15** (1999), 81–90.

[6] F. Piazzese, A quasi-classical description of particle's dynamics, submitted to the *International Journal of Theoretical Physics.*

Franco Israel Piazzese
Department of Physics, Politecnico
Corso Duca degli Abruzzi 24
10129 Torino, Italy
E-mail: piazzese@polito.it

Received: September 30, 1999; Revised: December 12, 1999

Clifford-Valued Cliffoms: A Geometric Language for Dirac Equations

Jose G. Vargas and Douglas G. Torr

To David Hestenes, for daring to ask truly fundamental questions in physics in the second half of the twentieth century.

ABSTRACT Arguments from Cartan's writings are shown to bear upon the related issues of "whether to use tangent or cotangent algebras to represent physical quantities" and of "the distinction, if any exists, between exterior forms and exterior differential forms." The three series of indices that enter the *tensor-valued differential forms* of the Kähler calculus are then interpreted.

When endowed with a vector-valued potential (the potential and wave-function of electrodynamics being scalar-valued), the wave function for the Kähler-Dirac equation acquires inhomogeneous valuedness from zero to infinity. We show how this unwanted explosion of valuedness is tamed by the introduction of *Clifford-valued clifforms*, the term "clifform" being used here to refer to differential forms whose underlying algebraic structure is Clifford rather than Grassmann. This double Clifford algebra, however, has limitations. These are taken care of by extending the double algebra to a Kaluza-Klein space associated with the given spacetime, where a canonical connection and interior covariant derivative immediately arise.

Keywords: Differential forms, interior calculus, Kähler calculus, Dirac equations.

1 Introduction

The Fifth International Conference on Clifford Algebras evidences the tremendous progress made in using these algebras to solve problems in cybernetics, robotics, image processing, and engineering. The field of Clifford analysis is vibrant and growing. Interesting Clifford-algebraic results continue to appear. However, it is fair to say that no major result has been obtained in theoretical physics through the use of Clifford algebras, con-

---

AMS Subject Classification: 53B99, 58A99, 53Z05.

trary to expectations.

We here submit that this lack of results has to do with the wrong use that has been made of Clifford algebras to represent the Dirac equation, and even classical equations. With respect to the latter–suffice it to notice that, although Maxwell's equations become very simple in Clifford algebra language, this language does not suggest any new, superseding set of electrodynamic equations, in spite of the fact that, as stated by Einstein, " ... no reasonable person believes that Maxwell's equations can hold rigorously. They are, in suitable cases, first approximations for weak fields" [1]. The use of Clifford algebras also has not shown, so far, any potential for unification of gravitation with other interactions. It is in the realm of Dirac equations where Clifford algebras may have the most to offer as we shall show.

Versions of Dirac equations that explicitly incorporate Clifford algebras are due to Kähler [2] and Hestenes [3]. Kähler's version employs a cotangent Clifford algebra (to be later defined) and a tangent tensor algebra. Hestenes' version involves just a tangent Clifford algebra, which results from an incorrect identification of tangent and cotangent spaces, a point already made by Oziewicz [4]. In this paper, we argue at length why this identification is incorrect, namely because of the use of a field of exterior forms where an exterior differential form should be used. (This is a common error in the literature and will be dealt with in Section 3.) We further show the way in which Clifford algebras can contribute to radically new developments in the Kähler theory.

The paper is organized as follows. In Section 2, we compare the Hestenes and Kähler versions of the Dirac equation. In Section 3, the relations among exterior forms, fields of exterior forms, and exterior differential forms are explored. In Section 4, we describe several *mathematical viruses*, a concept first introduced by Hestenes [5], that impede a clear understanding of the Kähler theory of Dirac equations . In Section 5, Kähler's tensor-valued clifforms are made to evolve into Clifford-valued clifforms. The term "clifform" refers to the fact that the differential forms have a Clifford algebra structure in Kähler's work and in this paper. This term appears to have been coined by Dimakis and Müller-Hoissen [6] with a different meaning since they use it to refer to Clifford-valuedness rather than to a separate Clifford structure of differential forms; these authors retained the usual Grassmann structure. In Section 6, we introduce Clifford-valued clifforms in a Kaluza-Klein space. In Section 7, we show that this space is endowed with a canonical teleparallel connection. The interior derivative for the teleparallel canonical connection is introduced in Section 8,where the new features of this calculus are exhibited. In the concluding remarks of Section 9, the results obtained are interpreted in the light of Finsler geometry, as this will open the way for new developments in Kähler's theory of Dirac equations.

2 The use of Clifford algebra in Dirac equations

Hestenes [3] represented the well-known equation

$$i\hbar\gamma^\mu\partial_\mu\psi = (mc + \frac{e}{c}\gamma^\mu A_\mu)\psi \tag{2.1}$$

as

$$\hbar\,\square\psi = (m + eA)\psi i. \tag{2.2}$$

The operator "$\square$" (the four-dimensional gradient) is formally a vector in Hestenes' Clifford algebra. In Equation (2.2), the factor i represents the unit pseudo-scalar in the same algebra. It plays the role of the unit imaginary when it multiplies the wave function on the right. Though the meaning of the factor i differs in Equations (2.1) and (2.2), its position in the respective equations is equivalent, as can be seen by transferring it from one side to another in any of these equations.

On the other hand, the Kähler-Dirac equation is

$$\partial u = a \vee u, \tag{2.3}$$

where ∂ represents the sum of the interior and exterior covariant derivatives. For minimal electromagnetic coupling, this equation becomes

$$-i\hbar\partial u = (im + eA)u, \tag{2.4}$$

where the Clifford product is understood and where i is the usual unit imaginary of the complex algebra. To distinguish between Equations (2.3) and (2.4), we shall refer to them as the Kähler-Dirac and Dirac-Kähler equations, respectively. Equation (2.2) is not equivalent to the Dirac-Kähler equation in the following sense: the position of the unit imaginary in Equation (2.2) differs from its position in Equation (2.4). From an additional perspective, the non-equivalence of the two equations has been discussed by Hestenes [7], who stated that the Kähler derivative differs from the derivative in the usual form of the Dirac equation (and, thus, in the Dirac-Hestenes equation). In particular, it couples minimal left ideals.

For our purposes, a more relevant difference between the Dirac-Hestenes equation and Kähler-Dirac equation is that, in the latter, both a and u in Equation (2.3) are non-homogeneous tensor-valued differential forms whose terms have three, rather than two, series of indices:

$$a^{i_1\ldots i_p}_{j_1\ldots j_q} = a^{i_1\ldots i_p}_{j_1\ldots j_q k_1\ldots k_r}\, dx^{k_l} \wedge \ldots \wedge dx^{k_r}. \tag{2.5}$$

Kähler did not exhibit the basis elements to match the i and j indices or explain why he took the unusual step of using a second series of subscripts. It is clear, however, that the difference is essential for, at least, the following feature of his calculus. Whereas a scalar-valued r-form (quantities with just a k series of indices) has an interior derivative in the sense of Kähler, a tensor-valued differential 0-form or tensor field (quantities with only i and j series of indices) has zero interior derivative, regardless of any considerations of antisymmetry in the j indices.

For reasons that will be clear in the next section, the Kähler-Dirac equation is the way to go (at least, if one has to choose one of the two options) except that Kähler's is just the first step in the right direction. His work has to be corrected for the following "infinite inflation of valuedness." Let u_s be the component of u of valuedness s. Let $val(x)$ denote the valuedness of x. The action of the operator ∂ is such that $val(\partial u_s) = val(u_s)$, whereas $val(a \vee u)$ is not the same as $val(u)$. More to the point, let x_{max} be the term of highest valuedness in x. If a is not scalar-valued, $val(a \vee u)_{max}$ is greater than $val(\partial u)_{max}$. This causes u to be of inhomogeneous valuedness extending all the way up to infinity, so that a matching of terms on both sides of the Kähler-Dirac equation may always be possible. Otherwise, one would have to match to zero the terms with the m highest valuedness in $a \vee u$, where m is $val(a)_{max}$. This is not the most natural way of avoiding infinite valuedness since the pervasiveness of the dot product of tangent vectors in physics and geometry suggests that we replace the tangent tensor algebra structure by a tangent Clifford algebra structure. This represents, in principle, a great increase in the algebraic richness in the theory of the Kähler-Dirac equation, with enormous potential implications for theoretical physics.

In the Kähler calculus, the k series of indices is used for the differential forms, but these are not the antisymmetric *covariant tensors*, or multilinear functions of tangent vectors. Indeed, if the differential forms were (antisymmetric) multilinear functions of vectors, the existence of a metric (which is always assumed in the Kähler calculus) would allow us to identify a basis (F^i) of linear functions of vectors with the reciprocal basis $(\mathbf{e}^i)$ of the tangent basis $(\mathbf{e}_i)$, i.e., such that $\mathbf{e}^i \cdot \mathbf{e}_j = \delta^i_j$. Differential forms would then be identifiable with and, thus, have the same derivatives as the antisymmetric tangent tensors. Since this is not the case in the Kähler calculus, his differential forms are not identifiable with multilinear functions of vectors. This runs contrary to the most common interpretation of the quantities that enter the Kähler calculus. We devote Section 4 to a discussion of mathematical viruses that impede seeing this calculus in the proper light. We should finally mention that, in spite of the differences among the foregoing equations, all of them have the same solution for the hydrogen atom.

3 Forms, fields of forms, and differential forms

What are differential forms if not antisymmetric multilinear functions of vectors? In at least one of his books on analysis [8], Rudin defines a differential form of order $k \geq 1$ as a function ω, symbolically represented by the sum

$$\omega = \sum a_{i_1 \dots i_k}(x) dx_{i_1} \wedge \dots \wedge dx_{i_k}, \tag{3.1}$$

which assigns to each k-surface Φ in an open set E in $\mathbb{R}^n$ a number $\omega(\phi) = \int_\Phi \omega$, according to the rule

$$\int_\Phi \omega = \int_D \sum a_{i_1 \dots i_k}(\Phi(u)) \frac{\partial(x_{i_1}, \dots, x_{i_k})}{\partial(u_1, \dots, u_k)} du, \tag{3.2}$$

where D is the parameter domain of Φ. The differential k-forms of this paper will be Rudin's, which are also Cartan's and Kähler's. They are not the differential k-forms that are prevalent in the literature, i.e., fields of k-forms, where the k-forms themselves are antisymmetric multilinear functions of vectors. Although Rudin's definition was made for open sets of $\mathbb{R}^n$, it is clear that it equally applies to any differentiable manifold.

For an illustration of the relations between these three concepts (differential forms, fields of forms, and forms), consider order one. As per the previous paragraph, a 1-form F is a linear map of tangent vectors which assigns to a vector $\mathbf{v}$ a real number. A basis of 1-forms F^i is defined by $F^i(\mathbf{e}_j) = \delta^j_i$. When we approximate the integral $\int_{x^i(a)}^{x^i(b)} A_j(x) dx^j$ by $A_j[x(a)][x^j(b) - x^j(a)]$, we are replacing Rudin's differential 1-form $A_j(x)dx^j$ by the 1-form $A_j[x(a)]F^j$. Whereas differential 1-forms are functions of curves, which are objects that live on the manifold, 1-forms are functions of vectors of a given tangent vector space. In our example, the vector where the form $A_j[x(a)]F^j$ is evaluated is $[x^j(b) - x^j(a)]\mathbf{e}_j$. As stated above, a metric permits one to identify (non-differential) 1-forms with vectors through the identification of F^i with the element $\mathbf{e}^i$ of the reciprocal tangent basis.

In the same way, as 1-forms may be viewed as linear approximations to differential 1-forms, k-forms may be viewed as linear approximations to differential k-forms. We shall refer to the fields of antisymmetric multilinear functions of vectors, i.e., to the fields of k-forms, simply as k-forms, without the qualification "*differential*"! Our differential k-forms (i.e., Rudin's) may still be referred to as cotangent k-tensors if one wants to emphasize their transformation properties. In this case, and to avoid confusion, one should not use the term cotangent tensors to refer to multilinear functions of tangent vectors.

In that literature, where a differential k-form is defined as a field of k-forms, our differential k-form is defined as a functional on fields of k-forms [9], also called *current*. In works on exterior differential systems (any partial differential equation or system thereof can be written as an exterior system), our concept of differential k-form is perhaps more pervasive, though the definition of this concept is not always clearly and explicitly made.

The key point is that the action of the derivative on differential k-forms is different from its action on fields of k-forms. It is worth recalling here that the exterior derivative, covariant derivative, and exterior covariant derivatives are, in essence, the same derivative operators acting upon scalar-valued differential k-forms, tangent tensors, and tensor-valued k-forms [10], respectively. We may, therefore, refer to the operator itself as the exterior derivative. In this light, the exterior derivative of dx^i is $ddx^i (= 0)$. The exterior derivative of $\omega^i (= A^i_j dx^j)$ is $A^i_{j,k} dx^k \wedge dx^j$, which, thus, is connection independent.

The foregoing considerations have to be qualified as follows. The "exterior derivative $d\mathbf{e}^i$" is introduced by definition as $d\mathbf{e}^i = -\omega^i_j \mathbf{e}^j$. Notice our use of "exterior derivative $d\mathbf{e}^i$" rather than "'exterior derivative' of $\mathbf{e}^i$" since a function $\mathbf{e}^i$ may not exist such that $d\mathbf{e}^i = -\omega^j_i \mathbf{e}^j$. In other words, the 1-form $d\mathbf{e}^i$ may be non-integrable. In the following, we shall abuse the language and refer to the "exterior derivative" of $\mathbf{e}^i$ since, after all, formal differentiations of expressions containing $\mathbf{e}^i$ as a factor are meaningful. The exterior derivative $d\psi^i$ of the element i of the field (ψ^i) of bases (F^i) of 1-forms is connection dependent as becomes obvious in particular when there is a metric (through the identification of F^i with $\mathbf{e}^i$ and of ψ^i with the corresponding field of $\mathbf{e}^i$). Similarly, the exterior derivative of (fields of) k-forms is connection-dependent, and the exterior derivative of differential k-forms is not. Of course, any ambiguities as to which of these derivatives we are using are removed by the use of bases throughout as, for example, $B^i_{jk} \mathbf{e}_i \psi^j dx^k$. If there is a metric, we could rewrite this as the equivalent object $B^i_{jk} \mathbf{e}_i \otimes \mathbf{e}^j dx^k$.

The derivative of a differential 1-form is connection independent, and the derivative of a field of 1-forms is connection dependent. Notice that the linear approximation of a differential 1-form is evaluated on a vector, not a vector field. Since the "approximating object" is a 1-form, not a field of 1-forms, the concept of exterior derivative does not apply to it: a 1-form does not have an exterior derivative. Fields of 1-forms, like vector fields, have exterior derivatives; here they are called covariant derivatives. So, it is not a question of the differential 1-form having a connection-independent exterior derivative and its linear approximation having a connection-dependent exterior derivative.

4 Mathematical viruses

Differential geometers of the Cartan mold will use the symbol ω^i to represent both a differential 1-form and a field of 1-forms (and similarly for k-forms). In such cases, ω^i represents the differential 1-form most of the time. This abuse of notation (a minor virus) is actually beneficial since shifting to the other meaning of the same symbol without further ado helps to shorten the arguments (when differentiation is not involved!).

A pathological mutant of this virus takes place when a physical magnitude which is a differential form is treated as if it were a tangent tensor field. It will be called the *transmutation virus.* It is highly pathological because it affects the important decision of whether to use Clifford algebras *à* la Kähler or *à* la Hestenes. (The conserved current of the Dirac-Kähler and Kähler-Dirac equations is an $(n-1)$-form, and the conserved current of the Dirac-Hestenes equation is a vector.) This point deserves discussion from a most general perspective: What is the true nature of a physical current? (not to be confused with De Rham's mathematical current).

To discuss this issue, we start by noticing that Equation (3.1) should simply be written as

$$\int \omega = \int \sum a_{i_1 \ldots i_k}(x) dx_{i_1} \wedge \ldots \wedge dx_{i_k}. \tag{4.1}$$

The $\int$ symbol only means the actual evaluation of the ω function, given by Equation (3.1). The use of exterior products makes it unnecessary to use Jacobians when changing coordinates since they automatically appear through those products. For example, the density of charge should be written as $\rho(x^i, t) dx \wedge dy \wedge dz$. Conservation of charge means that, when we integrate (read evaluate) $\rho(x^i, t) dx \wedge dy \wedge dz$ on some domain at constant time, the result is independent of the time chosen. An expression like $\int\int\int \rho(x^i, t) dx \wedge dy \wedge dz$ is, then, called a *Poincaré absolute integral invariant* although we should use this term to refer to the function given by the symbol on the right-hand side of equation (3.1) rather than to its evaluation, represented on the right-hand side of Equation (4.1). It is called *Poincaré* because it is a *constant time integration*, or integration extended to a set of simultaneous points [11]. It is called *absolute* because the integration can be extended to any domain of integration (integral invariants are called *relative* if the integration is independent of domain only for closed domains).

Poincaré's absolute integral invariants can be given a form which gives the right conserved quantity when integrated over domains which are not constituted by simultaneous points. The rule then is: replace dx, dy, dz by $dx - u_x dt$, $dy - u_y dt$, and $dz - u_z dt$ in the expression for the Poincaré integral invariant. The result of this replacement yields

$$\rho(x^i,t)dx \wedge dy \wedge dz - \rho(x^i,t)u_x dt \wedge dy \wedge dz \\ - \rho(x^i,t)u_y dx \wedge dt \wedge dz - \rho(x^i,t)u_z dx \wedge dy \wedge dt.$$

The new integrand is a new function. It is a 3-form with components ρ, $-\rho u_x$, $-\rho u_y$, $-\rho u_z$. When integrated on arbitrary cross sections of the world-hypertube of the given distribution of charge, it gives the same value as the old function did at constant time.

The result just obtained regarding the nature of the current is independent of the particular kinematics. It is valid in Newtonian mechanics if $\rho(x^i,t)dx \wedge dy \wedge dz$ is the density of some conserved scalar quantity. It is certainly the case that in Einstein's relativities one gets a vector field with components $\rho, \rho u_x, \rho u_y, \rho u_z$. This is a special result which does not have the general validity of the current 3-form. Thus, conserved currents are differential forms by nature, not vector fields. This becomes even more obvious in dealing with surface densities in a higher number of dimensions. Since the true nature of the current is an $(n-1)$-form (a "volume" density is given by an $(n-1)$-form in a space of $n-1$ spatial dimensions, plus time), the quantity that is conserved from the Dirac equation must be a (differential) 3-form. This is the case with Kähler-Dirac equations (see, for instance, the very first pages of Kähler's 1961 paper, [2]), and not with the Hestenes version of the Dirac equation. The Hestenes-Dirac equation is infected with the transmutation virus.

Another manifestation of the transmutation virus consists in treating the two indices of the electromagnetic field as (tangent) tensor indices rather than as differential form indices. The cotangent nature of these indices was first pointed out by Cartan [12], who, on the basis of this fact, went on to conclude that Maxwell's equations are independent of the affine connection of spacetime. This conclusion is correct only if one assumes, as Cartan did, that the electromagnetic field is represented by two independent differential forms, namely, $B_i dx^i \wedge dx^k + E_i dx^i \wedge dt$ and $D_i dx^i \wedge dx^k + H_i dx^i \wedge dt$. The two 2-forms are, of course, related, and because of this the argument is faulty. His point remains valid, however, that whether the connection of spacetime enters the generalizations of a system of physical equations depends on the nature of the indices that enter the physical quantities. The related point of whether such generalizations depend on the metric was recently discussed at length by us in Section 4.2 of [13].[1]

Another virus of relevance for this paper, the *bachelor algebra virus*, consists in neglecting one of the two algebras that enter the fundamental

[1]In [13], an error was inadvertently introduced in the reviewing process, where the words " ... two independent 2-forms, one for the fields and one for the sources ... " were introduced in lieu of " ... two independent 2-forms, one for the fields and one for the inductions ... ".

equations of the physics. The two algebras are in essence a tangent and a cotangent algebra (in the sense of cotangent defined above, i.e., pertaining to our differential r-forms). This neglect is a necessity in the standard form of the tensor calculus, where only two series of indices appear in components, these being unaccompanied by the respective bases. The virus, however, infects even equations where bases are used. Take, for instance, the metric tensor field. When obtained from the translation differential form $d\mathbf{P}(=dx^\mu \mathbf{e}_\mu)$, the process goes as follows:

$$ds^2 = d\mathbf{P} \cdot d\mathbf{P} = \mathbf{e}_\mu \cdot \mathbf{e}_\nu dx^\mu dx^\nu = g_{\mu\nu} dx^\mu dx^\nu. \tag{4.2}$$

It is well known, however, that $g_{\mu\nu}dx^\mu dx^\nu$ actually means $g_{\mu\nu}dx^\mu \otimes dx^\nu$. A formulation of Equation (4.2) in a way which shows that two algebras are involved here, as well as their nature, would read as follows:

$$ds^2 = d\mathbf{P}(\otimes, .)d\mathbf{P} = g_{\mu\nu}dx^\mu \otimes dx^\nu, \tag{4.3}$$

where the first product in the parentheses refers to the cotangent algebra and the second product refers to the tangent algebra.

The failure by practitioners to identify and do something about this virus is largely responsible for the neglect of the following fundamental problem: one constructs the so-called energy-momentum tensor from the electromagnetic field. The first is vector-valued and given by contracted products of the second. It is a mystery how the cotangent indices of the electromagnetic field become the value or tangent index of the energy-momentum tensor. If one were to claim (incorrectly) that the indices of the electromagnetic field are value indices, the problem would still persist since one has to generate the cotangent indices of the energy-momentum vector-valued form.

5 Clifford-valued clifforms in spacetime

Kähler introduced a Clifford algebra of differential forms through the condition

$$dx^i \vee dx^k + dx^k \vee dx^i = 2g^{ik}. \tag{5.1}$$

Since the left-hand side of Equation (5.1) is twice the formal symmetric part of $dx^i \vee dx^k$, it is pertinent to write this equation as

$$dx^i \cdot dx^k = g^{ik} \tag{5.2}$$

so that the nature of the Kähler algebra as a Clifford algebra becomes more transparent. This justifies referring to the differential forms of the Kähler calculus as clifforms, meaning, again, that the Grassmann algebra of differential forms is replaced by a Clifford algebra.

Equation (5.2) replaces the action of a certain operator e^i on dx^k. Consequently, many equations of the original Kähler calculus have to be rewritten, though without change in content. Details have been provided elsewhere [14]. Since the Clifford algebra is a quotient algebra of the general tensor algebra, it is immediately obvious that the Kähler calculus of tensor-valued clifforms gives rise to a calculus of Clifford-valued clifforms, based on the well-known equation $\mathbf{e}_i \cdot \mathbf{e}_j = g_{ij}$. The Clifford-valuedness on 4-dimensional vector spaces ends at order four, and the explosion of valuedness of the Dirac equation for vector-valued potentials with tangent-valued differential forms does not take place. A major problem has been solved, but two new problems have been created.

Consider the product $d\mathbf{P}(\vee,\vee)d\mathbf{P}$:

$$\begin{aligned} d\mathbf{P}(\vee,\vee)d\mathbf{P} &= dx^\mu \vee dx^\nu \mathbf{e}_\mu \vee \mathbf{e}_\nu \\ &= dx^\mu \wedge dx^\nu \mathbf{e}_\mu \wedge \mathbf{e}_\nu + 0 + 0 + g_{\mu\nu} g^{\mu\nu} \\ &= dx^\mu \wedge dx^\nu \mathbf{e}_\mu \wedge \mathbf{e}_\nu + 4. \end{aligned} \tag{5.3}$$

Of course, since 4 is a differential 0-form, it is a function of a 0-surface, or point. In a physical context, it would have to be evaluated on any instantly appearing point particle in physical theories incorporating these Clifford-valued clifforms. A second problem lies in that this product is not even able to accommodate the ds^2. The issue is not that one does not obtain Equation (4.3), which will not be possible without a cotangent tensor algebra, but rather that one does not obtain some alternative form of the ds^2.

In standard Clifford algebra, a quadratic form $\mathbf{Q}$ on vectors is given. It defines the square of a vector $\mathbf{a}^2$ as $\mathbf{Q}(\mathbf{a})$, with juxtaposition denoting the Clifford product. It thus follows that

$$\mathbf{a} \cdot \mathbf{b} = \frac{1}{2}[(\mathbf{a}+\mathbf{b})^2 - \mathbf{a}^2 - \mathbf{b}^2]. \tag{5.4}$$

An extension of the concept of Clifford algebra is easily achieved if one defines $\mathbf{a} \cdot \mathbf{b}$ in a way other than by using $\mathbf{a} \cdot \mathbf{a} = \mathbf{a}^2 = \mathbf{Q}(\mathbf{a})$ in Equation (5.4), especially if $\mathbf{a}^2$ is other than a number, when not zero. A requirement, though, is that $\mathbf{a} \cdot \mathbf{b}$ be symmetric, so that the Clifford product of vectors still be expressed as a sum

$$\mathbf{a} \vee \mathbf{b} = \mathbf{a} \wedge \mathbf{b} + \mathbf{a} \cdot \mathbf{b} \tag{5.5}$$

of antisymmetric and symmetric parts. Before doing this, let us justify it.

Consider the quantum mechanical description of a charged particle in an electromagnetic field. This is, by nature, a problem belonging to a dualistic description of the physics, i.e., a description of particles in fields. In a pure Dirac-field description, as the present theory is, particles are in principle absent from the equations representing the system in question. Particles will be specific solutions of the sought Kähler-Dirac equation without a mass term and without external field (other than a fundamental, stochastic background field). These solutions are, therefore, intrinsic to the particles. They have to be extricated in appropriate ways from general solutions of the Kähler-Dirac equation so as to produce a dualistic description. This means the following. The "total" wave function, i.e., wave function representing the pure Dirac-field solution, should be suitably "broken" into parts. If this theory conforms to reality, the standard quantum-mechanical description of a particle in an external field should result in the following. One of the parts should represent the particle's intrinsic wave function. Another part should represent the particles that create the external field of the dualistic description and should be reformulated as such. A last part, the actual wave function of quantum mechanics, is the remnant wave function interfacing the other two parts.

The first part should be dominant only in a very small region around the particle, the latter then being represented just by its global properties (mass, charge, etc). Retrospectively, this first part should contribute the mass term when dealing with the situation that is known as "minimal electromagnetic coupling." In the reformulation of the Kähler-Dirac equation for the physical situation represented by this coupling, the factor a re-emerges as the external electromagnetic potential times the charge of the particle. All these manipulations are, in essence, manipulations with the evaluations, i.e., integrations, of the different inhomogeneous differential forms. Each k-term of these inhomogeneous forms has to be integrated over a k-surface. The aforementioned first part of the pure Dirac-field wave function is to be considered as practically non-null only at the world line of the point particle. This motivates considering the arena of the physics for present purposes as the direct sum, $M^4 \oplus M^1$, of spacetime and the 1-dimensional space that plays the role of the world line of each individual particle. This is precisely the point where we wanted to get as justification for the course of action that follows. The issue now is: what could possibly be the a factor in the Kähler-Dirac equation? The remainder of the paper will take us towards a canonical answer to this question.

6 Clifford-valued clifforms in Kaluza-Klein space

The construction that follows was first obtained as a natural evolution of work done on Finsler geometry [15]. We here present the same results without resorting to this geometry. In this way, the present work can be seen as a discovery of an alternative form of the Finslerian theory of moving frames. In other words, the natural evolution of the theory of clifforms leads us to Finsler geometry as we shall later discuss.

Consider either Minkowski spacetime or, more generally, a 4-manifold M^4 endowed with a pseudo-Riemannian metric of Lorentzian signature and a compatible teleparallel connection. Both options allow us to identify the tangent spaces at different points on M^4, so that they constitute just one vector space V^4. Let $\mathbf{a}_\mu$ be a constant (pseudo)-orthonormal basis of V^4, and let ω^μ be the basis of differential 1-forms dual to $\mathbf{a}_\mu$. Let $\mathbf{u}$ be a unit tangent vector in the tangent space V^1 to a differentiable manifold M^1. Let s be the coordinate on M^1 dual to the unit tangent vector $\mathbf{u}$. On $V^4 \oplus V^1$, let $\mathbf{u}$ be the fifth element $\mathbf{a}_4$ of bases $(\mathbf{a}_A)$, with the indices A, B, ... running from zero to four. Needless to say that $\omega^4 = ds$. The $\mathbf{a}_\mu$ are defined as above.

A translation element $d\wp$ is defined on $M^4 \oplus M^1$ by

$$d\wp = \omega^\mu \mathbf{a}_\mu + ds\, \mathbf{u}.$$

Because of the definitions, the elements g_{AB} are

$$g_{AB} = \mathbf{a}_A \cdot \mathbf{a}_B = \begin{pmatrix} 1 & 0 & 0 & 0 & g_{40} \\ 0 & -1 & 0 & 0 & g_{41} \\ 0 & 0 & -1 & 0 & g_{42} \\ 0 & 0 & 0 & -1 & g_{43} \\ g_{40} & g_{41} & g_{42} & g_{43} & -1 \end{pmatrix}.$$

The $g_{\mu\nu}$ components are constant because of our choice of an orthonormal basis (or frame). The $g_{4\mu}$ are functions of the spacetime coordinates only. The basis $\mathbf{a}_\mu$ is not a canonical basis—i.e., a (pseudo)-orthonormal basis—of the tangent Clifford algebra defined by this dot product.

We shall now complete the orthonormalization to obtain a canonical basis $\mathbf{e}_A$ of the Clifford algebra. Let the function G be defined by

$$G \equiv 1 + (g_{40})^2 - (g_{41})^2 - (g_{42})^2 - (g_{43})^2.$$

One easily shows that, with $\mathbf{e}_\mu \equiv \mathbf{a}_\mu$ and $\mathbf{e}_4$ defined by

$$\mathbf{e}_4 = G^{-\frac{1}{2}}(\mathbf{u} - g_{40}\mathbf{e}_0 + g_{4i}\mathbf{e}_i),$$

one obtains a canonical basis.

We postulate

$$\omega^A \cdot \omega^B = 0, \quad \text{for} \quad A \neq B, \tag{6.1}$$

so that the double dot product of $d\wp$ with itself becomes

$$d\wp(\cdot,\cdot)d\wp = \omega^A \cdot \omega^B \mathbf{a}_A \cdot \mathbf{a}_B = \omega^0 \cdot \omega^0 - \omega^1 \cdot \omega^1 - \ldots - ds \cdot ds, \tag{6.2}$$

where we have used Equations (5.4) and (5.5). We further postulate

$$d\wp(\cdot\,,\,\cdot)d\wp = 0,$$

which yields

$$ds \cdot ds = \omega^0 \cdot \omega^0 - \omega^1 \cdot \omega^1 - \omega^2 \cdot \omega^2 - \omega^3 \cdot \omega^3. \tag{6.3}$$

With c equal to one, Equation (6.3) justifies referring to ds and $\mathbf{u}$, respectively, as the differential of proper time and as the 4-velocity even if the latter is not contained in spacetime (i.e., it is not a linear combination of just the four $\mathbf{a}_\mu$, but also of $d\mathbf{u}$). The basis of 1-forms dual to the $(\mathbf{e}_A)$ is readily obtained by means of

$$d\wp = \mathbf{a}_A \omega^A = \mathbf{e}_A \omega'^A. \tag{6.4}$$

The result is

$$\omega'^\mu = \omega^\mu + g_4{}^\mu ds, \quad \omega'^4 = G^{\frac{1}{2}} ds \tag{6.5}$$

with $g_4{}^0$ and $g_4{}^i$ defined as g_{40} and $-g_{4i}$, respectively. It is important to notice that, whereas $\omega^A \cdot \omega^B$ is zero for $A \neq B$, the same is not the case for $\omega'^A \cdot \omega'^B$.

Equation (6.2) relates the distance ds on the trajectories of particles to the structure of the spacetime manifold. We still need to define what the functions that constitute the different terms on the right-hand side of Equation (6.2) are, which we will now do. Given the spacetime differential 1-forms α and β, we define $\alpha \cdot \beta$ as a real map on a set of spacetime curves. The evaluation of $\alpha \cdot \beta$ on a curve γ is denoted as $(\alpha \cdot \beta)_\gamma$ and is defined by $(\alpha \cdot \beta)_\gamma = [(\sqrt{(\alpha \cdot \beta)}_\gamma]^2$, with $[\sqrt{(\alpha \cdot \beta)}]_\gamma$ in turn defined by

$$(\sqrt{\alpha \cdot \beta})_\gamma = \int_{\lambda_1}^{\lambda_2} \sqrt{\frac{\alpha \cdot \beta}{d\lambda^2}} d\lambda$$

and where, obviously, the symbol $\alpha \cdot \beta / d\lambda^2$ denotes $f_\mu g_\nu (dx^\mu / d\lambda)(dx^\nu / d\lambda)$ with $\alpha / d\lambda$ equal to $f_\mu (dx^\mu / d\lambda)$. Because of Equation (6.1), we need only be concerned with what $\omega^0 \cdot \omega^0, \omega^1 \cdot \omega^1, \omega^2 \cdot \omega^2$ and $\omega^3 \cdot \omega^3$ are. They are determined in the process of reducing the spacetime metric in terms of the coordinates $g'_{\mu\nu} dx^\mu dx^\nu$ to the form

$$\omega^0 \otimes \omega^0 - \omega^1 \otimes \omega^1 - \omega^2 \otimes \omega^2 - \omega^3 \otimes \omega^3,$$

also written as $\omega^0 \omega^0 - \omega^1 \omega^1 - \omega^2 \omega^2 - \omega^3 \omega^3$. The latter also has become $\omega^0 \cdot \omega^0 - \omega^1 \cdot \omega^1 - \omega^2 \cdot \omega^2 - \omega^3 \cdot \omega^3$.

To summarize, the double Clifford algebra is defined by the fields $g_{4\mu}$ and the functions $\omega^0 \cdot \omega^0, \omega^1 \cdot \omega^1, \omega^2 \cdot \omega^2$ and $\omega^3 \cdot \omega^3$ of spacetime curves. These are in turn given by the $g'_{\mu\nu}$ which, like the $g'_{4\mu}s$, are dot products of elements of bases of tangent vectors. The problem of representing the metric without tensor products, which appeared not to have a solution in spacetime, has a solution in Kaluza-Klein space. This has been done in Equation (6.3), obtained with the help of the equation $d\wp(\cdot,\cdot)d\wp = 0$. This equation also implies that

$$d\wp(\vee, \vee)d\wp = d\wp(\wedge, \wedge)d\wp = \omega^A \wedge \omega^B \mathbf{a}_A \wedge \mathbf{a}_B. \tag{6.6}$$

Since there is now no constant term on the right-hand side of Equation (6.6), the other problem that we faced with the Kähler algebra of Clifford-valued clifforms in spacetime (see discussion of Equation (5.3) is not present either.

7 The canonical connection of Kaluza-Klein space

The space that we have just constructed admits a *canonical connection.* It was derived elsewhere in (5.5). We summarize the results. With ω_{AB} defined by

$$\omega_{AB} \equiv \omega_A{}^C g_{CB}, \tag{7.1}$$

the condition of metric compatibility becomes the standard equation

$$\omega_{AB} + \omega_{BA} = dg_{AB} \tag{7.2}$$

which, together with the requirement that

$$\omega_\rho^\sigma = 0, \tag{7.3}$$

implies that all the ω_{AB} and $\omega_A{}^C$ are zero except for

$$\omega_{4\rho} = dg_{4\rho}, \quad \omega_4^\rho = \eta_\rho(\omega_{4\rho} - g_{4\rho}\omega_4^4), \quad \omega_4^4 = G^{-1}\eta_\rho g_{4\rho} dg_{4\rho}, \tag{7.4}$$

where η_ρ is $(+1, -1, -1, -1)$ and provided that all $g_{4\mu}$ are assumed to be different from zero. The mismatch of indices in the last equations and other equations to follow has to do with the fact that, in obtaining them, we have expressed the g^{AB} explicitly in terms of the g_{CD}. It should be obvious that there is no summation over repeated indices in the middle equation (7.4), which can also be written as

$$\omega_4^\rho = -G^{-1}\eta_\rho\eta_\phi g_{4\rho} g_{4\phi} dg_{4\phi}, \tag{7.5}$$

where there is obviously summation over ϕ but not over ρ. From the fact that there are only four linearly independent 1-forms, $dg_{4\rho}$, one can expect that ω_4^4 is a linear combination of the ω_4^ρ, namely,

$$\omega_4^4 = \omega_4^\rho g_{4\rho} \tag{7.6}$$

which results from

$$0 = \omega_{44} = \omega_4^A g_{A4}. \tag{7.7}$$

To tie this together, I must make the following statement. In obtaining the connection in (7.4), special "tricks" were used to quickly arrive at the result. The "tricks" were dependent on the assumption that all the $g_{4\rho}$ were different from zero (the "tricks" become cumbersome, otherwise). The result obtained, however, is simply the solution of a linear system in a large number of variables, the ω_A^B and ω_{AB}. Once the solution has been obtained in the case stated, that is the solution. If any of the $g_{4\rho}$ are zero, one need only substitute this in the solution obtained for the case where all of them are different from zero.

We use the opportunity to call attention to a typo in [15]. The components g^{4A} of the inverse matrix were given as $(G)^{-1}\eta_A g_{4A}$ with $\eta_A = (1,-1,-1,-1,-1)$. This is incorrect. For the given equation to be correct, one has to take $\eta_A = (1,-1,-1,-1,+1)$. Notice that this η_A is not the diagonal of the orthonormalized metric, which is $(1,-1,-1,-1,-1)$.

Finally, our connection is the canonical connection of $M^4 \oplus M^1$ when M^4 is Minkowski spacetime itself. The torsion of the structure, which is $d(d\mathbf{P})$, still makes mathematical sense as a difference between the changes of $d\wp$ along two different paths between two nearby points of $M^4 \oplus M^1$. Everything has a clear interpretation because s and $\mathbf{u}$ are geometric objects.

8 The interior derivative for the canonical connection

We proceed to apply the rule for obtaining Kähler's covariant derivative in the form that applies to spaces with torsion [14]. (As explained in this reference, Kähler worked out the case of the Levi-Civita connection only.) After describing the result obtained in [14], we adapt the result to the Kaluza-Klein (K-K) structure.

The interior derivative of an n-form is an $(n-1)$-form. Hence, the interior derivative of a 0-form, and of a vector field in particular, is zero. Let α be a scalar-valued 1-form, $\alpha = a_\mu \omega^\mu$, where (ω^μ) is some basis of 1-forms. We have

$$\begin{aligned} d\alpha &= a_{\mu,\nu}\omega^\mu \wedge \omega^\nu + a_\rho(\Omega^\rho + \omega^\mu \wedge \omega_\mu^\rho) \\ &= \omega^\nu \wedge [a_{\mu;\nu}\,\omega^\mu + a_\rho(R^\rho_{\nu\mu}\omega^\mu - \Gamma^\rho_{\mu\nu}\omega^\mu)] \\ &= \omega^\nu \wedge d_\nu(a_\mu\omega^\mu), \end{aligned} \tag{8.1}$$

where $d_\nu(a_\mu\omega^\mu)$ is defined as the expression in the square bracket in Equation (8.1). Following Kähler, the interior derivative is obtained by dot product: $\delta\alpha = \omega^\nu \cdot d_\nu(a_\mu\omega^\mu)$. Since $\omega^\nu \cdot \omega^\mu$ is symmetric and $R^\rho_{\nu\mu}$ is antisymmetric in (ν,μ), the torsion term in $d_\nu\alpha$ does not contribute to the interior derivative. One readily gets $\delta\alpha = a^\mu{}_{;\mu}$ (where ";" is the symbol for the usual covariant derivative). The expression for the interior derivative of a 1-form goes in most of the literature by the name of divergence of a vector field. Notice that the interior derivative of a 1-form does not depend explicitly on the torsion, but only implicitly through the connection. Consider now a 2-form, $\beta = \alpha_{\mu\nu}(\omega^\mu \wedge \omega^\nu)$, where the parentheses around $\omega^\mu \wedge \omega^\nu$ indicates that the summation over repeated indices extends only to $\mu < \nu$. We rewrite this as $\frac{1}{2}a_{\mu\nu}\omega^\mu \wedge \omega^\nu$, with summation over all μ and ν. Easy calculations yield

$$d\beta = \frac{1}{2}a_{\mu\nu;\eta}\omega^\eta \wedge \omega^\nu \wedge \omega^\mu + a_{\mu\nu}\omega^\eta \wedge R^\mu_{\eta\rho}\omega^\rho \wedge \omega^\nu. \tag{8.2}$$

Hence,

$$d_\eta\beta = \frac{1}{2}a_{\mu\nu;\eta}\omega^\nu \wedge \omega^\mu \ + a_{\mu\nu}R^\mu_{\eta\rho}\omega^\rho \wedge \omega^\nu, \tag{8.3}$$

and, therefore,

$$\delta\beta = \omega^\eta \cdot d_\eta\beta = a_{\nu\mu;}{}^\mu\omega^\nu + 2a_{\mu\nu}R^{\mu\rho}_\rho\omega^\nu. \tag{8.4}$$

The factor "2" in the last term has to do with our defining the components of the torsion Ω through $\Omega = R^\mu_{\eta\kappa}\omega^\eta \wedge \omega^\kappa \mathbf{e}_\mu$ rather than through $\Omega = R'^\mu_{\eta\kappa}\omega^\eta \wedge \omega^\kappa \mathbf{e}_\mu$, which is $\frac{1}{2}R'^\mu_{\eta\kappa}\omega^\kappa \wedge \omega^\nu \mathbf{e}_\mu$. In other words, the factor disappears in terms of the $R'^\mu_{\eta\kappa}$ components.

Let us proceed to explain the general technique for obtaining interior covariant derivatives in K-K space with canonical connection. Since the physics cannot depend on the history along the path, nothing depends on the coordinate s. Furthermore, since the ω^ν are linear combinations of the dx^μ, but not of ds, a linear combination of the dx^μ is a linear combination of the ω^ν also. We shall thus have

$$df = f_{,B}\omega^B = f_{,\mu}\omega^\mu, \tag{8.5}$$

where a quantity "f" with a subscript " , " denotes the coefficients with respect to some basis of 1-forms of the differential of "f". This, of course, contains the usual meaning where these components are the partial derivatives if the basis of 1-forms is constituted by the differentials of the coordinates. Notice that since df is a linear combination of the ω^μ, it is a linear combination of the ω'^μ *and* the ds, as implied by Equation (6.3).

We shall need the torsion of the K-K space, which we shall denote as Ω'. Formal differentiation of $d\wp$ yields

$$\Omega' = d(d\mathbf{P} + ds\mathbf{u}) = \Omega - ds \wedge d\mathbf{u}, \tag{8.6}$$

where $\Omega = d\omega^\mu \mathbf{a}_\mu$ is the spacetime torsion expressed in terms of constant frame fields.

We proceed to obtain the interior derivative of a 1-form. We now have

$$d\alpha = a_{A,\mu}\omega^\mu \wedge \omega^A + a_B(\Omega'^B + \omega^C \wedge \omega^B_C) = a_{A,\mu}\omega^\mu \wedge \omega^A + a_\nu\Omega^\nu. \tag{8.7}$$

For the first equality, we have used Equation (8.5) and the first equation of structure. For the second step, we have used Equation (8.7). In terms of a constant frame field, $\Omega^\mu = d\omega^\mu = R^\mu_{\kappa\nu}\omega^\kappa \wedge \omega^\nu$. Hence,

$$d\alpha = \omega^\nu \wedge [a_{A,\nu}\omega^A + a_\rho R^\rho_{\nu\mu}\omega^\mu] \tag{8.8}$$

and

$$d_B\alpha = a_{A,\mu}\omega^A + a_\rho R^\rho_{\mu\nu}\omega^\nu, \tag{8.9}$$

where the subscript μ on the right corresponds to the subscript B on the left. To facilitate the next step, let us rewrite $d_B\alpha$ as $a_{A,B}\omega^A + a_\rho R^\rho_{B\nu}\omega^\nu$, with the understanding that we later set $a_{A,4}$ and $R^\rho_{4\nu}$ equal to zero if need be. The interior derivative is then given by

$$\begin{aligned}
\delta\alpha &= \omega^B \cdot d_B\alpha \\
&= a_{A,B}\omega^B \cdot \omega^A + a_\rho R^\rho_{B,\nu}\omega^B \cdot \omega^\nu \\
&= a_{B,B}\omega^B \cdot \omega^B + a_\rho R^\rho_{\nu\nu}\omega^\nu \cdot \omega^\nu \\
&= a_{0,0}\omega^0 \cdot \omega^0 + a_{1,1}\omega^1 \cdot \omega^1 + a_{2,2}\omega^2 \cdot \omega^2 + a_{3,3}\omega^3 \cdot \omega^3.
\end{aligned} \tag{8.10}$$

One should notice that we have obtained the coefficients $a_{\mu,\nu}$, calculating with constant frame fields. Hence, these coefficients can also be written as $a_{\mu;\nu}$. It should be obvious that, in terms of general bases, $a_{\mu,\nu}$ would have to be replaced with $a_{\mu;\nu}$. The same ideas that we have implemented here to obtain the interior derivative of 1-forms can easily be extended to r-forms.

9 Closing remarks

On the basis of the tenets of the theory of integral invariants, which are exterior differential forms, and the difference existing between exterior differential forms and fields of exterior forms, we have concluded that Hestenes' theory of Dirac equations is infected by viruses, denoted as the *transmutation* and *bachelor algebra* viruses. Although Kähler's theory of Dirac equations is not similarly afflicted, it is deficient in other respects. These deficiencies are, however, easily corrected after recognizing that the introduction of a metric on a manifold M^4 endowed with an affine structure is equivalent to the attaching to it of a manifold M^1 representing curves on M^4. The metric then specifies how arbitrary differential curves on the manifold stretch out over "the measuring rod" M^1.

Our K-K space shares with the standard K-K space the independence of functions on the fifth coordinate. Given the meaning of s, it is the only option that makes sense in our case. The usual K-K space, on the other hand, is not a direct sum of manifolds but is rather a 5-dimensional

Riemannian manifold, therefore, differentiable. Hence, the non-dependence on the fifth coordinate has to be introduced by hand in the standard K-K theory, but not in ours.

Another difference is that, in our K-K space, the fifth dimension is not some additional dimension which happens to curl over a very small distance. The fifth dimension is simply the distance on curves. It is present at ordinary scales, just like time was present as a fourth coordinate before we learned to see it in this light in 1905. It simply happens that the difference as a coordinate between s and the spacetime coordinates is far deeper than the difference between the space and time coordinates themselves. Still another difference is that we do not have a metric in five dimensions which would be a 5-dimensional version of the $ds^2 = g_{\mu\nu}dx^\mu dx^\nu$. The fifth dimension is spanned by s itself. We, however, have a dot product in the 5-dimensional tangent vector space, which gives rise to $g_{AB} = \mathbf{a}_A \cdot \mathbf{a}_B$ and which is instrumental in getting the traditional spacetime metric as in Equation (6.3).

The most important difference, perhaps, is that the connection is canonical and yet is not the Levi-Civita connection of the 5-dimensional metric. It is canonical. If we assume Equation (7.3), which expresses spacetime teleparallelism, the connection is just determined by metric compatibility, Equation (7.3). Yet, this connection is not the Levi-Civita connection because, as one can easily verify, the equation

$$d\omega^A = \omega^B \wedge \omega_B^A \tag{9.1}$$

that defines the Levi-Civita connection is not satisfied by our connection. This should also be obvious from Equation (8.6) as the K-K torsion is not zero even if the spacetime torsion is. Notice, finally, that our canonical connection does not depend on the coefficients $g'_{\mu\nu}$ of the spacetime metric $g'_{\mu\nu}dx^\mu dx^\nu$.

Once we have a canonical connection, the Kähler derivative, or sum of the exterior and interior (covariant) derivatives, is defined. In other words, we have the left-hand side of the Kähler-Dirac equation. An important point now is that the differential invariants that characterize the K-K space can be connected with a structure of classical differential geometry. When such contact is made, one can in principle still extend further the Kähler theory of Dirac equations, as we now explain.

The differential invariants of our Kaluza-Klein structure are given by $(\omega^\mu, \omega_4^\lambda)$, which define $d\mathbf{P}$ and $d\mathbf{u}$, since $d\mathbf{P} = \omega^\mu \mathbf{a}_\mu$ and $d\mathbf{u} = \omega_4^\lambda \mathbf{a}_\lambda + \omega_4^4 \mathbf{u}$, the ω_4^4 being a linear combination of the ω_4^λ. But these differential invariants also are, in essence, the differential invariants that define the base space of the Finsler bundle (ω^μ, ω_0^i), where we have $d\mathbf{P} = \omega^\mu \mathbf{a}_\mu$ and $d\mathbf{u} = \omega_0^i \mathbf{a}_i$ [16]. Hence, our K-K space, which was obtained without resort to Finsler geometry, may be viewed as a reformulation of this geometry after the appropriate translations between Finslerian equations and K-K equations have been performed. Notice that although the canonical connection of our K-K space was motivated by Finslerian considerations [15], the actual construction did not involve any Finsler geometry.

The realization of this connection of Finsler geometry with this brand of K-K theory can now be used in the following way. It can be shown that the different pieces of the torsion in the Finsler bundle can be put in correspondence with the different non-gravitational interactions [17]. If this is an indication of some deeper truth, the potential A in the Dirac-Kähler equation should be replaced by the potential of the torsion, which is $d\wp$ in the K-K space. Still more enticing is the possibility that one may only need to replace a with $d\wp$ in the Kähler-Dirac equation to give

$$\partial\psi = d\wp \vee \psi. \tag{9.2}$$

Like the derivative operator on the left-hand side of this equation, the $d\wp$ and its derivatives depend only on the differential invariants that define the K-K structure, equivalently the horizontal space of the Finslerian structure. Hence, Equation (9.2) is determined by the structure. It remains to be seen whether this equation is a canonical invariant equation or whether, to be so, the right-hand side has to be multiplied by some factor (say, a factor whose square is minus one). A discussion of these issues has been given in [16].

To conclude, the inner dynamics of the Kähler theory of Dirac equations has led us to a few possibilities for fully geometric "Dirac equations," and, potentially, to a canonical option. Given the elegance and intrinsic relevance of anything geometric and canonical, it would be foolish to neglect what this offspring of the Kähler theory of Dirac equations has to offer to the program of unification.

References

[1] R. Debever, ed., *Elie Cartan-Albert Einstein, Letters on Absolute Parallelism,* Princeton University Press, Princeton, 1979.

[2] E. Kähler, Innerer und äußerer differentialkalkül, *Abh. Dtsch. Akad. Wiss. Berlin, Kl. für Math. Phys. Tech.* No. **4** (1960); Die Dirac-gleichung, *Abh. Dtsch. Akad. Wiss. Berlin, Kl. für Math. Phys. Tech.* No. **1** (1961); Der innere differentialkalkül, *Rendiconti di Matematica* **21** (1962), 3–4, 425.

[3] D. Hestenes, *Space-Time Algebra,* Gordon and Breach, 1966.

[4] Z. Oziewicz, in the Foreword of *Spinors, Twistors, Clifford Algebras and Quantum Deformations,* Kluwer Academic Publishers, Dordrecht, 1993.

[5] D. Hestenes, Mathematical viruses, in *Clifford Algebras and their Applications in Mathematical Physics,* A. Micali, R. Boudet, and J. Helmstetter, eds., Kluwer, Dordrecht, 1992, 3.

[6] A. Dimakis and F. Müller-Hoissen, Clifform calculus with applications to classical field theories, *Class. Quantum Grav.* **8** (1991), 2093.

[7] D. Hestenes, in Appendix of Clifford algebra and the interpretation of quantum mechanics, in *Clifford Algebras and Their Applications in Mathematical Physics,* J. S. R. Chisholm and A. K. Common, eds., Reidel Publishing Company, Dordrecht, 1986, 3, 321–346.

[8] W. Rudin, *Principles of Mathematical Analysis,* McGraw Hill, New York, 1976.

[9] G. De Rham, *Variétées Différentiables,* Hermann, Paris, 1960.

[10] J. G. Vargas and D. G. Torr, Conservation of vector-valued forms and the question of existence of gravitational energy-momentum tensors in general relativity, *Gen. Rel. Grav.* **23** (1991), 713.

[11] É. Cartan, *Leçons sur les Invariants Intégraux,* Hermann, Paris, 1922, Chapter 3.

[12] É. Cartan, Sur les variétés à connexion affine et la théorie de la relativité généralisée, *Ann. Ec. Normale* **41 1** (1924). Reprinted in *Oeuvres Completes, Vol. III, Part 2,* Éditions du CNRS, Paris, 1984.

[13] J. G. Vargas and D. G. Torr, The Cartan-Einstein unification with teleparallelism and the discrepant measurements of Newton's gravitational constant G, *Found. Phys.* **29** (1999), 145.

[14] J. G. Vargas and D. G. Torr, Teleparallel Kähler calculus for spacetime, *Found. Phys.* **28** (1998), 931.

[15] J. G. Vargas and D. G. Torr, The emergence of Kaluza-Klein microgeometry from the invariants of optimally Euclidean Lorenzian spaces, *Found Phys.* **27** (1997), 533.

[16] J. G. Vargas and D. G. Torr, The marriage of Clifford algebra and Finsler geometry, preprint, 1999.

[17] J. G. Vargas and D. G. Torr, The theory of acceleration within its context of differential invariants: the root of the problem with cosmological models?, *Found. Phys.* **29** (1999), 1543.

Jose G. Vargas
Center for Science Education and
Department of Physics and Astronomy
University of South Carolina
Columbia, SC 29208
E-mail: Vargas@mail.psc.sc.edu

Douglas G. Torr
Department of Physics and Astronomy
University of South Carolina
Columbia, SC 29208
E-mail: dtorr@mindspring.com

Received: September 27, 1999; Revised: March 20, 2000

3.

GEOMETRY AND LOGIC

The Principle of Duality in Clifford Algebra and Projective Geometry

Oliver Conradt

ABSTRACT A completely dual approach to Clifford algebra is presented in this article. It leads to the introduction of two new products, the dual geometric product $*$ and the dual inner product $\circ$, and sheds new light on the duality relation between the progressive and regressive outer products of the Clifford algebra. On the firm base of this dual approach, the projective principle of duality is formulated in this completeness for the first time in the language of Clifford algebra. In order to provide mathematical concepts which are close to possible applications in theoretical physics, section 5 is devoted to projective coordinate systems. The incidence relations between the primitive geometric forms, the linear complex, and Desargues' theorem are discussed as well.

Keywords: Principle of duality, projective geometry, Clifford algebra, projective coordinate systems, incidence relations, dual Clifford product, Hodge duality.

1 Introduction

The present article originates in the quest of the role played by the projective principle of duality in physics. Projective geometry and its principle of duality was applied to mechanics in the 19th century by several mathematicians and physicists ([11]). One of the first among them was the French geometer M. Chasles who considered the principle of duality as an important instrument not only in geometry but also in mechanics: "L'application des mêmes idées de dualité peut s'étendre à la Mécanique. En effet, l'élément primitif des corps auquel on applique d'abord les premiers principes de cette science, est, comme dans la Géométrie ancienne, le *point* mathématique. Ne sommes-nous pas autorisés à penser, maintenant, qu'en prenant le *plan* pour l'élément de l'étendue, et non plus le *point*, on sera conduit à d'autres doctrines, faisant pour ainsi dire une nouvelle science? Et s'il existe un principe unique pour passer de cette science à l'ancienne, comme le théorème de Géométrie qui établit la corrélation des propriétés de l'étendue figurée, ce principe sera la base d'une *dualité* semblable, dans la science du mouvement des corps." ([11], p. 21) One of the

AMS Subject Classification: 15A66, 51A25, 08A99.

most advanced approaches to 'geometric mechanics'—a notion coined by R. Ziegler in [11] for the application of methods from projective geometry to mechanics—certainly, is the dissertation of F. Lindemann [6].

Nevertheless, the significance of the principle of duality in modern physics, especially in relativistic quantum mechanics, has not yet been worked out. Since Clifford algebra is an excellent language for projective geometry and physics as well, it was necessary to find the equivalent formulation of the principle of duality in Clifford algebra. In this article we limit our intention to the formulation of the projective principle of duality in Clifford algebra. It will serve as the mathematical backbone in further investigations.

The Clifford algebra was introduced by the English mathematician W. K. Clifford (1878) in his analysis of Grassmann's papers. W. K. Clifford called the new object "geometric algebra." In addition, Clifford algebra appears to be an ideal tool for the formulation of geometric concepts. For this reason we will use the term "geometric algebra" instead of "Clifford algebra" from now on throughout this article.

After a short introduction to the concepts of geometric algebra in section 2, we describe the completely dual approach to geometric algebra in section 3. Section 4 introduces the primitive geometric forms of synthetic projective geometry. Since coordinate systems are important tools in physics, we present in section 5 projective coordinate systems in different dimensions. Special emphasis is put on the dual construction of the projective coordinate systems. Section 6 is devoted to the incidence relations between the primitive geometric forms of section 4. The linear complex and the operations "intersection" and "connection" are discussed as well. On the base of the completely dual approach, section 7 provides the formulation of the principle of duality in the language of geometric algebra. The principle of duality is applied to Desargues' theorem.

2 Geometric Algebra

In this section we review some definitions and notations of geometric algebra. For a thorough introduction to the topic, we refer to [2].

A generic element of a geometric algebra is called a *multivector*. Throughout this article uppercase letters $A, B, \ldots$ and lowercase letters $a, b, \ldots$ denote multivectors, and lowercase Greek letters $\alpha, \beta, \ldots$ denote real numbers. A *real geometric algebra* $\mathcal{G}_n$ is obtained by viewing the total of all multivectors as a linear space over the reals (with dimension 2^n) endowed with a *geometric product* defined by the following properties holding for all multivectors.

(2.1a)–(2.1c) (see below) require the geometric algebra to be associative and distributive with respect to addition. (2.1d) guarantees a one-element for the geometric product, and (2.1e) expresses the commutativity of the scalar multiplication:

$$(AB)C = A(BC), \tag{2.1a}$$
$$A(B+C) = AB + AC, \tag{2.1b}$$
$$(B+C)A = BA + CA, \tag{2.1c}$$
$$1A = A, \tag{2.1d}$$
$$\lambda A = A\lambda. \tag{2.1e}$$

A real geometric algebra $\mathcal{G}_n$ is generated through continued geometric multiplication out of the vectors of a real n-dimensional vector space $\mathcal{V}$ ($\neq \mathcal{G}_n$). The *contraction rule* distinguishes the geometric algebra from a mere associative algebra by requiring the square of any vector of $\mathcal{V}$ to be a *scalar*,

$$A^2 = \lambda 1, \quad \forall A \in \mathcal{V}, \tag{2.2}$$

i.e., a multivector proportional to 1.

The geometric product of any two homogeneous multivectors $A_{\bar{r}}$ and $B_{\bar{s}}$ decomposes into a sum,

$$A_{\bar{r}}B_{\bar{s}} = \langle A_{\bar{r}}B_{\bar{s}}\rangle_{|r-s|} + \langle A_{\bar{r}}B_{\bar{s}}\rangle_{|r-s|+2} + \ldots + \langle A_{\bar{r}}B_{\bar{s}}\rangle_{\mathcal{D}_n(r+s)}$$
$$= \sum_{k=0}^{m} \langle A_{\bar{r}}B_{\bar{s}}\rangle_{|r-s|+2k}, \tag{2.3}$$

of k-vectors with $m := 2^{-1}(\mathcal{D}_n(r+s) - |r-s|)$ and the index function

$$\mathcal{D}_n(i) = \begin{cases} i, & 0 \leq i \leq n \\ 2n - i, & n \leq i \leq 2n. \end{cases} \tag{2.4}$$

From the geometric product we will define two more products. The *inner product*

$$A_{\bar{r}} \cdot B_{\bar{s}} := \langle A_{\bar{r}}B_{\bar{s}}\rangle_{|r-s|}, \tag{2.5}$$

and the *outer product*

$$A_{\bar{r}} \wedge B_{\bar{s}} := \langle A_{\bar{r}}B_{\bar{s}}\rangle_{r+s}, \quad r+s \leq n, \tag{2.6a}$$

for any two homogeneous multivectors $A_{\bar{r}}$ and $B_{\bar{s}}$. In the case $r+s > n$ the outer product vanishes,

$$A_{\bar{r}} \wedge B_{\bar{s}} = 0, \quad r+s > n \tag{2.6b}$$

because the multivectors contain linearly dependent factors. In case of 1-vectors ($r = s = 1$), the geometric product decomposes into a sum of an inner and an outer product

$$AB = A \cdot B + A \wedge B, \quad \forall A, B \in \mathcal{V}. \tag{2.7}$$

	Basis elements	Direct subspace	Dim.
scalar	1	$\mathcal{G}^0_{p,q}$	$\binom{n}{0}$
vector	$P_1, P_2, P_3, \ldots, P_n$	$\mathcal{G}^1_{p,q} \equiv \mathcal{V}$	$\binom{n}{1}$
bivector	$P_iP_j,\ \forall i<j$	$\mathcal{G}^2_{p,q}$	$\binom{n}{2}$
trivector	$P_iP_jP_k,\ \forall i<j<k$	$\mathcal{G}^3_{p,q}$	$\binom{n}{3}$
$\vdots$	$\vdots$	$\vdots$	$\vdots$
k-vector	$P_{i_1}P_{i_2}\cdots P_{i_k},\ \forall i_1<i_2<\ldots<i_k$	$\mathcal{G}^k_{p,q}$	$\binom{n}{k}$
$\vdots$	$\vdots$	$\vdots$	$\vdots$
$(n-1)$-vector	$P_{i_1}P_{i_2}\cdots P_{i_{n-1}},\ \forall i_1<i_2<\ldots<i_{n-1}$	$\mathcal{G}^{n-1}_{p,q}$	$\binom{n}{n-1}$
pseudoscalar	$\mathrm{I} := P_1P_2\cdots P_{n-1}P_n$	$\mathcal{G}^n_{p,q}$	$\binom{n}{n}$

TABLE 1.1. Basis of the geometric algebra $\mathcal{G}_{p,q}$.

The total of all k-vectors forms a linear space denoted by $\mathcal{G}^k_n$. In particular the one-dimensional linear space of all scalars is denoted by $\mathcal{G}^0_n$, the n-dimensional space of all vectors by $\mathcal{G}^1_n \equiv \mathcal{V}$, and the one-dimensional linear space of all pseudoscalars by $\mathcal{G}^n_n$.

To get the dimension of any linear subspace $\mathcal{G}^k_n$, we explicitly choose a basis in $\mathcal{G}^1_n$. For a non-degenerate geometric algebra there is always a basis P_i, $i = 1, \ldots, n = p+q$, where the inner products take the values

$$P_i \cdot P_j = \eta_{ij}1, \tag{2.8}$$

with

$$\eta_{ij} = \begin{cases} 0, & \text{if } i \neq j, \\ 1, & \text{if } 1 \leq i \leq p, \\ -1, & \text{if } p+1 \leq i \leq p+q = n. \end{cases} \tag{2.9}$$

The geometric algebra $\mathcal{G}_{p,q} \equiv \mathcal{G}_n$ is said to have a *signature* (p,q). Following the notation for the whole algebra, the linear subspaces $\mathcal{G}^k_n$ are equivalently denoted by $\mathcal{G}^k_{p,q}$. A basis for the whole geometric algebra is shown in Table 1.1. $M^\dagger$ denotes the *reverse* of the multivector M.

3 A completely dual approach to Geometric Algebra

So far the approach to the geometric algebra $\mathcal{G}_{p,q}$ was based on the vectors P_i of the vector space $\mathcal{G}^1_{p,q}$, thus, generating the whole algebra through

I	$:=$	$P_1P_2\cdots P_{n-1}P_n$	I^{-1}	$=$	$(-1)^{\frac{n(n-1)}{2}}(-1)^q\mathrm{I}$
$\mathrm{I}^\dagger$	$=$	$(-1)^{\frac{n(n-1)}{2}}\mathrm{I}$	I^2	$=$	$[\mathrm{I}^{-1}]^2 = [\mathrm{I}^\dagger]^2 = (-1)^{\frac{n(n-1)}{2}}(-1)^q 1$
$P_i\mathrm{I}$	$=$	$(-1)^{n-1}\mathrm{I}P_i$	$A_{\bar{k}}\mathrm{I}$	$=$	$(-1)^{k(n-1)}\mathrm{I}A_{\bar{k}}$

TABLE 1.2. Properties of the unit pseudoscalar I.

continued geometric multiplication. From this development grew a space $\mathcal{G}_{p,q}$, the subspaces $\mathcal{G}^k_{p,q}$ of which are "left-right" symmetrically distributed with regard to the dimensions.

$$\dim(\mathcal{G}^k_{p,q}) = \binom{n}{k} = \binom{n}{n-k} = \dim(\mathcal{G}^{n-k}_{p,q}) \tag{3.1}$$

In the light of this symmetry the vectors P_i lose their central role, and it is obvious that one would try to generate the *same* geometric algebra $\mathcal{G}_{p,q}$ from the $(n-1)$-vectors of the n-dimensional linear space $\mathcal{G}^{n-1}_{p,q}$ as well. This second approach will be implemented in this section.

3.1 Duality

The geometric product of any homogeneous multivector $A_{\bar{r}}$ with any pseudoscalar $N = \mu\mathrm{I}$ is a homogeneous multivector

$$B = \langle B\rangle_{n-r} = A_{\bar{r}}N = \mu A_{\bar{r}}\mathrm{I} \tag{3.2}$$

of grade $(n-r)$. Thus, geometric multiplication with a pseudoscalar mediates between the left-right symmetric subspaces $\mathcal{G}^r_{p,q}$ and $\mathcal{G}^{n-r}_{p,q}$. Following D. Hestenes and R. Ziegler [3] we define the *dual* $\tilde{A}$ of a multivector A by geometric multiplication from the right side with the unit-n-blade I^{-1}.

$$\tilde{A} := A\mathrm{I}^{-1} = A\cdot\mathrm{I}^{-1} \tag{3.3}$$

Table 1.2 lists frequently used properties of the unit pseudoscalar I.

3.2 Dual geometric product

With a commutative diagram of mappings, we introduce a new $*$-*product*,

$$\begin{array}{ccc}
\mathcal{G}_{p,q}\otimes\mathcal{G}_{p,q} & \text{geometric product} & \mathcal{G}_{p,q} \\
(A,B) & \longrightarrow & AB \\
\Big\downarrow \mathrm{I}^{-1}\otimes\mathrm{I}^{-1} & & \Big\downarrow \mathrm{I}^{-1} \\
\mathcal{G}_{p,q}\otimes\mathcal{G}_{p,q} & *\text{-product} & \mathcal{G}_{p,q} \\
(\tilde{A},\tilde{B}) & \longrightarrow & \tilde{A}*\tilde{B} := (AB)^\sim
\end{array} \tag{3.4}$$

which translates into the given geometric product according to

$$\tilde{A} * \tilde{B} = (AB)^{\sim}, \tag{3.5a}$$

$$A * B = (\tilde{A}\tilde{B})^{\sim}. \tag{3.5b}$$

The $*$-product is associative,

$$\begin{aligned}(A * B) * C &= (\tilde{A}\tilde{B})^{\sim} * C = [\mathrm{I}^2(\tilde{A}\tilde{B})\tilde{C}]^{\sim} \\ &= [\mathrm{I}^2\tilde{A}(\tilde{B}\tilde{C})]^{\sim} = A * (\tilde{B}\tilde{C})^{\sim} \\ &= A * (B * C),\end{aligned} \tag{3.6}$$

distributive with respect to addition,

$$\begin{aligned}A * (B + C) &= [\tilde{A}(\tilde{B} + \tilde{C})]^{\sim} = (\tilde{A}\tilde{B})^{\sim} + (\tilde{A}\tilde{C})^{\sim} \\ &= A * B + A * C,\end{aligned} \tag{3.7a}$$

$$\begin{aligned}(B + C) * A &= [(\tilde{B} + \tilde{C})\tilde{A}]^{\sim} = (\tilde{B}\tilde{A})^{\sim} + (\tilde{C}\tilde{A})^{\sim} \\ &= B * A + C * A,\end{aligned} \tag{3.7b}$$

and the pseudoscalar $\tilde{\mathrm{I}} = \mathrm{I}^{-1}$ plays the role of the one-element

$$\tilde{\mathrm{I}} * A = \mathrm{I}^{-1} * A = \mathrm{I}^4 A = A. \tag{3.8}$$

In addition, there is a dual contraction rule with respect to (2.2). The square of any $(n-1)$-vector A is a pseudoscalar,

$$A * A = \tilde{A}^2\mathrm{I}^{-1} = \lambda\mathrm{I}^{-1} = \lambda\tilde{\mathrm{I}}, \tag{3.9}$$

i.e., a multivector proportional to the one-element of the $*$-product. Thus, the $*$-product satisfies the axioms (2.1) and (2.2) and represents a new geometric product.

The original geometric product and the *dual geometric product* $*$ act in the same geometric algebra $\mathcal{G}_{p,q}$ (viewed as a linear space). The first one generates the geometric algebra from the vectors of the vector space $\mathcal{G}^1_{p,q}$, and the second one does so from the $(n-1)$-vectors of the linear space $\mathcal{G}^{n-1}_{p,q}$; see Table 1.3. Thus, we have shown how to generate the same geometric algebra by continued dual geometric multiplication starting with the homogeneous multivectors of the n-dimensional linear space $\mathcal{G}^{n-1}_{p,q}$.

3.3 One space with two aspects – two spaces for one aspect

Left-right symmetry demands another consequence. If the geometric algebra $\mathcal{G}_{p,q}$ is developed from the vectors P_i^- (Table 1.3) as well as from the vectors P_i (Table 1.1), it should also be possible to adjust the grade of the homogeneous multivectors to the actual approach in the first case. For this reason we introduce a *plus-minus notation*. It will distinguish between the left and the right approach to the geometric algebra. From now on we supply the vectors of the vector space $\mathcal{G}^1_{p,q}$ with an extra plus sign, $A^+_{\mathrm{I}} \equiv A_{\mathrm{I}}$, as

Basis elements	**Direct subspace**	**Dim.**
$1^- := \tilde{1} = \mathrm{I}^{-1}$	$\mathcal{G}^{0-}_{p,q} \equiv \mathcal{G}^{n}_{p,q}$	$\binom{n}{0}$
$P_i^- := \tilde{P}_i = P_i\mathrm{I}^{-1}, \quad \forall i \in \{1,2,\ldots,n\}$ $= \mathrm{I}^2(-1)^{i-1}\eta_{ii}P_1\cdots P_{i-1}P_{i+1}\cdots P_n$	$\mathcal{G}^{1-}_{p,q} \equiv \mathcal{G}^{n-1}_{p,q}$	$\binom{n}{1}$
$P_i^- * P_j^- = \tilde{P}_i * \tilde{P}_j = P_iP_j\mathrm{I}^{-1}, \ \forall i < j$ $= \mathrm{I}^2(-1)^{(i+j)-2}\eta_{ii}\eta_{jj}P_1\cdots P_{i-1}P_{i+1}\cdots P_{j-1}P_{j+1}\cdots P_n$	$\mathcal{G}^{2-}_{p,q} \equiv \mathcal{G}^{n-2}_{p,q}$	$\binom{n}{2}$
$P_i^- * P_j^- * P_k^- = \quad \forall i < j < k$ $= \tilde{P}_i * \tilde{P}_j * \tilde{P}_k$ $= (P_iP_j\mathrm{I}^{-1}) * (P_k\mathrm{I}^{-1})$ $= P_iP_jP_k\mathrm{I}^{-1}$ $= \mathrm{I}^2(-1)^{(i+j+k)-3}\eta_{ii}\eta_{jj}\eta_{kk}P_1\cdots P_{i-1}\times$ $P_{i+1}\cdots P_{j-1}P_{j+1}\cdots P_{k-1}P_{k+1}\cdots P_n$	$\mathcal{G}^{3-}_{p,q} \equiv \mathcal{G}^{n-3}_{p,q}$	$\binom{n}{3}$
$\vdots$	$\vdots$	$\vdots$
$P_{i_1}^- * P_{i_2}^- * \cdots * P_{i_k}^- = \quad \forall i_1 < i_2 < \ldots < i_k$ $= \tilde{P}_{i_1} * \tilde{P}_{i_2} * \cdots * \tilde{P}_{i_{k-1}} * \tilde{P}_{i_k}$ $= (P_{i_1}P_{i_2}\cdots P_{i_{k-1}}\mathrm{I}^{-1}) * (P_{i_k}\mathrm{I}^{-1})$ $= P_{i_1}P_{i_2}\cdots P_{i_{k-1}}P_{i_k}\mathrm{I}^{-1}$ $= \mathrm{I}^2(-1)^{\sum_{l=1}^{k}(i_l-1)}(\prod_{l=1}^{k}\eta_{i_li_l})P_1\cdots P_{i_1-1}P_{i_1+1}\cdots\cdots P_{i_k-1}P_{i_k+1}\cdots P_n$	$\mathcal{G}^{k-}_{p,q} \equiv \mathcal{G}^{n-k}_{p,q}$	$\binom{n}{k}$
$\vdots$	$\vdots$	$\vdots$
$P_{i_1}^- * P_{i_2}^- * \cdots * P_{i_{n-1}}^- = \quad \forall i_1 < i_2 < \ldots < i_{n-1}$ $= \tilde{P}_{i_1} * \tilde{P}_{i_2} * \cdots * \tilde{P}_{i_{n-1}}$ $= P_{i_1}P_{i_2}\cdots P_{i_{n-2}}P_{i_{n-1}}\mathrm{I}^{-1}$	$\mathcal{G}^{(n-1)-}_{p,q} \equiv \mathcal{G}^{1}_{p,q}$	$\binom{n}{n-1}$
$\mathrm{I}^- := P_1^- * P_2^- * \cdots * P_{n-1}^- * P_n^-$ $= P_1P_2\cdots P_{n-1}P_n\mathrm{I}^{-1}$ $= 1$	$\mathcal{G}^{n-}_{p,q} \equiv \mathcal{G}^{0}_{p,q}$	$\binom{n}{n}$

TABLE 1.3. Basis of the geometric algebra $\mathcal{G}^-_{p,q}$. The new "minus" basis elements are translated into the basis of Table 1.1.

well as the vector space itself, $\mathcal{G}^{1+}_{p,q} \equiv \mathcal{G}^1_{p,q}$. The same applies to any k-vector $A_{\bar{k}}^+ \equiv A_{\bar{k}}$, multivector $A^+ \equiv A$, subspace $\mathcal{G}^{k+}_{p,q} \equiv \mathcal{G}^k_{p,q}$, and the geometric algebra $\mathcal{G}^+_{p,q} \equiv \mathcal{G}_{p,q}$. A plus sign indicates that the geometric algebra $\mathcal{G}^+_{p,q}$ is generated by the vectors P_i^+ of the vector space $\mathcal{G}^{1+}_{p,q}$. Equivalently, a minus sign points out that the geometric algebra $\mathcal{G}^-_{p,q}$ is generated from the vectors P_i^- of the vector space $\mathcal{G}^{1-}_{p,q}$. If the geometric algebras $\mathcal{G}^+_{p,q}$ and $\mathcal{G}^-_{p,q}$ are considered as *linear spaces*, there is no difference between them, i.e., we may speak of *one space*,

$$\mathcal{G}^+_{p,q} \equiv \mathcal{G}_{p,q} \equiv \mathcal{G}^-_{p,q}. \tag{3.10}$$

The linear space $\mathcal{G}_{p,q}$ is endowed with two geometric products denoted by juxtaposition and $*$. If the geometric product denoted by the juxtaposition is taken as the original geometric product, the linear space $\mathcal{G}_{p,q}$ turns into the geometric algebra $\mathcal{G}^+_{p,q}$. If the $*$-product represents the original geometric product, $\mathcal{G}_{p,q}$ turns into the geometric algebra $\mathcal{G}^-_{p,q}$. Thus, the geometric algebras $\mathcal{G}^+_{p,q}$ and $\mathcal{G}^-_{p,q}$ represent *two* different *aspects* of the same linear space $\mathcal{G}_{p,q}$.

By virtue of the isomorphisms ($\psi^- = (\psi^+)^{-1}$)

$$\begin{aligned} \psi^+ : \quad \mathcal{G}^-_{p,q} &\longrightarrow \mathcal{G}^+_{p,q} \qquad (3.11a)\\ A^-_{\bar{k}} &\longmapsto \langle A^-_{\bar{k}}\rangle^+_{n-k}, \\ \psi^- : \quad \mathcal{G}^+_{p,q} &\longrightarrow \mathcal{G}^-_{p,q} \qquad (3.11b)\\ A^+_{\bar{k}} &\longmapsto \langle A^+_{\bar{k}}\rangle^-_{n-k}, \end{aligned}$$

the geometric algebra $\mathcal{G}^+_{p,q}$ contains $\mathcal{G}^-_{p,q}$ (3.11a), and the geometric algebra $\mathcal{G}^-_{p,q}$ contains $\mathcal{G}^+_{p,q}$ (3.11b). Each basis element, homogeneous multivector, and subspace of the linear space $\mathcal{G}_{p,q}$ appears in two different grades.

$$\langle P^-_i\rangle^+_{n-1} \equiv P^-_i \tag{3.12a}$$

$$P^+_i \equiv \langle P^+_i\rangle^-_{n-1} \tag{3.12b}$$

$$A^+_{\bar{k}} \equiv \langle A^+_{\bar{k}}\rangle^-_{n-k} \tag{3.13a}$$

$$\langle A^-_{\bar{k}}\rangle^+_{n-k} \equiv A^-_{\bar{k}} \tag{3.13b}$$

$$\mathcal{G}^{k+}_{p,q} \equiv \mathcal{G}^{(n-k)-}_{p,q} \tag{3.14a}$$

$$\mathcal{G}^{(n-k)+}_{p,q} \equiv \mathcal{G}^{k-}_{p,q} \tag{3.14b}$$

On the other hand we may stress the fact that the geometric algebras $\mathcal{G}^+_{p,q}$ and $\mathcal{G}^-_{p,q}$ are equal regarding their form and their construction, i.e., we may speak of *one aspect*, namely,

$$\mathcal{G}^+_{p,q} \simeq \mathcal{G}^-_{p,q}. \tag{3.15}$$

Thus, the structure of geometric algebra is realized in *two spaces* $\mathcal{G}^+_{p,q}$ and $\mathcal{G}^-_{p,q}$. By virtue of the isomorphisms

$$\begin{aligned} \phi^+ : \quad \mathcal{G}^-_{p,q} &\longrightarrow \mathcal{G}^+_{p,q} \qquad (3.16a)\\ A^- &\longmapsto \langle A^- * (\mathrm{I}^-)^{-1}\rangle^+, \end{aligned}$$

$$\begin{aligned} \phi^- : \quad \mathcal{G}^+_{p,q} &\longrightarrow \mathcal{G}^-_{p,q} \qquad (3.16b)\\ A^+ &\longmapsto \langle A^+(\mathrm{I}^+)^{-1}\rangle^-, \end{aligned}$$

each grade is represented by two different homogeneous multivectors and subspaces:

$$A_{\bar{k}}^{+} \simeq \langle A_{\bar{k}}^{+}(\mathrm{I}^{+})^{-1}\rangle_{\bar{k}}^{-} \tag{3.17a}$$

$$\langle A_{\bar{k}}^{-} * (\mathrm{I}^{-})^{-1}\rangle_{\bar{k}}^{+} \simeq A_{\bar{k}}^{-} \tag{3.17b}$$

$$\mathcal{G}_{p,q}^{k+} \simeq \mathcal{G}_{p,q}^{k-}. \tag{3.18}$$

We will not always apply the plus-minus notation rigorously. If plus and minus signs are absent, the explanations and statements hold for *both* signs. For instance, the basis elements and subspaces of Table 1.1 can be supplied with a plus as well as with a minus sign. Both cases are correct. All the concepts introduced so far hold rigorously for both aspects.

Explicit computation of the square and the inverse of the pseudoscalar I^{-},

$$(\mathrm{I}^{-})^2 = \mathrm{I}^{-} * \mathrm{I}^{-} = (-1)^{\frac{n(n-1)}{2}}(-1)^q 1^{-}, \tag{3.19}$$

$$(\mathrm{I}^{-})^{-1} = (-1)^{\frac{n(n-1)}{2}}(-1)^q \mathrm{I}^{-}, \tag{3.20}$$

confirm the equivalence to the corresponding formulas in Table 1.2. Duality in the geometric algebra $\mathcal{G}_{p,q}^{-}$ complies with the definition (3.3) in the geometric algebra $\mathcal{G}_{p,q}^{+}$,

$$A * (\mathrm{I}^{-})^{-1} = \tilde{A}, \tag{3.21}$$

and for the translation of the original geometric product in $\mathcal{G}_{p,q}^{+}$ into the original geometric product in $\mathcal{G}_{p,q}^{-}$, we obtain

$$\tilde{A}\tilde{B} = (\mathrm{I}^{-})^2 (A * B)^{\sim}, \tag{3.22a}$$

$$AB = (\mathrm{I}^{-})^2 (\tilde{A} * \tilde{B})^{\sim}, \tag{3.22b}$$

thus, providing the inverse relations of (3.5).

3.4 Inner and outer products

In the geometric algebra $\mathcal{G}_{p,q}^{+}$ as well as in the geometric algebra $\mathcal{G}_{p,q}^{-}$, there is a geometric product, an outer product, and an inner product. To distinguish between these *six* different products we introduce, in addition to the geometric product, a new notation for the inner and outer products in the geometric algebra $\mathcal{G}_{p,q}^{-}$; see Table 1.4. According to (2.5) and (2.6), the inner and outer products in the geometric algebra $\mathcal{G}_{p,q}^{-}$ are defined by

$$A_{\bar{r}}^{-} \circ B_{\bar{s}}^{-} = \langle A_{\bar{r}}^{-} * B_{\bar{s}}^{-}\rangle_{\overline{|r-s|}}^{-} \tag{3.23}$$

and

$$A_{\bar{r}}^{-} \vee B_{\bar{s}}^{-} = \langle A_{\bar{r}}^{-} * B_{\bar{s}}^{-}\rangle_{\overline{r+s}}^{-}, \quad r+s \leq n, \tag{3.24a}$$

$$A_{\bar{r}}^{-} \vee B_{\bar{s}}^{-} = 0, \quad r+s > n, \tag{3.24b}$$

	Geometric product	Inner product	Outer product
$\mathcal{G}^+_{p,q}$	(juxtaposition)	$\cdot$	$\wedge$
$\mathcal{G}^-_{p,q}$	$*$	$\circ$	$\vee$

TABLE 1.4. Notation of the products in $\mathcal{G}^+_{p,q}$ and $\mathcal{G}^-_{p,q}$.

because the homogeneous multivectors $A^-_{\bar{r}}$ and $B^-_{\bar{s}}$ contain linearly dependent factors in this case. Dual to (2.3), the geometric product $*$ between two homogeneous multivectors is decomposed into a sum of homogeneous multivectors,

$$\begin{aligned} A^-_{\bar{r}} * B^-_{\bar{s}} &= \langle A^-_{\bar{r}} * B^-_{\bar{s}} \rangle^-_{\overline{|r-s|}} + \langle A^-_{\bar{r}} * B^-_{\bar{s}} \rangle^-_{\overline{|r-s|+2}} + \dots \\ &\qquad \dots + \langle A^-_{\bar{r}} * B^-_{\bar{s}} \rangle^-_{\overline{\mathcal{D}_n(r+s)}} \\ &= \sum_{k=0}^{m} \langle A^-_{\bar{r}} * B^-_{\bar{s}} \rangle^-_{\overline{|r-s|+2k}}, \end{aligned} \tag{3.25}$$

with $m := 2^{-1}(\mathcal{D}_n(r+s) - \mid r-s \mid)$ and the index function (2.4). In the case $r = s = 1$ we obtain the dual expression to (2.7).

$$A^-_{\bar{1}} * B^-_{\bar{1}} = A^-_{\bar{1}} \circ B^-_{\bar{1}} + A^-_{\bar{1}} \vee B^-_{\bar{1}} \tag{3.26}$$

The inner products in $\mathcal{G}^-_{p,q}$ between two basis elements P^-_i,

$$\begin{aligned} P^-_i \circ P^-_j &= \langle P^-_i * P^-_j \rangle^-_{\bar{0}} = \langle P^-_i (\mathrm{I}^+)^{-1} P^-_j (\mathrm{I}^+)^{-1} (\mathrm{I}^+)^{-1} \rangle^+_{\bar{n}} \\ &= \langle P^+_i P^+_j \rangle^+_{\bar{0}} (\mathrm{I}^+)^{-1} = P^+_i \cdot P^+_j (\mathrm{I}^+)^{-1} \\ &= \eta_{ij} 1^-, \end{aligned} \tag{3.27}$$

are in exact correspondence to (2.8). From the translation law for the geometric products (3.5) we get the duality of the outer products of $\mathcal{G}^+_{p,q}$ and $\mathcal{G}^-_{p,q}$,

$$\begin{aligned} (A^+_{\bar{r}})^\sim \vee (B^+_{\bar{s}})^\sim &= \langle (A^+_{\bar{r}})^\sim \rangle^-_{\bar{r}} \vee \langle (B^+_{\bar{s}})^\sim \rangle^-_{\bar{s}} \\ &= \langle (A^+_{\bar{r}})^\sim * (B^+_{\bar{s}})^\sim \rangle^-_{\overline{r+s}} \\ &= \langle A^+_{\bar{r}} B^+_{\bar{s}} \mathrm{I}^{-1} \rangle^+_{\overline{n-(r+s)}} \\ &= \langle A^+_{\bar{r}} B^+_{\bar{s}} \rangle^+_{\overline{r+s}} \mathrm{I}^{-1} \\ &= (A^+_{\bar{r}} \wedge B^+_{\bar{s}})^\sim, \end{aligned} \tag{3.28}$$

with $r+s \leq n$. For the complementary grades $r+s \geq n$ the equation (3.28) holds trivially. Thus, for generic multivectors A and B we have

$$\tilde{A} \vee \tilde{B} = (A \wedge B)^\sim, \tag{3.29a}$$

$$A \vee B = (\tilde{A} \wedge \tilde{B})^\sim, \tag{3.29b}$$

and equivalently from the translation law (3.22)

$$\tilde{A} \wedge \tilde{B} = (\mathrm{I}^-)^2 (A \vee B)^\sim, \tag{3.30a}$$

$$A \wedge B = (\mathrm{I}^-)^2 (\tilde{A} \vee \tilde{B})^\sim. \tag{3.30b}$$

The relation (3.29) is a natural consequence of the completely dual approach and expresses the duality of the outer products $\wedge$ and $\vee$. In the usual one-sided construction of the geometric algebra, the relation (3.29) is taken as the definition for the product $\vee$, and one may then show that it also has the structure of an outer product, see for example [2, 3]. The latter step is unnecessary in our approach, thus, already indicating the power of the dual approach.

Another new result is the duality relation between the inner products of $\mathcal{G}^+_{p,q}$ and $\mathcal{G}^-_{p,q}$,

$$\begin{aligned}(A^+_{\bar{r}})^\sim \circ (B^+_{\bar{s}})^\sim &= \langle (A^+_{\bar{r}})^\sim \rangle^-_r \circ \langle (B^+_{\bar{s}})^\sim \rangle^-_s \\ &= \langle (A^+_{\bar{r}})^\sim * (B^+_{\bar{s}})^\sim \rangle^-_{|r-s|} \\ &= \langle A^+_{\bar{r}} B^+_{\bar{s}} \mathrm{I}^{-1} \rangle^+_{n-|r-s|} \\ &= \langle A^+_{\bar{r}} B^+_{\bar{s}} \rangle^+_{|r-s|} \mathrm{I}^{-1} \\ &= (A^+_{\bar{r}} \cdot B^+_{\bar{s}})^\sim .\end{aligned} \tag{3.31}$$

Thus, for generic multivectors A and B we have

$$\tilde{A} \circ \tilde{B} = (A \cdot B)^\sim, \quad A \circ B = (\tilde{A} \cdot \tilde{B})^\sim, \tag{3.32}$$

and equivalently

$$\tilde{A} \cdot \tilde{B} = (\mathrm{I}^-)^2 (A \circ B)^\sim, \quad A \cdot B = (\mathrm{I}^-)^2 (\tilde{A} \circ \tilde{B})^\sim. \tag{3.33}$$

The inner and outer products of $\mathcal{G}^+_{p,q}$ and separately of $\mathcal{G}^-_{p,q}$ are connected by the *dual conjugation*:

$$\begin{aligned}(A^+_{\bar{r}} \wedge B^+_{\bar{s}})^\sim &= \langle A^+_{\bar{r}} B^+_{\bar{s}} \rangle_{r+s} \mathrm{I}^{-1} \\ &= \langle A^+_{\bar{r}} B^+_{\bar{s}} \mathrm{I}^{-1} \rangle_{n-(r+s)} \\ &= (-1)^{s(n-1)} \langle A^+_{\bar{r}} \mathrm{I}^{-1} B^+_{\bar{s}} \rangle_{n-(r+s)} \\ &= \begin{cases} A^+_{\bar{r}} \cdot (B^+_{\bar{s}})^\sim, & r+s \le n \\ (-1)^{s(n-1)} (A^+_{\bar{r}})^\sim \cdot B^+_{\bar{s}}, & r+s \le n \end{cases}\end{aligned} \tag{3.34}$$

$$(A_{\bar{r}}^{-} \vee B_{\bar{s}}^{-})^{\sim} = \begin{cases} A_{\bar{r}}^{-} \circ (B_{\bar{s}}^{-})^{\sim}, & r+s \leq n \\ (-1)^{s(n-1)}(A_{\bar{r}}^{-})^{\sim} \circ B_{\bar{s}}^{-}, & r+s \leq n \end{cases} \tag{3.35}$$

$$\begin{aligned} (A_{\bar{r}}^{+} \cdot B_{\bar{s}}^{+})^{\sim} &= \langle A_{\bar{r}}^{+} B_{\bar{s}}^{+} \rangle_{|r-s|} \mathrm{I}^{-1} \\ &= \langle A_{\bar{r}}^{+} B_{\bar{s}}^{+} \mathrm{I}^{-1} \rangle_{n-|r-s|} \\ &= (-1)^{s(n-1)} \langle A_{\bar{r}}^{+} \mathrm{I}^{-1} B_{\bar{s}}^{+} \rangle_{n-|r-s|} \\ &= \begin{cases} A_{\bar{r}}^{+} \wedge (B_{\bar{s}}^{+})^{\sim}, & r \leq s \\ (-1)^{s(n-1)}(A_{\bar{r}}^{+})^{\sim} \wedge B_{\bar{s}}^{+}, & r \geq s \end{cases} \end{aligned} \tag{3.36}$$

$$(A_{\bar{r}}^{-} \circ B_{\bar{s}}^{-})^{\sim} = \begin{cases} A_{\bar{r}}^{-} \vee (B_{\bar{s}}^{-})^{\sim}, & r \leq s \\ (-1)^{s(n-1)}(A_{\bar{r}}^{-})^{\sim} \vee B_{\bar{s}}^{-} & r \geq s. \end{cases} \tag{3.37}$$

In the geometric algebra $\mathcal{G}_{p,q}^{+}$, the outer product $\wedge$ between two homogeneous multivectors $A_{\bar{r}}^{+}$ and $B_{\bar{s}}^{+}$ is a vector $A_{\bar{r}}^{+} \wedge B_{\bar{s}}^{+}$ of grade $r+s$ $(r, s \leq r+s \leq n)$. It is therefore said to be *progressive.* The outer product $A_{\bar{r}}^{+} \vee B_{\bar{s}}^{+}$ forms a homogeneous multivector of grade $r+s-n$ $(r+s-n \leq r, s \leq n)$ and is thus *regressive.* From the point of view of the geometric algebra $\mathcal{G}_{p,q}^{-}$ the outer product $\vee$ is progressive and $\wedge$ is regressive. The specifying terms for the six different products are compiled in Table 1.5.

	(juxtaposition)	$*$	$\cdot$	$\circ$	$\wedge$	$\vee$
$\mathcal{G}_{p,q}^{+}$	original	dual	original	dual	progressive	regressive
$\mathcal{G}_{p,q}^{-}$	dual	original	dual	original	regressive	progressive

TABLE 1.5. Specifying terms for geometric, inner, and outer products depending on the chosen aspect.

4 Primitive geometric forms

There are three different elements in the geometry of space: *point*, *line*, and *plane*. Synthetic projective geometry treats these elements equally, i.e., on an elementary level point, line, and plane are thought of as *pure* elements. A pure point is thought to exist without lines and planes passing through it, a pure line is thought to exist without points lying in it and planes passing through it, and a pure plane is thought to exist without lines and points lying in it. Thus, a pure element doesn't enter into an incidence relation with either of the two other elements. The equal handling of the elements has far reaching consequences for geometry. A first one is that the set of all pure planes, the three dimensional *space of planes*, opens up an alternative illustration for a three dimensional manifold besides the three

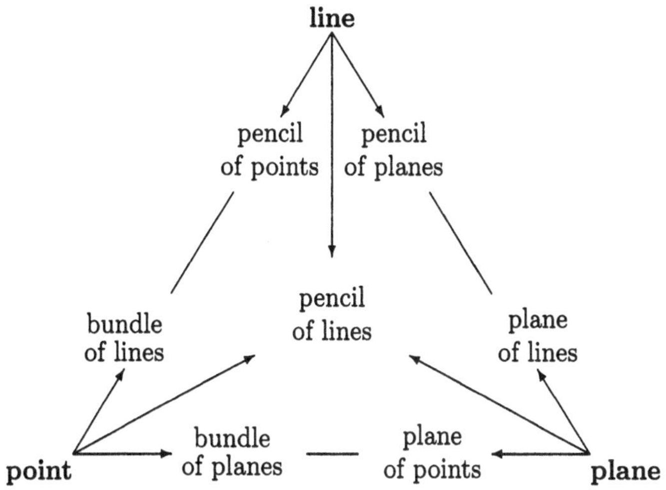

FIGURE 1. The incidence relations between the pure elements.

dimensional *space of points*, the set of all pure points. The set of all pure lines, the *space of lines*, forms a four dimensional manifold.

Having rejected all the incidence relations between the elements, we reintroduce them again systematically. There are infinitely many points lying in and infinitely many planes passing through a line. The line appears under the aspect "point" as a *pencil of points* (the set of all points lying in the line) and under the aspect "plane" as a *pencil of planes* (the set of all planes passing through the line). There are infinitely many lines and infinitely many points lying in a plane. The plane appears under the aspect "line" as a *plane of lines* (the set of all lines lying in the plane) and under the aspect "point" as a *plane of points* (the set of all points lying in the plane). There are infinitely many lines and infinitely many planes passing through a point. The point appears under the aspect "line" as a *bundle of lines* (the set of all lines passing through the point) and under the aspect "plane" as a *bundle of planes* (the set of all planes passing through the point). There are infinitely many lines lying in the plane and passing through the point of an incident plane point pair. The plane point pair appears under the aspect "line" as a *pencil of lines* (the set of all lines which are at once on the same point and the same plane). A complete overview over the incidence relations between the pure elements is given in the diagram of Figure 1. The seven incidence relations, together with the space of planes, the space of points, and the space of lines, constitute the ten *primitive geometric forms* listed in Table 1.6 with their dimensions.

Dim.	Primitive geometric form	Proj. space
-	point, line, plane	-
1	pencil of points, pencil of lines, pencil of planes	P^1
2	{ planar field: plane of points, plane of lines centric bundle: bundle of planes, bundle of lines }	P^2
3	space: space of points, space of planes	P^3
4	space of lines	-

TABLE 1.6. Primitive geometric forms of synthetic projective geometry.

5 The order of pure elements in primitive geometric forms. Projective coordinate systems

The order of pure elements can be treated within synthetic projective geometry as shown by O. Veblen and J. W. Young (chapter I and II of volume II in [10]) and by L. Locher (chapter I in [7] and p. 37ff of [8]). We will follow another direction and establish one-to-one correspondences between primitive geometric forms and geometric algebras. These correspondences lead to projective coordinate systems. The order of the pure elements then is reflected by the order of the homogeneous multivectors. We will not consider the order relations in geometric algebra here. This section aims only at developing projective coordinate systems in primitive geometric forms. As a natural product from the introduction of a projective coordinate system in space, the linear complex, a generalized line in space, will be included into the set of geometric objects illustrating homogeneous multivectors.

The projective interpretation for homogeneous multivectors developed by D. Hestenes and R. Ziegler in section 3 of [3] is unique to within a scale factor. Thus, a real geometric algebra $\mathcal{G}_{n+1}$ of dimension 2^{n+1} contains the real projective space P^n. We take this as a prerequisite to our introduction of projective coordinate systems. Special emphasis is put on duality, thereby yielding two dual coordinate systems in every dimension.

Before we begin a detailed development, a warning is in order: A projective coordinate system *does not determine the signature* (or the metric) in the corresponding primitive geometric form. It *only reflects the order* of the pure elements. The choice of signature remains still open after introducing a projective coordinate system.

5.1 Pencils

The pencil of points, the pencil of lines, and the pencil of planes are three different examples for a one-dimensional projective space. Each of them is represented algebraically by the four-(= 2^2)-dimensional geometric algebra $\mathcal{G}_2$. The homogeneous multivectors, together with their geometric interpre-

$\mathcal{G}_2^+$	Pencil of points	Pencil of lines	Pencil of planes	$\mathcal{G}_2^-$
$\rho 1^+$				$\rho \mathrm{I}^-$
$\left.\begin{array}{l}\rho A_{\bar{1}}^+ = \mu_1^+ P_1^+ + \\ \quad + \mu_2^+ P_2^+\end{array}\right\}$	point	line	plane	$\left\{\begin{array}{l}\rho A_{\bar{1}}^- = \mu_1^- P_1^- + \\ \quad + \mu_2^- P_2^-\end{array}\right.$
$\rho \mathrm{I}^+$				$\rho 1^-$

TABLE 1.7. The geometric interpretation for a 1-vector of $\mathcal{G}_2^+$ and $\mathcal{G}_2^-$ depends on the chosen type of pencil.

tation, are shown in Table 1.7. According to Table 1.3, the translation of the basis elements of the geometric algebra $\mathcal{G}_2^+$ into the basis elements of the geometric algebra $\mathcal{G}_2^-$ is given by

$$1^+ = \mathrm{I}^-, \tag{5.1a}$$

$$\{\eta_{11}P_1^+, \eta_{22}P_2^+\} = \{P_2^-, -P_1^-\}, \tag{5.1b}$$

$$\eta_{11}\eta_{22}\mathrm{I}^+ = -1^-. \tag{5.1c}$$

A projective coordinate system in a pencil is a continuous one-to-one correspondence between the elements of the pencil and the 1-vectors of the geometric algebra $\mathcal{G}_2$. It is well defined by any choice of the fundamental elements 0, 1, and ∞; see, for example, chapter V in [5] or chapter VI of volume I in [10]. We may choose any three elements and assign to them the *homogeneous* coordinates $\rho(0,1)$, $\rho(1,1)$, and $\rho(1,0)$ in any order. Thus, the fundamental element $\rho(0,1)$ is represented by the second basis vector ρP_2, the fundamental element $\rho(1,1)$ by the 1-vector $\rho(P_1 + P_2)$, and the fundamental element $\rho(1,0)$ by the first basis vector ρP_1.

Any 1-vector $\rho A_{\bar{1}} = \mu_1 P_1 + \mu_2 P_2$ belongs to the element with homogeneous coordinates $\rho(\mu_1, \mu_2)$ or the projective coordinate $\frac{\mu_1}{\mu_2}$.

This manner of introducing a coordinate system applies in both approaches to the geometric algebra $\mathcal{G}_2$. In the first case the 1-vectors and coordinates are supplied with plus signs, in the second case with minus signs. The connection between these dual coordinate systems is given by the equations (5.1). From

$$\mu_1^+ P_1^+ + \mu_2^+ P_2^+ = \mu_1^- P_1^- + \mu_2^- P_2^- \tag{5.2}$$

we obtain

$$\mu^- := \frac{\mu_1^-}{\mu_2^-} = -\frac{\eta_{22}}{\eta_{11}}\frac{\mu_2^+}{\mu_1^+} =: \frac{\eta_{22}}{\eta_{11}}\frac{1}{\mu^+}. \tag{5.3}$$

Thus, the projective coordinate of a fixed element in one coordinate system equals the positive or negative reciprocal of the coordinate in the dual

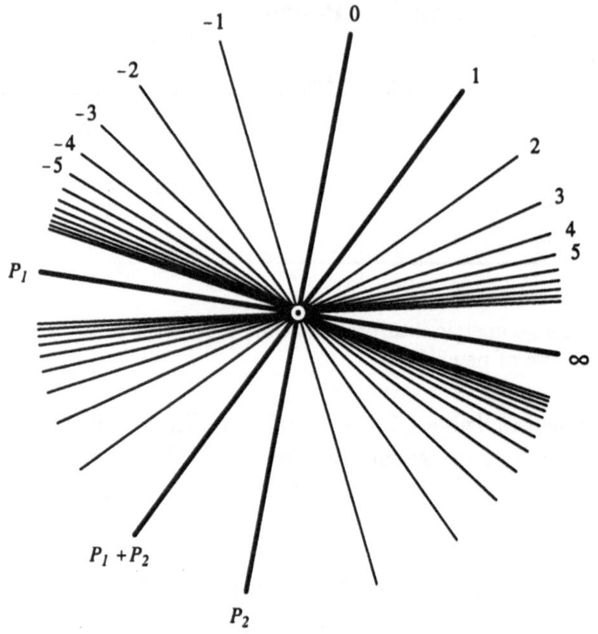

FIGURE 2. Projective line coordinates of a pencil of lines.

coordinate system. The sign of the reciprocal depends on the signature of the geometric algebra $\mathcal{G}_2$. It is positive for $\eta_{11} = -\eta_{22}$ and negative for $\eta_{11} = \eta_{22}$.

Pencil of lines

In a pencil of lines, a generic multivector M of $\mathcal{G}_2$ decomposes into two different types of numbers and a line.

$$M = \langle \text{scalar} \rangle_0 + \langle \text{line} \rangle_1 + \langle \text{pseudoscalar} \rangle_2 \tag{5.4}$$

Figure 2 shows an example for the coordinatization of a pencil of lines.

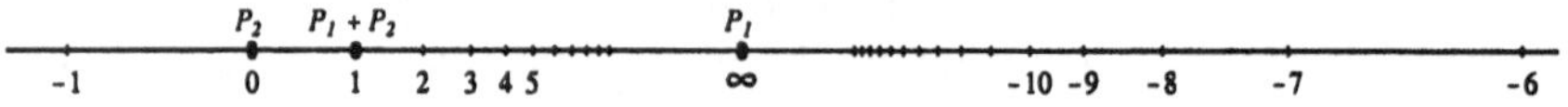

FIGURE 3. Projective point coordinates of a pencil of points. The fundamental points 0, 1, and ∞ are arbitrarily chosen. The remaining projective coordinates are determined by the logarithm of the cross-ratio; see [5].

Pencil of points

In a pencil of points, a generic multivector M of $\mathcal{G}_2$ decomposes into two different types of numbers and a point.

$$M = \langle \text{scalar} \rangle_0 + \langle \text{point} \rangle_1 + \langle \text{pseudoscalar} \rangle_2. \tag{5.5}$$

Figure 3 shows an example for the coordinatization of a pencil of points.

Pencil of planes

In a pencil of planes, a generic multivector M of $\mathcal{G}_2$ decomposes into two different types of numbers and a plane.

$$M = \langle\text{scalar}\rangle_0 + \langle\text{plane}\rangle_1 + \langle\text{pseudoscalar}\rangle_2 \tag{5.6}$$

Figure 4 shows an example for the coordinatization of a pencil of planes.

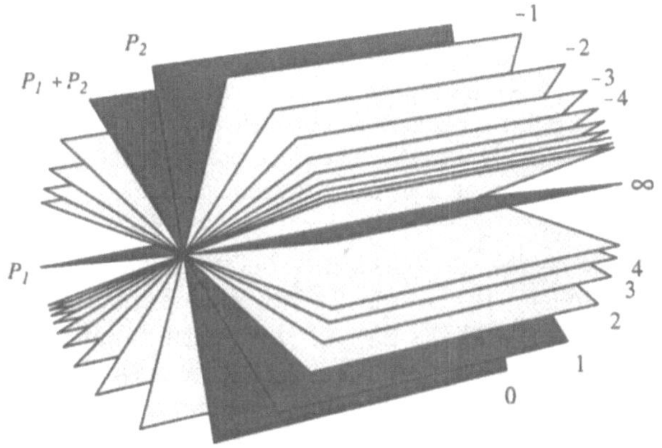

FIGURE 4. Projective plane coordinates of a pencil of planes.

5.2 Bundle and field

The plane of points, the plane of lines, the bundle of lines, and the bundle of planes are four different examples of two-dimensional projective spaces P^2. The planar field and the centric bundle are represented algebraically by the eight ($= 2^3$)-dimensional geometric algebra $\mathcal{G}_3$. To simplify expressions we introduce a new notation for the basis elements of grade 2 in *both* approaches to the geometric algebra $\mathcal{G}_3$.

$$l_1 := P_2P_3, \quad l_2 := P_3P_1, \quad l_3 := P_1P_2. \tag{5.7}$$

The homogeneous multivectors, together with their geometric interpretation, are shown in Table 1.8. According to Table 1.3, the translation of the basis elements of the geometric algebra $\mathcal{G}_3^+$ into the basis elements of the geometric algebra $\mathcal{G}_3^-$ is given by

$$1^+ = \mathrm{I}^-, \tag{5.8a}$$

$$\{\eta_{11}P_1^+, \eta_{22}P_2^+, \eta_{33}P_3^+\} = \{l_1^-, l_2^-, l_3^-\}, \tag{5.8b}$$

$$\{\eta_{22}\eta_{33}l_1^+, \eta_{11}\eta_{33}l_2^+, \eta_{11}\eta_{22}l_3^+\} = \{-P_1^-, -P_2^-, -P_3^-\}, \tag{5.8c}$$

$$\eta_{11}\eta_{22}\eta_{33}\mathrm{I}^+ = -1^-. \tag{5.8d}$$

$\mathcal{G}_3^+$	Planar field	Centric bundle	$\mathcal{G}_3^-$
$\rho 1^+$			$\rho \mathrm{I}^-$
$\rho A_{\bar{1}}^+ = \mu_1^+ P_1^+ + \mu_2^+ P_2^+ + \mu_3^+ P_3^+$	point	plane	$\rho A_{\bar{2}}^- = \mu_1^- l_1^- + \mu_2^- l_2^- + \mu_3^- l_3^-$
$\rho A_{\bar{2}}^+ = \lambda_1^+ l_1^+ + \lambda_2^+ l_2^+ + \lambda_3^+ l_3^+$	line	line	$\rho A_{\bar{1}}^- = \lambda_1^- P_1^- + \lambda_2^- P_2^- + \lambda_3^- P_3^-$
$\rho \mathrm{I}^+$			$\rho 1^-$

TABLE 1.8. Geometric interpretation for the homogeneous multivectors of grade 1 and 2 in the planar field and the centric bundle. The grade of a geometric object depends on the chosen aspect to the geometric algebra $\mathcal{G}_3$. For example, in the planar field a point appears as a geometric object of grade 1 in $\mathcal{G}_3^+$, and as a geometric object of grade 2 in $\mathcal{G}_3^-$. The same holds for the plane in the centric bundle, and analogous for the lines in both geometries.

A continuous one-to-one correspondence between the points and the lines of the planar field and the 1- and 2-vectors of the geometric algebra $\mathcal{G}_3$ or between the planes and the lines of the centric bundle and the 1- and 2-vectors of the geometric algebra $\mathcal{G}_3$ is a two-dimensional projective coordinate system. Any choice of *four* generic fundamental elements defines the projective coordinate system (chapter VII of volume I in [10]). In the "plus" approach we start with the plane of points or the bundle of planes, choose any four generic elements, and assign to them in any order the homogeneous coordinates $\rho(0,0,1)^+$, $\rho(0,1,0)^+$, $\rho(1,0,0)^+$, and $\rho(1,1,1)^+$. Any 1-vector $\rho A_{\bar{1}}^+ = \mu_1^+ P_1^+ + \mu_2^+ P_2^+ + \mu_3^+ P_3^+$ belongs to the point or plane with homogeneous coordinates $(\mu_1, \mu_2, \mu_3)^+$ or projective coordinates $(\frac{\mu_1}{\mu_3}, \frac{\mu_2}{\mu_3})^+$, and any 2-vector $\rho A_{\bar{2}}^+ = \lambda_1^+ l_1^+ + \lambda_2^+ l_2^+ + \lambda_3^+ l_3^+$ belongs to the line with homogeneous line coordinates $(\lambda_1, \lambda_2, \lambda_3)^+$ or projective line coordinates $(\frac{\lambda_1}{\lambda_3}, \frac{\lambda_2}{\lambda_3})^+$. In the dual approach to the same planar field or the same centric bundle, we start with the plane of lines or bundle of lines, respectively, choose any four generic lines, and assign to them in any order the homogeneous coordinates $\rho(0,0,1)^-$, $\rho(0,1,0)^-$, $\rho(1,0,0)^-$, and $\rho(1,1,1)^-$. Any 1-vector $\rho A_{\bar{1}}^- = \lambda_1^- P_1^- + \lambda_2^- P_2^- + \lambda_3^- P_3^-$ belongs to the line with homogeneous line coordinates $(\lambda_1, \lambda_2, \lambda_3)^-$ or projective line coordinates $(\frac{\lambda_1}{\lambda_3}, \frac{\lambda_2}{\lambda_3})^-$, and any 2-vector $\rho A_{\bar{2}}^- = \mu_1^- l_1^- + \mu_2^- l_2^- + \mu_3^- l_3^-$ belongs to the point or plane with homogeneous coordinates $(\mu_1, \mu_2, \mu_3)^-$ or projective coordinates $(\frac{\mu_1}{\mu_3}, \frac{\mu_2}{\mu_3})^-$. The projective coordinates in the dual geometric algebras differ only by a sign depending on the signature chosen. From

$$\mu_1^+ P_1^+ + \mu_2^+ P_2^+ + \mu_3^+ P_3^+ = \mu_1^- l_1^- + \mu_2^- l_2^- + \mu_3^- l_3^- \tag{5.9}$$

we obtain the relations for the point or plane coordinates,

$$\mu_x^- := \frac{\mu_1^-}{\mu_3^-} = \frac{\eta_{11}}{\eta_{33}} \frac{\mu_1^+}{\mu_3^+} =: \frac{\eta_{11}}{\eta_{33}} \mu_x^+, \quad \mu_y^- := \frac{\mu_2^-}{\mu_3^-} = \frac{\eta_{22}}{\eta_{33}} \frac{\mu_2^+}{\mu_3^+} =: \frac{\eta_{22}}{\eta_{33}} \mu_y^+. \tag{5.10}$$

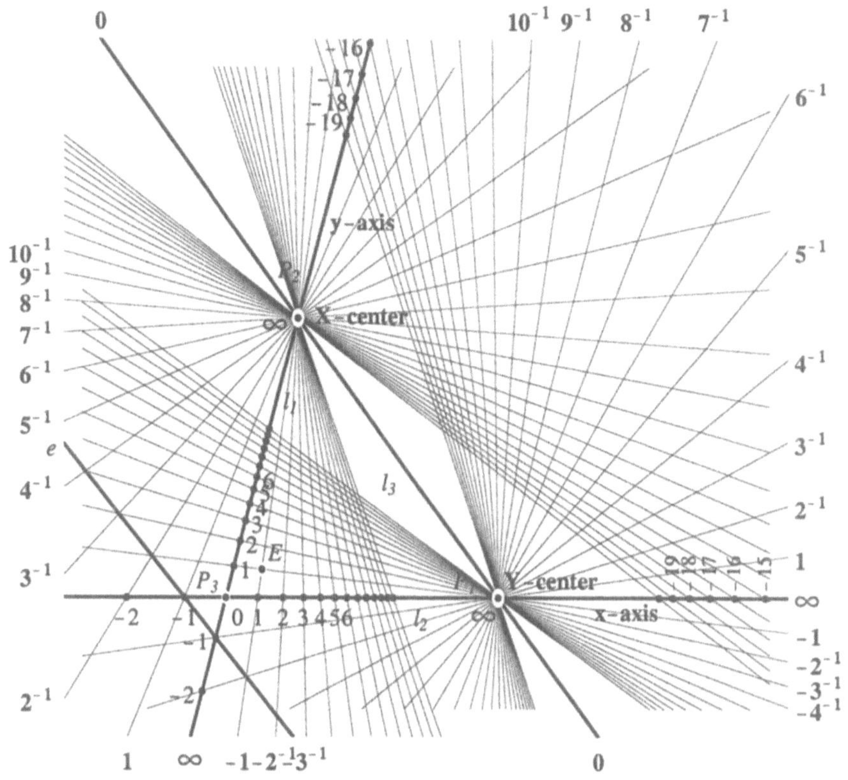

FIGURE 5. Projective coordinate system in a planar field. The projective point coordinates are indicated along the x-axis (l_2) and the y-axis (l_1). The projective line coordinates are indicated for the X-center (P_2) on the left side and the bottom, for the Y-center (P_1) on the right side and the top of the figure. The projective point coordinates of E are $(1,1)^+$, and the projective line coordinates of e are $(1,1)^+$. This coordinate system refers to the plus approach only.

And from

$$\lambda_1^+ l_1^+ + \lambda_2^+ l_2^+ + \lambda_3^+ l_3^+ = \lambda_1^- P_1^- + \lambda_2^- P_2^- + \lambda_3^- P_3^- \tag{5.11}$$

we get the relations for the line coordinates,

$$\lambda_x^- := \frac{\lambda_1^-}{\lambda_3^-} = \frac{\eta_{33}}{\eta_{11}} \frac{\lambda_1^+}{\lambda_3^+} =: \frac{\eta_{33}}{\eta_{11}} \lambda_x^+, \quad \lambda_y^- := \frac{\lambda_2^-}{\lambda_3^-} = \frac{\eta_{33}}{\eta_{22}} \frac{\lambda_2^+}{\lambda_3^+} =: \frac{\eta_{33}}{\eta_{22}} \lambda_y^+. \tag{5.12}$$

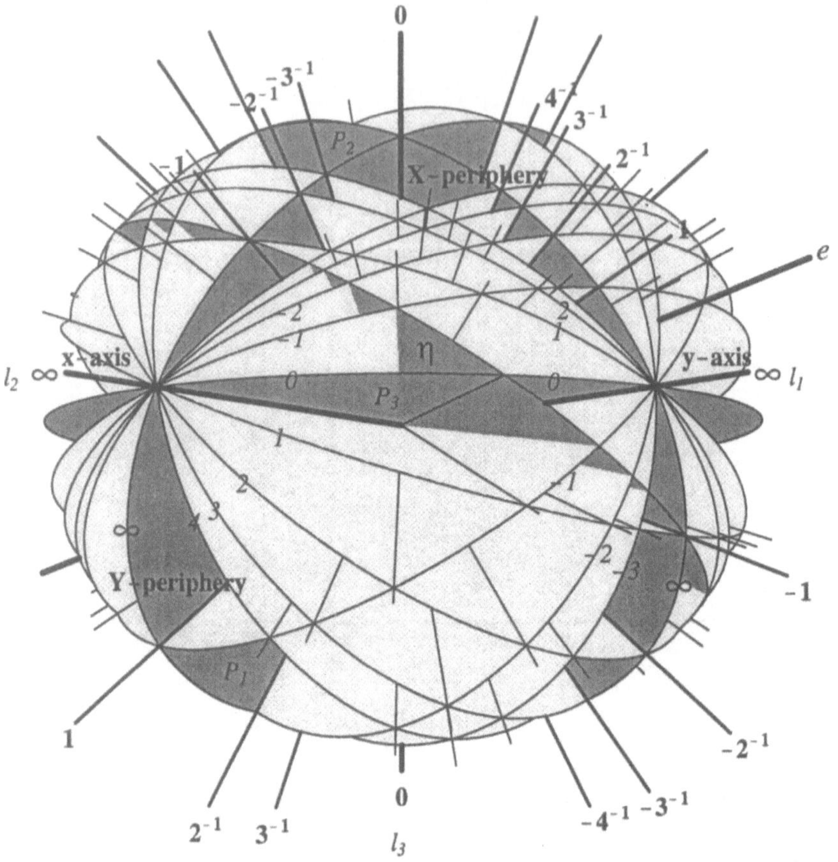

FIGURE 6. Projective coordinate system in a centric bundle. The projective plane coordinates are indicated around the x-axis (l_2) and the y-axis (l_1). The projective line coordinates are indicated along the circumferences of the X-periphery (P_2) and the Y-periphery (P_1). The projective plane coordinates of η are $(1,1)^+$, and the projective line coordinates of e are $(1,1)^+$. This coordinate systems only shows the plus approach.

Planar field

In a planar field, a generic multivector M of $\mathcal{G}_3$ decomposes into two different types of numbers, a point, and a line.

$$M^+ = \langle\text{scalar}\rangle_0 + \langle\text{point}\rangle_1 + \langle\text{line}\rangle_2 + \langle\text{pseudoscalar}\rangle_3 \tag{5.13a}$$

$$M^- = \langle\text{scalar}\rangle_0 + \langle\text{line}\rangle_1 + \langle\text{point}\rangle_2 + \langle\text{pseudoscalar}\rangle_3 \tag{5.13b}$$

Figure 5 shows an example for the coordinatization of a planar field.

Centric bundle

In a centric bundle, a generic multivector M of $\mathcal{G}_3$ decomposes into two different types of numbers, a plane, and a line.

$$M^+ = \langle\text{scalar}\rangle_0 + \langle\text{plane}\rangle_1 + \langle\text{line}\rangle_2 + \langle\text{pseudoscalar}\rangle_3 \tag{5.14a}$$

$$M^- = \langle\text{scalar}\rangle_0 + \langle\text{line}\rangle_1 + \langle\text{plane}\rangle_2 + \langle\text{pseudoscalar}\rangle_3 \tag{5.14b}$$

Figure 6 shows an example for the coordinatization of a centric bundle.

5.3 Space

The space of points and the space of planes provide two illustrations for the three-dimensional projective space; the *space of linear complexes* is an example for a *five*-dimensional projective space containing the space of lines as a four-dimensional (quadratic) sub-manifold. All of them are represented algebraically by the 16-($= 2^4$)-dimensional geometric algebra $\mathcal{G}_4$. To simplify expressions we introduce a new notation for the basis elements of grade 2 in both approaches to the geometric algebra $\mathcal{G}_4$,

$$\begin{aligned} l_1 &:= P_2P_3, \quad l_2 := P_3P_1, \quad l_3 := P_1P_2, \\ l_4 &:= P_4P_1, \quad l_5 := P_4P_2, \quad l_6 := P_4P_3, \end{aligned} \tag{5.15}$$

and also for the basis elements of grade 3.

$$\mathbf{P}_1 := P_2P_3P_4,\ \mathbf{P}_2 := P_3P_4P_1,\ \mathbf{P}_3 := P_4P_1P_2,\ \mathbf{P}_4 := P_1P_2P_3. \tag{5.16}$$

The homogeneous multivectors, together with their geometric interpretation, are shown in Table 1.9. According to Table 1.3, the translation of the basis elements of the geometric algebra $\mathcal{G}_4^+$ into the basis elements of the

$\mathcal{G}_4^+$	**Space**	$\mathcal{G}_4^-$
$\rho 1^+$		ρI^-
$\rho A_{\bar 1}^+ = \mu_1^+ P_1^+ + \mu_2^+ P_2^+ + \mu_3^+ P_3^+ + \mu_4^+ P_4^+$	point	$\rho A_{\bar 3}^- = \mu_1^- \mathbf{P}_1^- + \mu_2^- \mathbf{P}_2^- + \mu_3^- \mathbf{P}_3^- + \mu_4^- \mathbf{P}_4^-$
$\rho A_{\bar 2}^+ = \lambda_1^+ l_1^+ + \lambda_2^+ l_2^+ + \lambda_3^+ l_3^+ + \lambda_4^+ l_4^+ + \lambda_5^+ l_5^+ + \lambda_6^+ l_6^+$	linear complex or a line	$\rho A_{\bar 2}^- = \lambda_1^- l_1^- + \lambda_2^- l_2^- + \lambda_3^- l_3^- + \lambda_4^- l_4^- + \lambda_5^- l_5^- + \lambda_6^- l_6^-$
$\rho A_{\bar 3}^+ = \sigma_1^+ \mathbf{P}_1^+ + \sigma_2^+ \mathbf{P}_2^+ + \sigma_3^+ \mathbf{P}_3^+ + \sigma_4^+ \mathbf{P}_4^+$	plane	$\rho A_{\bar 1}^- = \sigma_1^- P_1^- + \sigma_2^- P_2^- + \sigma_3^- P_3^- + \sigma_4^- P_4^-$
ρI^+		$\rho 1^-$

TABLE 1.9. Projective interpretation for homogeneous multivectors in space.

geometric algebra $\mathcal{G}_4^-$ is given by

$$1^+ = \mathrm{I}^-, \tag{5.17a}$$

$$\{\eta_{11}P_1^+, -\eta_{22}P_2^+\} = \{\mathbf{P}_1^-, \mathbf{P}_2^-\},$$
$$\{\eta_{33}P_3^+, -\eta_{44}P_4^+\} = \{\mathbf{P}_3^-, \mathbf{P}_4^-\}, \tag{5.17b}$$

$$\{\eta_{22}\eta_{33}l_1^+, \eta_{11}\eta_{33}l_2^+\} = \{l_4^-, l_5^-\},$$
$$\{\eta_{11}\eta_{22}l_3^+, \eta_{11}\eta_{44}l_4^+\} = \{l_6^-, l_1^-\},$$
$$\{\eta_{22}\eta_{44}l_5^+, \eta_{33}\eta_{44}l_6^+\} = \{l_2^-, l_3^-\}, \tag{5.17c}$$

$$\{\eta_{22}\eta_{33}\eta_{44}\mathbf{P}_1^+, \eta_{11}\eta_{33}\eta_{44}\mathbf{P}_2^+\} = \{P_1^-, -P_2^-\},$$
$$\{\eta_{11}\eta_{22}\eta_{44}\mathbf{P}_3^+, \eta_{11}\eta_{22}\eta_{33}\mathbf{P}_4^+\} = \{P_3^-, -P_4^-\}, \tag{5.17d}$$

$$\eta_{11}\eta_{22}\eta_{33}\eta_{44}\mathrm{I}^+ = 1^-. \tag{5.17e}$$

A continuous one-to-one correspondence between the points, linear complexes, and planes and the 1-, 2-, and 3-vectors of the geometric algebra $\mathcal{G}_4$ provide a projective coordinate system in space. Any choice of either *five* generic fundamental points or *five* generic fundamental planes defines a projective coordinate system (chapter VII of volume I in [10]). In the "plus" approach we start with the space of points, choose any five generic points, and assign to them in any order the homogeneous coordinates $\rho(0,0,0,1)^+$, $\rho(0,0,1,0)^+$, $\rho(0,1,0,0)^+$, $\rho(1,0,0,0)^+$, and $\rho(1,1,1,1)^+$.

Any 1-vector

$$\rho A_1^+ = \mu_1^+ P_1^+ + \mu_2^+ P_2^+ + \mu_3^+ P_3^+ + \mu_4^+ P_4^+$$

belongs to the point with homogeneous point coordinates $(\mu_1, \mu_2, \mu_3, \mu_4)^+$ or the projective point coordinates $(\frac{\mu_1}{\mu_4}, \frac{\mu_2}{\mu_4}, \frac{\mu_3}{\mu_4})^+$, any 2-vector

$$\rho A_2^+ = \lambda_1^+ l_1^+ + \lambda_2^+ l_2^+ + \lambda_3^+ l_3^+ + \lambda_4^+ l_4^+ + \lambda_5^+ l_5^+ + \lambda_6^+ l_6^+$$

belongs to the linear complex with the homogeneous complex coordinates $(\lambda_1, \lambda_2, \lambda_3, \lambda_4, \lambda_5, \lambda_6)^+$, and any 3-vector

$$\rho A_3^+ = \sigma_1^+ \mathbf{P}_1^+ + \sigma_2^+ \mathbf{P}_2^+ + \sigma_3^+ \mathbf{P}_3^+ + \sigma_4^+ \mathbf{P}_4^+$$

belongs to the plane with homogeneous plane coordinates $(\sigma_1, \sigma_2, \sigma_3, \sigma_4)^+$ or projective plane coordinates $(\frac{\sigma_1}{\sigma_4}, \frac{\sigma_2}{\sigma_4}, \frac{\sigma_3}{\sigma_4})^+$; see Figure 7. In the "minus" approach, we start with the space of planes, choose any five generic planes, and assign to them in any order the homogeneous coordinates $\rho(0,0,0,1)^-$, $\rho(0,0,1,0)^-$, $\rho(0,1,0,0)^-$, $\rho(1,0,0,0)^-$, and $\rho(1,1,1,1)^-$.

Any 1-vector

$$\rho A_1^- = \mu_1^- P_1^- + \mu_2^- P_2^- + \mu_3^- P_3^- + \mu_4^- P_4^-$$

belongs to the plane with homogeneous plane coordinates $(\mu_1, \mu_2, \mu_3, \mu_4)^-$ or the projective plane coordinates $(\frac{\mu_1}{\mu_4}, \frac{\mu_2}{\mu_4}, \frac{\mu_3}{\mu_4})^-$, any 2-vector

$$\rho A_2^- = \lambda_1^- l_1^- + \lambda_2^- l_2^- + \lambda_3^- l_3^- + \lambda_4^- l_4^- + \lambda_5^- l_5^- + \lambda_6^- l_6^-$$

belongs to the linear complex with the homogeneous complex coordinates $(\lambda_1, \lambda_2, \lambda_3, \lambda_4, \lambda_5, \lambda_6)^-$, and any 3-vector

$$\rho A_3^- = \sigma_1^- \mathbf{P}_1^- + \sigma_2^- \mathbf{P}_2^- + \sigma_3^- \mathbf{P}_3^- + \sigma_4^- \mathbf{P}_4^-$$

belongs to the point with homogeneous point coordinates $(\sigma_1, \sigma_2, \sigma_3, \sigma_4)^-$ or projective point coordinates $(\frac{\sigma_1}{\sigma_4}, \frac{\sigma_2}{\sigma_4}, \frac{\sigma_3}{\sigma_4})^-$. The projective coordinates of

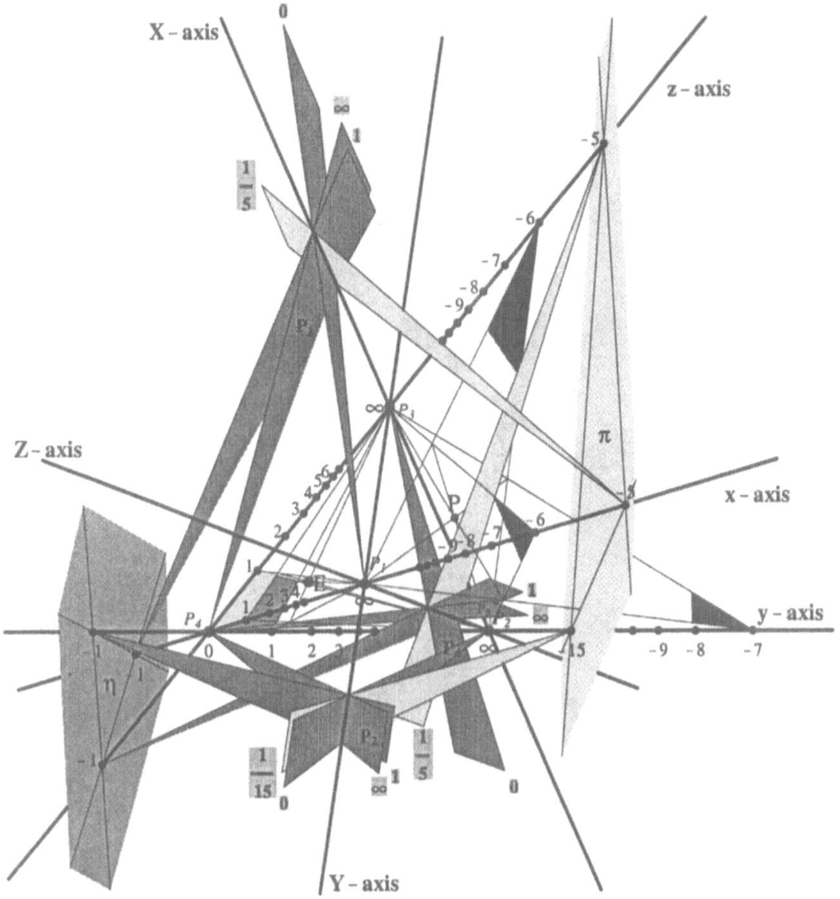

FIGURE 7. Projective coordinate system in space. The projective point coordinates are indicated along the x-axis ($l_4 = P_4P_1$), the y-axis ($l_5 = P_4P_2$), and the z-axis ($l_6 = P_4P_3$). $\mathbf{P}_4$ represents the plane at infinity in the space of points. The projective plane coordinates are indicated around the X-axis ($l_1 = P_2P_3$), the Y-axis ($l_2 = P_3P_1$), and the Z-axis ($l_3 = P_1P_2$), and are put on a gray background. P_4 represents the point at infinity in the space of planes. The projective point coordinates of E and P are $(1, 1, 1)^+$ and $(-6, -7, -6)^+$, respectively. The projective plane coordinates of η and π are $(1, 1, 1)^+$ and $(\frac{1}{5}, \frac{1}{15}, \frac{1}{5})^+$, respectively. This figure refers to the plus approach only.

points and planes in the dual geometric algebras differ only by a sign which depends on the chosen signature. From

$$\begin{aligned}\mu_1^+ P_1^+ + \mu_2^+ P_2^+ + \mu_3^+ P_3^+ + \mu_4^+ P_4^+ \\ = \mu_1^- \mathbf{P}_1^- + \mu_2^- \mathbf{P}_2^- + \mu_3^- \mathbf{P}_3^- + \mu_4^- \mathbf{P}_4^-,\end{aligned} \tag{5.18}$$

we get the relations for the point coordinates

$$\begin{aligned}\mu_x^- &:= \frac{\mu_1^-}{\mu_4^-} = -\frac{\eta_{11}}{\eta_{44}} \frac{\mu_1^+}{\mu_4^+} = -\frac{\eta_{11}}{\eta_{44}} \mu_x^+, \\ \mu_y^- &:= \frac{\mu_2^-}{\mu_4^-} = \frac{\eta_{22}}{\eta_{44}} \frac{\mu_2^+}{\mu_4^+} = \frac{\eta_{22}}{\eta_{44}} \mu_y^+, \\ \mu_z^- &:= \frac{\mu_3^-}{\mu_4^-} = -\frac{\eta_{33}}{\eta_{44}} \frac{\mu_3^+}{\mu_4^+} = -\frac{\eta_{33}}{\eta_{44}} \mu_z^+.\end{aligned} \tag{5.19}$$

And from

$$\begin{aligned}\sigma_1^+ \mathbf{P}_1^+ + \sigma_2^+ \mathbf{P}_2^+ + \sigma_3^+ \mathbf{P}_3^+ + \sigma_4^+ \mathbf{P}_4^+ \\ = \sigma_1^- P_1^- + \sigma_2^- P_2^- + \sigma_3^- P_3^- + \sigma_4^- P_4^-,\end{aligned} \tag{5.20}$$

we obtain the relations for the plane coordinates

$$\begin{aligned}\sigma_x^- &:= \frac{\sigma_1^-}{\sigma_4^-} = -\frac{\eta_{11}}{\eta_{44}} \frac{\sigma_1^+}{\sigma_4^+} = -\frac{\eta_{11}}{\eta_{44}} \sigma_x^+, \\ \sigma_y^- &:= \frac{\sigma_2^-}{\sigma_4^-} = \frac{\eta_{22}}{\eta_{44}} \frac{\sigma_2^+}{\sigma_4^+} = \frac{\eta_{22}}{\eta_{44}} \sigma_y^+, \\ \sigma_z^- &:= \frac{\sigma_3^-}{\sigma_4^-} = -\frac{\eta_{33}}{\eta_{44}} \frac{\sigma_3^+}{\sigma_4^+} = -\frac{\eta_{33}}{\eta_{44}} \sigma_z^+.\end{aligned} \tag{5.21}$$

For the geometric interpretation of the line coordinates and the coordinates of linear complexes, we refer to the book of H.-J. Stoß [9].

In space, a generic multivector M of $\mathcal{G}_4$ decomposes into two different types of numbers, a point, a (special) linear complex, and a plane

$$M^+ = \langle\text{sc.}\rangle_0 + \langle\text{point}\rangle_1 + \langle\text{complex}\rangle_2 + \langle\text{plane}\rangle_3 + \langle\text{p.-sc.}\rangle_4, \tag{5.22a}$$

$$M^- = \langle\text{sc.}\rangle_0 + \langle\text{plane}\rangle_1 + \langle\text{complex}\rangle_2 + \langle\text{point}\rangle_3 + \langle\text{p.-sc.}\rangle_4. \tag{5.22b}$$

6 Basic relations and operations

In the first part of this section we review the incidence relations of homogeneous multivectors in space, introduce the linear complex, and apply the incidence relation also to the non-homogeneous multivector of a point plane pair. The second part reviews the operations of connection and intersection.

6.1 Incidence

The incidence relations are formulated as identities in geometric algebra.

Definition 1. *Two multivectors A and B are* incident *if and only if $A \wedge B = 0$ and $A \vee B = 0$.*

$$A^+_{\bar{1}} \wedge B^+_{\bar{1}} = 0, \quad A^+_{\bar{1}} \vee B^+_{\bar{1}} = 0 \quad \Leftrightarrow \quad A^+_{\bar{1}} = \beta B^+_{\bar{1}} \tag{6.1a}$$

$$A^-_{\bar{1}} \vee B^-_{\bar{1}} = 0, \quad A^-_{\bar{1}} \wedge B^-_{\bar{1}} = 0 \quad \Leftrightarrow \quad A^-_{\bar{1}} = \beta B^-_{\bar{1}} \tag{6.1b}$$

The incidence relation between points (6.1a) and planes (6.1b), respectively, is reflexive. Thus, a point lies in itself and a plane passes through itself.

Theorem 1. *The following statements are equivalent: i) $A_{\bar{2}}$ represents a line; ii) $A_{\bar{2}} \wedge A_{\bar{2}} = 0$; iii) $A_{\bar{2}} \vee A_{\bar{2}} = 0$; and iv) $A_{\bar{2}}$ is incident with itself.*

Proof. **i) ⇒ ii)** Any line may be represented by the connection of two distinct points $B^+_{\bar{1}}$ and $C^+_{\bar{1}}$: $A^+_{\bar{2}} = B^+_{\bar{1}} \wedge C^+_{\bar{1}}$. It follows that $A^+_{\bar{2}} \wedge A^+_{\bar{2}} = 0$. In the dual approach any line may be represented by the intersection of two distinct planes $B^-_{\bar{1}}$ and $C^-_{\bar{1}}$: $A^-_{\bar{2}} = B^-_{\bar{1}} \vee C^-_{\bar{1}}$. It follows that $A^-_{\bar{2}} \wedge A^-_{\bar{2}} = 0$.

ii) ⇒ iii) $A_{\bar{2}} \vee A_{\bar{2}} = (-1)^q (A_{\bar{2}} \wedge A_{\bar{2}})^\sim, \quad \forall A_{\bar{2}} \in \mathcal{G}^2_{p,q}$.

iii) ⇒ iv) Definition 1.

iv) ⇒ i): The solutions of the incidence equations $A_{\bar{2}} \wedge X_{\bar{1}} = 0$ and $A_{\bar{2}} \vee X_{\bar{1}} = 0$ have the form of a pencil of points or of a pencil of planes: $X_{\bar{1}} = \mu_1 B_{\bar{1}} + \mu_2 C_{\bar{1}}$. Thus, $A_{\bar{2}}$ represents a line. □

The incidence relation between 2-vectors is *not* reflexive in general. A direct geometric interpretation for 2-vectors $A_{\bar{2}}$ with $A_{\bar{2}} \wedge A_{\bar{2}} \neq 0$ and $A_{\bar{2}} \vee A_{\bar{2}} \neq 0$ does not exist. Nevertheless, any 2-vector is incident with certain lines in space and is, thus, characterized by them.

Definition 2. *The locus of all lines that are incident with a given 2-vector $A_{\bar{2}}$ is called* a linear complex of $A_{\bar{2}}$ *if $A_{\bar{2}} \wedge A_{\bar{2}} \neq 0$ and $A_{\bar{2}} \vee A_{\bar{2}} \neq 0$, and* a special linear complex of $A_{\bar{2}}$ *if $A_{\bar{2}} \wedge A_{\bar{2}} = 0$ and $A_{\bar{2}} \vee A_{\bar{2}} = 0$.*

Instead of the correct expression "(special) linear complex of $A_{\bar{2}}$" we will use the shorter but less precise expression "(special) linear complex $A_{\bar{2}}$."

Theorem 2. *A (special) linear complex is a three-dimensional subset in the space of lines.*

Proof. A line $B_{\bar{2}}$ is incident with a (special) linear complex $A_{\bar{2}}$ if and only if $A_{\bar{2}} \wedge B_{\bar{2}} = 0$ and $A_{\bar{2}} \vee B_{\bar{2}} = 0$. Thus, three degrees of freedom are left to the line $B_{\bar{2}}$. □

Theorem 3.

a) Given a linear complex $A_{\bar{2}}$. Every bundle in space contains one and only one pencil of lines belonging to the linear complex $A_{\bar{2}}$, and every field in space contains one and only one pencil of lines belonging to the linear complex $A_{\bar{2}}$.

b) Given a special linear complex $A_{\bar{2}}$. The same statement as in a) holds for all bundles and fields whose points and planes, respectively, are not incident with the line $A_{\bar{2}}$. In case they are incident, all the lines of the bundle and all the lines of the field belong to the special linear complex $A_{\bar{2}}$.

Proof. a) Given any point $B_{\bar{1}}^{+}$ and the linear complex $A_{\bar{2}}$, solutions of the system $A_{\bar{2}}^{+} \wedge X_{\bar{2}}^{+} = 0$, $A_{\bar{2}}^{+} \vee X_{\bar{2}}^{+} = 0$, $B_{\bar{1}}^{+} \wedge X_{\bar{2}}^{+} = 0$, $B_{\bar{1}}^{+} \vee X_{\bar{2}}^{+} = 0$ form a pencil of lines with center $B_{\bar{1}}^{+}$. Given any plane $B_{\bar{1}}^{-}$ and the linear complex $A_{\bar{2}}$, solutions of the system $A_{\bar{2}}^{-} \vee X_{\bar{2}}^{-} = 0$, $A_{\bar{2}}^{-} \wedge X_{\bar{2}}^{-} = 0$, $B_{\bar{1}}^{-} \vee X_{\bar{2}}^{-} = 0$, $B_{\bar{1}}^{-} \wedge X_{\bar{2}}^{-} = 0$ form a pencil of lines in the plane $B_{\bar{1}}^{-}$.
b) In case the point $B_{\bar{1}}^{+}$ or the plane $B_{\bar{1}}^{-}$ belongs to the line $A_{\bar{2}}$, the incidence equations $A_{\bar{2}} \wedge X_{\bar{2}} = 0$ and $A_{\bar{2}} \vee X_{\bar{2}} = 0$ hold trivially for all $X_{\bar{2}}$ passing through $B_{\bar{1}}^{+}$ or lying in $B_{\bar{1}}^{-}$, respectively. Otherwise see case a). □

Null polarities ([1], p. 69f) are intimately connected with non-special linear complexes.

Theorem 4. *Every null polarity determines one and only one non-special linear complex, and every non-special linear complex determines one and only one null polarity. The lines of the non-special linear complex are the only invariant elements of the corresponding null polarity.*

Proof. See chapter XI of volume I in [10]. □

Definition 3. *Two non-special linear complexes $A_{\bar{2}}$ and $B_{\bar{2}}$ are* null invariant to each other *if and only if the linear complex $A_{\bar{2}}$ is invariant under the null polarity of $B_{\bar{2}}$ or, which is the same, if and only if the linear complex $B_{\bar{2}}$ is invariant under the null polarity of $A_{\bar{2}}$. In case at least one of the linear complexes is special, say $A_{\bar{2}}$, the linear complexes $A_{\bar{2}}$ and $B_{\bar{2}}$ are null invariant, if and only if the line $A_{\bar{2}}$ belongs to the (special) linear complex $B_{\bar{2}}$.*

$$A_{\bar{2}} \wedge B_{\bar{2}} = 0, \quad A_{\bar{2}} \vee B_{\bar{2}} = 0. \tag{6.1c}$$

The equations (6.1c) express the incidence relations for lines and linear complexes. In detail we have

1. If $A_{\bar{2}} = \beta B_{\bar{2}}$, then $A_{\bar{2}}$ represents a line.
2. If $A_{\bar{2}} \neq \beta B_{\bar{2}}$, then we distinguish the following cases:

   (a) If $A_{\bar{2}} \wedge A_{\bar{2}} = 0$, $B_{\bar{2}} \wedge B_{\bar{2}} = 0$, and equations (6.1c) are satisfied, then the lines $A_{\bar{2}}$ and $B_{\bar{2}}$ meet, i.e., they pass through a common point and lie in a common plane.

(b) $A_{\bar{2}} \wedge A_{\bar{2}} \neq 0$ and $B_{\bar{2}} \wedge B_{\bar{2}} = 0$.
The line $B_{\bar{2}}$ belongs to the linear complex $A_{\bar{2}}$.

(c) $A_{\bar{2}} \wedge A_{\bar{2}} = 0$ and $B_{\bar{2}} \wedge B_{\bar{2}} \neq 0$.
The line $A_{\bar{2}}$ belongs to the linear complex $B_{\bar{2}}$.

(d) $A_{\bar{2}} \wedge A_{\bar{2}} \neq 0$ and $B_{\bar{2}} \wedge B_{\bar{2}} \neq 0$.
The linear complexes $A_{\bar{2}}$ and $B_{\bar{2}}$ are null invariant.

$$A_{\bar{2}}^{+} \wedge B_{\bar{1}}^{+} = 0, \quad A_{\bar{2}}^{+} \vee B_{\bar{1}}^{+} = 0. \tag{6.1d}$$

The second equation of (6.1d) is trivially satisfied.

1. If the 2-vector $A_{\bar{2}}^{+}$ represents a line, the equations (6.1d) require the point $B_{\bar{1}}^{+}$ and the line $A_{\bar{2}}^{+}$ to be incident. Thus, the set of all points lying in a fixed line forms an one-dimensional *pencil of points*. The set of all lines passing through a fixed point forms a two-dimensional *bundle of lines.*

2. Direct computation shows that the equations (6.1d) have no solutions other than 0 in case $A_{\bar{2}}^{+}$ represents a linear complex. Thus, there are no points lying in a linear complex!

$$A_{\bar{2}}^{-} \vee B_{\bar{1}}^{-} = 0, \quad A_{\bar{2}}^{-} \wedge B_{\bar{1}}^{-} = 0. \tag{6.1e}$$

The second equation of (6.1e) is trivially true.

1. If the 2-vector $A_{\bar{2}}^{-}$ represents a line, the equations (6.1e) require the plane $B_{\bar{1}}^{-}$ and the line $A_{\bar{2}}^{-}$ to be incident. Thus, the set of all planes passing through a fixed line forms a one-dimensional *pencil of planes.* The set of all lines lying in a fixed plane form a two-dimensional *field of lines.*

2. Direct computation shows that the equations (6.1e) do not have solutions other than 0 in case $A_{\bar{2}}^{-}$ represents a linear complex. Thus, there are no planes passing through a linear complex!

$$A_{\bar{3}} \wedge B_{\bar{1}} = 0, \quad A_{\bar{3}} \vee B_{\bar{1}} = 0. \tag{6.1f}$$

The equations (6.1f) require the plane $A_{\bar{3}}^{+}$, $B_{\bar{1}}^{-}$ and the point $B_{\bar{1}}^{+}$, $A_{\bar{3}}^{-}$, respectively, to be incident. The set of all points lying in a fixed plane forms a two-dimensional *plane of points*, and the set of all planes passing through a fixed point forms a two-dimensional *bundle of planes.*

Let us now consider incident multivectors. What is the condition for two generic point plane pairs $A = \langle A\rangle_1 + \langle A\rangle_3$ and $B = \langle B\rangle_1 + \langle B\rangle_3$ to be incident?

$$\left.\begin{array}{l} A \wedge B = 0 \\ A \vee B = 0 \end{array}\right\} \quad \Leftrightarrow \quad \left\{\begin{array}{l} \langle A\rangle_1 = \alpha\langle B\rangle_1 \\ \langle A\rangle_3 = \beta\langle B\rangle_3 \\ \langle A\rangle_1 \wedge \langle A\rangle_3 = \langle B\rangle_1 \vee \langle B\rangle_3 = 0. \end{array}\right. \tag{6.1g}$$

They are incident if and only if A and B represent the same incident point plane pair. Thus, the incidence relation between incident point plane pairs is reflexive as for points (6.1a) and planes (6.1b).

An incident point plane pair $A = \langle A\rangle_1 + \langle A\rangle_3$ and a line $B_{\bar{2}}$ are incident,

$$(\langle A\rangle_1 + \langle A\rangle_3) \wedge B_{\bar{2}} = 0, \quad (\langle A\rangle_1 + \langle A\rangle_3) \vee B_{\bar{2}} = 0, \tag{6.1h}$$

if and only if the line $B_{\bar{2}}$ passes through the point $\langle A\rangle_{\bar{1}}^{+}$ ($\langle A\rangle_{\bar{3}}^{-}$) and lies in the plane $\langle A\rangle_{\bar{3}}^{+}$ ($\langle A\rangle_{\bar{1}}^{-}$) simultaneously. The set of all lines that are incident with a fixed incident point plane pair forms a one-dimensional *pencil of lines*.

The equations (6.1a), (6.1b), (6.1c), and (6.1g) express the incidence relations for pure elements. The equations (6.1d), (6.1e), (6.1f), and (6.1h) show the incidence relations between different pure elements according to the diagram of Figure 1.

6.2 Connection and intersection

Definition 4. *The* connection *S of two multivectors A and B is $S = A \wedge B$ if A and B are not incident. The* intersection *S of two multivectors A and B is $S = A \vee B$ if A and B are not incident.*

The equations (6.2) review the relations of connection and intersection in space. We limit our attention to homogeneous multivectors.

$$S_{\bar{2}}^{+} = A_{\bar{1}}^{+} \wedge B_{\bar{1}}^{+} \neq 0, \quad S_{\bar{2}}^{-} = A_{\bar{1}}^{-} \vee B_{\bar{1}}^{-} \neq 0. \tag{6.2a}$$

The connection of two distinct points $A_{\bar{1}}^{+}$ and $B_{\bar{1}}^{+}$ is one and only one line $S_{\bar{2}}^{+}$ ($S_{\bar{2}}^{+} \wedge S_{\bar{2}}^{+} = 0$). The intersection of two distinct planes $A_{\bar{1}}^{-}$ and $B_{\bar{1}}^{-}$ is one and only one line $S_{\bar{2}}^{-}$ ($S_{\bar{2}}^{-} \vee S_{\bar{2}}^{-} = 0$).

$$S_{\bar{4}}^{+} = A_{\bar{2}}^{+} \wedge B_{\bar{2}}^{+} \neq 0, \quad S_{\bar{4}}^{-} = A_{\bar{2}}^{-} \vee B_{\bar{2}}^{-} \neq 0. \tag{6.2b}$$

The connection of two not necessarily distinct (special) linear complexes $A_{\bar{2}}^{+}$ and $B_{\bar{2}}^{+}$, that are not null invariant to each other, is one and only one pseudoscalar $S_{\bar{4}}^{+}$. The intersection of two not necessarily distinct (special) linear complexes $A_{\bar{2}}^{-}$ and $B_{\bar{2}}^{-}$, that are not null invariant to each other, is one and only one pseudoscalar $S_{\bar{4}}^{-}$.

$$A_{\bar{3}}^{+} \wedge B_{\bar{3}}^{+} = 0, \quad A_{\bar{3}}^{-} \vee B_{\bar{3}}^{-} = 0, \quad \forall A_{\bar{3}}, B_{\bar{3}} \in \mathcal{G}_{p,q}^{3}. \tag{6.2c}$$

There is no connection between any two planes $A_{\bar{3}}^{+}$ and $B_{\bar{3}}^{+}$ and no intersection between any two points $A_{\bar{3}}^{-}$ and $B_{\bar{3}}^{-}$.

$$S_{\bar{3}}^{+} = A_{\bar{2}}^{+} \wedge B_{\bar{1}}^{+} \neq 0, \quad S_{\bar{3}}^{-} = A_{\bar{2}}^{-} \vee B_{\bar{1}}^{-} \neq 0. \tag{6.2d}$$

The connection of a (special) linear complex $A^+_{\bar{2}}$ and a point $B^+_{\bar{1}}$ is one and only one plane $S^+_{\bar{3}}$. The intersection of a (special) linear complex $A^-_{\bar{2}}$ and a plane $B^-_{\bar{1}}$ is one and only one point $S^-_{\bar{3}}$.

$$S^+_{\bar{4}} = A^+_{\bar{3}} \wedge B^+_{\bar{1}} \neq 0, \quad S^-_{\bar{4}} = A^-_{\bar{3}} \vee B^-_{\bar{1}} \neq 0. \tag{6.2e}$$

The connection of a plane $A^+_{\bar{3}}$ and a point $B^+_{\bar{1}}$ is one and only one pseudoscalar $S^+_{\bar{4}}$. The intersection of a point $A^-_{\bar{3}}$ and a plane $B^-_{\bar{1}}$ is one and only one pseudoscalar $S^-_{\bar{4}}$.

$$A^+_{\bar{3}} \wedge B^+_{\bar{2}} = 0, \quad A^-_{\bar{3}} \vee B^-_{\bar{2}} = 0, \quad \forall A_{\bar{3}} \in \mathcal{G}^3_{p,q}, \quad \forall B_{\bar{2}} \in \mathcal{G}^2_{p,q}. \tag{6.2f}$$

There is no connection defined between a plane $A^+_{\bar{3}}$ and a (special) linear complex $B^+_{\bar{2}}$. Equivalently there is no intersection between a point $A^-_{\bar{3}}$ and a (special) linear complex $B^-_{\bar{2}}$.

7 Principle of duality

The principle of duality was first stated by the French mathematicians J. V. Poncelet and J. D. Gergonne (1826) in the first quarter of the 19th century. In Germany, A. F. Möbius rediscovered the principle of duality independently and used the homogeneous coordinates in his book *Das barycentrische Calcul* (1827) for the first time. J. Plücker discovered the line and plane coordinates that gave a firm analytic base to the principle of duality. F. Klein removed the last vestiges of Euclidean geometry by showing every metric geometry—non-Euclidean or Euclidean—with constant curvature to be embedded in projective geometry ([5]). He also provided the algebraic foundation for projective geometry. F. Klein considered the principle of duality as a significant step in the development of mathematics (and natural sciences). "Die Auffindung des Dualitätsprinzips, das von unserem heutigen Standpunkt aus nicht allzu tiefliegend erscheint, stellte eine wesentliche wissenschaftliche Leistung dar. Man erkennt dies am besten daran, dass rund 150 Jahre nach der Auffindung des Pascalschen Satzes vergangen sind, ehe der Satz des Brianchon gefunden wurde, der sich doch mit Hilfe des Dualitätsprinzips durch eine unmittelbare Übertragung aus dem ersten Satz ableiten lässt." ([5], p. 38)

To justify the principle of duality in synthetic projective geometry, the *method of formal inference* is used. Starting from a system of explicitly stated axioms that imply their own duals, half of the theorems appear self-evident. If one theorem is proved by tracing it back to some axioms, the dual theorem is immediately true by formal inference, i.e., it traces back to the dual axioms belonging to the system of explicitly stated axioms. This is one of the important advantages of this method. The books of O. Veblen and J. W. Young [10], L. Locher [7, 8], and H. S. M. Coxeter [1] all approach projective geometry from the vantage point of formal inference.

Plane of points		Plane of lines	$\mathcal{G}_3^+$		$\mathcal{G}_3^-$
			$A^+_{\bar{0}}$	$\leftrightarrow$	$A^-_{\bar{0}}$
point	$\leftrightarrow$	line	$A^+_{\bar{1}}$	$\leftrightarrow$	$A^-_{\bar{1}}$
line	$\leftrightarrow$	point	$A^+_{\bar{2}}$	$\leftrightarrow$	$A^-_{\bar{2}}$
			$A^+_{\bar{3}}$	$\leftrightarrow$	$A^-_{\bar{3}}$
lying in	$\leftrightarrow$	passing through	$A^+ \wedge B^+ = 0$ and $A^+ \vee B^+ = 0$	$\leftrightarrow$	$A^- \vee B^- = 0$ and $A^- \wedge B^- = 0$
passing through	$\leftrightarrow$	lying in	$A^+ \wedge B^+ = 0$ and $A^+ \vee B^+ = 0$	$\leftrightarrow$	$A^- \vee B^- = 0$ and $A^- \wedge B^- = 0$
connect	$\leftrightarrow$	intersect	$\wedge$	$\leftrightarrow$	$\vee$
intersect	$\leftrightarrow$	connect	$\vee$	$\leftrightarrow$	$\wedge$
			$\cdot$	$\leftrightarrow$	$\circ$
			$\circ$	$\leftrightarrow$	$\cdot$
			(juxtaposition)	$\leftrightarrow$	$*$
			$*$	$\leftrightarrow$	(juxtaposition)

TABLE 1.10. The principle of duality in the field $\mathcal{G}_3$.

"The principle of duality in the *plane* affirms that every definition remains significant, and every theorem remains true, when we interchange 'point' and 'line', and make a few consequent alterations in wording" ([1], p. 26). The "consequent alterations in wording" are shown on the left side in Table 1.10. To illustrate the principle of duality in the field, we use the famous *theorem of Desargues* (Theorem 5) which is self-dual (Theorem 6).

Let $\Delta = \{A, B, C\}$ and $\Delta' = \{A', B', C'\}$ be two generic triangles with the sides $\delta = \{a, b, c\}$ and $\delta' = \{a', b', c'\}$.

Theorem 5. *If and only if Δ and Δ' are perspective from a point Z (i.e., the three connecting lines $p := AA'$, $q := BB'$, and $r := CC'$ pass through the common point Z), the three intersecting points $P := aa'$, $Q := bb'$, and $R := cc'$ are perspective from a line z (i.e., the points P, Q, and R lie on the common line z).*

Theorem 6. *If and only if δ and δ' are perspective from a line z (i.e., the three intersecting points $P := aa'$, $Q := bb'$, and $R := cc'$ lie on the common line z), the three lines $p := AA'$, $q := BB'$, and $r := CC'$ are perspective from a point Z (i.e., the lines p, q, and r pass through the common point Z.)*

The principle of duality in the *bundle* allows for the analogous interchange of "plane" and "line" and a few consequent alterations shown on the left side in Table 1.11. Theorem 7 and 8 provide an example for the

Bundle of planes		Bundle of lines	$\mathcal{G}_3^+$		$\mathcal{G}_3^-$
			$A_{\bar{0}}^{+}$	$\leftrightarrow$	$A_{\bar{0}}^{-}$
plane	$\leftrightarrow$	line	$A_{\bar{1}}^{+}$	$\leftrightarrow$	$A_{\bar{1}}^{-}$
line	$\leftrightarrow$	plane	$A_{\bar{2}}^{+}$	$\leftrightarrow$	$A_{\bar{2}}^{-}$
			$A_{\bar{3}}^{+}$	$\leftrightarrow$	$A_{\bar{3}}^{-}$
lying in	$\leftrightarrow$	passing through	$A^+ \wedge B^+ = 0$ and $A^+ \vee B^+ = 0$	$\leftrightarrow$	$A^- \vee B^- = 0$ and $A^- \wedge B^- = 0$
passing through	$\leftrightarrow$	lying in	$A^+ \wedge B^+ = 0$ and $A^+ \vee B^+ = 0$	$\leftrightarrow$	$A^- \vee B^- = 0$ and $A^- \wedge B^- = 0$
connect	$\leftrightarrow$	intersect	$\wedge$	$\leftrightarrow$	$\vee$
intersect	$\leftrightarrow$	connect	$\vee$	$\leftrightarrow$	$\wedge$
			$\cdot$	$\leftrightarrow$	$\circ$
			$\circ$	$\leftrightarrow$	$\cdot$
			(juxtaposition)	$\leftrightarrow$	$*$
			$*$	$\leftrightarrow$	(juxtaposition)

TABLE 1.11. The principle of duality in the bundle $\mathcal{G}_3$.

duality in the bundle. Let $\Delta = \{A, B, C\}$ and $\Delta' = \{A', B', C'\}$ be two generic "triplanes" with the edges $\delta = \{a, b, c\}$ and $\delta' = \{a', b', c'\}$.

Theorem 7. *If and only if Δ and Δ' are perspective from a plane Z (i.e., the three intersecting lines $p := AA'$, $q := BB'$, and $r := CC'$ lie in the common plane Z), the three connecting planes $P := aa'$, $Q := bb'$, and $R := cc'$ are perspective from a line z (i.e., the planes P, Q, and R pass through the common line z).*

Theorem 8. *If and only if δ and δ' are perspective from a line z (i.e., the three connecting planes $P := aa'$, $Q := bb'$, and $R := cc'$ pass through the common line z), the three lines $p := AA'$, $q := BB'$, and $r := CC'$ are perspective from a plane Z (i.e., the lines p, q, and r lie in the common plane Z).*

The principle of duality in *space* affords interchanging of "plane" and "point". The notion "line" interchanges with itself. The necessary alterations in space are shown on the left side in Table 1.12. Since the Theorems 5 to 8 are true in space too, Theorem 5 and 7, as well as Theorem 6 and 8, are revealed to be dual to each other in space.

We will now define the principle of duality in geometric algebra $\mathcal{G}_n$ using the completely dual approach of section 3. In a natural way one is led to distinguish between a major and a minor form of the principle of duality

Space of points		Space of planes	$\mathcal{G}_4^+$		$\mathcal{G}_4^-$
			$A_{\bar{0}}^+$	↔	$A_{\bar{0}}^-$
point	↔	plane	$A_{\bar{1}}^+$	↔	$A_{\bar{1}}^-$
line	↔	line	$A_{\bar{2}}^+$	↔	$A_{\bar{2}}^-$
plane	↔	point	$A_{\bar{3}}^+$	↔	$A_{\bar{3}}^-$
			$A_{\bar{4}}^+$	↔	$A_{\bar{4}}^-$
lying in	↔	passing through	$A^+ \wedge B^+ = 0$ and $A^+ \vee B^+ = 0$	↔	$A^- \vee B^- = 0$ and $A^- \wedge B^- = 0$
passing through	↔	lying in	$A^+ \wedge B^+ = 0$ and $A^+ \vee B^+ = 0$	↔	$A^- \vee B^- = 0$ and $A^- \wedge B^- = 0$
connect	↔	intersect	$\wedge$	↔	$\vee$
intersect	↔	connect	$\vee$	↔	$\wedge$
			$\cdot$	↔	$\circ$
			$\circ$	↔	$\cdot$
			(juxtaposition)	↔	$*$
			$*$	↔	(juxtaposition)

TABLE 1.12. The principle of duality in space $\mathcal{G}_4$.

depending on whether the sign of the plus-minus notation is changed or not.

Definition 5. *The* major dual *S' of any equation, theorem, or definition S in $\mathcal{G}_n^+$ or $\mathcal{G}_n^-$ is obtained by interchanging juxtaposition with $*$, $\wedge$ with $\vee$, $\cdot$ with $\circ$, and by reversing the sign of the plus-minus notation. Then S' is also an equation, theorem, or definition in $\mathcal{G}_n^-$ or $\mathcal{G}_n^+$, respectively.*

Theorem 9 (The Major Principle of Duality). *Any statement S, deduced in the frame of the geometric algebra $\mathcal{G}_n^+$, is true if and only if the major dual statement S' is true in the geometric algebra $\mathcal{G}_n^-$.*

Proof. The isomorphism (3.16) guarantees the simultaneous validity of statement S and S'. □

Definition 6. *The* minor dual *S' of any equation, theorem, or definition S in $\mathcal{G}_n^+$ or $\mathcal{G}_n^-$ is obtained by interchanging juxtaposition with $*$, $\wedge$ with $\vee$, $\cdot$ with $\circ$, and by replacing the grades of a multivector with the corresponding dual grades ($k \to n-k$). Then S' is also an equation, theorem, or definition in $\mathcal{G}_n^+$ or $\mathcal{G}_n^-$, respectively.*

Theorem 10 (The Minor Principle of Duality). *Any statement S, deduced in the frame of the geometric algebra $\mathcal{G}_n^+$ or $\mathcal{G}_n^-$, is true if and only if the minor dual statement S' is true in the geometric algebra $\mathcal{G}_n^+$ or $\mathcal{G}_n^-$, respectively.*

Proof. The isomorphisms (3.16) and (3.11) guarantee the simultaneous validity of statement S and S'. □

The Tables 1.10, 1.11, and 1.12 show that the principle of duality in synthetic projective geometry (in its major form) corresponds one-to-one to the major principle of duality in geometric algebra.

To close this section we apply the principle of duality to Desargues' theorem formulated in geometric algebra. The Theorems 11 to 14 provide the conditions for three points in the plane $\mathcal{G}_3$ to be collinear, and three lines in the plane $\mathcal{G}_3$ to be concurrent.

Theorem 11. *The following statements are equivalent:*

i) The points $P_{\bar{1}}^+$, $Q_{\bar{1}}^+$, $R_{\bar{1}}^+$ are collinear, i.e., they lie on a common line.

ii) $(P_{\bar{1}}^+ \wedge Q_{\bar{1}}^+) \vee R_{\bar{1}}^+ = 0.$

iii) $\langle P_{\bar{1}}^+ * Q_{\bar{1}}^+ * R_{\bar{1}}^+ \rangle_3^+ = 0.$

Proof. **i) ⇔ ii)** From $0 = (P_{\bar{1}}^+ \wedge Q_{\bar{1}}^+) \vee R_{\bar{1}}^+ = [(P_{\bar{1}}^+ \wedge Q_{\bar{1}}^+)^\sim \wedge (R_{\bar{1}}^+)^\sim]^\sim = (\mathrm{I}^+)^2 \langle (P_{\bar{1}}^+ \wedge Q_{\bar{1}}^+) R_{\bar{1}}^+ \rangle_3^+ (\mathrm{I}^+)^{-1} = (\mathrm{I}^+)^2 [(P_{\bar{1}}^+ \wedge Q_{\bar{1}}^+) \wedge R_{\bar{1}}^+]^\sim \Leftrightarrow (P_{\bar{1}}^+ \wedge Q_{\bar{1}}^+) \wedge R_{\bar{1}}^+ = 0$ it follows $(P_{\bar{1}}^+ \wedge Q_{\bar{1}}^+) \vee R_{\bar{1}}^+ = 0 \Leftrightarrow (P_{\bar{1}}^+ \wedge Q_{\bar{1}}^+) \vee R_{\bar{1}}^+ = 0$ and $(P_{\bar{1}}^+ \wedge Q_{\bar{1}}^+) \wedge R_{\bar{1}}^+ = 0 \Leftrightarrow$ The points $P_{\bar{1}}^+$, $Q_{\bar{1}}^+$, $R_{\bar{1}}^+$ are collinear.

ii) ⇔ iii) $0 = (P_{\bar{1}}^+ \wedge Q_{\bar{1}}^+) \vee R_{\bar{1}}^+ = \langle (P_{\bar{1}}^+ \wedge Q_{\bar{1}}^+) * R_{\bar{1}}^+ \rangle_3^- = \langle [(P_{\bar{1}}^+)^\sim \vee (Q_{\bar{1}}^+)^\sim]^\sim * R_{\bar{1}}^+ \rangle_3^- = \langle \langle P_{\bar{1}}^+ * (\mathrm{I}^-)^{-1} * Q_{\bar{1}}^+ * (\mathrm{I}^-)^{-1} \rangle_2^- * (\mathrm{I}^-)^{-1} * Q_{\bar{1}}^+ \rangle_3^- = (\mathrm{I}^-)^2 \langle P_{\bar{1}}^+ * Q_{\bar{1}}^+ * R_{\bar{1}}^+ \rangle^- * (\mathrm{I}^-)^{-1} \Leftrightarrow \langle P_{\bar{1}}^+ * Q_{\bar{1}}^+ * R_{\bar{1}}^+ \rangle_3^+ = 0.$ □

Theorem 12. *The following statements are equivalent:*

i) The lines $p_{\bar{2}}^+$, $q_{\bar{2}}^+$, $r_{\bar{2}}^+$ are concurrent, i.e., they pass through a common point.

ii) $(p_{\bar{2}}^+ \vee q_{\bar{2}}^+) \wedge r_{\bar{2}}^+ = 0.$

iii) $\langle p_{\bar{2}}^+ q_{\bar{2}}^+ r_{\bar{2}}^+ \rangle^+ = 0.$

Proof. Apply minor duality to Theorem 11. □

Theorem 13. *The following statements are equivalent:*

i) The lines $P_{\bar{1}}^-$, $Q_{\bar{1}}^-$, $R_{\bar{1}}^-$ are concurrent, i.e., they pass through a common point.

ii) $(P_{\bar{1}}^- \vee Q_{\bar{1}}^-) \wedge R_{\bar{1}}^- = 0.$

iii) $\langle P_{\bar{1}}^- Q_{\bar{1}}^- R_{\bar{1}}^- \rangle_{\bar{3}}^- = 0.$

Proof. Apply major duality to Theorem 11. □

Theorem 14. *The following statements are equivalent:*

i) The points $p_{\bar{2}}^-$, $q_{\bar{2}}^-$, $r_{\bar{2}}^-$ *are collinear, i.e., they lie in a common line.*

ii) $(p_{\bar{2}}^- \wedge q_{\bar{2}}^-) \vee r_{\bar{2}}^- = 0.$

iii) $\langle p_{\bar{2}}^- * q_{\bar{2}}^- * r_{\bar{2}}^- \rangle^- = 0.$

Proof. Apply major duality to Theorem 12. □

Let $\Delta = \{A_{\bar{1}}^+, B_{\bar{1}}^+, C_{\bar{1}}^+\}$ and $\Delta' = \{A_{\bar{1}}'^+, B_{\bar{1}}'^+, C_{\bar{1}}'^+\}$ be two generic triangles in the field $\mathcal{G}_3^+$ with

$$J^+ = \langle J^+ \rangle_3 = A_{\bar{1}}^+ \wedge B_{\bar{1}}^+ \wedge C_{\bar{1}}^+ = \mu(\mathrm{I}^+)^{-1}, \tag{7.1a}$$

$$J'^+ = \langle J'^+ \rangle_3 = A_{\bar{1}}'^+ \wedge B_{\bar{1}}'^+ \wedge C_{\bar{1}}'^+ = \mu'(\mathrm{I}^+)^{-1}, \tag{7.1b}$$

and

$$(J^+)^2 = -(A_{\bar{1}}^+)^2(B_{\bar{1}}^+)^2(C_{\bar{1}}^+)^2, \quad (J'^+)^2 = -(A_{\bar{1}}'^+)^2(B_{\bar{1}}'^+)^2(C_{\bar{1}}'^+)^2. \tag{7.2}$$

The corresponding sides,

$$a_{\bar{2}}^+ := B_{\bar{1}}^+ \wedge C_{\bar{1}}^+, \qquad b_{\bar{2}}^+ := C_{\bar{1}}^+ \wedge A_{\bar{1}}^+, \qquad c_{\bar{2}}^+ := A_{\bar{1}}^+ \wedge B_{\bar{1}}^+, \tag{7.3a}$$

$$a_{\bar{2}}'^+ := B_{\bar{1}}'^+ \wedge C_{\bar{1}}'^+, \quad b_{\bar{2}}'^+ := C_{\bar{1}}'^+ \wedge A_{\bar{1}}'^+, \quad c_{\bar{2}}'^+ := A_{\bar{1}}'^+ \wedge B_{\bar{1}}'^+, \tag{7.3b}$$

the connecting lines between homologous angles, and the intersecting points between homologous sides,

$$p_{\bar{2}}^+ := A_{\bar{1}}^+ \wedge A_{\bar{1}}'^+, \qquad q_{\bar{2}}^+ := B_{\bar{1}}^+ \wedge B_{\bar{1}}'^+, \qquad r_{\bar{2}}^+ := C_{\bar{1}}^+ \wedge C_{\bar{1}}'^+, \tag{7.4a}$$

$$P_{\bar{1}}^+ := a_{\bar{2}}^+ \vee a_{\bar{2}}'^+, \qquad Q_{\bar{1}}^+ := b_{\bar{2}}^+ \vee b_{\bar{2}}'^+, \qquad R_{\bar{1}}^+ := c_{\bar{2}}^+ \vee c_{\bar{2}}'^+, \tag{7.4b}$$

transform under multiplication from the right side with the pseudoscalar J^+ or J'^+ into the following homogeneous multivectors:

$$\begin{aligned} a_{\bar{2}}^+ J^+ &= -(B_{\bar{1}}^+ \wedge C_{\bar{1}}^+)(C_{\bar{1}}^+ \wedge B_{\bar{1}}^+ \wedge A_{\bar{1}}^+) \\ &= -\langle B_{\bar{1}}^+ C_{\bar{1}}^+ \rangle_{\bar{2}}^+ \langle C_{\bar{1}}^+ B_{\bar{1}}^+ A_{\bar{1}}^+ \rangle_{\bar{3}}^+ \\ &= -(B_{\bar{1}}^+)^2 (C_{\bar{1}}^+)^2 A_{\bar{1}}^+, \\ b_{\bar{2}}^+ J^+ &= -(A_{\bar{1}}^+)^2 (C_{\bar{1}}^+)^2 B_{\bar{1}}^+, \\ c_{\bar{2}}^+ J^+ &= -(A_{\bar{1}}^+)^2 (B_{\bar{1}}^+)^2 C_{\bar{1}}^+, \end{aligned} \tag{7.5a}$$

$$\begin{aligned} a_{\bar{2}}'^+ J'^+ &= -(B_{\bar{1}}'^+)^2 (C_{\bar{1}}'^+)^2 A_{\bar{1}}'^+, \\ b_{\bar{2}}'^+ J'^+ &= -(A_{\bar{1}}'^+)^2 (C_{\bar{1}}'^+)^2 B_{\bar{1}}'^+, \\ c_{\bar{2}}'^+ J'^+ &= -(A_{\bar{1}}'^+)^2 (B_{\bar{1}}'^+)^2 C_{\bar{1}}'^+, \end{aligned} \tag{7.5b}$$

$$\begin{aligned}
P_{\bar{1}}^{+}J^{+} &= \frac{(\mathrm{I}^{+})^2}{\mu'}\langle a_{\bar{2}}^{+}J^{+}a_{\bar{2}}'^{+}J'^{+}\rangle_{\bar{2}}^{+} \\
&= \frac{(\mathrm{I}^{+})^2}{\mu'}(B_{\bar{1}}^{+})^2(C_{\bar{1}}^{+})^2(B_{\bar{1}}'^{+})^2(C_{\bar{1}}'^{+})^2\langle A_{\bar{1}}^{+}A_{\bar{1}}'^{+}\rangle_{\bar{2}}^{+} \\
&= \frac{(\mathrm{I}^{+})^2}{\mu'}(B_{\bar{1}}^{+})^2(C_{\bar{1}}^{+})^2(B_{\bar{1}}'^{+})^2(C_{\bar{1}}'^{+})^2 p_{\bar{2}}^{+}, \\
Q_{\bar{1}}^{+}J^{+} &= \frac{(\mathrm{I}^{+})^2}{\mu'}(C_{\bar{1}}^{+})^2(A_{\bar{1}}^{+})^2(C_{\bar{1}}'^{+})^2(A_{\bar{1}}'^{+})^2 q_{\bar{2}}^{+}, \\
R_{\bar{1}}^{+}J^{+} &= \frac{(\mathrm{I}^{+})^2}{\mu'}(A_{\bar{1}}^{+})^2(B_{\bar{1}}^{+})^2(A_{\bar{1}}'^{+})^2(B_{\bar{1}}'^{+})^2 r_{\bar{2}}^{+}.
\end{aligned} \tag{7.6}$$

We can now state the Desargues' theorem in the plane $\mathcal{G}_3$.

Theorem 15. $\langle P_{\bar{1}}^{+} * Q_{\bar{1}}^{+} * R_{\bar{1}}^{+}\rangle_{\bar{3}}^{+} = 0 \Leftrightarrow \langle p_{\bar{2}}^{+}q_{\bar{2}}^{+}r_{\bar{2}}^{+}\rangle^{+} = 0.$

Proof.

$$\begin{aligned}
\langle P_{\bar{1}}^{+} * Q_{\bar{1}}^{+} * R_{\bar{1}}^{+}\rangle_{\bar{3}}^{+} &= \frac{(\mathrm{I}^{+})^2}{\mu^3}\langle P_{\bar{1}}^{+}J^{+}Q_{\bar{1}}^{+}J^{+}R_{\bar{1}}^{+}J^{+}\rangle^{+}(\mathrm{I}^{+})^{-1} \\
&= \frac{(J^{+}J'^{+})^2}{\mu^3\mu'^3}\langle p_{\bar{2}}^{+}q_{\bar{2}}^{+}r_{\bar{2}}^{+}\rangle^{+}(\mathrm{I}^{+})^{-1} \\
&= \frac{(\mathrm{I}^{+})^2}{J^{+}J'^{+}}\langle p_{\bar{2}}^{+}q_{\bar{2}}^{+}r_{\bar{2}}^{+}\rangle^{+}(\mathrm{I}^{+})^{-1}.
\end{aligned}$$

□

Theorem 16. $\langle P_{\bar{1}}^{-}Q_{\bar{1}}^{-}R_{\bar{1}}^{-}\rangle_{\bar{3}}^{-} = 0 \Leftrightarrow \langle p_{\bar{2}}^{-} * q_{\bar{2}}^{-} * r_{\bar{2}}^{-}\rangle^{-} = 0.$

Proof. Apply major duality to Theorem 15. □

Theorem 15 and 16 are clearly self-dual and may also represent Desargues' theorem in the bundle $\mathcal{G}_3$.

For the proof of Desargues' theorem in *space* $\mathcal{G}_4$, we start with two triangles $\Delta = \{A_{\bar{1}}^{+}, B_{\bar{1}}^{+}, C_{\bar{1}}^{+}\}$ and $\Delta' = \{A_{\bar{1}}'^{+}, B_{\bar{1}}'^{+}, C_{\bar{1}}'^{+}\}$ that lie perspective from a center $Z_{\bar{1}}^{+}$. Thus, we may write

$$\alpha' A_{\bar{1}}'^{+} = Z_{\bar{1}}^{+} + \alpha A_{\bar{1}}^{+},\ \beta' B_{\bar{1}}'^{+} = Z_{\bar{1}}^{+} + \beta B_{\bar{1}}^{+},\ \gamma' C_{\bar{1}}'^{+} = Z_{\bar{1}}^{+} + \gamma C_{\bar{1}}^{+}, \tag{7.7}$$

and

$$\begin{aligned}
\alpha'\beta'\gamma' A_{\bar{1}}'^{+} \wedge B_{\bar{1}}'^{+} \wedge C_{\bar{1}}'^{+} &= Z_{\bar{1}}^{+} \wedge (\beta\gamma B_{\bar{1}}^{+} \wedge C_{\bar{1}}^{+} + \alpha\gamma C_{\bar{1}}^{+} \wedge A_{\bar{1}}^{+} \\
&\quad + \alpha\beta A_{\bar{1}}^{+} \wedge B_{\bar{1}}^{+}) + \alpha\beta\gamma A_{\bar{1}}^{+} \wedge B_{\bar{1}}^{+} \wedge C_{\bar{1}}^{+}.
\end{aligned} \tag{7.8}$$

Hence, the 2-vector

$$z_{\bar{2}}^{+} := \beta\gamma B_{\bar{1}}^{+} \wedge C_{\bar{1}}^{+} + \alpha\gamma C_{\bar{1}}^{+} \wedge A_{\bar{1}}^{+} + \alpha\beta A_{\bar{1}}^{+} \wedge B_{\bar{1}}^{+} \tag{7.9}$$

represents the line of intersection of the plane $A'^{+}_{\bar{1}} \wedge B'^{+}_{\bar{1}} \wedge C'^{+}_{\bar{1}}$ and the plane $A^{+}_{\bar{1}} \wedge B^{+}_{\bar{1}} \wedge C^{+}_{\bar{1}}$ if the two planes are different. In any case $z^{+}_{\bar{2}}$ is the common line of

$$P^{+}_{\bar{1}} := \gamma C^{+}_{\bar{1}} - \beta B^{+}_{\bar{1}},\; Q^{+}_{\bar{1}} := \alpha A^{+}_{\bar{1}} - \gamma C^{+}_{\bar{1}},\; R^{+}_{\bar{1}} := \beta B^{+}_{\bar{1}} - \alpha A^{+}_{\bar{1}}, \tag{7.10}$$

whose expressions can be deduced from the equations

$$\begin{aligned}
\alpha'\beta' A'^{+}_{\bar{1}} \wedge B'^{+}_{\bar{1}} &= Z^{+}_{\bar{1}} \wedge (\beta B^{+}_{\bar{1}} - \alpha A^{+}_{\bar{1}}) + \alpha\beta A^{+}_{\bar{1}} \wedge B^{+}_{\bar{1}},\\
\beta'\gamma' B'^{+}_{\bar{1}} \wedge C'^{+}_{\bar{1}} &= Z^{+}_{\bar{1}} \wedge (\gamma C^{+}_{\bar{1}} - \beta B^{+}_{\bar{1}}) + \beta\gamma B^{+}_{\bar{1}} \wedge C^{+}_{\bar{1}},\\
\gamma'\alpha' C'^{+}_{\bar{1}} \wedge A'^{+}_{\bar{1}} &= Z^{+}_{\bar{1}} \wedge (\alpha A^{+}_{\bar{1}} - \gamma C^{+}_{\bar{1}}) + \gamma\alpha C^{+}_{\bar{1}} \wedge A^{+}_{\bar{1}}.
\end{aligned} \tag{7.11}$$

Direct computation shows

$$\begin{aligned}
z^{+}_{\bar{2}} \wedge P^{+}_{\bar{1}} &= 0 \quad \text{and} \quad z^{+}_{\bar{2}} \vee P^{+}_{\bar{1}} = 0,\\
z^{+}_{\bar{2}} \wedge Q^{+}_{\bar{1}} &= 0 \quad \text{and} \quad z^{+}_{\bar{2}} \vee Q^{+}_{\bar{1}} = 0,\\
z^{+}_{\bar{2}} \wedge R^{+}_{\bar{1}} &= 0 \quad \text{and} \quad z^{+}_{\bar{2}} \vee R^{+}_{\bar{1}} = 0.
\end{aligned} \tag{7.12}$$

For the other direction of Desargues' theorem in space, we start with two triangles Δ and Δ' where the intersecting points $P^{+}_{\bar{1}}$, $Q^{+}_{\bar{1}}$, and $R^{+}_{\bar{1}}$ lie on a common line $z^{+}_{\bar{2}}$. Hence, we can write

$$\begin{aligned}
\beta B^{+}_{\bar{1}} &= \alpha A^{+}_{\bar{1}} + R^{+}_{\bar{1}}, & \beta' B'^{+}_{\bar{1}} &= \alpha' A'^{+}_{\bar{1}} + R^{+}_{\bar{1}},\\
\alpha A^{+}_{\bar{1}} &= \gamma C^{+}_{\bar{1}} + Q^{+}_{\bar{1}}, & \alpha' A'^{+}_{\bar{1}} &= \gamma' C'^{+}_{\bar{1}} + Q^{+}_{\bar{1}},\\
\gamma C^{+}_{\bar{1}} &= \beta B^{+}_{\bar{1}} + P^{+}_{\bar{1}}, & \gamma' C'^{+}_{\bar{1}} &= \beta' B'^{+}_{\bar{1}} + P^{+}_{\bar{1}}.
\end{aligned} \tag{7.13}$$

The 1-vector

$$Z^{+}_{\bar{1}} := \alpha' A'^{+}_{\bar{1}} - \alpha A^{+}_{\bar{1}} = \beta' B'^{+}_{\bar{1}} - \beta B^{+}_{\bar{1}} = \gamma' C'^{+}_{\bar{1}} - \gamma C^{+}_{\bar{1}} \tag{7.14}$$

clearly represents the common intersection point of the lines $A^{+}_{\bar{1}} \wedge A'^{+}_{\bar{1}}$, $B^{+}_{\bar{1}} \wedge B'^{+}_{\bar{1}}$, and $C^{+}_{\bar{1}} \wedge C'^{+}_{\bar{1}}$, i.e., it is the center of perspectivity of the triangles Δ and Δ'. Thus, we have proved Desargues' theorem in space.

By major duality in space, we find that two "triplanes" $\Delta = \{A^{-}_{\bar{1}}, B^{-}_{\bar{1}}, C^{-}_{\bar{1}}\}$ and $\Delta' = \{A'^{-}_{\bar{1}}, B'^{-}_{\bar{1}}, C'^{-}_{\bar{1}}\}$ are perspective from a plane $Z^{-}_{\bar{1}}$, i.e., the lines of intersection $p^{-}_{\bar{2}}$, $q^{-}_{\bar{2}}$, and $r^{-}_{\bar{2}}$ lie in the common plane $Z^{-}_{\bar{1}}$ if and only if the planes of connection $P^{-}_{\bar{1}}$, $Q^{-}_{\bar{1}}$, and $R^{-}_{\bar{1}}$ pass through a common line $z^{-}_{\bar{2}}$.

Acknowledgments

I would like to thank D. Trautman, H.-J. Stoß, P. Gschwind, T. Heim, F. Sommen, P. Lounesto, B. Fauser, G. Sobczyk, and D. Hestenes for their support and many fruitful discussions.

REFERENCES

[1] H. S. M. Coxeter, *Non-Euclidean Geometry, Fifth Edition,* University of Toronto Press, Toronto, 1978.

[2] D. Hestenes and G. Sobczyk, *Clifford Algebra to Geometric Calculus,* D. Reidel Publishing Company, Dordrecht, 1985.

[3] D. Hestenes and R. Ziegler, Projective geometry with Clifford algebra, *A*cta Appl. Math. **23** (1991), 25–63.

[4] D. Hestenes, The design of linear algebra and geometry, *Acta Appl. Math.* **23** (1991), 65–93.

[5] F. Klein, *Vorlesungen über Nicht-Euklidische Geometrie,* Springer, Berlin, 1968.

[6] F. Lindemann, Über unendlich kleine bewegungen und über kraftsysteme bei allgemeiner projectivischer massbestimmung, *Math. Ann.* **7** (1874), 56–144.

[7] L. Locher, *Projektive Geometrie und die Grundlagen der Euklidischen und Polareuklidischen Geometrie,* **2**, Auflage, Verlag am Goetheanum, Dornach (Schweiz), 1980.

[8] L. Locher, *Raum und Gegenraum,* **2**. Auflage, Verlag am Goetheanum, Dornach (Schweiz), 1970.

[9] H.-J. Stoß, *Treffgeraden und Nullinvarianz,* Verlag am Goetheanum, Dornach (Schweiz), 1995.

[10] O. Veblen and J. W. Young, *Projective Geometry,* Volume I and II, Ginn and Company, Boston, 1910 and 1918.

[11] R. Ziegler, *Die Geschichte der Geometrischen Mechanik im 19. Jahrhundert,* Franz Steiner Verlag Wiesbaden, Stuttgart, 1985.

Oliver Conradt
Department of Physics and Astronomy
University of Basel
Klingelbergstrasse 82, CH-4056 Basel, Switzerland
E-mail: oliver.conradt@unibas.ch

Received: September 20, 1999; Revised: October 29, 1999

Doing Geometric Research with Clifford Algebra

Hongbo Li

ABSTRACT Clifford algebra is a coordinate-free approach to geometric research and applications. In this paper we present some basic techniques that we have been developing and employing in our research on a Clifford algebra approach to hyperbolic geometry and automated geometry theorem proving. Our work shows that, following this approach, not only can we reformulate existing results with improvements and generalizations, but we also can make new discoveries.
Keywords: Hyperbolic geometry, conformal geometry, automated theorem-proving, Grassmann-Cayley algebra.

1 Introduction

Algebraic methods for geometry date back to Descartes' analytic geometry in the 17th century. Although coordinate methods are fundamental, their lack of geometric invariance often makes the translation from algebra to geometry impossible. Moreover, coordinate methods often require a huge amount of computational work, which may not even be carried out on modern computers. These phenomena occur in both theoretical and applied research and can only be overcome by coordinate-free algebraic methods. Coordinate-free algebraic methods for geometry date back to Grassmann's extension theory in the 19th century. Clifford algebra is one of the algebras that grew out of Grassmann's work. It is a universal algebraic language not only for geometry, but also for mechanics, physics, and a large portion of mathematics (Hestenes, 1987).

Stimulated by the work of Fenchel (1989), Iversen (1992) applied Clifford algebra in a hybrid formalism of both vectors and matrices to hyperbolic geometry, and presented a beautiful algebraic description of that geometry.

The version of Clifford algebra described in (Hestenes and Sobczyk, 1984) is used by Hestenes (1987), and Hestenes and Ziegler (1991) to reformulate spherical and projective geometries. The model of Euclidean geometry in a Minkowski space proposed by Wachter (see Seidel, 1952) is reformulated in the language of Clifford algebra by Hestenes (1991) and Havel (1991, 1995)

AMS Subject Classification: 14N10, 51B10, 51B20, 68T15.

in the more general framework of conformal geometry. These reformulations not only simplify algebraic descriptions and computations, but also help gain new insights into geometries. In geometric applications, a branch of Clifford algebra, called Grassmann-Cayley algebra, is applied to automated geometry theorem proving by Crapo and Richter-Gebert (1995), Mourrain (1999), and to computer vision and robotics by Faugeras and Mourrain (1995), Bayro-Corrochano, Lasenby and Sommer (1996), et al. For the treatment of many problems arising from applications, significant improvements can be achieved by using Clifford algebra.

Influenced by the work of Hestenes and helped by Hestenes himself, we started to apply Clifford algebra in geometric research in 1993. Our research focused on two aspects: hyperbolic geometry and automated geometry theorem proving. In both aspects we have been able to not only reformulate existing theorems and methods with improvements and generalizations but also discover new theorems. One example is that in hyperbolic plane geometry we generalize the existing two formulas on triangles which represent the area of a triangle with its three side lengths (Greenberg, 1980) and represent the perimeter of a triangle with its three inner angles (Fenchel, 1989) to two formulas on convex n-gons. The classical formulas can be formulated in the Clifford algebra $\mathcal{G}_{2,1}$ as follows (see later in this section for explanations of the symbols):

Theorem 1.1 (Old Result). *Let ABC be a triangle. Then*

$$\left\{\begin{array}{rcl} \cos\dfrac{K_{ABC}}{2} & = & 2\dfrac{1-A\cdot B-B\cdot C-C\cdot A}{|A+B||B+C||C+A|}, \\[2ex] \sin\dfrac{K_{ABC}}{2} & = & 2\dfrac{|A\wedge B\wedge C|}{|A+B||B+C||C+A|}. \end{array}\right.$$

Let

$$a_1=\frac{(A\wedge B)^{\sim}}{|A\wedge B|},\ a_2=\frac{(B\wedge C)^{\sim}}{|B\wedge C|},\ a_3=\frac{(C\wedge A)^{\sim}}{|C\wedge A|}.$$

Then

$$\left\{\begin{array}{rcl} \cosh\dfrac{L_{ABC}}{2} & = & 2\dfrac{1+a_1\cdot a_2+a_2\cdot a_3+a_3\cdot a_1}{|a_1+a_2||a_2+a_3||a_3+a_1|}, \\[2ex] \sinh\dfrac{L_{ABC}}{2} & = & 2\dfrac{|a_1\wedge a_2\wedge a_3|}{|a_1+a_2||a_2+a_3||a_3+a_1|}. \end{array}\right.$$

Let a convex n-gon have vertices $A_1,\ldots,A_n$. Let $K_{A_1\cdots A_n}$ be its area and let $L_{A_1\cdots A_n}$ be its perimeter. By means of spinor representations of Lorentz transformations and a group of Gauss equalities, we obtain the following generalizations (Li, 1997):

Theorem 1.2 (New Result).

$$\begin{cases} \cos\frac{K_{A_1A_2\cdots A_n}}{2} &= 2\frac{1+\sum\limits_{k=1}^{[\frac{n}{2}]}\sum\limits_{P_k}(-1)^k < \prod\limits_{l=1}^{2k} A_{P_k(l)} >_0}{\prod\limits_{i=1}^{n}|A_i+A_{i+1}|}, \\ \sin\frac{K_{A_1A_2\cdots A_n}}{2} &= 2\lambda\frac{\sum\limits_{k=1}^{[\frac{(n-1)}{2}]}\sum\limits_{\sigma_k}(-1)^{k-1} < \prod\limits_{l=1}^{2k+1} A_{\sigma_k(l)} >_3^{\sim}}{\prod\limits_{i=1}^{n}|A_i+A_{i+1}|}, \end{cases}$$

where $\lambda = \frac{(A_1\wedge A_2\wedge A_3)^{\sim}}{|A_1\wedge A_2\wedge A_3|}$, *and the indices are understood to be modulo* n. P_k *is a list of* $2k$ *elements from* $\{1,2,\ldots,n\}$, *and* $P_k(i)$ *denotes the* i*-th element in the list.* σ_k *is any list of* $2k+1$ *elements from* $\{1,2,\ldots,n\}$ *and* $\sigma_k(i)$ *denotes the* i*-th element in the list. Also,* $P_k(i) < P_k(i+1), \sigma_k(i) < \sigma_k(i+1)$.

Theorem 1.3 (New Result). *Let* $a_i = \frac{(A_i\wedge A_{i+1})^{\sim}}{|A_i\wedge A_{i+1}|}$, *where the indices are understood to be modulo* n. *Then*

$$\begin{cases} \cosh\frac{L_{A_1A_2\cdots A_n}}{2} &= 2\frac{1+\sum\limits_{k=1}^{[\frac{n}{2}]}\sum\limits_{P_k} < \prod\limits_{l=1}^{2k} a_{P_k(l)} >_0}{\prod\limits_{i=1}^{n}|a_i+a_{i+1}|}, \\ \sinh\frac{L_{A_1A_2\ldots A_n}}{2} &= 2\frac{|\sum\limits_{k=1}^{[\frac{(n-1)}{2}]}\sum\limits_{\sigma_k} < \prod\limits_{l=1}^{2k+1} a_{\sigma_k(l)} >_3^{\sim}|}{\prod\limits_{i=1}^{n}|a_i+a_{i+1}|}, \end{cases}$$

where P_k *is a list of* $2k$ *elements from* $\{1,2,\ldots,n\}$ *and* $P_k(i)$ *denotes the* i*-th element in the list;* σ_k *is a list of* $2k+1$ *elements from* $\{1,2,\ldots,n\}$ *and* $\sigma_k(i)$ *denotes the* i*-th element in the list. Also,* $P_k(i) < P_k(i+1), \sigma_k(i) < \sigma_k(i+1)$.

Another example is the generalization of Wachter's model of Euclidean geometry to a universal homogeneous model for Euclidean, spherical and hyperbolic geometries (Li, Hestenes and Rockwood, 1999a, b, c). Within the universal model, a natural correspondence is established among theorems in the three geometries, i.e., a theorem in one geometry can be translated into two theorems in the other two geometries by interpreting rescaled null vectors differently.

We have successfully applied Clifford algebra in automated geometry theorem proving using a general method combining Clifford algebra with Wu's method of characteristic sets (Li, 1994, 1996a; Li and Cheng, 1997). The method can be used to prove theorems in Euclidean, affine, projective, non-Euclidean, and differential geometries, and it can also be used in mechanics and robotics. The proofs produced are often short and have geometric interpretations. In particular, in the local theory of space curves and surfaces in differential geometry, by integrating E. Cartan's moving frames and the calculus of exterior differential forms with Wu's method, we can produce proofs that are similar to those given in textbooks and sometimes even simpler.

The general automated geometry theorem proving method can also be used to study sophisticated geometric problems. Below is a theorem first proposed by E. Cartan as a conjecture and later proved by S. S. Chern (1985).

Theorem 1.4 (Chern's Theorem). *A non-trivial family of isometric surfaces having the same principal curvatures is either a family of surfaces of constant mean curvature or a family of Weingarten surfaces of non-constant mean curvature, assuming that they do not contain umbilics and are* c^5.

Our computer-generated proof (Li, 1997a) uses only three rules in the computation:

1. substitution;
2. solving vectorial equations: if ω_1, ω_2 are independent 1-forms defined over a patch of space surface, then a 1-form ω can be solved from
$$\omega \wedge \omega_1 = \Omega_1, \quad \omega \wedge \omega_2 = \Omega_2,$$
where Ω_1, Ω_2 are 2-forms, and the solution is
$$\omega = \frac{\Omega_2}{\omega_1 \wedge \omega_2}\omega_1 - \frac{\Omega_1}{\omega_1 \wedge \omega_2}\omega_2;$$
3. generating integrability equations: if the leading variable of a vectorial equation is df, where f is a function, then the equation can be substituted into $d(df) = 0$ to get an integrability condition.

The proof is straightforward computing compared with the original proof using techniques like complex differential forms.

As the final example, we have used the general method, together with some techniques from algebraic geometry, to tackle a conjecture proposed by Erdös, et al., in 1994, and obtained the best result so far. The original problem is from work of Erdös, Jackson and Mauldlin.

For ten points A_{ij}, $1 \leq i < j \leq 5$ on a plane, if there are five points A_k, $1 \leq k \leq 5$ on the plane, including points at infinity, of which at least two points are different, such that A_i, A_j, A_{ij} are collinear, $1 \leq i < j \leq 5$, we say the five points form a consistent 5-tuple. Now assume that no three of the A_{ij}'s are collinear. Is it true that there are only finitely many consistent 5-tuples?

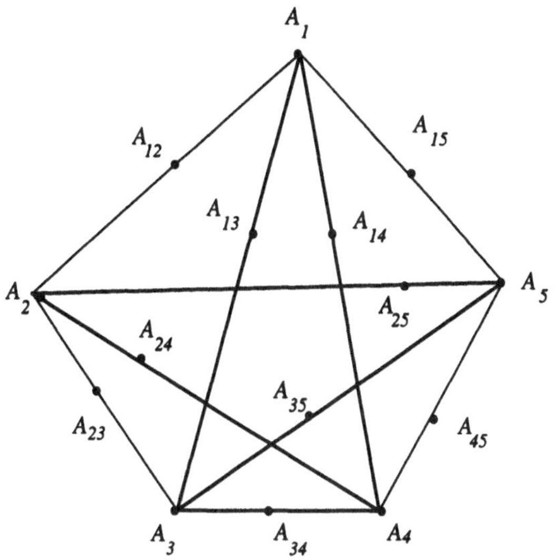

FIGURE 1. A conjecture by Erdös, et al., around 1994.

Erdös, et al., proved that if there are only finitely many solutions, then there are at most 49. They conjectured that there are always finitely many solutions but could not prove it. They sent the problem to Boyer asking if their prover could solve it. Boyer forwarded the problem to Chou and Gao, and, in turn, Chou and Gao sent it to us. Chou and Gao proved that, generically, there are only finitely many solutions. In (Li and Shi, 1997), we proved that

Theorem 1.5 (New Result). *For 10 general points A_{ij} on the plane, there are at most 6 solutions.*

Our work shows that we can do geometric research by a Clifford algebra approach. The present paper intends to attract attention. For this purpose, we present some basic techniques that we have been developing and using in our research on a Clifford algebra approach to hyperbolic geometry and Clifford algebraic methods for automated geometry theorem proving. The paper is divided into two sections. The first section is on Clifford algebraic representations of hyperbolic geometry, and the second is on a general method combining Clifford algebra with Wu's method for automated geometry theorem proving. The notations on Clifford algebra are from (Hestenes and Sobczyk, 1984); the terminology on hyperbolic geometry is from (Iversen, 1992); and the terminology on automated theorem proving is from (Wu, 1978, 1979).

To facilitate reading the material, we present below a short description of symbols and notions for the Clifford algebra used in this paper.

A real *Clifford algebra* generated by an n-dimensional real inner product space $\mathcal{V}^n$ of signature[1] (p,q,r), is the quotient algebra of the tensor algebra $\otimes(\mathcal{V}^n)$ by the two-sided ideal generated by elements of the form $x\otimes x - x\cdot x$ where $x \in \mathcal{V}^n$ and the dot denotes the inner product in $\mathcal{V}^n$. The Clifford algebra is denoted by $\mathcal{G}(\mathcal{V}^n)$, or $\mathcal{G}_{p,q,r}$ when the signature of the space is stressed. An element of a Clifford algebra is called a *multivector*, and the multiplication of the algebra induced from the tensor product is called the *geometric product.*

The *inner product* of the space $\mathcal{V}^n$ can be extended to the whole of $\mathcal{G}(\mathcal{V}^n)$. The geometric product also induces a product, called the *outer product*, in the algebra. The vector space $\mathcal{G}(\mathcal{V}^n)$ together with the outer product forms a *Grassmann algebra.* The outer product is denoted by "$\wedge$". There is another important product in Clifford algebra called the *cross product.* The cross product of two multivectors x and y is defined as $x\times y = \frac{1}{2}(xy - yx)$ where the juxtaposition of multivectors represents the geometric product. The *square* of a multivector x is $x^2 = xx$.

For two multivectors x, y, if $xy = yx = 1$, they are called *invertible*, denoted by $x = y^{-1}$ and $y = x^{-1}$. Not all multivectors are invertible. For two multivectors x, y, if $xy^{-1} = y^{-1}x$, the following symbol is used: $x/y = xy^{-1} = y^{-1}x$.

A fundamental concept in Clifford algebra is an *r-blade* (blade of *grade* r). Let $a_1, \ldots, a_r$ be vectors of $\mathcal{V}^n$. Then $a_1 \wedge \cdots \wedge a_r$ is called an r-blade if it is nonzero. The square of a blade is a scalar that equals the inner product of the blade with itself. The *magnitude* of a blade x is $|x| = \sqrt{|x^2|}$. The *reverse* of an r-blade x is $x^\dagger = (-1)^{r(r-1)/2}x$. The *main anti-automorphism* of an r-blade x is $x^* = (-1)^r x$. The latter two concepts can be extended to any multivector by linearity.

Let x be an invertible blade. The *projection* of a multivector y onto x is $P_x(y) = (y\cdot x)x^{-1}$. The *dual* of a multivector y with respect to x is $yx^\dagger$. In particular, when x is an n-blade, called a *pseudoscalar*, the dual is denoted by $y^\sim$. The *meet* of two multivectors y, z is $y \vee z = y^\sim \cdot z$.

The set of all r-blades where r is even, together with $\mathbb{R}$, generates a vector subspace $\mathcal{G}^+(\mathcal{V}^n)$ of $\mathcal{G}(\mathcal{V}^n)$. It is a subalgebra of $\mathcal{G}(\mathcal{V}^n)$ called the *even subalgebra.*

Let $a_1, \ldots, a_r$ be invertible vectors of $\mathcal{V}^n$. Then $a_1\cdots a_r$ is called a *versor.* When r is even, $a_1\cdots a_r$ is called a *spinor.* The *adjugation* of a multivector y by a versor x is defined by $Ad^*_x(y) = x^{*-1}yx$. When x is a spinor, since $Ad^*_x(y) = x^{-1}yx$, a different symbol $Ad_x(y)$ can be used to replace $Ad^*_x(y)$.

[1]The *signature* of a real inner product space is a triplet of nonnegative integers whose sum equals the dimension of the space; the first (second, third) integer equals the dimension of a maximal subspace where every vector x satisfies $x\cdot x > 0\,(<0, =0)$. A real inner product space with signature (p,q,r) is often denoted by $\mathbb{R}^{p,q,r}$; when $r=0$, it is denoted by $\mathbb{R}^{p,q}$; when $q=r=0$, it is denoted by $\mathbb{R}^p$; when $p=r=0$, it is denoted by $\mathbb{R}^{-q}$.

2 Clifford algebraic representations of hyperbolic geometry

2.1 The hyperboloid model

For hyperbolic n-space, there are five important analytic models: the Poincaré ball model, the Poincaré half-space model, the Klein ball model, the hemisphere model and the hyperboloid model. In the Minkowski space $\mathbb{R}^{n,1}$, the set

$$\mathcal{D}^n = \{X \in \mathbb{R}^{n,1} | X \cdot X = -1\} \tag{2.1}$$

is called a double hyperbolic n-space. It has two components, denoted by $\mathcal{H}^n$ and $-\mathcal{H}^n$, respectively. The component $\mathcal{H}^n$ is called the hyperboloid model of the hyperbolic n-space.

Definition 2.1. *A generalized point of $\mathcal{D}^n$ is either a point, an end, or a direction. A point is an element of $\mathcal{D}^n$. An end is a one-dimensional null half-space of $\mathbb{R}^{n,1}$; a direction is a one-dimensional Euclidean half-space of $\mathbb{R}^{n,1}$. A point at infinity refers to a pair of antipodal ends.*

Let A, B be two distinct points in $\mathcal{D}^n$. Then they are on the same component if and only if $A \cdot B < 0$. Since the blade $A \wedge B$ is Minkowski, $(A \wedge B)^2 = (A \cdot B)^2 - 1 > 0$. So A, B are on the same component if and only if $A \cdot B < -1$.

Proposition 2.1. *Let A, B be two points in $\mathcal{H}^n$, and let $d(A, B)$ be their hyperbolic distance. Then*

$$A \cdot B = -\cosh d(A, B). \tag{2.2}$$

Definition 2.2. *A generalized point of $\mathcal{H}^n$ is either a point, a point at infinity, or an imaginary point. A point is an element of $\mathcal{H}^n$. A point at infinity is a null 1-space of $\mathbb{R}^{n,1}$; an imaginary point is a Euclidean 1-space of $\mathbb{R}^{n,1}$.*

Algebraically, a point can be represented by a vector of square -1; an end or a point at infinity can be represented by a null vector; a direction or an imaginary point can be represented by a unit vector. The set of points at infinity of $\mathcal{D}^n$ (or $\mathcal{H}^n$) is called the sphere at infinity. The set of ends of $\mathcal{D}^n$ has two components, similar to $\mathcal{D}^n$ itself.

Definition 2.3. *A hyperbolic r-plane of $\mathcal{D}^n$ (or $\mathcal{H}^n$) is the intersection of $\mathcal{D}^n$ (or $\mathcal{H}^n$) with an $(r+1)$-space of $\mathbb{R}^{n,1}$, where $0 \leq r \leq n-1$.*

Algebraically, a hyperbolic r-plane can be represented by the Minkowski $(r+1)$-blade representing the $(r+1)$-space of $\mathbb{R}^{n,1}$. By this representation, any r-plane is oriented. When $r = 1$, a hyperbolic 1-plane is a line. The tangent direction of a line L_2 at a point A is the direction of vector $A \cdot L_2$, where L_2 is a 2-blade representing the line.

Definition 2.4. *An r-sphere at infinity of $\mathcal{D}^n$ (or $\mathcal{H}^n$) is the intersection of the sphere at infinity with an $(r+1)$-plane of $\mathcal{D}^n$ (or $\mathcal{H}^n$).*

Now we introduce the concept of generalized spheres.

Definition 2.5. *A generalized sphere of $\mathcal{D}^n$ (or $\mathcal{H}^n$) is either a sphere, or a horosphere, or a hypersphere. It is determined by a pair (C, ρ), where C is a vector in $\mathbb{R}^{n,1}$ representing a generalized point, called the center of the generalized sphere, and $\rho > 0$ is called the generalized radius. When C is a point, the set $\{P \in \mathcal{D}^n \text{ (or } \mathcal{H}^n)|P\cdot C = -(1+\rho)\}$ is called a sphere; when C is an end, the set $\{P \in \mathcal{D}^n|P\cdot C = -\rho\}$ is called a horosphere of $\mathcal{D}^n$; when C is a point at infinity, the set $\{P \in \mathcal{H}^n||P\cdot C| = \rho\}$ is called a horosphere of $\mathcal{H}^n$; when C is a direction, the set $\{P \in \mathcal{D}^n \text{ (or } \mathcal{H}^n)|P\cdot C = -\rho\}$ is called a hypersphere. The hyperplane of $\mathcal{D}^n$ (or $\mathcal{H}^n$) represented by $C^\sim$ is called the axis of the hypersphere.*

Definition 2.6. *A generalized r-sphere of $\mathcal{D}^n$ (or $\mathcal{H}^n$) is a generalized sphere in an $(r+2)$-plane, when taking the $(r+2)$-plane as a double hyperbolic (or hyperbolic) $(r+1)$-space.*

Definition 2.7. *A doublesphere of $\mathcal{D}^n$ (or $\mathcal{H}^n$) is a hypersphere together with its reflection with respect to the axis. An r-doublesphere is an r-hypersphere together with its reflection with respect to the axis.*

A more algebraic definition of generalized spheres is as follows.

Definition 2.8. *A generalized r-sphere of $\mathcal{D}^n$ (or $\mathcal{H}^n$) is the intersection of $\mathcal{D}^n$ (or $\mathcal{H}^n$) with an affine $(r+1)$-plane of $\mathbb{R}^{n,1}$, assuming that the intersection is not empty. When the affine $(r+1)$-plane does not go through the origin, the generalized r-sphere is called an r-dimensional sphere, horosphere or hypersphere, if the space of displacements of the affine $(r+1)$-plane is Euclidean, degenerate or Minkowski, respectively.*

Definition 2.9. *A total sphere of $\mathcal{D}^n$ refers to a generalized sphere, or a hyperplane, or the sphere at infinity. A total r-sphere of $\mathcal{D}^n$ is an r-dimensional generalized sphere, plane, or sphere at infinity.*

The hyperboloid model has some unique features in representing geometric entities and transformations.

- The model is isotropic in that, at every point of the model, the metric of the tangent space is the same.
- A hyperbolic r-plane can be identified with a projective r-plane when viewed from the origin. This enables us to study hyperbolic r-planes in the framework of linear subspaces of $\mathbb{R}^{n,1}$.
- The angle of two intersecting lines is the Euclidean angle between their tangent directions at the intersection. This is the conformal property of the model.

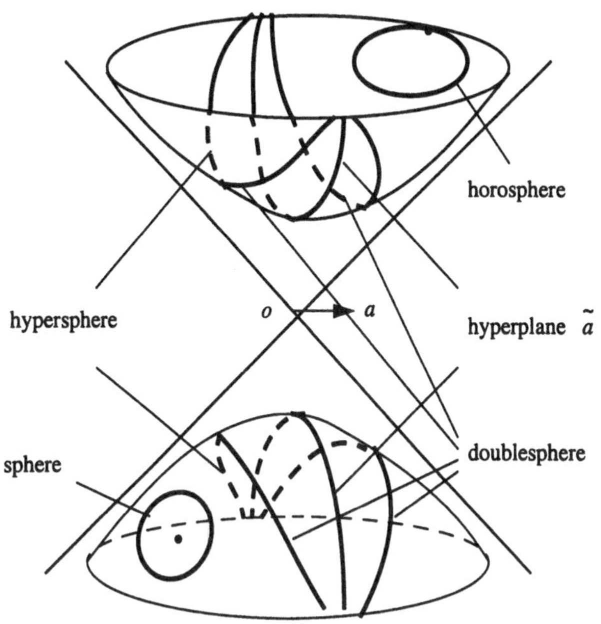

FIGURE 2. Total spheres and a doublesphere.

- Formula (2.2) reduces a problem on distances to a problem on inner products.
- Definition 2.8 enables us to study generalized r-spheres in the framework of affine $(r+1)$-planes of $\mathbb{R}^{n,1}$.
- The hyperbolic isometries are those orthogonal transformations of $\mathbb{R}^{n,1}$ keeping $\mathcal{H}^n$ invariant. In particular, they are all linear transformations.
- The model is quite similar to the model of spherical n-space in $\mathbb{R}^{n+1}$.

These features decide that it is natural to apply Clifford algebra in hyperbolic geometry, just as the applications of Clifford algebra in projective geometry (Hestenes and Ziegler, 1991) and spherical geometry (Hestenes, 1987). Indeed, some applications of Clifford algebra in hyperbolic 3-space have been worked out by Iversen (1992).

2.2 Generalized triangles

In this subsection, we present a fundamental geometric concept in hyperbolic plane geometry called generalized triangles. This concept unifies and generalizes the classical hyperbolic triangles, right pentagons and right hexagons. The Clifford algebra used here is $\mathcal{G}_{2,1}$, and the model for the hyperbolic plane is the hyperboloid model.

Definition 2.10. *A generalized triangle in $\mathcal{H}^2$ is composed of three non-collinear generalized points and the three lines connecting them, assuming that the lines exist.*

Proposition 2.2. *Let A, B, C be three generalized points. Then they form a generalized triangle if and only if $A \wedge B \wedge C \neq 0$, $(A \cdot B)(B \cdot C)(C \cdot A) \neq 0$ and the three blades $A \wedge B, B \wedge C, A \wedge C$ are all Minkowski.*

We can easily recognize that $(A \wedge B \wedge C)^{\sim}$ is the magnitude (Greenberg, 1980) of the triangle ABC when A, B, C are points. What is the geometric meaning of $(A \cdot B)(B \cdot C)(C \cdot A)$? We shall see that its sign characterizes the convexity of ABC.

Definition 2.11. *A generalized triangle is said to be convex if any two sides of it are on the same side of its third side.*

Proposition 2.3. *Let ABC be a generalized triangle.*

1. *When A, B, C are points or points at infinity, then ABC is convex.*
2. *When A is an imaginary point, and B, C are points or points at infinity, then ABC is convex if and only if B, C are on the same side of line $A^{\sim}$.*
3. *When A, B are imaginary points, and C is a point or a point at infinity, then ABC is convex if and only if C is between the lines $A^{\sim}, B^{\sim}$.*
4. *When A, B, C are imaginary points, then ABC is convex if and only if any of the three lines $A^{\sim}, B^{\sim}, C^{\sim}$ is between the other two.*

Theorem 2.4 (New Result). *Let ABC be a generalized triangle. Then it is convex if and only if $(A \cdot B)(B \cdot C)(C \cdot A) < 0$.*

Proof. When A, B, C are points or points at infinity, the conclusion is trivial. Now we assume that A is an imaginary point. When B, C are points or points at infinity, let $B \cdot C < 0$. B, C are on the same side of line $A^{\sim}$ if and only if $(B \wedge A^{\sim})^{\sim} (C \wedge A^{\sim})^{\sim} = (A \cdot B)(A \cdot C) > 0$, i.e., if and only if $(A \cdot B)(B \cdot C)(C \cdot A) < 0$. When B is an imaginary point, since $A \wedge B$ is Minkowski, $|A \cdot B| > 1$. Let $A \cdot B < -1$; then the vector $A + B$ corresponds to the midpoint of the rectilinear segment between the two feet of the common perpendicular of lines $A^{\sim}, B^{\sim}$. When C is a point or a point at infinity between lines $A^{\sim}, B^{\sim}$, then

$$(C \cdot A)(A \cdot (A+B))((A+B) \cdot C) < 0, \quad (C \cdot B)(B \cdot (A+B))((A+B) \cdot C) < 0.$$

Since $A \cdot (A+B) < 0$, $B \cdot (A+B) < 0$, we have $(C \cdot A)(C \cdot B)((A+B) \cdot C)^2 > 0$, i.e., $(A \cdot B)(B \cdot C)(C \cdot A) < 0$. When C is not between the lines $A^{\sim}, B^{\sim}$, then $(A \cdot B)(B \cdot C)(C \cdot A) > 0$. When B, C are both imaginary points, let $A \cdot B < -1$. When $(C \cdot A)(C \cdot B) > 0$, since

$$\{A \cdot (C \cdot (B \wedge C))\}\{B \cdot (C \cdot (B \wedge C))\} = ((A \cdot C)(B \cdot C) - A \cdot B)(B \wedge C)^2 > 0,$$

the point P_1 on $\mathcal{H}^2$ corresponding to the vector $C \cdot (B \wedge C)$ is between lines $A^\sim, B^\sim$. Since line $C^\sim$ passes through point P_1 and does not intersect with any of lines $A^\sim, B^\sim$, it is also between the two lines. Similarly, line $A^\sim$ (or $B^\sim$) is between lines $C^\sim$, $B^\sim$ (or $C^\sim$, $A^\sim$). Conversely, if any of the three lines $A^\sim, B^\sim, C^\sim$ is between the other two, then point P_1 is between lines $A^\sim, B^\sim$, the point P_2 corresponding to vector $B \cdot (B \wedge C)$ is between lines $A^\sim, C^\sim$. Let $A \cdot C < -1$. Then

$$(A \cdot C)(B \cdot C) - A \cdot B > 0, \quad (A \cdot B)(B \cdot C) - A \cdot C > 0.$$

So $B \cdot C < \min(\frac{A \cdot C}{A \cdot B}, \frac{A \cdot B}{A \cdot C}) \leq 1$. Since $|B \cdot C| > 1$, it follows that $B \cdot C < -1$. So $(A \cdot B)(B \cdot C)(C \cdot A) < 0$. □

In Euclidean plane geometry, we have the concept right triangles. In hyperbolic plane geometry we have a similar concept.

Definition 2.12. *A generalized triangle is called a right triangle if at least one of its vertices is a point and the inner angle at a point vertex is* 90°.

Proposition 2.5. *Let ABC be a generalized triangle. Then it is right if and only if* $((A \wedge B) \cdot (B \wedge C))\,((B \wedge C) \cdot (C \wedge A))\,((C \wedge A) \cdot (A \wedge B)) = 0$.

The sign of $((A \wedge B) \cdot (B \wedge C))\,((B \wedge C) \cdot (C \wedge A))\,((C \wedge A) \cdot (A \wedge B))$ characterizes another geometric invariant which is described in the following definition and theorem.

Definition 2.13. *A generalized triangle is called an acute triangle if it is convex and its inner angle at every point vertex is acute.*

Theorem 2.6 (New Result). *Let ABC be a generalized triangle. Then it is acute if and only if* $((A \wedge B) \cdot (B \wedge C))\,((B \wedge C) \cdot (C \wedge A))\,((C \wedge A) \cdot (A \wedge B)) < 0$.

Analogous to Euclidean plane geometry, various properties can be established for generalized circles and double cycles associated with generalized triangles; for example, generalized circles and double cycles circumscribing a generalized triangle, inscribed or escribed in a convex generalized triangle (see Li, 1997).

2.3 The homogeneous model

Since generalized r-spheres can be studied in the framework of affine $(r+1)$-planes of $\mathbb{R}^{n,1}$, and since affine $(r+1)$-planes can be conveniently studied when embedded in a one-dimensional higher space as intersections of linear

$(r+2)$-spaces with an affine hyperplane, to facilitate the study of generalized r-spheres, we will embed $\mathbb{R}^{n,1}$ into $\mathbb{R}^{n+1,1}$ and embed $\mathcal{D}^n$ into the null cone of $\mathbb{R}^{n+1,1}$. Then we obtain the homogeneous model of double hyperbolic space which is very convenient for the study of hyperbolic conformal geometry. Let a_0 be a fixed unit vector of $\mathbb{R}^{n+1,1}$. The space represented by $a_0^\sim$ is a Minkowski $(n+1)$-space which we denote by $\mathbb{R}^{n,1}$. The set $\mathcal{D}^n$ is in the space $\mathbb{R}^{n,1}$. The mapping

$$X \mapsto X - a_0, \text{ for } X \in \mathcal{D}^n, \tag{2.3}$$

maps $\mathcal{D}^n$ in a one-to-one manner onto the set

$$\mathcal{N}_{a_0}^n = \{X \in \mathbb{R}^{n+1,1} | X \cdot X = 0, X \cdot a_0 = -1\}.$$

The set $\mathcal{N}_{a_0}^n$, together with the mapping (2.3), defines a model for $\mathcal{D}^n$, called the homogeneous model of the double hyperbolic n-space. The following theorem says that not only generalized r-spheres, but all total r-spheres have beautiful algebraic representations in the homogeneous model.

Theorem 2.7 (New Result). *Let $B_{r-1,1}$ be a Minkowski r-blade in $\mathcal{G}_{n+1,1}$, $2 \le r \le n+1$. Then $B_{r-1,1}$ represents a total $(r-2)$-sphere. If*

1. *$a_0 \cdot B_{r-1,1} = 0$, then $B_{r-1,1}$ represents an $(r-2)$-sphere at infinity.*
2. *$a_0 \cdot B_{r-1,1}$ is Euclidean, then $B_{r-1,1}$ represents an $(r-2)$-sphere.*
3. *$a_0 \cdot B_{r-1,1}$ is degenerate, then $B_{r-1,1}$ represents an $(r-2)$-horosphere.*
4. *$a_0 \cdot B_{r-1,1}$ is Minkowski, but $a_0 \wedge B_{r-1,1} \neq 0$, then $B_{r-1,1}$ represents an $(r-2)$-hypersphere.*
5. *$a_0 \wedge B_{r-1,1} = 0$, then $B_{r-1,1}$ represents an $(r-2)$-plane.*

When $r = n+1$, the dual form of the above theorem is as follows.

Theorem 2.8 (New Result). *Let s be a vector of positive square in $\mathbb{R}^{n+1,1}$. Then $s^\sim$ represents a total sphere. If*

1. *$a_0 \wedge s = 0$, then $s^\sim$ represents the sphere at infinity. The sphere at infinity is represented by $a_0^\sim$.*
2. *$a_0 \wedge s$ is Minkowski, then $s^\sim$ represents a sphere. The sphere with center C and generalized radius ρ is represented by $(C - \rho a_0)^\sim$ where C is the representation of the point in the homogeneous model.*
3. *$a_0 \wedge s$ is degenerate, then $s^\sim$ represents a horosphere. The horosphere with center C and generalized radius ρ is represented by $(C - \rho a_0)^\sim$.*
4. *$a_0 \wedge s$ is Euclidean, but $a_0 \cdot s \neq 0$, then $s^\sim$ represents a hypersphere. The hypersphere with center C and generalized radius ρ is represented by $(C - \rho a_0)^\sim$.*

5. $a_0 \cdot s = 0$, *then* $s^\sim$ *represents a hyperplane. A hyperplane with normal direction* C *is represented by* $C^\sim$.

Proof of Theorem 2.8. If

1. $s \wedge a_0 = 0$, then any null vector in the space $s^\sim$ represents a point at infinity, and vice versa.

2. $s \wedge a_0$ is Minkowski, then $s \cdot a_0 \neq 0$. Let ϵ be the sign of $s \cdot a_0$. Let

$$\mathbf{C} = -\epsilon \frac{P^{\perp}_{a_0}(s)}{|P^{\perp}_{a_0}(s)|}, \quad \rho = \frac{|a_0 \cdot s|}{|a_0 \wedge s|} - 1. \tag{2.4}$$

Then $\mathbf{C}$ is a point, $\rho > 0$, as $|a_0 \wedge s|^2 = (a_0 \cdot s)^2 - s^2 < (a_0 \cdot s)^2$. Let $s' = -\epsilon s / |a_0 \wedge s|$. Then

$$s' = \mathbf{C} - (1+\rho)a_0 = C - \rho a_0, \tag{2.5}$$

where $C = \mathbf{C} - a_0$. A point represented by a null vector P is on the sphere $(\mathbf{C}, \rho)$ if and only if $P \cdot s' = 0$, which is equivalent to $P \wedge s^\sim = 0$.

3. $s \wedge a_0$ is degenerate, then $|s \cdot a_0| = |s| \neq 0$. Let ϵ be the sign of $s \cdot a_0$. Let

$$C = -\epsilon P^{\perp}_{a_0}(s), \quad \rho = |a_0 \cdot s| = |s|. \tag{2.6}$$

Then C is an end, $\rho > 0$. Let $s' = -\epsilon s$. Then

$$s' = C - \rho a_0. \tag{2.7}$$

A point represented by a null vector P is on the horosphere (C, ρ) if and only if $P \wedge s^\sim = 0$.

4. $s \wedge a_0$ is Euclidean, but $s \cdot a_0 \neq 0$, let ϵ be the sign of $s \cdot a_0$. Let

$$C = -\epsilon \frac{P^{\perp}_{a_0}(s)}{|P^{\perp}_{a_0}(s)|}, \quad \rho = \frac{|a_0 \cdot s|}{|a_0 \wedge s|}. \tag{2.8}$$

Then C is a direction, $\rho > 0$. Let $s' = -\epsilon s / |a_0 \wedge s|$. Then

$$s' = C - \rho a_0. \tag{2.9}$$

A point represented by a null vector P is on the hypersphere (C, ρ) if and only if $P \wedge s^\sim = 0$.

5. $s \cdot a_0 = 0$, a point represented by a null vector P is on the hyperplane normal to s if and only if $P \wedge s^\sim = 0$.

This finishes the proof of Theorem 2.8. □

Proof of Theorem 2.7. If $a_0 \wedge B_{r-1,1} = 0$, then $B_{r-1,1} = a_0 \wedge C_{r-2,1}$ where $C_{r-2,1}$ is a Minkowski blade in the space $\mathbb{R}^{n,1}$ represented by $a_0^{\sim}$. Let $p \in \mathcal{D}^n$. Then p is in the $(r-2)$-plane of $\mathcal{D}^n$ represented by $C_{r-2,1}$ in the hyperboloid model if and only if $p \wedge C_{r-2,1} = 0$, or equivalently, if and only if $(p - a_0) \wedge B_{r-1,1} = 0$. So in the homogeneous model, $B_{r-1,1}$ represents an $(r-2)$-plane of $\mathcal{D}^n$. If $a_0 \wedge B_{r-1,1} \neq 0$, then $a_0 \wedge B_{r-1,1}$ is a Minkowski blade. It represents a Minkowski $(r+1)$-space $\mathbb{R}^{r,1}$ of $\mathbb{R}^{n+1,1}$, at the same time it represents an $(r-1)$-plane of $\mathcal{D}^n$ in the homogeneous model. By Theorem 2.8, since $s = B_{r-1,1} \cdot (a_0 \wedge B_{r-1,1})$ is a vector of positive square of $\mathbb{R}^{r,1}$, its dual $B_{r-1,1}$ in $\mathcal{G}(\mathbb{R}^{r,1})$ represents a total sphere of the $(r-1)$-plane which is a total $(r-2)$-sphere of $\mathcal{D}^n$. The classification is determined by s, or dually, by $B_{r-1,1}$. □

In the homogeneous model, beautiful algebraic representations are available not only for single total spheres, but also for various collections of total spheres.

Definition 2.14. *A bundle of total spheres determined by r total spheres, which are represented by r Minkowski $(n+1)$-blades $B_1, \ldots, B_r$, is the set of total spheres represented by $\lambda_1 B_1 + \ldots + \lambda_r B_r$ where the λ's are scalars.*

When $B_1 \vee \cdots \vee B_r \neq 0$, the integer $r-1$ is called the dimension of the bundle. A one-dimensional bundle is also called a pencil. The dimension of a bundle is allowed to be between 1 and $n-1$. The blade $A_{n-r+2} = B_1 \vee \cdots \vee B_r$ can be used to represent the bundle. There are five classes of bundles:

1. When $a_0 \cdot A_{n-r+2} = 0$, the bundle is called a concentric bundle. It is composed of the sphere at infinity and the generalized spheres whose centers lie in the subspace $(a_0 \wedge A_{n-r+2})^{\sim}$ of $\mathbb{R}^{n,1}$.

2. When A_{n-r+2} is Minkowski and $a_0 \cdot A_{n-r+2}$, $a_0 \wedge A_{n-r+2} \neq 0$, the bundle is called a concurrent bundle. It is composed of total spheres containing the generalized $(n-r)$-sphere A_{n-r+2}. In particular, when A_{n-r+2} represents an $(n-r)$-hypersphere, the bundle is composed of hyperspheres only. $a_0 \wedge (a_0 \cdot A_{n-r+2})$ represents the axis of the $(n-r)$-hypersphere, and it is the intersection of all axes of the hyperspheres in the bundle.

3. When A_{n-r+2} is degenerate and $a_0 \cdot A_{n-r+2}$, $a_0 \wedge A_{n-r+2} \neq 0$, the bundle is called a tangent bundle. Any two non-intersecting total spheres in the bundle are tangent to each other. The tangency occurs at the point or point at infinity corresponding to the unique null 1-space in the space A_{n-r+2}.

4. When A_{n-r+2} is Euclidean and $a_0 \cdot A_{n-r+2}, a_0 \wedge A_{n-r+2} \neq 0$, the bundle is called a Poncelet bundle. $A_{n-r+2}^{\sim}$ represents a generalized

$(r-2)$-sphere, called the Poncelet sphere, which is self-inversive with respect to every total sphere in the bundle.

5. When $a_0 \wedge A_{n-r+2} = 0$, the bundle is called a hyperplane bundle. It is composed of hyperplanes (1) perpendicular to the $(r-1)$-plane represented by $a_0 \wedge A^{\sim}_{n-r+2}$, or (2) whose representations pass through the space A_{n-r+2}, or (3) passing through the $(n-r)$-plane represented by A_{n-r+2}, if the blade A_{n-r+2} is (1) Euclidean, or (2) degenerate, or (3) Minkowski, respectively.

2.4 Lorentz and conformal transformations

By a Lorentz transformation of $\mathcal{H}^n$, we understand an orthogonal transformation of $\mathbb{R}^{n,1}$ preserving $\mathcal{H}^n$. A Lorentz transformation is called even if its determinant equals 1; it is called odd if its determinant equals -1. In Clifford algebra, the orthogonal group $O(n,1)$ is represented by the projective pin group $\mathbf{Pin}(n,1)$ which is the quotient of the versor group of $\mathbb{R}^{n,1}$ by $\mathbb{R} \setminus \{0\}$. $\mathbf{Pin}(n,1)$ is isomorphic to $O(n,1)$. It has four connected components:

$E_+(n,1)$, the set of versors which are geometric products of an even number of positive-signatured vectors and an even number of negative-signatured vectors. It is a subgroup of $\mathbf{Pin}(n,1)$ isomorphic to the even Lorentz group $\mathbf{Lor}^+(n+1)$.

$E_-(n,1)$, the set of versors which are geometric products of an odd number of positive-signatured vectors and an odd number of negative-signatured vectors. $E_-(n,1)$ and $E_+(n,1)$ form a subgroup of the group $\mathbf{Pin}(n,1)$ isomorphic to the special orthogonal group $SO(n,1)$.

$O_+(n,1)$, the set of versors which are geometric products of an odd number of positive-signatured vectors and an even number of negative-signatured vectors. $O_+(n,1)$ and $E_+(n,1)$ form a subgroup of the group $\mathbf{Pin}(n,1)$ isomorphic to the Lorentz group $\mathbf{Lor}(n+1)$.

$O_-(n,1)$, the set of versors which are geometric products of an even number of positive-signatured vectors and an odd number of negative-signatured vectors. $O_-(n,1)$ and $E_+(n,1)$ form a subgroup of the group $\mathbf{Pin}(n,1)$ isomorphic to the subgroup $\mathbf{Lor}^-(n+1)$ of $O(n,1)$ which is composed of even Lorentz transformations and non-special orthogonal transformations mapping $\mathcal{H}^n$ to $-\mathcal{H}^n$.

Lorentz transformations are characterized by the versors inducing them. For example, in hyperbolic plane geometry, Lorentz transformations are classified as follows.

Theorem 2.9 (Old Result). *Any Lorentz transformation of $\mathcal{H}^2$ is one of the following types. The first three are even transformations and the fourth is odd:*

1. *Rotation: induced by the spinor ab, where a, b are vectors and $a \wedge b$ is Euclidean.*
2. *Horolation: induced by the spinor ab, where a, b are vectors and $a \wedge b$ is degenerate.*
3. *Translation: induced by the spinor ab, where a, b are vectors, $a \wedge b$ is Minkowski and $a^2b^2 > 0$.*
4. *Glide-reflection: induced by the versor $abI_{2,1}$, where a, b are vectors, $a \wedge b$ is Minkowski and $a^2b^2 < 0$.*

For hyperbolic conformal transformations, it is a well-known fact that the orthogonal group $O(n+1,1)$ of $\mathbb{R}^{n+1,1}$ is a double covering of the conformal group $M(n)$ of $\mathcal{D}^n : M(n) = O(n+1,1)/\{\pm 1\}$. Let $I_{n+1,1}$ be a unit pseudoscalar of $\mathcal{G}_{n+1,1}$. Then the versor action of $I_{n+1,1}$ maps X to $-X$, for any X in $\mathbb{R}^{n+1,1}$. Therefore,

$$M(n) = \mathbf{Pin}(n+1,1)/\{I_{n+1,1}\}. \tag{2.10}$$

Theorem 2.10 (Old Result). *In the homogeneous model, any hyperbolic conformal transformation of $\mathcal{D}^n$ can be realized by the versor action of a versor in $\mathcal{G}_{n+1,1}$, and vice versa. Any two versors realize the same conformal transformation if and only if they are equal up to a nonzero scalar or a pseudoscalar factor.*

The pseudoscalar $I_{n+1,1}$ also induces a duality in $\mathcal{G}_{n+1,1}$, under which $O_+(n+1,1)$ and $O_-(n+1,1)$ are interchanged when n is even, and $O_+(n+1,1)$ and $E_-(n+1,1)$ are interchanged when n is odd. Therefore,

$$M(n) = \begin{cases} \mathbf{Lor}(n+2) = \mathbf{Lor}^-(n+2) & \text{when } n \text{ is even;} \\ \mathbf{Lor}(n+2) = SO(n+1,1) & \text{when } n \text{ is odd.} \end{cases}$$

We obtain once again the classical result that $M(n)$ is isomorphic to the Lorentz group of $\mathcal{H}^{n+1}$.

3 Clifford algebra approach to automated geometry theorem proving

In modern algebraic methods for automated geometry theorem proving, Wu's characteristic set method (Wu, 1978, 1994; Chou, 1988) and the Gröbner basis method (Buchberger, Collins and Kutzler, 1988; Kutzler and

Stifter, 1986; Kapur, 1986) are two basic ones. In these methods, the first step is to set up a coordinate system, and represent the geometric entities and constraints in the hypothesis of a theorem by coordinates, polynomial equalities and inequalities. The second step is to compute a characteristic set or Gröbner basis by algebraic manipulations among the polynomial equalities. The third step is to verify the conclusion of the theorem using the characteristic set or Gröbner basis.

Since coordinate representations do not keep geometric meaning, and since algebraic manipulations of polynomials in the coordinates are sometimes too complicated to be carried out on modern computers, in recent years, there has been a trend to use geometric invariants for algebraic representations, and combine Wu's method or the Gröbner basis method with an algebra of geometric invariants for algebraic manipulations. Among these methods, there are the method of Crapo and Richter-Gebert (1995) integrating Grassmann-Cayley algebra with the Gröbner basis method, the method of Mourrain (1999) combining Grassmann-Cayley algebra with Wu's method, and the area method (Chou, Gao and Zhang, 1994; Yang, Gao, Chou and Zhang, 1996) combining a set of high-level geometric invariants with Wu's method and logic methods. Proofs based on these methods are often shorter, more readable, and have better geometric meaning.

Because of the universal applicability of Clifford algebra in geometries, it is natural to study combining Clifford algebra with Wu's method or the Gröbner basis method. In (Li, 1994, 1996a; Li and Cheng, 1997), a general method which combines Clifford algebra with Wu's method is proposed. Various applications of the method are carried out in automated theorem proving in Euclidean geometry, non-Euclidean geometry and differential geometry by Li (1995, 1997a, 1997b, 1999), Li and Cheng (1997, 1998a, 1998b), Li and Shi (1997). Also, research and applications are carried out by Wang (1996, 1998) in combining Clifford algebra with the Gröbner basis method and term rewriting techniques for automated geometry theorem proving.

Our general method can be described as follows:

1. Triangulate the hypothesis using substitutions, pseudo-divisions and vectorial equation-solving. The result is called a triangular sequence. Prove the conclusion with the triangular sequence and continue if the proof fails.

2. Triangulate the triangular sequence with vectorial equation-solving, parametric equation-solving, substitutions and pseudo-divisions. The result is called a parametric triangular sequence. Prove the conclusion with the parametric triangular sequence and continue if the proof fails.

3. Select a coordinate system and translate all expressions in the parametric triangular sequence into polynomials. Use Wu's method to

compute a characteristic set and prove the conclusion with the characteristic set.

We illustrate the method with a theorem in solid geometry.

Theorem 3.1 (Old Result). *Let ABCD be a tetrahedron. Let P be a plane intersecting lines AB, AC, DC, DB at M, N, E, F, respectively. If for different positions of P, MNEF is a parallelogram, then the center O of the parallelogram is on a fixed straight line.*

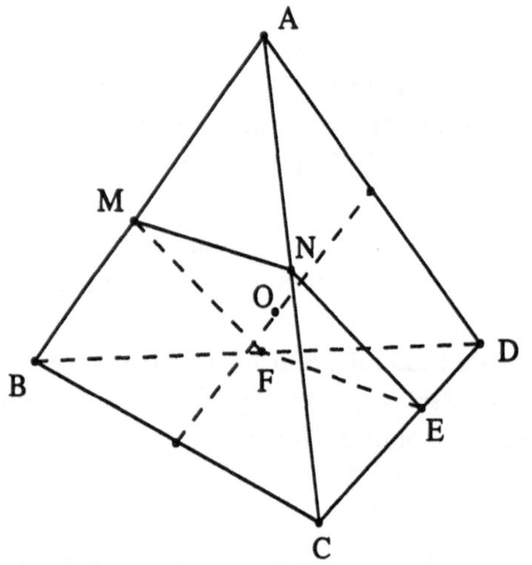

FIGURE 3. A theorem in solid geometry.

In the Grassmann model $\mathcal{G}_4$ of the space, the hypothesis is

$$\begin{cases} M - N = F - E, & MNEF \text{ is a parallelogram} \\ A \wedge B \wedge M = 0, & A, B, M \text{ are collinear} \\ A \wedge C \wedge N = 0, & A, C, N \text{ are collinear} \\ C \wedge D \wedge E = 0, & C, D, E \text{ are collinear} \\ B \wedge D \wedge F = 0, & B, D, F \text{ are collinear} \\ O = \dfrac{M+E}{2}, & O \text{ is the midpoint of } ME \\ A \wedge B \wedge C \wedge D \neq 0, & A, B, C, D \text{ are not coplanar.} \end{cases} \tag{3.1}$$

The conclusion cannot be algebraized. The order of variables for triangulation is: $A \prec B \prec C \prec D \prec M \prec N \prec E \prec F \prec O$. There are two vectorial equation sets to be solved during the triangulation of the hypothesis. The first is solving

$$\begin{cases} C \wedge D \wedge E = 0 \\ B \wedge D \wedge E + B \wedge D \wedge M - B \wedge D \wedge N = 0 \end{cases} \tag{3.2}$$

for E, where the second equation comes from the fifth of (3.1), and the solution is

$$\begin{cases} (E-D)(A\wedge B\wedge C\wedge D)^{\sim}+(C-D)(A\wedge B\wedge D\wedge(N-M))^{\sim}=0 \\ B\wedge C\wedge D\wedge N-B\wedge C\wedge D\wedge M=0 \end{cases}$$

with the nondegeneracy condition $A\wedge B\wedge C\wedge D\neq 0$. The second is solving

$$\begin{cases} A\wedge C\wedge N=0 \\ B\wedge C\wedge D\wedge N-B\wedge C\wedge D\wedge M=0 \end{cases} \tag{3.3}$$

for N, and the solution is

$$(N-C)(A\wedge B\wedge C\wedge D)^{\sim}+(A-C)(B\wedge C\wedge D\wedge M)^{\sim}=0 \tag{3.4}$$

with the nondegeneracy condition $A\wedge B\wedge C\wedge D\neq 0$. After triangulation, we get the following triangular sequence:

$$\begin{array}{l} 2O-M-E=0, \\ F-E-M+N=0, \\ E(A\wedge B\wedge C\wedge D)^{\sim}-D(A\wedge B\wedge C\wedge D)^{\sim} \\ \quad -D(A\wedge B\wedge D\wedge N)^{\sim}+C(A\wedge B\wedge D\wedge N)^{\sim}=0, \\ N(A\wedge B\wedge C\wedge D)^{\sim}-C(A\wedge B\wedge C\wedge D)^{\sim} \\ \quad -C(B\wedge C\wedge D\wedge M)^{\sim}+A(B\wedge C\wedge D\wedge M)^{\sim}=0, \\ A\wedge B\wedge M=0. \end{array} \tag{3.5}$$

The nondegeneracy condition is $A\wedge B\wedge C\wedge D\neq 0$, which is in the original hypothesis. The conclusion cannot be obtained from the triangular sequence as M does not have an explicit expression. There is no vectorial equation to be solved, and only one vectorial equation to be solved parametrically during further triangulation of (3.5). Getting M from

$$A\wedge B\wedge M=0, \tag{3.6}$$

the solution is

$$M=(1-\lambda)A+\lambda B \tag{3.7}$$

where λ is a parameter. The nondegeneracy condition is $A\neq B$. After the triangulation, we obtain the following parametric triangular sequence:

$$\begin{array}{l} 2O-A-D+\lambda A+\lambda D-\lambda B-\lambda C=0, \\ F-D+\lambda D-\lambda B=0, \\ E-D+\lambda D-\lambda C=0, \\ N-A+\lambda A-\lambda C=0, \\ M-A+\lambda A-\lambda B=0. \end{array} \tag{3.8}$$

The nondegeneracy conditions are

$$A\wedge B\wedge C\wedge D,\ \ A\neq B,$$

which are all guaranteed by the original hypothesis. The conclusion is obvious from the first equality of (3.8):

$$O = (1-\lambda)\frac{A+D}{2} + \lambda\frac{B+C}{2},$$

i. e., O is on the line passing through the midpoint of line segment AD and the midpoint of line segment BC.

Acknowledgments

This paper is supported partially by a 973 project of China, the Qiu Shi Science and Technology Foundations of Hong Kong, the DFG and the AvH Foundations of Germany.

References

[1] E. Bayro-Corrochano, J. Lasenby, and G. Sommer, Geometric algebra: a framework for computing point and line correspondences and projective structure using n-uncalibrated cameras, in *Proc. ICPR'96* Vol. **1**, Vienna, 1996, 334–338.

[2] E. Bayro-Corrochano and J. Lasenby, A unified language for computer vision and robotics, in *Algebraic Frames for the Perception-Action Cycle,* G. Sommer and J. J. Koenderink, eds., LNCS **1315** (1997), 219–234.

[3] B. Buchberger, G. E. Collins, and B. Kutzler, Algebraic methods for geometric reasoning, *Ann. Rev. Comput. Sci.* **3** (1988), 85–119.

[4] J. W. Cannon, W. J. Floyd, R. Kenyon, and W. R. Parry, Hyperbolic geometry, in *Flavors of Geometry,* S. Levy, ed., Cambridge, 1997.

[5] S. S. Chern, Deformation of surfaces preserving principal curvatures, in *Differential Geometry and Complex Analysis, Volume in Memory of H. Rauch,* Springer, New York, 1984, 155–163.

[6] S.-C. Chou, *Mechanical Geometry Theorem Proving,* D. Reidel, Dordrecht, 1988.

[7] S.-C. Chou, X.-S. Gao, and J.-Z. Zhang, *Machine Proofs in Geometry,* World Scientific, Singapore, 1994.

[8] H. Crapo and J. Richter-Gebert, Automatic proving of geometric theorems, in *Invariant Methods in Discrete and Computational Geometry,* N. L. White, ed., Kluwer, Dordrecht, 1995, 167–196.

[9] A. Crumeyrolle, *Orthogonal and Symplectic Clifford Algebras,* D. Reidel, Dordrecht, 1990.

[10] R. Delanghe, F. Sommen, and V. Soucek, *Clifford Algebra and Spinor-Valued Functions,* D. Reidel, Dordrecht, 1992.

[11] O. Faugeras and B. Mourrain, On the geometry and algebra of the point and line correspondences between N images, in *Proc. of Europe-China Workshop on Geometrical Modeling and Invariants for Computer Vision,* R. Mohr and C. Wu, eds., 1995, 102–109.

[12] W. Fenchel, *Elementary Geometry in Hyperbolic Space,* Walter de Gruyter and Company, Berlin, New York, 1989.

[13] M. J. Greenberg, *Euclidean and Non-Euclidean Geometries, Second Edition,*, W. H. Freeman and Company, San Francisco, 1980.

[14] T. Havel, Some examples of the use of distances as coordinates for Euclidean geometry, *J. Symbolic Computat.* **11** (1991), 579–593.

[15] T. Havel, Geometric algebra and Möbius sphere geometry as a basis for Euclidean invariant theory, in *Invariant Methods in Discrete and Computational Geometry,* N. L. White, ed., D. Reidel, Dordrecht, 1995, 245–256.

[16] D. Hestenes and G. Sobczyk, *Clifford Algebra to Geometric Calculus,* D. Reidel, Dordrecht, Boston, 1984.

[17] D. Hestenes, *New Foundations for Classical Mechanics,* D. Reidel, Dordrecht, Boston, 1987.

[18] D. Hestenes and R. Ziegler, Projective geometry with Clifford algebra, *Acta Appl. Math* **23** (1991), 25–63.

[19] D. Hestenes, The design of linear algebra and geometry, *Acta Appl. Math.* **23** (1991), 65–93.

[20] D. Hestenes, H. Li, and A. Rockwood, New algebraic tools for classical geometry, in *Geometric Computing with Clifford Algebra,* G. Sommer, ed., Springer, 1999.

[21] B. Iversen, *Hyperbolic Geometry,* Cambridge Univ. Press, Cambridge, 1992.

[22] D. Kapur, Using Gröbner base to reason about geometry problems, *J. Symbolic Computat.* **2** (1986), 399–408.

[23] B. Kutzler and S. Stifter, On the application of Buchberger's algorithm to automated geometry theorem proving, *J. Symbolic Computat.* **2** (1986), 389–397.

[24] H. Li, New explorations in automated theorem proving in geometries, Ph. D. Thesis, Peking University, Beijing, 1994.

[25] H. Li, Automated reasoning with differential forms, in *Proc. ASCM '95,* Beijing, 1995, 29–32.

[26] H. Li, Clifford algebra and area method, *MMRC Research Report,* Institute of Systems Science, Academia Sinica, Beijing **14** (1996), 37–69.

[27] H. Li, Clifford algebra and Lobachevsky geometry, in *Clifford Algebras and Their Applications in Mathematical Physics,* V. Dietrich, et al., ed., D. Reidel, Dordrecht, 1996, 239–245.

[28] H. Li, *On Mechanical Theorem Proving in Differential Geometry – Local Theory of Surfaces,* Science in China, Series A, **40** (4) (1997), 350–356.

[29] H. Li, *Ordering in Mechanical Geometry Theorem Proving,* Science in China, Series A, **40** (3) (1997), 225–233.

[30] H. Li, Hyperbolic geometry with Clifford algebra, *Acta Appl. Math.* **48** (3) (1997), 317–358.

[31] H. Li, Some applications of Clifford algebra to geometries, in *Proc. ADG'98,* Beijing, 1998.

[32] H. Li, Vectorial equation-solving for mechanical geometry theorem proving, *J. of Automated Reason,* to appear, 1999.

[33] H. Li and M. T. Cheng, Proving theorems in elementary geometry with Clifford algebraic method, *Chinese Math. Progress* **26** (4) (1997), 357–371.

[34] H. Li and M. T. Cheng, Clifford algebraic reduction method for mechanical theorem proving in differential geometry, *J. Automated Reasoning* **21** (1998), 1–21.

[35] H. Li and M. T. Cheng, Ordering in mechanical theorem proving in differential geometry, *Acta Math. Appl. Sinica* **14** (4)(1998), 358–362.

[36] H. Li and H. Shi, On Erdös' ten-point problem, *Acta Mathematica Sinica, New Series* **13** (2) (1997), 221–230.

[37] H. Li, D. Hestenes, and A. Rockwood, Generalized homogeneous coordinates for computational geometry, in *Geometric Computing with Clifford Algebra,* G. Sommer, ed., Springer, 1999.

[38] H. Li, D. Hestenes, and A. Rockwood, A universal model for conformal geometries of Euclidean, spherical, and double hyperbolic spaces, in *Geometric Computing with Clifford Algebra,* G. Sommer, ed., Springer, 1999.

[39] H. Li, D. Hestenes, and A. Rockwood, Spherical conformal geometry with geometric algebra, in *Geometric Computing with Clifford Algebra,* G. Sommer, ed., Springer, 1999.

[40] B. Mourrain and N. Stolfi, Computational symbolic geometry, in *Invariant Methods in Discrete and Computational Geometry,* N. L. White, ed., D. Reidel, Dordrecht, Boston, 1995, 107–139.

[41] B. Mourrain, New aspects of geometrical calculus with invariants. `http://www-sop.inria.fr/safir/whoswho/Bernard.Mourrain/publi.html`, 1999.

[42] J. Seidel, Distance-geometric development of two-dimensional Euclidean, hyperbolic and spherical geometry I, II. *Simon Stevin* **29** (1952), 32–50, 65–76.

[43] G. Sommer, E. Bayro-Corrochano, and T. Bülow, Geometric algebra as a framework for the perception–action cycle, *Workshop on Theoretical Foundation of Computer Vision,* Dagstuhl, 1996.

[44] D. Wang, Clifford algebraic calculus for geometric reasoning, with application to computer vision, in *Automated Deduction in Geometry,* D. Wang, ed., LNAI **1360** (1996), 115–140.

[45] D. Wang, Gröbner Bases applied to geometric theorem proving and discovering, in *Gröbner Bases and Application,* B. Buchberger and F. Winker, eds., 1998, 281–301.

[46] W. T. Wu, On the decision problem and the mechanization of theorem proving in elementary geometry, *Scientia Sinica* **21** (1978).

[47] W. T. Wu, On the mechanization of theorem proving in elementary and differential geometry, *Scientia Sinica, Math. Supplement (I)* (1979), 94–102.

[48] W. T. Wu, *Mechanical Theorem Proving in Geometries: Basic Principle* (translated from Chinese edition 1984), Springer, Vienna, 1994.

[49] L. Yang, X. -S. Gao, S. -C. Chou, and J. -Z. Zhang, Automated production of readable proofs for theorems in non-Euclidean geometries, in *Proc. ADG'96*, LNAI **1360** (1996), 171–188.

Hongbo Li
Institute of Systems Science
Academia Sinica
Beijing 100080, P. R. China
E-mail: hli@mmrc.iss.ac.cn

Received: October 5, 1999; Revised: November 29, 1999

Clifford Algebra of Quantum Logic

Bernd Schmeikal

ABSTRACT We shall show how elementary logic calculus can be represented within the Clifford algebras $C\ell_{1,1}, C\ell_{3,1} \simeq C\ell_{2,2}$ and $C\ell_{m,m}$, in general. Thus, we shall also understand why logic is connected with the standard model of physics and the quantum structure of spacetime, respectively. "It is not the substance which is in space," as Alfred North Whitehead had pointed out, "but the attributes". Accordingly, as we consider the Dirac algebra $\mathrm{Mat}(4, \mathbb{C})$, annihilation and time reversal turn out to be logic operations acting on Dirac spinors. This is remarkable as three generations of physicists have calculated time reverted wave functions. In a complete Clifford algebraic image of the Dirac equation, those wave functions do not exist because time reversal, when carried out in the whole algebra, annihilates any Dirac spinor. It is then, essentially in the quantum chromodynamics context of the Majorana algebra, that classical logic becomes incomplete with regard to definite truth. That is, color rotations cannot be represented as products of logic operations. It also follows that once we connect spinor spaces with probability distributions, classical logic naturally turns into quantum logic. Future developments may show how logic derivative procedures can be represented synchronically as elements in neutral signature and, further, how logic quantum computations can be generated practically by reflections of waves in subnuclear arrays.

Keywords: Quantum logic, Clifford algebra, idempotent lattice, spinors, Dirac equation.

1 Prologue

As a novice in theoretical physics, I came upon an introductory booklet on what was then called quantum logic and published by Professor Mittelstädt in the Bibliographic Institute (pocket-edition)[1]. This fascinating investigation emphasized the determinism-probabilism dichotomy and pointed at the fact that in quantum mechanics "tertium datur". That is, *"A or not A"* does not cover the whole of truth. Meanwhile, the situation has changed

---

AMS Subject Classification: 03G10, 03G12, 3G25, 15A66, 14L35.

[1]BI-Hochschultaschenbücher, Bibliographisches Institut Mannheim: Peter Mittelstaedt, 1963, *"Philosophische Probleme der Modernen Physik"*.

considerably. The above duality has lost much of its persuasive power, and the historic idea comes along in different garments – disguised, for example, as concrete logic, as matrix logic, as quantum information and so forth. All those different mathematical objects, to my taste, are not fully convincing. It would be better to work out a consistent approach conscious of the historical concepts which will show consequences of engineering applicability at the same time.

The following rigor began with the idea that there is an interface between matter and mind. This idea first led me to some special inquiries in paleontology, anthropology and the genetic structuralism. But next it led me to Charles S. Peirce's iconic notation for the logic relatives (Peirce 1870, 1902, 1903). Peirce's logic icons can be represented in at least two ways, namely, either as multivectors or as primitive idempotents of some Clifford algebra $C\ell_{m,m}$. The quarks can be conceived as representants of fundamental logic relations, and the establishment of logic transformation rules turns out as a matter of high energy physics and Clifford algebra, respectively. In this paper, I wish to make you familiar with some sound basis of quantum logic which will perhaps allow you to go deeper into the subject matter.

2 Basic reflections of Clifford algebra and the syntagmatic group

First, I will give you an idea of a fundamental concept which we shall use throughout this lecture and which I denote a *"basic reflection"* of Clifford algebra. Consider the grade involution $\hat{}$, the reversion $\tilde{}$, and the Clifford conjugation $\bar{}$ in some real Clifford algebra $C\ell_{p,q}$. Let $\{\mathbf{e}_i\}, i = 1, \ldots, n = p+q$ be an orthonormal basis of $\mathbb{R}^{p,q}$ and the set of multivectors $\mathbf{G}_{p,q} = \{\pm 1, \pm\mathbf{e}_i, \pm\mathbf{e}_{ij}, \ldots, \pm\mathbf{w}\}$ the Dirac group of $C\ell_{p,q}$. Then, we define elements $s_\alpha, s_\eta, s_\gamma, \ldots \in C\ell_{p,q}$, which act like generating reflections. We define:

D1: The elements $s_\eta \in C\ell_{p,q}$ as *basic reflections* in $C\ell_{p,q}$ such that

$$s_\eta^2 = +1, \ s_\eta \text{ is invertible, of period two, and } s_\eta \tilde{s}_\eta \in \mathbf{G}_{p,q}. \quad (2.1)$$

That is, by reversion $\tilde{}$, *the basic reflections* are normalized to units in the Dirac group

C1: If s_η sends $y \in \mathbf{G}_{p,q}$ to $\mathbf{G}_{p,q}$, that is,

$$s_\eta y s_\eta \in \mathbf{G}_{p,q}, \quad (2.2)$$

then we say that the s_η carries out a transposition in $\mathbf{G}_{p,q}$. The s_η generate a group. This is a group of elements normalized by reversion onto the units of the Dirac group

$$\mathbf{S}_{p,q} = \{s \mid s\tilde{s} \in \mathbf{G}_{p,q}\}. \tag{2.3}$$

Let us call $\mathbf{S}_{p,q}$ the ***syntagmatic group*** of $C\ell_{p,q}$. Elements of the syntagmatic group shall be used as tools to transpose idempotents and their corresponding minimal left ideals.

3 The iconic notation of Charles S. Peirce

Consider two prime propositions A and B of the elementary logic, briefly called attributes. Both combine to the frame of a truth table having four entries (TT, TF, FT, FF).

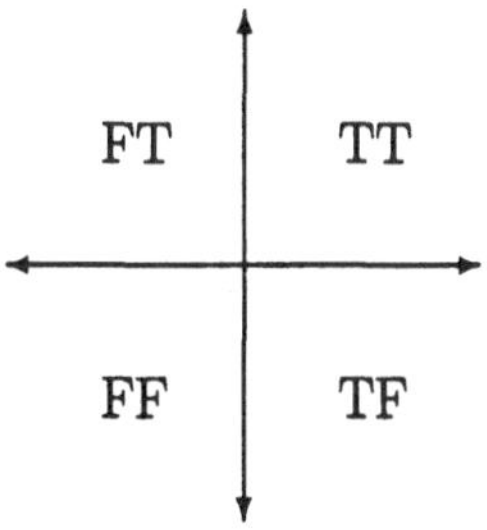

A logic binary connective such as $A \wedge B$ assigns to each of these four entries a definite value of either T(rue) or F(alse). The connective $A \wedge B$ projects onto the entry TT the value T and gives an F to all other entries. It has the truth table (TFFF). Following McCulloch and Zellweger (both in Zellweger 1992), we put a dot into the upper right quadrant which results in the following figure:

$A \wedge B$

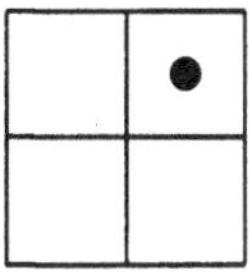

Charles S. Peirce had chosen a different arrangement. He put the quadrants as follows and left open the quarters where T's (dots) are located. So he made the right icon denote the *logic conjunction* $A \wedge B$.

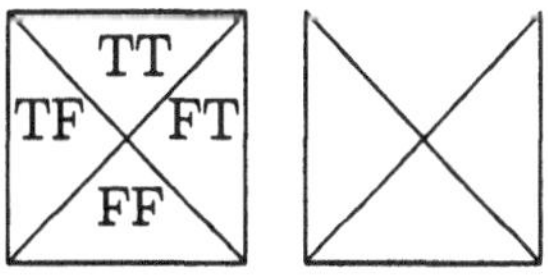

Now, we may take a step into the Clifford algebra $C\ell_{2,0} \simeq C\ell_{1,1}$, generated by the vector space $\mathbb{R}^2$, which combines the algebra of the real vector plane with that of the complex plane. We observe that the dot is located in the upper right quadrant of the real vector plane; this figure can very well be represented by the unit $\frac{1}{\sqrt{2}}(\mathbf{e}_1 + \mathbf{e}_2)$:

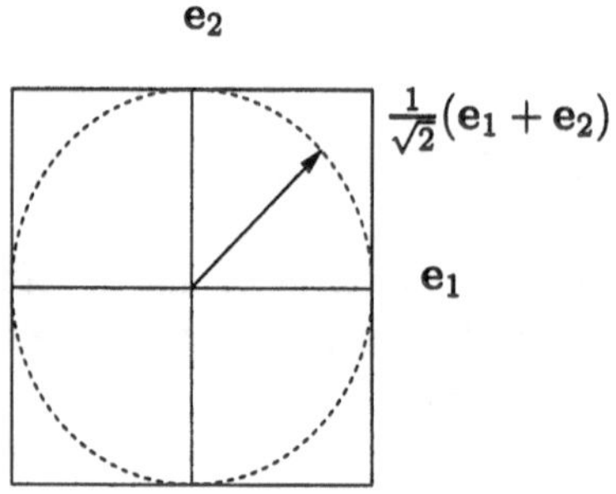

For convenience, we may use such units of the Clifford algebra to represent the binary connectives. Yet, we shall at first prefer some representation by primitive idempotents in order to learn something about the quantum geometric nature of logic transformations. Note that we have four entries, each of which can take the value of T or F (or equivalently 1 or 0). There is a total of $2^4 = 16$ logic binary connectives. One of these has only zero entries. This denotes the identically false or "contradictory" proposition $(A \wedge \neg A)$ with truth table (FFFF), symbolized by the following notation:

	de Morgan, Frege, Boole	*McCulloch, Parry*	*Peirce*	*Zellweger*	*idempotent*
code	truth table	dot-chart	box-X	letter shape	in $C\ell_{3,1}$
$L_1\,(A \wedge \neg A)$	(FFFF)	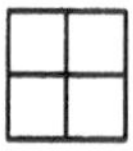			f

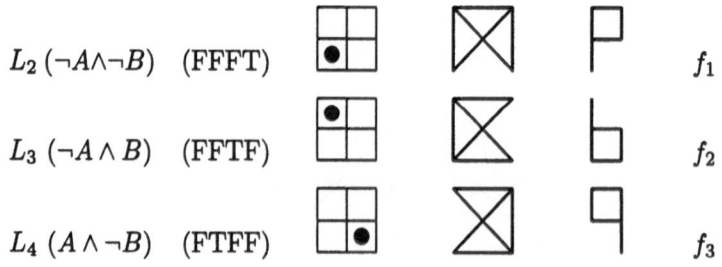

Those binary connectives having one dot only are represented by four conjunctions:

$L_2\,(\neg A \wedge \neg B)$	(FFFT)	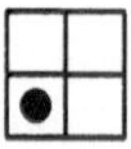			f_1
$L_3\,(\neg A \wedge B)$	(FFTF)	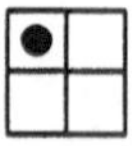			f_2
$L_4\,(A \wedge \neg B)$	(FTFF)	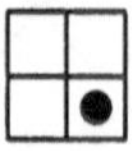			f_3

$L_5\ (A \wedge B)$ (TFFF)  f_4

Next, there follow $\binom{4}{2} = 6$ binary connectives with 2 dots in their dot-charts, with the first two representing the single attributes:

$L_6\ (A)$ (TTFF) 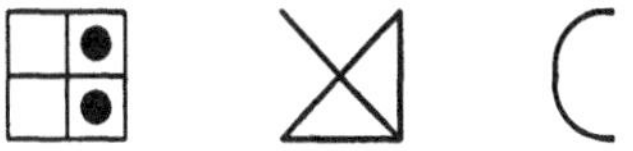 $f_3 + f_4$

$L_7\ (B)$ (TFTF)  $f_2 + f_4$

The next connective signifies James S. Coleman's (1974) famous "interactive attribute" which takes the value "True" if and only if the truth values of A and B are equal. This is the logical identity $A \equiv B$ often called "equivalence" and written as $(A \wedge B) \vee (\neg A \wedge \neg B)$.

L_8 (A identical B)

$A \equiv B$ (TFFT)  $f_1 + f_4$

L_9 (A or else B)

$A \neq B$ (FTTF)  $f_2 + f_3$

$L_{10}\ (\neg B)$ (FTFT)  $f_1 + f_3$

$L_{11}\ (\neg A)$ (FFTT)  $f_1 + f_2$

There follow $\binom{4}{3} = 4$ binary connectives having 3 dots in their charts. Their construction involves three primitive idempotents

$L_{12}\ (\neg A \vee \neg B)$ (FTTT)  $f_1 + f_2 + f_3$

$L_{13}\ (\neg A \vee B)$ (TFTT)  $f_1 + f_2 + f_4$

$L_{14}\ (A \vee \neg B)$ (TTFT)  $f_1 + f_3 + f_4$

$L_{15}\ (A \vee B)$ (TTTF) 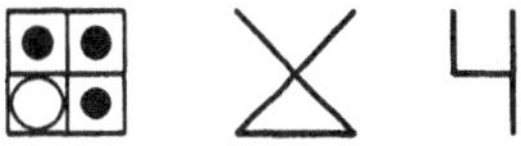 $f_2 + f_3 + f_4$

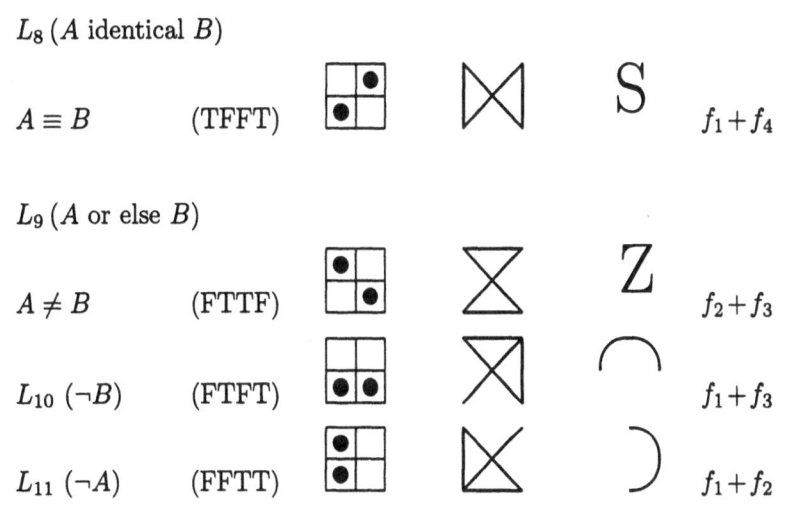

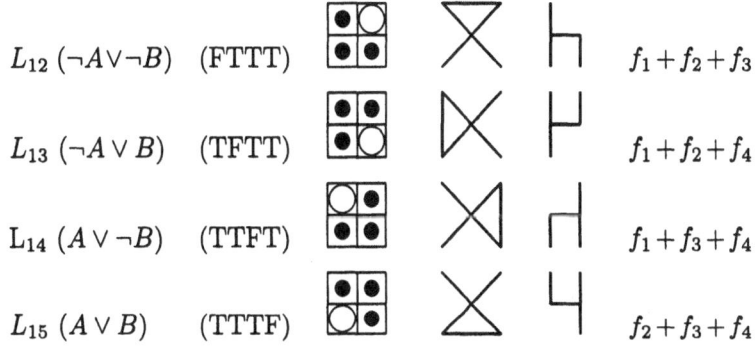

The last connective L_{16} represents the "True" statement. It corresponds with the Clifford unity **1** and, as a geometric letter shape, has the symmetry $\mathbf{D}_4$ which is not by fortune.

$L_{16}\ (A \vee \neg A)$ (TTTT)  $f_1+f_2+f_3+f_4$

4 Primitive idempotents and basic reflections in $C\ell_{3,1}$

Consider the Minkowski space in the opposite metric $\mathbb{R}^{3,1}$ with its Clifford algebra $C\ell_{3,1} \simeq \mathrm{Mat}(4,\mathbb{R})$, the Majorana algebra. The Dirac equation in this space is

$$\partial\psi + ie\mathbf{A}\psi = \mathrm{m}\psi. \tag{4.1}$$

It is possible to represent $C\ell_{3,1}$ either by real matrices (Lounesto 1996, p. 15) or in a complexified structure given by $\mathbb{C}\otimes\mathrm{Mat}(4,\mathbb{R}) \simeq \mathrm{Mat}(4,\mathbb{C})$. One possible complex representation is based on the unit vectors

$$\begin{aligned} \mathbf{e}_1 &= \begin{pmatrix} \sigma_3 & 0 \\ 0 & -\sigma_3 \end{pmatrix}, \quad \mathbf{e}_2 = \begin{pmatrix} \sigma_2 & 0 \\ 0 & -\sigma_2 \end{pmatrix}, \\ \mathbf{e}_3 &= \begin{pmatrix} 0 & \sigma_{123} \\ -\sigma_{123} & 0 \end{pmatrix}, \quad \mathbf{e}_4 = \begin{pmatrix} 0 & -\sigma_{123} \\ -\sigma_{123} & 0 \end{pmatrix}, \end{aligned} \tag{4.2}$$

where $\sigma_1, \sigma_2, \sigma_3$ are Pauli matrices and $-\sigma_{123}$ has diagonal entries $i = \sqrt{-1}$. This representation is obtained by the following choice of the primitive idempotent

$$f_1 = \frac{1}{2}(1+\mathbf{e}_1)(1+\mathbf{e}_{34}). \tag{4.3}$$

Together with

$$f_2 = \frac{1}{2}(1-\mathbf{e}_1)(1+\mathbf{e}_{34}),\ f_3 = \frac{1}{2}(1-\mathbf{e}_1)(1-\mathbf{e}_{34}),\ f_4 = \frac{1}{2}(1+\mathbf{e}_1)(1-\mathbf{e}_{34})$$

we have a maximal set $F_1 = \{f_i\}$, $i = 1,2,3,4$ of mutually annihilating primitive idempotents which sum up to unity. The f_i are represented by diagonal matrices with 1 at the location (i,i) and zeros elsewhere. These are used for two purposes, namely (a) to represent the lattice of logic binary connectives as was done in section 3 and (b) to build up the basic reflections from which we calculate subspaces invariant under SU(3).

Consider ***basic reflections*** constructed by f_1, f_2, f_3, f_4 and bivectors $\mathbf{e}_{ij}$ with $i, j \neq 4$.

$$\begin{aligned}
s_{11} &= f_1 + f_2 + \mathbf{e}_{12}(f_4 - f_3), & s_{12} &= f_1 + f_2 + \mathbf{e}_{12}(f_3 - f_4),\\
s_{21} &= f_1 + f_3 + \mathbf{e}_{13}(f_2 - f_4), & s_{22} &= f_1 + f_3 + \mathbf{e}_{13}(f_4 - f_2),\\
s_{31} &= f_1 + f_4 + \mathbf{e}_{23}(f_3 - f_2), & s_{32} &= f_1 + f_4 + \mathbf{e}_{23}(f_2 - f_3),\\
s_{41} &= f_2 + f_3 + \mathbf{e}_{23}(f_1 - f_4), & s_{42} &= f_2 + f_3 + \mathbf{e}_{23}(f_4 - f_1),\\
s_{51} &= f_2 + f_4 + \mathbf{e}_{13}(f_3 - f_1), & s_{52} &= f_2 + f_4 + \mathbf{e}_{13}(f_1 - f_3),\\
s_{61} &= f_3 + f_4 + \mathbf{e}_{12}(f_2 - f_1), & s_{62} &= f_3 + f_4 + \mathbf{e}_{12}(f_1 - f_2),\\
s_{71} &= g_1 + g_2 + \mathbf{e}_{12}(g_4 - g_3), & s_{72} &= g_1 + g_2 + \mathbf{e}_{12}(g_3 - g_4),\\
s_{81} &= g_1 + g_3 + \mathbf{e}_{13}(g_2 - g_4), & s_{82} &= g_1 + g_3 + \mathbf{e}_{13}(g_4 - g_2),\\
s_{91} &= g_1 + g_4 + \mathbf{e}_{23}(g_3 - g_2), & s_{92} &= g_1 + g_4 + \mathbf{e}_{23}(g_2 - g_3),\\
s_{101} &= g_2 + g_3 + \mathbf{e}_{23}(g_1 - g_4), & s_{102} &= g_2 + g_3 + \mathbf{e}_{23}(g_4 - g_1),\\
s_{111} &= g_2 + g_4 + \mathbf{e}_{13}(g_3 - g_1), & s_{112} &= g_2 + g_4 + \mathbf{e}_{13}(g_1 - g_3),\\
s_{121} &= g_3 + g_4 + \mathbf{e}_{12}(g_2 - g_1), & s_{122} &= g_3 + g_4 + \mathbf{e}_{12}(g_1 - g_2).
\end{aligned} \tag{4.4}$$

Here, the g_i are obtained from the f_i by a unitary transformation or by regarding the isomorphism $C\ell_{3,0} \simeq C\ell_{3,1}^+$ determined by the correspondences $\mathbf{e}_1, \mathbf{e}_2, \mathbf{e}_3 \simeq \mathbf{e}_{14}, \mathbf{e}_{24}, \mathbf{e}_{34}$. The Pauli algebra $C\ell_{3,0}$ contains central invertible elements having the form $x + y\mathbf{e}_{123} \in C\ell_{3,0}^-$ with $x, y \in \mathbb{R}$. A subset of these forms a group

$$\mathbf{W}_1 = \{x + y\mathbf{e}_{123} | \, x, y \in \mathbb{R}; x^2 + y^2 = 1\} \simeq \mathrm{U}(1),$$

isomorphic to the unitary group $\mathrm{U}(1)$. Being aware that the expressions of the form

$$(1 + \mathbf{e}_i), \quad (1 + \mathbf{e}_{i4}), \quad (1 + \mathbf{e}_{ik4})$$

(for $i, k \neq 4$) are not invertible, there remain five more groups of the type

$$\begin{aligned}
\mathbf{W}_2 &= \{x + y\mathbf{e}_{12} | x, y \in \mathbb{R}; x^2 + y^2 = 1\} \simeq \mathrm{U}(1),\\
\mathbf{W}_3 &= \{x + y\mathbf{e}_{13} | x, y \in \mathbb{R}; x^2 + y^2 = 1\} \simeq \mathrm{U}(1),\\
\mathbf{W}_4 &= \{x + y\mathbf{e}_{23} | x, y \in \mathbb{R}; x^2 + y^2 = 1\} \simeq \mathrm{U}(1),\\
\mathbf{W}_5 &= \{x + y\mathbf{e}_{4} | x, y \in \mathbb{R}; x^2 + y^2 = 1\} \simeq \mathrm{U}(1),\\
\mathbf{W}_6 &= \{x + y\mathbf{e}_{1234} | x, y \in \mathbb{R}; x^2 + y^2 = 1\} \simeq \mathrm{U}(1).
\end{aligned}$$

Thus, the g_i can also be calculated with the following unitary transformation:

$$u_2 = \frac{1}{\sqrt{2}}(1 - \mathbf{e}_4) \in \mathbf{W}_5. \tag{4.5}$$

For the set $G = \{g_i\}$, we have

$$G = u_2 F \bar{u}_2, \tag{4.6}$$

where $\bar{u}_2$ is the Clifford conjugate of u_2. The 24 reflections $s_{j\alpha}$ with $j = 1, 2, \ldots, 12$ and $\alpha = 1, 2$ are close relatives of the Schönfließ symbols σ'_d, σ''_d (Schmeikal 1996). They have the properties required by **D1**. They are invertible, and their squares equal the unity

$$s_{j\alpha}^2 = +1, \quad \forall j, \alpha. \tag{4.7}$$

By reversion $\tilde{}$, they are normalized to unit vectors in the basis of $\mathbb{R}^3$. Precisely, they satisfy the 24 equations

$$\begin{array}{llll} s_{11}\tilde{s}_{11} = -\mathbf{e}_2, & s_{12}\tilde{s}_{12} = +\mathbf{e}_2, & s_{21}\tilde{s}_{21} = +\mathbf{e}_3, & s_{22}\tilde{s}_{22} = -\mathbf{e}_3, \\ s_{31}\tilde{s}_{31} = +\mathbf{e}_1, & s_{32}\tilde{s}_{32} = +\mathbf{e}_1 & s_{41}\tilde{s}_{41} = -\mathbf{e}_1, & s_{42}\tilde{s}_{42} = -\mathbf{e}_1, \\ s_{51}\tilde{s}_{51} = +\mathbf{e}_3, & s_{52}\tilde{s}_{52} = -\mathbf{e}_3 & s_{61}\tilde{s}_{61} = +\mathbf{e}_2, & s_{62}\tilde{s}_{62} = -\mathbf{e}_2, \\ s_{71}\tilde{s}_{71} = -\mathbf{e}_3, & s_{72}\tilde{s}_{72} = -\mathbf{e}_3, & s_{81}\tilde{s}_{81} = -\mathbf{e}_1, & s_{82}\tilde{s}_{82} = +\mathbf{e}_1, \\ s_{91}\tilde{s}_{91} = +\mathbf{e}_2, & s_{92}\tilde{s}_{92} = -\mathbf{e}_2, & s_{101}\tilde{s}_{101} = -\mathbf{e}_2, & s_{102}\tilde{s}_{102} = +\mathbf{e}_2, \\ s_{111}\tilde{s}_{111} = +\mathbf{e}_1, & s_{112}\tilde{s}_{112} = +\mathbf{e}_1, & s_{121}\tilde{s}_{121} = +\mathbf{e}_3, & s_{122}\tilde{s}_{122} = +\mathbf{e}_3. \end{array}$$

Notice that the $s_{j\alpha}$ do not belong to $\mathbf{Pin}(3,1)$ because, in this case, by reversion they would be normalized to ± 1. They are composed by even and odd components. Yet, they do not belong to the Lipschitz group $\Gamma_{3,1}$ because vectors $\mathbf{x} \in \mathbb{R}^{3,1}$ would be transformed into vectors $s\mathbf{x}\hat{s}^{-1} = s\mathbf{x}\hat{s} \in \mathbb{R}^{3,1}$. However, this is not the case with the $s_{j\alpha}$. Basic reflections are mathematical tools to transpose the primitive idempotents. We use them to generate the orientation of space-time and the symmetry SU(3) of the interaction space. Similarly, as in Coxeter's finite reflection groups and in our case, the basic reflections are generators not only of the coordinate units $\mathbf{e}_i$ but also of the octahedral symmetries which are responsible for the orientation of the Minkowski space. Further, they bring forth the tetrahedral rotations of quantum chromodynamics. Another important feature of the basic reflections $s_{j\alpha}$ is in the multiplication rules

$$s_{j1}s_{j2} = c_{j2}, \quad \text{and} \quad c_{j2}^2 = 1, \tag{4.8}$$

the meaning of which is well known in traditional geometry. Quantities s_{j1}, s_{j2} act just like the Schönfließ symbols of rotatory reflections. That is, the product of rotatory reflections $\sigma'_d, \sigma''_d \in \mathbf{D}_{2d}$ in the dihedral group bring forth a half turn C_2 and, thus, a non-trivial centre of the octahedral symmetry. Precisely, the $s_{j\alpha}$ are generators of hyperoctahedral symmetries acting on the hypercubes of idempotent lattices.

5 Transpositions of idempotents by basic reflections

There exist, indeed, 6 different sets of primitive idempotents in $C\ell_{3,1}$. These are composed from the ordered pairs $ik, i, k \neq 4$ by four units

$1, \mathbf{e}_i, \mathbf{e}_{k4}, \mathbf{e}_{ik4}$ with the appropriate signatures. We have used only two of them, namely the f_i built up by $\{1, \mathbf{e}_1, \mathbf{e}_{34}, \mathbf{e}_{134}\}$ and the g_i which stem from the set $\{1, \mathbf{e}_3, \mathbf{e}_{34}, \mathbf{e}_{134}\}$. Note that all those basis units squared give $+1$. Now consider the first 12 *basic reflections* from s_{11} to s_{61}. They are very similar to the Schönfließ symbols σ_d', σ_d'' representing rotatory reflections in the spatial dihedral group D_{2d}. Consider, for example, σ_d'' which results in a flip of the $\{\mathbf{e}_1, \mathbf{e}_2\}$ plane by π about the main diagonal $y = x$. By this flip, $\mathbf{e}_1$ is transposed onto $\mathbf{e}_2$, and vice versa. Clearly σ_d'' can be conceived as an element of the rotation group SO(3). In the Pauli algebra SU(2), this element has two representations by $s'' = \frac{1}{\sqrt{2}}(\mathbf{e}_{13} - \mathbf{e}_{23})$ and its double $s''^{-1} = -s''$.

Recall that a Pauli spinor $\psi \to s''^{-1}\psi$ turns only half the angle of a vector $\mathbf{r}$ obeying the transformation rule $\mathbf{r} \to s''^{-1}\mathbf{r}s''$. Thus, we obtain the transpositions $\mathbf{e}_1 \to s''^{-1}\mathbf{e}_1 s'' = \mathbf{e}_2$ and $\mathbf{e}_2 \to s''^{-1}\mathbf{e}_2 s'' = \mathbf{e}_1$. For such a transposition or commutation of elements, we write $\mathbf{e}_1 \longleftrightarrow \mathbf{e}_2$. A similar statement can be made about the basic reflections which obey the identity $s_{j\alpha}^{-1} = s_{j\alpha}$.

Acting on the set of primitive idempotents $\mathrm{F} = \{f_1, f_2, f_3, f_4\}$, each of those basic reflections leaves two f_i invariant but transposes the remaining two. Suppose that f_i is transposed onto f_k, and the other way around; then we write $f_i \longleftrightarrow f_k$. A transposition is carried out by multiplying an f_i with an $s_{j\alpha}$ from the left and right. For example, we have $s_{12}f_3s_{12} = f_4$ and $s_{12}f_4s_{12} = f_3$; therefore, we put $f_3 \longleftrightarrow f_4$. Altogether, we observe the following transpositions

$$\begin{array}{lll} & \text{invariant} & \text{transposed} \\ s_{11}, s_{12}: & f_1, f_2 & f_3 \longleftrightarrow f_4, \\ s_{21}, s_{22}: & f_1, f_3 & f_2 \longleftrightarrow f_4, \\ s_{31}, s_{32}: & f_1, f_4 & f_2 \longleftrightarrow f_3, \\ s_{41}, s_{42}: & f_2, f_3 & f_1 \longleftrightarrow f_4, \\ s_{51}, s_{52}: & f_2, f_4 & f_1 \longleftrightarrow f_3, \\ s_{61}, s_{62}: & f_3, f_4 & f_1 \longleftrightarrow f_2. \end{array} \tag{5.1}$$

The $s_{j\alpha}$ also carry out transpositions between different sets of primitive idempotents. Consider, for example, the set H built up by components $\{1, \mathbf{e}_1, \mathbf{e}_{24}, \mathbf{e}_{124}\}$:

$$\begin{aligned} h_1 &= \tfrac{1}{2}(1+\mathbf{e}_1)(1+\mathbf{e}_{24}), \quad & h_2 &= \tfrac{1}{2}(1-\mathbf{e}_1)(1+\mathbf{e}_{24}), \\ h_3 &= \tfrac{1}{2}(1+\mathbf{e}_1)(1-\mathbf{e}_{24}), \quad & h_4 &= \tfrac{1}{2}(1-\mathbf{e}_1)(1-\mathbf{e}_{24}). \end{aligned} \tag{5.2}$$

We have

$$\begin{aligned} s_{11}: &\quad g_1 \longleftrightarrow h_1, \; g_2 \longleftrightarrow h_2, \; g_3 \longleftrightarrow h_3, \; g_4 \longleftrightarrow h_4, \\ s_{12}: &\quad g_1 \longleftrightarrow h_3, \; g_2 \longleftrightarrow h_4, \; g_3 \longleftrightarrow h_1, \; g_4 \longleftrightarrow h_2. \end{aligned} \tag{5.3}$$

Acting on F, the reflection s_{71} generates still another set of primitive idempotents, and so on. Finally, we obtain six sets of primitive idempotents F, G, H, K, M, N which are interconnected by basic reflections. Each set generates a lattice with 16 idempotents which can be pictured as a 4-dimensional hypercube with a 3-dimensional projection being a rhombidodecahedron. In each of those hypercubes, there exist four tetrahedral operators, together with their inverses, which belong to an octahedral orientation symmetry **O**.

6 The color rotations of QCD

Consider reflections s_{11} and s_{31} from equations (5.1). They carry out transpositions $f_3 \longleftrightarrow f_4$ and $f_2 \longleftrightarrow f_3$, but do not alter f_1. It is, therefore, that the product $c_{13} = s_{11}s_{31}$ color-rotates the idempotents f_2, f_3, f_4. The inverse of c_{13} is the product $s_{31}s_{11}$, and we have

$$c_{13}f_2c_{13}^{-1} = f_3, \quad c_{13}f_3c_{13}^{-1} = f_4, \quad c_{13}f_4c_{13}^{-1} = f_2. \tag{6.1}$$

Fixing either f_2 or f_3 or f_4, we obtain a total of four tetrahedral rotations together with their inverses. For each set of primitive idempotents F, G, H, K, M, N, there exists such a set of 8 tetrahedral rotations. We obtain the 24 quantities ${}_ic_{k3}$ with $i = 1, \ldots, 6$ and $k = 1, \ldots, 4$ and their inverse rotations ${}_ic_{k3}^{-1}$ where the index 3 merely indicates period 3. We can fix the primitive idempotents f_1, g_1, h_1, k_1, m_1 and n_1 and interpret them as leptons to 3 generations of quarks (d,u; s,c; b,t-quarks). The color rotations turn out as tetrahedral elements of the octahedral orientation symmetry of the idempotent lattices.

7 Octahedral symmetries of idempotent lattices

The color rotations are typical elements of the octahedral orientation symmetry of cubes and hypercubes. Each of the sets F, G, H, K, M, N of primitive idempotents generates a 16-element lattice of idempotents which can be pictured as a 4-dimensional cube or its 3-dimensional projection which is a rhombidodecahedron. The octahedral group **O** is isomorphic to the symmetric group S_4 having 24 elements. Consider **O**(F), the octahedral group of the first set of idempotents. It has a

$$\textbf{minimal generating set} \qquad \{s_{11}, s_{11}s_{12}s_{21}\} \tag{7.1}$$

and consists of the following elements:

$$\begin{array}{ll}
s_{11}, s_{12}, s_{21}, s_{22}, s_{31}, s_{32} & \text{basic reflections,} \\
c_{12} = s_{11}s_{12}, c_{22} = s_{21}s_{22}, c_{32} = s_{31}s_{32} & \text{main rotations,} \\
c_{14} = s_{31}s_{32}s_{11}, c_{24} = s_{11}s_{12}s_{21}, c_{34} = s_{21}s_{22}s_{31} & \text{period 4 rotations,} \\
c_{14}^{-1} = s_{11}s_{32}s_{31}, c_{24}^{-1} = s_{21}s_{12}s_{11}, c_{34}^{-1} = s_{31}s_{22}s_{21} & \text{their inverses,} \\
c_{13} = s_{11}s_{31}, c_{23} = s_{11}s_{22}, c_{33} = s_{12}s_{31}, c_{43} = s_{21}s_{32} & \text{color rotations,} \\
c_{13}^{-1} = s_{31}s_{11}, c_{23}^{-1} = s_{22}s_{11}, c_{33}^{-1} = s_{31}s_{12}, c_{43}^{-1} = s_{32}s_{21} & \text{their inverses,} \\
1 & \text{unity.}
\end{array} \tag{7.2}$$

8 Six copies of SU(3)

There are 6 hyperoctahedral lattices of idempotents and in each of them there exist four typical color operators. Fixing the first primitive idempotent in each lattice such that we obtain a spinor space invariant under the action of a color rotation, we should obtain six copies of the symmetric unitary group SU(3). The Gell-Mann matrices can be calculated as follows. Consider the lattice generated by the set F. The Gell-Mann matrices for F read as follows:[2]

$$\begin{aligned}
\lambda_1 &= \tfrac{1}{2}(s_{31} - s_{32}) &&= \tfrac{1}{2}(-\mathbf{e}_{24} + \mathbf{e}_{124}), \\
\lambda_2 &= \tfrac{i}{2}(s_{21}s_{22}s_{31} - s_{31}s_{22}s_{21}) &&= \tfrac{i}{2}(-\mathbf{e}_{23} + \mathbf{e}_{123}), \\
\lambda_3 &= \tfrac{1}{2}(s_{11}s_{12} - s_{21}s_{22}) &&= \tfrac{1}{2}(\mathbf{e}_{34} + \mathbf{e}_{134}), \\
\lambda_4 &= \tfrac{i}{2}(s_{11}s_{12}s_{21} - s_{21}s_{12}s_{11}) &&= \tfrac{i}{2}(-\mathbf{e}_{4} + \mathbf{e}_{13}), \\
\lambda_5 &= \tfrac{1}{2}(s_{21} - s_{22}) &&= \tfrac{1}{2}(\mathbf{e}_{3} + \mathbf{e}_{14}), \\
\lambda_6 &= \tfrac{i}{2}(s_{11}s_{32}s_{31} - s_{31}s_{32}s_{11}) &&= \tfrac{i}{2}(-\mathbf{e}_{12} + \mathbf{e}_{1234}), \\
\lambda_7 &= \tfrac{1}{2}(s_{11} - s_{12}) &&= \tfrac{1}{2}(-\mathbf{e}_{2} + \mathbf{e}_{234}), \\
\lambda_8 &= \tfrac{1}{\sqrt{12}}(4f_{22} - 1 - \mathbf{e}_1) &&= \tfrac{1}{\sqrt{12}}(-2\mathbf{e}_1 + \mathbf{e}_{34} + \mathbf{e}_{134}).
\end{aligned}$$

[2]In $C\ell_{3,1}$ there is a second representation to λ_8, which has the form

$$\lambda_8 = \frac{1}{\sqrt{12}}(-1 - 3s_{31}s_{32}) = -1 - 3\mathbf{e}_1$$

and differs from the first by a nonzero entry in the first row and column. This does not affect the group structure if we omit the first row and column.

To prove this is correct use basis (4.2) and calculate, for example,

$$\lambda_5 = \frac{1}{2}\begin{pmatrix} 0 & 0 & -i & 0 \\ 0 & 0 & 0 & -i \\ i & 0 & 0 & 0 \\ 0 & i & 0 & 0 \end{pmatrix} + \frac{1}{2}\begin{pmatrix} 0 & 0 & i & 0 \\ 0 & 0 & 0 & -i \\ -i & 0 & 0 & 0 \\ 0 & i & 0 & 0 \end{pmatrix} = \left(\begin{array}{c|ccc} 0 & 0 & 0 & 0 \\ \hline 0 & 0 & 0 & -i \\ 0 & 0 & 0 & 0 \\ 0 & i & 0 & 0 \end{array}\right).$$

Up till now we have used the basic reflections in order to

(i) generate the basis of Euclidean 3-space,

(ii) transpose the primitive idempotents,

(iii) generate the octahedral symmetries of idempotent lattices,

(iv) construct the generators of the interaction space SU(3).

We shall now use them to bring forth all possible ($256 = 16 \times 16$) transformations among the elementary logic relations. For this purpose, we first define the generalized logic operations of negation and attribute commutation.

9 Generalized logic operators

The present day logic calculi are not aware of the geometric nature of logic. But using Boole's notation, we proceed in a straight line following certain rules of derivation. We are not aware of the iconic pattern of logic transformations because this, at first, involves more than one dimension. The logician goes on from one definite, well-formed term or formula to another one which is identical to the first or which follows from it. If B can be derived from A by some rules of derivation or "laws of thought" – using Boole's words – B stands with A in a logic relation of implication $\rightarrow$ or identity $\equiv$. Thus, often the use of metalogical signs like $\xrightarrow{L}$ for deducibility is not metalogic at all, but stays within the logic framework entirely. Charles S. Peirce, William K. Clifford and in our time mathematicians, psychologists and sociologists like Glenn Clark, Jean Piaget, Shea Zellweger, and Claude Lévi-Strauss have realized the geometric nature of logic transformations. They used what I will call *"generalized logic operations."* Consider a well-formed Boolean term such as $\neg A \wedge B$. It consists of a proposition to the left, an operator or "junctor" $\wedge$ in the middle, and a second proposition to the right. There are essentially three locations in the term, and on each location there may act a generalized negation. Basically, we allow for three elementary negations of the form NOO, ONO and OON. We observe, for example, $\mathrm{NOO}(\neg A \wedge B) = A \wedge B$, $\mathrm{ONO}(\neg A \wedge B) = A \vee \neg B$, and $\mathrm{OON}(\neg A \wedge B) = \neg A \wedge \neg B$. For $\mathrm{ONO}(A \wedge B)$ we also write $A \neg\wedge B$, where $\neg\wedge$ means "not and". The generalized negation ONO denotes "negation of connective" while NOO and OON denote "negation of proposition". Often it is also accounted for the possibility to "commute" propositions. Then by "commutation" the term $\neg A \wedge B$ is turned into $A \wedge \neg B$ while $A \wedge B, \neg A \wedge \neg B, A \vee B$ (and so on) are not altered. Let us define in a loose manner at first

D2: Generalized logic operators

$$\nu_1 = \text{NOO},\ \nu_2 = \text{ONO},\ \nu_3 = \text{OON},$$

$$\tau = \text{transposition of propositions},$$

$$\gamma = \text{C(orrelation)},\ \epsilon = \text{OOO (the identity transformation)}.$$

Those operators can be multiplied so that one obtains, for example,[3]

$$\begin{aligned} \nu_1\nu_2\gamma(A \wedge \neg B) &= (\text{NOO})(\text{ONO})\tau(A \wedge \neg B) \\ &= (\text{NOO})(\text{ONO})(\neg A \wedge B) \\ &= \text{NOO}(A \vee \neg B) = \neg A \vee \neg B. \end{aligned}$$

Generally, we define

D3: **A binary connective** is a triple $A * B$ with an abstract junctor $*$, which has 16 possible values given by the idempotents in [1]. Note, each idempotent represents one definite truth table.[4]

D4a: **Negation of the left proposition:** a generalized negation ν_1 of $A * B$ involves a value change of the junctor $*$ which is determined by the two transpositions $f_1 \longleftrightarrow f_3$ and $f_2 \longleftrightarrow f_4$. Note, these transpositions reflect the desired rearrangement of the truth table.

4b: **Negation of connective:** a generalized negation ν_2 of the connective $*$ is given by a turnover from the idempotent L_i to the dual idempotent $1 - L_i$. Note, this means an exchange of 0 and 1 in the truth table of $*$.

4c: **Negation of the right proposition:** a generalized negation ν_3 of $A * B$ involves a value change of the junctor $*$ which is determined

---

[3]Clark and Zellweger (1993) have concatenated the operators and introduced expressions of the form OOOO, NOOO, ... , NOOC, ... NONC, ... , NNNC. There are 16 such concatenated operators depending on whether or not there is a negation of the first proposition, is a negation of the relation, is a negation of the second proposition and whether or not there is a transposition of propositions. But one must account for the succession of the operators because they do not commute, that is, ν_1 and ν_3 do not commute with τ.

[4]Such a binary logic relation $A * B$ requires considerably more locations in the Boole notation. For example, we have to define "identity" $A \equiv B$ by the 13-letter term $A \equiv B \div (A \wedge B) \vee (\neg A \wedge \neg B)$.

by the two transpositions $f_1 \longleftrightarrow f_2$ and $f_3 \longleftrightarrow f_4$. Note, these transpositions reflect the desired rearrangement of the truth table.

4d: Transposition of propositions: a transposition (or exchange) τ of propositions in $A*B$ involves a value change of the junctor $*$ which is determined by the transposition $f_2 \longleftrightarrow f_3$. Note, this transposition reflects the desired rearrangement of the truth table.[5]

4e: Correlation of junctions: correlation γ is given by an exchange of the dual operators $\wedge$ and $\vee$. In any connective as determined by its Boolean expression, the junctor $\wedge$ is replaced by the dual $\vee$ and the other way around.[6]

Definitions **D4a** to **4e** allow for all 256 possible transformations among the 16 connectives.

9.1 Negation ν_1 of the left proposition in $C\ell_{3,1}$

Being aware of the transformation properties (7.2) of the basic reflections $s_{i\alpha}$, we can be sure that a generalized negation ν_1 of $A * B$ is determined by the basic reflections s_{51} and s_{21} (alternatively s_{52} and s_{22}). The first proposition in any connective L_i, where $i = 1, \dots, 16$, can be negated by $L_i \rightarrow s_{51}s_{21}L_is_{21}s_{51}$ or, alternatively, $L_i \rightarrow s_{52}s_{21}L_is_{21}s_{52}$. This is the same as

$$L_i \rightarrow \mathbf{e}_3 L_i \mathbf{e}_3 \quad \text{or, alternatively,} \quad L_i \rightarrow \mathbf{e}_{14} L_i \mathbf{e}_{14}. \tag{9.1}$$

Example 1. *Consider $L_{15} = A \vee B$ which according to [1] is represented by the idempotent $f_2 + f_3 + f_4$. We have $\mathbf{e}_3(f_2 + f_3 + f_4)\mathbf{e}_3 = f_1 + f_2 + f_4$ which represents L_{13}, that is, in Boole's notation $\neg A \vee B$.*

9.2 Generalized negation ν_2 of the connective $A * B$ in $C\ell_{3,1}$

Generalized negation ν_2 of the connective $A * B$ in $C\ell_{3,1}$ is implicitly defined by the lattice complement, that is, negation of any connective L_i has to result in the idempotent $1 - L_i$:

$$L_i \rightarrow 1 - L_i. \tag{9.2}$$

---

[5]Exchange of proposition has been denoted "commutation" by Zellweger and symbolized by C. We should use another symbol because C was introduced by Piaget to denote "correlation". Also we use "commutation" in the algebraic sense.

[6]Example: $(A \wedge B) \vee (\neg A \wedge \neg B)$ is replaced by the term $(A \vee B) \wedge (\neg A \vee \neg B)$. A little calculation shows that this is equal to $(A \wedge \neg B) \vee \neg A \wedge B)$. This represents $A \neg \equiv B$ ('A or else B').

Example 2. *Consider* $L_2 = \neg A \wedge \neg B$ *represented by the idempotent* f_1. *Negation of the connective of* L_2 *is brought forth by*

$$\nu_2 L_2 \div 1 - L_2 = f_2 + f_3 + f_4$$

which according to [1] is representing L_{15}, *that is,* $A \vee B$.

9.3 Negation ν_3 of the right proposition in $C\ell_{3,1}$

By the transformation properties (5.1) of the basic reflections $s_{i\alpha}$, we find that a generalized negation ν_3 of $A * B$ is determined by the basic reflections s_{61} and s_{11} (alternatively s_{62} and s_{12}). The right proposition in any connective L_i, where $i = 1, \ldots, 16$, can be negated by $L_i \to s_{61}s_{11}L_i s_{11}s_{61}$ or alternatively $L_i \to s_{62}s_{11}L_i s_{11}s_{62}$. This is the same as

$$L_i \to \mathbf{e}_{234} L_i \mathbf{e}_{234} \quad \text{or alternatively} \quad L_i \to \mathbf{e}_2 L_i \mathbf{e}_2. \tag{9.3}$$

Example 3. *Consider the identity relation* $A \equiv B$ *which is* $L_8 = f_1 + f_4$. *We calculate* $\nu_3(A \equiv B) \div \mathbf{e}_{234} L_8 \mathbf{e}_{234} = f_2 + f_3$ *which is* L_9 *or* $A\neg \equiv B$. *Clearly, negating one of the two propositions negates the identity relation.*

9.4 Exchange τ of propositions

As an exchange (or transposition) τ of propositions in $A * B$ is determined by the transposition $f_2 \longleftrightarrow f_3$, this is brought forth by the application of s_{31}. In any connective L_i the two propositions can be transposed by multiplying the representative idempotent by s_{31} from the left and right.

$$L_i \to s_{31} L_i s_{31}. \tag{9.4}$$

Example 4. *The identity relation* $A \equiv B$ *is not affected by an exchange of propositions. In* $C\ell_{3,1}$ *this is expressed by the equation*

$$s_{31}(f_1 + f_4)s_{31} = f_1 + f_4.$$

9.5 Correlation γ of junctions

Correlation γ has been introduced as a symmetry operation by Jean Piaget (1985) and is given by an exchange of the dual operators $\wedge$ and $\vee$. It is not independent of the other operations of generalized negation, but it can be brought about by a product of $\nu_2\nu_1\nu_3$ in this sequence. We can define $\gamma L_i = \nu_2\nu_1\nu_3 L_i$. The product $\nu_1\nu_3$ has four possible representations because each ν_1 and ν_3 have two. Therefore, correlation γ has four representations

$$\begin{aligned}
\gamma L_i &= 1 - \mathbf{e}_{32} L_i \mathbf{e}_{23} = 1 + \mathbf{e}_{23} L_i \mathbf{e}_{23}, \\
\gamma L_i &= 1 - \mathbf{e}_{124} L_i \mathbf{e}_{124}, \\
\gamma L_i &= 1 - \mathbf{e}_{24} L_i \mathbf{e}_{24}, \\
\gamma L_i &= 1 - (-\mathbf{e}_{123} L_i \mathbf{e}_{123}) = 1 + \mathbf{e}_{123} L_i \mathbf{e}_{123}.
\end{aligned} \tag{9.5}$$

Example 5. *Consider* $L_5 = f_4$, *the logic conjunction* $A \wedge B$. *We have* $\gamma f_4 = 1 - \mathbf{e}_{24} f_4 \mathbf{e}_{24} = f_2 + f_3 + f_4$ *which correctly represents* L_{15}.

10 De Morgan's laws

In mathematical logic we learn that, for any logic propositions A and B in the elementary logic calculus L_1, we have to have

(i) $\neg(A \wedge B) = \neg A \vee \neg B$,

(ii) $\neg(A \vee B) = \neg A \wedge \neg B$.

Those are known as de Morgan's laws of elementary logic or Boole's algebra, respectively. To close the circle, let us go back to Peirce's notation to see what they mean. First, take L_5 which is $A \wedge B$ and L_{15} being $A \vee B$. In Peirce's iconic notation these are

$A \wedge B$

$A \vee B$

De Morgan's law (i) says that ν_2, when applied to the left figure $(A \wedge B)$, will result in the same figure as $\nu_1\nu_3$ applied to the right figure $(A \vee B)$. Now we know the result; namely, this is

$\nu_2(A \wedge B)$

$= \neg A \vee \neg B =$

$\nu_1\nu_3(A \vee B)$

This figure is obtained from the upper left by "mating". There are four boxes (primitive idempotents), and each one can be filled with a ball. Take the upper left figure, fill the empty boxes by a ball and empty the filled boxes, that is, form a complement. The "mated icon" on the left should equal the figure on the right rotated by π. We said that in $C\ell_{3,1}$ the mating operation is defined by the transformation $L_i \to 1 - L_i$. The rotation $\nu_1\nu_3$, we have seen, can be represented by $L_i \to -\mathbf{e}_{23} L_i \mathbf{e}_{23}$. This is in analogy with rotation of a vector by spin matrices. As a matter of fact, we have

$$1 - f_4 = -\mathbf{e}_{23}(f_2 + f_3 + f_4)\mathbf{e}_{23}. \tag{10.1}$$

Multiplying equation (10.1) with $\mathbf{e}_{32}$ from the left and right, we obtain

$$\mathbf{e}_{32}\mathbf{e}_{32} - \mathbf{e}_{32}f_4\mathbf{e}_{32} = -\mathbf{e}_{32}\mathbf{e}_{23}(f_2 + f_3 + f_4)\mathbf{e}_{23}\mathbf{e}_{32} = -(f_2 + f_3 + f_4).$$

With $\mathbf{e}_{32}\mathbf{e}_{32} = -1$ this gives us the second law

$$1 - (f_2 + f_3 + f_4) = -\mathbf{e}_{23}f_4\mathbf{e}_{23}. \tag{10.2}$$

That is, mating L_{15} gives the same as rotating L_5. In iconic notation

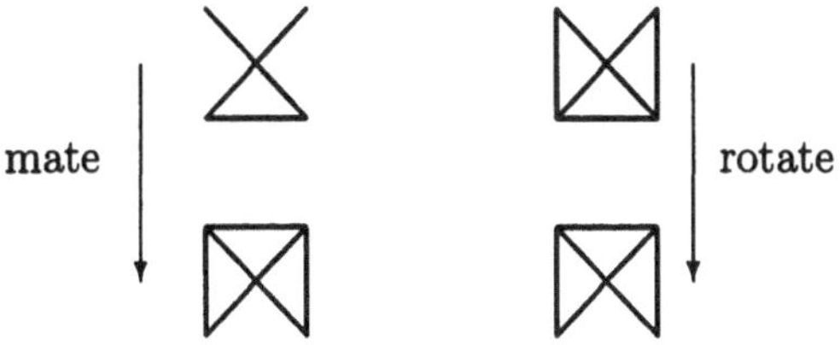

De Morgan's law has a wider range than is indicated by (10.1) and (10.2). In its general form, it says that the logic relation – which is dual to L_i – can be obtained by mating the rotated icon L_i, precisely by negating that connective (ν_2) which is obtained by negating both propositions ($\nu_1\nu_3$) in L_i :

$$\gamma L_i = 1 + \mathbf{e}_{23}L_i\mathbf{e}_{23}. \tag{10.3}$$

Thus, de Morgan's law is a definition of a duality-relation between connectives γL_i and L_i. Let $\Lambda = \{L_i \,|\, i = 1, \dots, 16\}$; then, the duality-relation $\Delta \subset \Lambda^2$ is given by

$$\Delta = \{L_i, \gamma L_i \,|\, \gamma L_i = 1 + \mathbf{e}_{23}L_i\mathbf{e}_{23}\}. \tag{10.4}$$

11 Logic of Dirac spinors

The Dirac algebra is a 2×2 matrix algebra with diagonal entries in the even part $C\ell^+_{3,1}$ and off-diagonal in $C\ell^-_{3,1}$. It is isomorphic to $\mathrm{Mat}(4, \mathbb{C})$. The Dirac spinors are elements of a minimal left ideal of the Dirac algebra and have the form

$$\psi = \begin{pmatrix} u^+ & 0 \\ u^- & 0 \end{pmatrix} f \quad \text{where} \quad f = \frac{1}{2}(1 + e_{34}) \quad \text{and} \quad u^\pm \in C\ell^\pm_{31}. \tag{11.1}$$

11.1 Negation ν_1 of first proposition

The quantity f is a primitive idempotent of $C\ell^\pm_{3,1}$ and, at the same time, a representative of the connective L_{11}, the negated proposition $\neg A$. Within the complex logic system of the $C\ell_{3,1}$, the logic of the Dirac spinors is disclosed as a rather simple restriction. In accordance with [1], a reversion of f brings about $\tilde{f} = \frac{1}{2}(1-\mathbf{e}_{34})$ which is L_{11} negated, that is, L_6 which represents A. In the basis (4.2), we have

$$L_{11} = f = \begin{pmatrix} 1 & 0 \\ 0 & 0 \end{pmatrix}, \quad \psi = \begin{pmatrix} u^+ & 0 \\ u^- & 0 \end{pmatrix}$$

and

$$L_6 = \tilde{f} = \frac{1}{2}(1 - \mathbf{e}_{34}) = \begin{pmatrix} 0 & 0 \\ 0 & 1 \end{pmatrix}.$$

Thus, by negation ν_1 of the first proposition, a spinor ψ is annihilated. Clearly, generalized logic operations, as were defined in Section 9, are acting on the Dirac spinor space. But ν_1 has to be excluded if one wants to avoid annihilation.

11.2 Logic Operations τ, ν_3 and γ on Dirac spinors

Transposition τ has to be excluded because it can lead to a negation of the first proposition. One may argue then that there remain ν_3 and γ. But ν_3 involves a transposition $f_1 \longleftrightarrow f_2$ which means that the lepton f_1 described by the Dirac spinor turns into a quark. Correlation γ does not alter $f = f_1 + f_2$, as can easily be verified. It sends u^- from $C\ell_{3,1}^-$ to $C\ell_{3,1}^- \cup C\ell_{3,1}^+$, as can be seen from $\gamma\mathbf{e}_1 = 1-\mathbf{e}_1$. Dirac spinors do not remain Dirac spinors, but they are turned into complexified Majorana spinors.

11.3 Time reversal as a logic operation

Time reversal is defined by a transition from $\mathbf{e}_4$ to $-\mathbf{e}_4$ which, because of equation (4.3), has the immediate consequence of transpositions of the primitive idempotents, namely $f_1 \longleftrightarrow f_4$ and $f_2 \longleftrightarrow f_3$. Note that ν_1 carries out transpositions (13)(24) while ν_3 does (12)(34). Thus, their product brings forth the desired cycles (14)(23). Product $\nu_1\nu_3$ has four representations only two of which turn $\mathbf{e}_4$ into $-\mathbf{e}_4$, namely, $\mathbf{e}_{24}$ and $\mathbf{e}_{123}$ Thus, take a look at the Dirac spinor

$$\psi = \begin{pmatrix} u^+ & 0 \\ u^- & 0 \end{pmatrix} L_{11} \quad \text{where} \quad L_{11} = f = \frac{1}{2}(1 + \mathbf{e}_{34}).$$

Although time reversal leaves $\mathbf{e}_{34}$ unaltered as it turns $\mathbf{e}_3, \mathbf{e}_4$ into $-\mathbf{e}_3, -\mathbf{e}_4$ (**PT**) (calculate, e.g., $\mathbf{e}_{24}\mathbf{e}_3\mathbf{e}_{24} = -\mathbf{e}_3$), it negates L_{11}. Again negation of L_{11} results in annihilation of the spinor. Next, consider a spin gauge model in which f_1 represents a lepton and c_{13} the color rotation operator. Also here, the color rotation breaks up $f = f_1 + f_2$ and turns it into $\tilde{f} = \frac{1}{2}(1 - \mathbf{e}_{34})$. The spinor is diminished.

We are now in a position to summarize what is resulting from our investigation into the Dirac algebra by the application of the internal idempotent logic of $C\ell_{3,1}$; namely, it tells us that (i) the space of Dirac spinors is not an appropriate image for a spinor space in quantum chromodynamics, and (ii) does not allow us to represent time reversal in lepton systems. The logic structure of the idempotent lattices of the Majorana algebra is such that it sends time reversed and color rotated wave functions to zero.

12 Quantum chromodynamics in the logic context

Consider one of the generators of the minimal basis of the orientation group O, namely, the period-4 operator $c_{24} = \nu_1\tau$ which carries out to the cycle (1243). The transformation $f_i \to \nu_1\tau f_i\tau\nu_1 = \mathbf{e}_3 s_{31} f_i s_{31}\mathbf{e}_3$ produces the figure

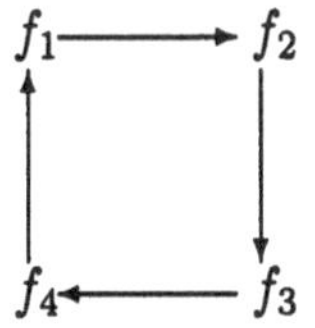

To transform the cycle (1243) into the color rotation (234), we need one more transposition, namely, (13). This can be generated by the reflection s_{51}. The color rotation operator c_{13} keeps f_1 unaltered while it turns

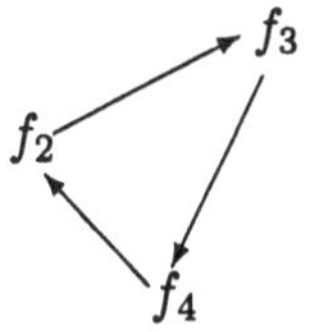

Although the color rotation is a logic operation inasmuch as it induces a rotation in the space of logic propositions, it cannot be represented by a product of generalized logic operators defined so far. It is a reflection of such an operator. Namely, we have

$$c_{13} = \tau\nu_1 s_{51} \quad \text{with the inverse} \quad c_{13}^{-1} = s_{51}\nu_1\tau, \tag{12.1}$$

and the f_i are transformed according to

$$f_i \to s_{51}\nu_1\tau f_i\tau\nu_1 s_{51} = s_{51}e_3 s_{31} f_i s_{31} e_3 s_{51} = c_{13}^{-1} f_i c_{13}. \tag{12.2}$$

13 Compound statements of higher complexity

Consider compound statements involving m prime propositions together with the Clifford algebra of neutral signature $C\ell_{m,m} = C\ell_{1,1} \otimes \ldots \otimes C\ell_{1,1}$. This constitutes the main factor in every real Clifford algebra of higher dimension and has a Dirac group given by products of m dihedral factors $\mathbf{D}_4$ (Bergtold 1996, Shaw 1995). We may denote $C\ell_{m,m}$ *as a "logic Clifford algebra of compound statements of* m *prime propositions."* $C\ell_{m,m}$ contains 2^m, mutually annihilating primitive idempotents $f_1, f_2, \ldots, f_{2^m}$ brought on by all possible sign combinations in

$$\frac{1}{2^m}(1 \pm \mathbf{e}_1\mathbf{e}_{m+1})(1 \pm \mathbf{e}_2\mathbf{e}_{m+2}) \cdots (1 \pm \mathbf{e}_m\mathbf{e}_{2m}). \tag{13.1}$$

These 2^m primitive idempotents f_i annihilate each other by Clifford multiplication and additively generate a lattice L_m of idempotents primitive in $C\ell_{m,m}$. Each element of L_m represents one and only one compound statement of the elementary logic calculus. The lattice

- L_m has $2^{(2^m)}$ elements.
- Compound statements have $2^{(2^m)}$ truth tables.
- $C\ell_{m,m} \simeq \mathrm{Mat}(2^m, \mathbb{R})$.

With the aid of the elements of L_m we construct a number of $2^{m-1}(2^m - 1)$ reflections of the form $s_{j\alpha}$ which carry out transpositions between the primitive idempotents f_i. There are $2^{m-1}(2^m - 1)$ such distinct pairs of $f_i \to f_k$ and reverse. It is enough to construct a set T of 2^m generating transpositions

$$f_1 \leftrightarrow f_2,\ f_2 \leftrightarrow f_3,\ \cdots,\ f_{2^m-1} \leftrightarrow f_{2^m}$$

which we abbreviate as τ_{jk}. Then, any product of arbitrary factors from T can be seen as a general logic transformation. Those transformations which signify "logic derivability" restrict the space to a series of identity relations and implications among terms analogous to terms L_8 and L_{13} in L_2. It should not be too difficult to construct and enumerate transpositions. For instance, consider any number $\alpha \leq m$ and index pair i, k with $i = 2\alpha - 1$ and $k = 2\alpha$; then, transposition $\tau_{i,k}$ is given by

$$\tau_{i,k} = (1 - f_i - f_k) + \mathbf{e}_{2m}(f_k - f_i). \tag{13.2}$$

Example 6. *Consider the Clifford algebra* $C\ell_{4,4}$. *The following expressions in* $C\ell_{4,4}$ *give transpositions* $\tau_{3,4}, \tau_{15,16}$, *and* $\tau_{6,7}$:

$$\begin{aligned}
\tau_{3,4} &= (1 - f_3 - f_4) + \mathbf{e}_8(f_4 - f_3),\\
\tau_{15,16} &= (1 - f_{15} - f_{16}) + \mathbf{e}_8(f_{16} - f_{15}),\\
\tau_{6,7} &= (1 - f_3 - f_4) + \mathbf{e}_{34}(f_7 - f_6).
\end{aligned}$$

14 The abstract algebraic form of iconic logic

There is a set GL which can strictly be denoted a *partial binary single-valued multiplicative system* also *called set of generalized logical operations.* Further, there is a set C of *logic connectives* and "$\bullet$" a binary relation $GL \bullet C$. We investigate mappings $GL \bullet C \to C$, especially when GL imposes on C the structure of a group. Considering the set $GL = \{\nu_1, \nu_2, \nu_3, \tau\}$ and binary connectives C, it can immediately be verified that GL imposes on C the group structure of $\mathbf{D}_4 \times Z_2$. The noncommutativity of the dihedral group $\mathbf{D}_4$ is brought in by the possibility to transpose statements A and B, that is, by the operator τ. In the Clifford algebra, this is expressed by anticommutativity.

15 Future developments

The Clifford algebra $C\ell_{3,1}$ decomposes into a direct sum of minimal left ideals with regard to its idempotents f_1, f_2, f_3, f_4. Interpreting these as quantum states and spinors subject to a Dirac-Fueter equation, it is clear that we need not to expect definite logical states to bring on definite truth values and definite connectives but rather stochastic states. Therefore, what seems to be classical logic at first encounter discloses the character of a quantum logic at the second. In this way, quantum logic naturally unfolds from classical logic. The future developments may show (1) how logic algebraic procedures of arbitrary complexity can be represented in neutral signature and (2) how logic processes such as quantum computation can be carried out in a subnuclear array.

REFERENCES

[1] R. Abłamowicz, P. Lounesto, J. M. Parra, *Clifford Algebras with Numeric and Symbolic Computations,* Birkhäuser, Boston, 1996.

[2] G. Bergdolt, Orthonormal basis sets in Clifford algebras, in *Clifford Algebras with Numeric and Symbolic Computations,* R. Abłamowicz, P. Lounesto, and J. M. Parra, eds., Birkhäuser, Boston, 1996, 269–284.

[3] H. W. Braden, Dimensional spinors; their properties in terms of finite groups, *J. Math. Phys.* **26** (1985), 613–620.

[4] J. Brent, *Charles Sanders Peirce: A Life,* Indiana University Press, Bloomington, 1993.

[5] J. Brunning, Peirce's development of the algebra of relations, Ph.D. Thesis, University of Toronto, 1981.

[6] J. S. R. Chisholm, Tetrahedral structure of idempotents of the Clifford algebra $C\ell_{3,1}$, in *Clifford Algebras and their Applications in Mathematical Physics,* A. Micali, et al., eds., Kluwer, Dordrecht, 1992, 27–32.

[7] G. Clark, New light on Peirce's iconic notation for the sixteen binary connectives, paper presented at the Peirce Sesquicentennial Congress at Harvard University, Sept. 1989, Mount Union College, Alliance, Ohio, 1989.

[8] G. Clark and Sh. Zellweger, Let the mirrors do the thinking, *Mount Union Magazine,* Spring, Alliance, 1993, 2–5.

[9] W. K. Clifford, *The Common Sense of the Exact Sciences,* K. Pearson, ed., Dover, New York, 1955.

[10] J. S. Coleman, Recent developments in American sociological methods, *The Polish Sociological Bulletin,* Vol. **30** (2) (1974), 11–23.

[11] K. Greider and T. Weiderman, *Generalised Clifford Algebras as Special Cases of Standard Clifford Algebras,* l'UCD preprint 16, 1988,(cited after Chisholm 1992).

[12] H. E. Hawkes, Estimate of Peirce's linear associative algebra, *Amer. J. Math.* **24** (1902), 87–95.

[13] D. Hestenes and G. Sobczyk, *Clifford Algebra to Geometric Calculus - A Unified Language for Mathematics and Physics,* Kluwer, Dordrecht, 1984.

[14] V. F. Lenzen, The contributions of Charles S. Peirce to linear algebra, in *Phenomenology and Natural Existence: Essays in Honor of Marvin Farber,* D. Riepe, ed., State University of New York at Albany, 1973, 239–254.

[15] P. Lounesto, Verifying and falsifying conjectures. Counterexamples in Clifford algebras with CLICAL, in *Clifford Algebras with Numeric and Symbolic Computations,* R. Abłamowicz, P. Lounesto, J. M. Parra, eds., Birkhäuser, Boston, 1996, 1–30.

[16] J. Maks, *Modulo (1,1)-Periodicity of Clifford Algebras and Generalized (Anti)-Möbius Transformations,* Proefschrift, Delft University of Technology, 1989.

[17] K. Menger, A group in the substitutive algebra of the calculus of propositions, *Archiv der Mathematik* **13** (1962), 471–478.

[18] A. Micali, et al., *Clifford Algebras and Their Applications in Mathematical Physics,* Kluwer, Dordrecht, 1992.

[19] P. Mittelstaedt, *Philosophische Probleme der Modernen Physik,* Bibliographisches Institut, Mannheim, 1963.

[20] Z. Oziewicz, Z., ed., *Spinors, Twistors, Clifford Algebras and Quantum Deformations,* Kluwer, Dordrecht, 1993, 97–106.

[21] Ch. S. Peirce, Description of a notation for the logic relatives, resulting from an amplification of the conceptions of Boole's calculus of logic, *Memoirs of the American Academy of Sciences* **9** (1870), 317–378.

[22] Ch. S. Peirce, A brief description of the algebra of relatives, 7th January 1882, with a postscript of 16th of January, privately printed brochure, 6 pages, Baltimore, *Collected Papers of Charles Sanders Peirce, Vol. 4,* Cambridge, 1882, 227–323.

[23] Ch. S. Peirce, The simplest mathematics, in *Collected Papers of Charles Sanders Peirce,* C. Hartshorne and P. Weiss, eds., Cambridge, 1902, 227–323.

[24] Ch. S. Peirce, Nomenclature and divisions of dyadic relations, (syllabus), *MS,* Robin catalog No. 539:00002 – 00025 (1903).

[25] Ch. S. Peirce, MS 431, in *Annotated Catalogue of the Papers of Charles S. Peirce,* R.S. Robin, ed., University of Massachusetts, 1967.

[26] Ch. S. Peirce, *Writings Edition Project: Writings of Charles Sanders Peirce: A Chronological Edition,* Indiana University, Indianapolis, 1986.

[27] J. Piaget, *Logic and Psychology,* Basic Books, New York, 1957.

[28] J. Piaget, *Jean Piaget über Jean Piaget,* Kindler, Munich, 1981.

[29] R. S. Robin, *Annotated Catalogue of the Papers of Charles S. Peirce,* University of Massachusetts, 1967.

[30] B. Schmeikal-Schuh, Soziogene der orientierung – ikonen der raumzeit, *WISDOM. Jg. III* **3/4** (1989), 1–18.

[31] B. Schmeikal-Schuh, Logic from space, *Quality and Quantity* **27** (1993), 117–137.

[32] B. Schmeikal, The generative process of space-time and strong interaction – quantum numbers of orientation, in *Clifford Algebras with Numeric and Symbolic Computations,* R. Abłamowicz, P. Lounesto, and J. M. Parra, eds., Birkhäuser, Boston, 1996, 83–100.

[33] B. Schmeikal, Reconstructions of science, *Institut für Höhere Studien. Reihe Soziologie* No. **33**, Vienna, 1996.

[34] R. Shaw, Finite geometry, Dirac groups and the table of real Clifford algebras, in *Clifford Algebras and Spinor Structures,* R. Abłamowicz and P. Lounesto, eds., Kluwer, Dordrecht, 1995, 59–99.

[35] A. Stern, *Matrix Logic,* North Holland-Elsevier, Amsterdam, 1991.

[36] A. Stern, *Matrix Logic and the Mind,* North Holland-Elsevier, Amsterdam, 1992.

[37] J. Tkadlec, Concrete Quantum Logics with generalised compatibility, *Mathematica Bohemica* **123** No. **2** (1998), 213–218.

[38] Sh. Zellweger, A Logical garnet as both a 3-D and a 4-D symmetry model of the 16 binary connectives, *6th International Congress of Logic, Methodology and Philosophy of Science,* abstracts, sections 13/14, Hannover, 1979.

[39] Sh. Zellweger, Sign-creation and man-sign engineering, *Semiotica* **38 1/2** (1982), 17–54.

[40] Sh. Zellweger, Cards, mirrors and hand held models that lead into elementary Logic, from the Sixteenth Annual Conference of the International Group for the Psychology of Mathematics Education, University of New Hampshire, August 7 – August 11, 1992, conference paper, Mount Union College, Alliance, Ohio, 1992.

Bernd Schmeikal
Biofield Laboratory
Kundmanngasse 26
A-1030 Vienna, Austria
Email: schmeika@isis.wu-wien.ac.at

Received: October 4, 1999; Revised: January 3, 2000

4.

MATHEMATICS: DEFORMATIONS

Hecke Algebra Representations in Ideals Generated by q-Young Clifford Idempotents

Rafał Abłamowicz and Bertfried Fauser

ABSTRACT It is a well-known fact from group theory that irreducible tensor representations of classical groups are suitably characterized by irreducible representations of the symmetric groups. However, due to their different nature, vector and spinor representations are only connected and not united in such descriptions.

Clifford algebras are an ideal tool with which to describe symmetries of multiparticle systems since they contain spinor and vector representations within the same formalism and, moreover, allow for a complete study of all classical Lie groups. In this work, together with an accompanying work also presented at this conference, an analysis of q-symmetry – for generic q's – based on the ordinary symmetric groups is given for the first time. We construct q-Young operators as Clifford idempotents and the Hecke algebra representations in ideals generated by these operators. Various relations, such as orthogonality of representations and completeness, are given explicitly, and the symmetry types of representations are discussed. Appropriate q-Young diagrams and tableaux are given. The ordinary case of the symmetric group is obtained in the limit $q \to 1$. All in all, a toolkit for Clifford algebraic treatment of multi-particle systems is provided. The distinguishing feature of this paper is that the Young operators of conjugated Young diagrams are related by Clifford reversion, connecting Clifford algebra and Hecke algebra features. This contrasts the purely Hecke algebraic approach of King and Wybourne, who do not embed Hecke algebras into Clifford algebras.
Keywords: Clifford algebras of multivectors, Clifford algebra representations, spinors, spinor representations, symmetric group, Hecke algebra, q-Young operators, q-Young diagrams and tableaux, q-deformation, multi-particle states, internal symmetries, quantum Clifford algebras.

1 Introduction

1.1 Motivation

We investigate a possibility of implementing the symmetric group S_n and its group algebra deformation, the Hecke algebra $H_{\mathbb{F}}(n,q)$, as a subalgebra

---

AMS Subject Classification: 15A66, 17B37, 20C30, 81R25.

of the Clifford algebra of multivectors which was called elsewhere quantum Clifford algebra. The latter algebra is defined as the Clifford algebra of a bilinear form with a suitably chosen anti-symmetric part. The presence of the antisymmetric part changes the structure of the corresponding Clifford algebra and allows one to introduce the needed deformation.

Our main interest in Clifford algebras arose from their ubiquitous appearance in mathematical physics as it has been demonstrated many times by D. Hestenes [22, 23]. Up to now, however, the main efforts have been devoted to the development of the real Dirac theory and other physical models such as, for example, the Weinberg–Salam theory [9] of electroweak interactions. Despite this enormous range of applicability, there exist problems in mathematical physics not yet formulated or discussed in the Clifford algebra framework.

One major unsolved problem is the proper formulation of multi-particle theories. Quantum field theory is a theory of infinitely many particles which causes, on one hand, great problems with renormalization, but on the other it provides one of the most precise formalisms developed so far in physics. There have been only a few attempts to tackle the multi-particle problem [10], while Hestenes uses matrices in this case [23]. On the other hand, we succeeded in showing that quantum field theory, when treated in terms of generating functionals, can be reformulated by Clifford algebras [12, 13, 17]. However, to treat such complicated theories correctly, one is *forced* to introduce non-symmetric bilinear forms in Clifford algebras, and there are at least two reasons why this needs to be done.

One reason is a problem of normal-ordering, which has to be performed in multi-particle quantum systems. The transition from time-ordered to normal-ordered generating functionals usually yields singularities, which can be seen as a calculational error if Clifford algebras are used [18]. The second reason is the connection of the vacuum structure of physical systems, which is intimately related to the antisymmetric part of the bilinear form in the Clifford algebra [14]. This is far beyond the abilities of other currently used methods.

The Clifford approach to such problems is very rigid. Treating Clifford numbers as single entities makes it easy to calculate with them, but it hides the internal structure of the involved objects. Since essentially all physical observables can be given in terms of two-spinors [32], we wish to have a mechanism which breaks up the Clifford numbers into smaller parts.

Taking an idea from group theory, it is possible to characterize tensor products of irreducible representations of classical groups by the irreducible representations of the symmetric group since both group actions commute. Such a situation, where one has a non-trivial action of S_n, is then considered to be an n-particle system. The symmetric group plays, hence, a dominant role not only in mathematics, e.g., in combinatorics, but also in physics. As an example, note that it was a group theoretical necessity to form mesons and hadrons from quarks after they had been postulated.

During the last two decades one has become aware that the deformed symmetric group algebra, i.e., the Hecke algebra, lies at the heart of the so-called quantized structures, see e.g., [6, 28]. In particular, quantum groups have provided a powerful tool for solving lattice problems in statistical physics. Jones polynomial and the knot theory are related to Hecke algebras, see the extensive discussion in [11]. It is, thus, a quite natural idea to bring the symmetric group and its deformation, the Hecke algebra, into the Clifford formalism. Furthermore, the symmetric group is the Coxeter group of the Dynkin diagram of the $\mathbf{A}_n$ complex Lie algebras [24] which explains the name given to the particular deformation discussed below.

To our knowledge, Clifford algebras have been used only marginally in the study of the symmetric groups. Only a spinor double cover is known, see [25]; however, D. Finkelstein discussed this quite extensively in his lecture.

In this paper we show explicit calculations for $H_{\mathbb{F}}(2,q)$ and $H_{\mathbb{F}}(3,q)$ where $\mathbb{F}$ is the base field of the group algebra in question.[1]

1.2 Definitions

Throughout this paper, we use quantum Clifford algebras with an arbitrary, not necessarily symmetric, bilinear form B. Such algebras can be constructed using Chevalley's approach [5]. However, he utilizes only symmetric forms thereafter. This issue was clearly addressed in [1, 16, 29, 30], while the connection to Hecke algebras was first discussed in [31]. A mathematically sound approach to such algebras including their applications can be found in [11, 17]. There is a need imposed by quantum physics to use this type of Clifford algebras, as shown in [14].

Spinors are usually defined to be elements of a minimal left ideal of a Clifford algebra. Such ideals can be constructed as linear spaces generated by a primitive idempotent. This means, that the spinor space $\mathcal{S} = < C\ell(B,V)\mathbf{f} >_{\mathbb{K}}$, where $\mathbf{f}$ is a primitive idempotent and $C\ell(B,V)$ is the full Clifford algebra of the pair (B,V), where B is the general bilinear form and V is a linear space over $\mathbb{F}$. It is well-known, that the smallest faithful representation of a Clifford algebra is a spinor representation of dimension 2^k, where k is related to a Radon-Hurwitz number. Comparing the dimension of the space of endomorphisms of the spinor space $\mathcal{S}$ with that of the Clifford algebra, one finds several cases of representations over the field $\mathbb{K} \cong \mathbb{R}$, $\mathbb{C}$, $\mathbb{H}$, $\mathbb{R} \oplus \mathbb{R}$ or $\mathbb{H} \oplus \mathbb{H}$. Moreover, since the primitive idempotents of a Clifford algebra decompose the unity $\mathbf{1} = \mathbf{f}_1 + \mathbf{f}_2 + \cdots + \mathbf{f}_k$, one ends up with a k-dimensional right-linear space $\mathcal{S}$ over the appropriate field $\mathbb{K}$.

The aim of this work is to provide a mechanism for breaking up the ordinary spinor representation of $C\ell_{n,n}$ into tensor products of smaller

[1] We distinguish the field $\mathbb{F}$ the algebras are built over and the (double) field $\mathbb{K}$, defined below, used in representations of Clifford algebras. They should not be confused.

representations using appropriate Young operators constructed as Clifford idempotents. This means that a suitable Clifford algebra is used as a carrier space for various tensor product representations.

The Young operators for various Young diagrams provide us with a set of idempotents which decompose the unity $\mathbf{1}$ of $C\ell_{n,n}$ as

$$\mathbf{1} = Y^{(\lambda_1)} + \cdots + Y^{(\lambda_n)}, \tag{1.1}$$

where (λ_i) is a partition of n characterizing the appropriate Young tableau, that is, a Young diagram (frame) with an allowed numbering. We denote a Young tableaux by $Y^{(\lambda_i)}_{i_1,\ldots,i_n}$, where (λ_i) is an ordered partition of n and $i_1,\ldots,i_n$ is an allowed numbering of the boxes in the Young diagram corresponding to (λ_i) as in [21, 27]. Furthermore, these Young operators are mutually annihilating idempotents

$$Y^{(\lambda_i)}Y^{(\lambda_j)} = \delta_{\lambda_i\lambda_j}Y^{(\lambda_j)}. \tag{1.2}$$

It appears natural to ask if these Young operators can be used to give representations of the symmetric group *within* the Clifford algebraic framework. The representation spaces which appear as a natural outcome of the embedding of the symmetric group, and its representations can then be looked at as multi-particle spinor states. However, these might not be spinors of the full Clifford algebra.

In order to be as general as possible, we give not only the representations of the symmetric group, but also of the Hecke algebra $H_{\mathbb{F}}(n,q)$. The Hecke algebra is the generalization of the group algebra of the symmetric group by adding the requirement that transpositions t_i of *adjacent* elements $i, i+1$ are no longer involutions s_i. We set $t_i^2 = (1-q)t_i + q$ which reduces to $s_i^2 = \mathbf{1}$ in the limit $q \to 1$.

Hecke algebras are 'truncated' braids since a further relation (see (1.3) below) is added to the braid group relations as in [4]. A detailed treatment of this topic with important links to physics may be found, for example, in [20, 34] and in the references of [11].

The defining relations of the Hecke algebra will be given according to Bourbaki [8]. Let $< \mathbf{1}, t_1, \ldots, t_n >$ be a set of generators which fulfill these relations:

$$t_i^2 = (1-q)t_i + q, \tag{1.3}$$

$$t_it_j = t_jt_i, \quad |i-j| \geq 2, \tag{1.4}$$

$$t_it_{i+1}t_i = t_{i+1}t_it_{i+1}. \tag{1.5}$$

Then their algebraic span is the Hecke algebra. Since we will compare our results with those of King and Wybourne [26] – hereafter denoted by KW – we give a transformation to their generators g_i, namely $g_i = -t_i$, which results in a new quadratic relation

$$g_i^2 = (q-1)g_i + q \tag{1.6}$$

while the other two remain unchanged. However, this small change in sign is responsible for great differences especially in the q-polynomials occurring in our formulas and in their formulas. One immediate consequence is that this transformation interchanges symmetrizers and antisymmetrizers. In particular, this replacement connects formula (3.4) in KW with our full symmetrizers while formula (3.3) in KW gives our full antisymmetrizers. Finally, the algebra morphism ρ which maps the Hecke algebra into the even part of an appropriate Clifford algebra can be found in [11].

Let $\{\mathbf{1}, \mathbf{e}_1, \dots, \mathbf{e}_{2n}\}$ be a set of generators of the Clifford algebra $C\ell(B,V)$ where the vector space $V = \{\mathbf{e}_i\} = <\mathbf{e}_i>$ is endowed with a non-symmetric $2n \times 2n$ bilinear form $B = [B(\mathbf{e}_i, \mathbf{e}_j)] = [B_{i,j}]$ defined as

$$B_{i,j} := \begin{cases} 0, & \text{if } 1 \leq i,j \leq n \text{ or } n < i,j \leq 2n, \\ q, & \text{if } i = j-n \text{ or } i-1-n = j, \\ -(1+q), & \text{if } i+1 = j-n \text{ or } i = j+1-n, \\ -1, & \text{if } |i-j-n| \geq 2 \text{ and } i > n, \\ 1, & \text{otherwise.} \end{cases} \tag{1.7}$$

The most general case would have $\nu_{ij} \neq 0$ in the last line of (1.7). For example, when $n = 4$, then

$$B = \begin{pmatrix} 0 & 0 & 0 & 0 & q & -1-q & 1 & 1 \\ 0 & 0 & 0 & 0 & -1-q & q & -1-q & 1 \\ 0 & 0 & 0 & 0 & 1 & -1-q & q & -1-q \\ 0 & 0 & 0 & 0 & 1 & 1 & -1-q & q \\ 1 & 1 & -1 & -1 & 0 & 0 & 0 & 0 \\ q & 1 & 1 & -1 & 0 & 0 & 0 & 0 \\ -1 & q & 1 & 1 & 0 & 0 & 0 & 0 \\ -1 & -1 & q & 1 & 0 & 0 & 0 & 0 \end{pmatrix}. \tag{1.8}$$

The bilinear form B in (1.8) is our particular choice that guarantees that the following equations hold:

$$\rho(t_i) = b_i := \mathbf{e}_i \wedge \mathbf{e}_{i+n}, \tag{1.9}$$

$$b_i b_j = b_j b_i, \quad \text{whenever } |i-j| \geq 2, \tag{1.10}$$

$$b_i b_{i+1} b_i = b_{i+1} b_i b_{i+1}. \tag{1.11}$$

This shows ρ to be a homomorphism of algebras implementing the Hecke algebra structure in the Clifford algebra $C\ell(B,V)$. One knows from [11] that ρ is not injective, and that its kernel contains all Young diagrams which are not L-shaped (that is, diagrams with at most one row and/or one column). The first instance, however – when this kernel is non-trivial – occurs in S_4 where the partition $4 = (2,2)$ gives a Young diagram of square form which is not L-shaped.

2 The case of $H_{\mathbb{F}}(2,q)$ and S_2

We begin with $H_{\mathbb{F}}(2,q)$ which reduces to S_2 in the limit $q \to 1$. $H_{\mathbb{F}}(2,q)$ is generated by $\{\mathbf{1}, b_1\}$. We have, thus, only one q-transposition, from which we can calculate a q-symmetrizer $R(12)$ and a q-antisymmetrizer $C(12)$.

Notice that in the limit $q \to 1$ we have the following relations for a set of new generators defined as $s_i := \lim b_i$ when $q \to 1$:

(i) $s_i^2 = 1$,

(ii) $s_i s_j = s_j s_i$, whenever $|i-j| \geq 2$,

(iii) $s_i s_{i+1} s_i = s_{i+1} s_i s_{i+1}$,

(iii)' $(s_i s_{i+1})^3 = \mathbf{1}$.

Property (iii)' follows from the fact that $s_i^2 = 1$ and $s_i^{-1} = s_i$. This is a presentation of the symmetric group according to Coxeter-Moser [7]. Now it is an easy matter to show that (ii) is valid for transpositions, and that (iii) can be calculated graphically using tangles as in Figure 1. In the cycle

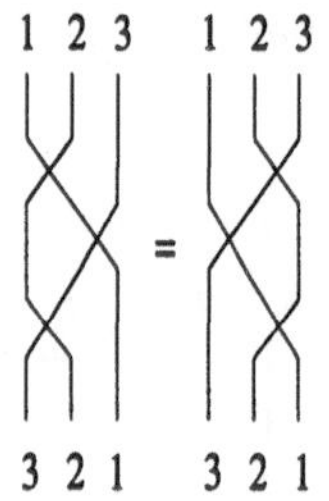

FIGURE 1. Tangles representing equation (1.11).

notation, (iii) can be written as: (12)(23)(12) = (23)(12)(23) = (13). In the Hecke case it does matter if one 'twists' the tangles one thread over or under the next one, which makes this relation quite non-trivial. It is sometimes called "the quantum-Yang-Baxter equation" in physics. In the case of the symmetric group, it does *not* matter which twist is used. In the crossings of the Hecke algebra we define the left-to-right moving tangle when reading from the top to the bottom as the *upper* one.

Thus, we define the q-symmetrizer $R(12)$ and the q-antisymmetrizer $C(12)$ (up to the normalization) as follows:

$$R(12) := q + b_1 = q + \mathbf{e}_1 \wedge \mathbf{e}_5, \tag{2.1}$$
$$C(12) := \mathbf{1} - b_1 = \mathbf{1} - \mathbf{e}_1 \wedge \mathbf{e}_5. \tag{2.2}$$

Notice that the q-antisymmetrizer is related to the q-symmetrizer by the operation of reversion denoted by tilde in the Clifford algebra $C\ell_{1,1}$, that is, $C(12)^{\tilde{}} = R(12)$ and $R(12)^{\tilde{}} = C(12)$. How do we know that $q + b_1$ gives the symmetrizer $R(12)$? Notice first that $R(12)$ is almost an idempotent since

$$R(12)R(12) = (1+q)\mathbf{e}_1 \wedge \mathbf{e}_5 + q(1+q) = (1+q)R(12). \tag{2.3}$$

Thus, when we normalize $R(12)$ by dividing it by $1+q$, the new element denoted as $R(12)_q$ will be an idempotent.

$$\begin{aligned} R(12)_q R(12)_q &= \frac{(1+q)\mathbf{e}_1 \wedge \mathbf{e}_5 + q(1+q)}{(1+q)^2} \\ &= \frac{\mathbf{e}_1 \wedge \mathbf{e}_5 + q}{1+q} = \frac{b_1 + q}{1+q} = R(12)_q. \end{aligned}$$

If we now take the limit of $R(12)_q$ as $q \to 1$, we obtain

$$\lim_{q \to 1} R(12)_q = \frac{1+s_1}{2} \tag{2.4}$$

with $s_1 = \mathbf{e}_1 \wedge \mathbf{e}_5$ squaring to $\mathbf{1}$ (in the limit $q \to 1$) in agreement with (i) above. Then the expression $\frac{1}{2}(1+s_1)$ acts as a symmetrizer on, for example, functions of two variables. Likewise, the normalized q-antisymmetrizer gives an idempotent element

$$C(12)_q = R(12)_q\bar{\ } = \frac{1-b_1}{1+q} \tag{2.5}$$

which in the limit $q \to 1$ gives the regular S_2 antisymmetrizer

$$\lim_{q \to 1} C(12)_q = \frac{1-s_1}{2}. \tag{2.6}$$

Notice that $R(12)_q$, $C(12)_q$ and their limits $\frac{1}{2}(1+s_1)$, $\frac{1}{2}(1-s_1)$ are pairwise mutually annihilating primitive idempotents in $C\ell_{1,1}$.

Therefore, following the standard theory of Young operators, we define the q-Young operator $Y_{1,2}^{(2)}$ as equal to the normalized q-symmetrizer $R(12)_q$ while the q-Young operator $Y_{1,2}^{(11)}$ is defined as equal to the normalized q-antisymmetrizer $C(12)_q$. We can conclude that $Y_{1,2}^{(2)}$ and $Y_{1,2}^{(11)}$ are mutually annihilating idempotents in $H_{\mathbb{F}}(2,q) \subset C\ell_{1,1}^{+} \subset C\ell_{1,1}$, and that they decompose the unity

$$\mathbf{1} = Y_{1,2}^{(2)} + Y_{1,2}^{(11)}. \tag{2.7}$$

In the Clifford algebra $C\ell_{1,1}$ we can construct various primitive idempotents. Since in our construction the Hecke algebra $H_{\mathbb{F}}(2,q)$ is a subalgebra of the even part $C\ell_{1,1}^{+}$ of $C\ell_{1,1}$, the idempotents of the Hecke algebra must be even Clifford elements.

It can be verified easily with CLIFFORD [2, 3] that the only two non-trivial mutually annihilating idempotents in the Hecke algebra $H_{\mathbb{F}}(2,q)$ are the Young operators found above. In this case, the Young operators happen to be the two even primitive idempotents in the Clifford algebra $C\ell_{1,1}$.[2]

With CLIFFORD we have also looked for any intertwining elements in the algebraic span of the two Young operators $Y^{(2)}_{1,2}$ and $Y^{(11)}_{1,2}$, and we have found none, as expected. That is, we have found no non-trivial element T in the algebraic span of the Hecke algebra such that $TY^{(2)}_{1,2} = Y^{(11)}_{1,2}T$. Since the representations are only one-dimensional here, there is no Garnir element (see definition below).

3 The case of $H_{\mathbb{F}}(3,q)$ and S_3

The Hecke algebra $H_{\mathbb{F}}(3,q)$ is spanned by the basis elements

$$\{\mathbf{1}, b_1, b_2, b_{12}, b_{21}, b_{121}\}$$

which are expressed in terms of Grassmann polynomials as

$$\begin{gathered}
b_1 = \mathbf{e}_1 \wedge \mathbf{e}_5,\; b_2 = \mathbf{e}_2 \wedge \mathbf{e}_6,\\
b_{12} = b_1 b_2 = -(1+q)\mathbf{1} + \mathbf{e}_1 \wedge \mathbf{e}_6 - \mathbf{e}_1 \wedge \mathbf{e}_2 \wedge \mathbf{e}_5 \wedge \mathbf{e}_6 + (1+q)\mathbf{e}_2 \wedge \mathbf{e}_5,\\
b_{21} = b_2 b_1 = -q(1+q)\mathbf{1} + (1+q)\mathbf{e}_1 \wedge \mathbf{e}_6 - \mathbf{e}_1 \wedge \mathbf{e}_2 \wedge \mathbf{e}_5 \wedge \mathbf{e}_6 + q\mathbf{e}_2 \wedge \mathbf{e}_5,\\
b_{121} = b_1 b_2 b_1 = q\mathbf{e}_1 \wedge \mathbf{e}_5 - (1+2q)\mathbf{1} + q\mathbf{e}_2 \wedge \mathbf{e}_6 + \mathbf{e}_1 \wedge \mathbf{e}_6\\
+(-1+q)\mathbf{e}_1 \wedge \mathbf{e}_2 \wedge \mathbf{e}_5 \wedge \mathbf{e}_6 - (-q-1+q^2)\mathbf{e}_2 \wedge \mathbf{e}_5.
\end{gathered}$$

We begin by defining our Young symmetrizer $Y^{(3)}_{1,2,3}$ as in [26]:

$$Y^{(3)}_{1,2,3} := \frac{q^3\mathbf{1} + q^2 b_1 + q^2 b_2 + q b_{12} + q b_{21} + b_{121}}{(1+q+q^2)(1+q)}. \tag{3.1}$$

By construction, our Young antisymmetrizer $Y^{(111)}_{1,2,3}$ is defined as the reversion of the symmetrizer $Y^{(3)}_{1,2,3}$, that is,

$$Y^{(111)}_{1,2,3} := Y^{(3)\,\tilde{}}_{1,2,3} = \frac{\mathbf{1} - b_1 - b_2 + b_{12} + b_{21} - b_{121}}{(1+q+q^2)(1+q)}. \tag{3.2}$$

Also in this case King and Wybourne's full antisymmetrizer is the reversion of their full symmetrizer. However, the KW Young operators corresponding to Young tableaux which are conjugate to each other in the sense of

---

[2]The Clifford algebra $C\ell_{1,1}$ is generated here by the 1-vectors $\mathbf{e}_1$ and $\mathbf{e}_5$. That is, we view $C\ell_{1,1}$ as being embedded in the Clifford algebra $C\ell_{4,4}$ of the 8×8 bilinear form B given in (1.8).

Mcdonald [27] (see also [19]) and generate representations of dimensions greater than one appear not to be related by the reversion. For example, King and Wybourne define

$$R(13) := q^3\mathbf{1} + b_{121}, \quad C(13) := \mathbf{1} - b_{121}, \tag{3.3}$$

yet $C(13) \neq R(13)\tilde{}$ since

$$\begin{aligned} C(13) - R(13)\tilde{} = {} & 2(q^2 - q)\mathbf{1} + (-1 - q^2 + 2q)b_1 + (-1 - q^2 + 2q)b_2 \\ & + (1 - q)b_{12} + (1 - q)b_{21}, \end{aligned} \tag{3.4}$$

which does equal 0 only in the limit $q \to 1$, that is, not for an arbitrary q. Therefore, the difference between the KW's antisymmetrizer $C(13)$ and their reversed symmetrizer $R(13)\tilde{}$ is the nonzero Hecke element (3.4) which vanishes automatically in the limit $q \to 1$. This is the reason for us to depart now from the KW formalism and introduce our own definitions of row-symmetrizers and column-antisymmetrizers and require that they be related through the reversion. By doing so, we realize the conjugation of all Young tableaux as the Clifford algebra reversion. This is a crucial fact: since the reversion is the distinguished anti-automorphism of the Clifford algebra, we are able to connect it here to the automorphism of the Hecke algebra which interchanges the roots in the quadratic Hecke relation $(q, -1) \stackrel{\sim}{\to} (-1, q)$. Such interchange is an element of the Galois group [33]. We have, thus, established a direct connection between Clifford anti-involutions and the Galois group elements acting in the Hecke algebra. We consider our setting, therefore, to be in some sense natural.

To be as general as possible, we do the construction of the Young operators in two steps. First, we split the unit element $\mathbf{1}$ in the Hecke algebra using the reversion into two non-primitive idempotents. Each of these idempotents generates a three-dimensional decomposable ideal. To achieve this goal, we can use any two of the following three equations since any two of them imply the third:

$$X + X\tilde{} = \mathbf{1}, \tag{3.5}$$

$$X^2 = X, \tag{3.6}$$

$$XX\tilde{} = 0. \tag{3.7}$$

By doing so, our goal is to find four Young operators known to exist from the general theory of the Hecke algebras for $n = 3$ [20, 34]. The four q-Young operators will still have only one parameter, and they will generalize four Young operators of S_3 described in Hamermesh [21] on p. 245. One of them will be a full symmetrizer, another one will be a full antisymmetrizer, and the other two will be of mixed symmetry.

In the first step, we will find the most general element

$$X = K_1\mathbf{1} + K_2b_1 + K_3b_2 + K_4b_{12} + K_5b_{21} + K_6b_{121} \tag{3.8}$$

in the Hecke algebra $H_{\mathbb{F}}(3,q)$ that satisfies (3.5). Upon substituting X into (3.5) we have found that X must have the following form:

$$X = (\frac{1}{2}qK_2 - \frac{1}{2}qK_6 + \frac{1}{2} + \frac{1}{2}qK_3 - \frac{1}{2}K_2 + \frac{1}{2}q^2K_6 - \frac{1}{2}K_3)\mathbf{1} \\ + K_2b_1 + K_3b_2 + K_4b_{12} + (-K_4 + qK_6 - K_6)b_{21} + K_6b_{121}. \quad (3.9)$$

The element X in (3.9) belongs to a family parameterized by four real or complex parameters K_2, K_3, K_4, K_6. Next we demand that X also satisfies (3.6).

After substituting X displayed in (3.9) into equation (3.6), we have found six sets of solutions. The solutions are parameterized by complex numbers satisfying two similar but different quadratic equations:

$$(1+q)z^2 + (-q^2K_4 + K_4 + qK_2 - 1 + K_2)z + K_4K_2 + K_2^2 \\ - K_4 - K_2 + qK_4 - q^2K_4^2 - qK_4^2 - q^2K_2K_4 + qK_2^2 = 0, \quad (3.10)$$

$$(1+q)z^2 + (-q^2K_4 + K_4 + qK_2 + 1 + K_2)z + K_4K_2 + K_2^2 \\ + K_4 + K_2 - qK_4 - q^2K_4^2 - qK_4^2 - q^2K_2K_4 + qK_2^2 = 0. \quad (3.11)$$

Let α be a root of equation (3.10) and κ be a root of equation (3.11). We obtain the following six representatives $r_i, i = 1, \ldots, 6$ of all solution families of (3.6) which we again express in the Hecke algebra basis:[3]

$$r_1 = \frac{1}{1+q} - K_4b_1 + qK_4b_2 + K_4b_{12} - \frac{(q^3K_4 + q + K_4 - 1)b_{21}}{q(1+q)} \\ - \frac{(-K_4 + q^2K_4 + 1)b_{121}}{q(1+q)},$$

$$r_2 = \frac{q\mathbf{1}}{1+q} - K_4b_1 + qK_4b_2 + K_4b_{12} - \frac{(q^3K_4 - q + K_4 + 1)b_{21}}{q(1+q)} \\ - \frac{(-K_4 + q^2K_4 - 1)b_{121}}{q(1+q)},$$

$$r_3 = \frac{1}{1+q} + qK_4b_1 - K_4b_2 + K_4b_{12} - \frac{(q^3K_4 + q + K_4 - 1)b_{21}}{q(1+q)} \\ - \frac{(-K_4 + q^2K_4 + 1)b_{121}}{q(1+q)},$$

$$r_4 = \frac{q\mathbf{1}}{1+q} + qK_4b_1 - K_4b_2 + K_4b_{12} - \frac{(q^3K_4 - q + K_4 + 1)b_{21}}{q(1+q)} \\ - \frac{(-K_4 + q^2K_4 - 1)b_{121}}{q(1+q)},$$

[3] One might notice that the roots α and κ provide us with two, possibly complex, solutions each, which yields 8 real solutions: 4 are linearly independent and 4 are related by the reversion.

$$r_5 = \frac{\mathbf{1}}{1+q} + K_2 b_1 + K_4 b_{12} - \frac{(\kappa + K_2 - qK_4 + K_4 - q^2 K_4^2 + qK_2^2) b_2}{(K_2 + K_4 - qK_4 + \kappa)(1+q)}$$
$$+ \frac{(q^2 K_2 K_4 + qK_4^2 - K_2^2 - K_4 K_2) b_2}{(K_2 + K_4 - qK_4 + \kappa)(1+q)}$$
$$- \frac{(q\kappa K_4 + qK_2 K_4 + q\kappa K_2 - \kappa K_2) b_{21}}{q(K_2 + K_4 - qK_4 + \kappa)} + \frac{(\kappa K_2 + qK_4^2) b_{121}}{q(-K_2 - K_4 + qK_4 - \kappa)},$$

$$r_6 = \frac{q\mathbf{1}}{1+q} + K_2 b_1 + K_4 b_{12} - \frac{(q^2 K_4^2 - qK_2^2 + q^2 K_2 K_4 + qK_4^2 + \alpha) b_2}{(-K_2 - K_4 + qK_4 - \alpha)(1+q)}$$
$$+ \frac{(qK_4 - K_2 - K_4 + K_2^2 + K_4 K_2) b_2}{(-K_2 - K_4 + qK_4 - \alpha)(1+q)}$$
$$+ \frac{(qK_2 K_4 + q\alpha K_4 + q\alpha K_2 - \alpha K_2) b_{21}}{(-K_2 - K_4 + qK_4 - \alpha) q} + \frac{(\alpha K_2 + qK_4^2) b_{121}}{q(-K_2 - K_4 + qK_4 - \alpha)}.$$

It can be checked with CLIFFORD that the rank of the set $\{r_i\}$, $i = 1, \ldots, 6$ is four. For our purpose we must select any four linearly independent elements, for example $\{r_1, r_2, r_3, r_5\}$, which we rename $\{f_1, f_2, f_3, f_4\}$. It can also be verified with CLIFFORD that the elements $\{f_1, f_2, f_3, f_4\}$ satisfy the required relations (3.5), (3.6), and (3.7).

We look for the Young operators obtained by one of the $f_i, i = 1, \ldots, 4$. Due to the fact that the representation spaces which correspond to the symmetric $Y_{1,2,3}^{(3)}$ and the antisymmetric $Y_{1,2,3}^{(111)}$ Young operators respectively are one-dimensional, they cannot have any free parameters besides q. The full symmetrizer can be given according to KW as the q-weighted sum of all six Hecke basis elements. However, in our construction the full antisymmetrizer is defined as the reversion of the full symmetrizer, that is, $Y_{1,2,3}^{(111)} := Y_{1,2,3}^{(3)\,\tilde{}}$, as it was done in dimension two. Then we have

$$Y_{1,2,3}^{(3)} := \frac{q^3 \mathbf{1} + q^2 b_1 + q^2 b_2 + q b_{12} + q b_{21} + b_{121}}{(1+q+q^2)(1+q)}, \tag{3.12}$$

$$Y_{1,2,3}^{(111)} := \frac{\mathbf{1} - b_1 - b_2 + b_{12} + b_{21} - b_{121}}{(1+q+q^2)(1+q)}. \tag{3.13}$$

Each of the four f_i elements defined above generates a three-dimensional one-sided ideal in the Hecke algebra, or, in other words, together with its reversion $\tilde{f}_i$ it decomposes the unity in $H_{\mathbb{F}}(3, q)$. Our objective is to further split each of these ideals into one- and two-dimensional vector spaces. The one-dimensional vector spaces will be generated by the full symmetrizer (3.12) and the full antisymmetrizer (3.13), respectively. The two-dimensional vector spaces will be generated by the mixed type Young operators. So each of the f_i elements has to be a sum of a full (anti)symmetrizer and a Young operator of the mixed type. If we pick f_1, we notice that it must contain the full antisymmetrizer; because when the parameter K_4 is replaced with $1/(q+1)$, the re-defined f_1 (or the r_1 defined above) reduces to an expression with alternating signs in the Hecke basis

$$\frac{1}{1+q}-\frac{b_1}{1+q}+\frac{qb_2}{1+q}+\frac{b_{12}}{1+q}-\frac{qb_{21}}{1+q}-\frac{b_{121}}{1+q}.$$

Therefore, by subtracting the full antisymmetrizer $Y^{(111)}_{1,2,3}$ from f_1 we find our first Young operator $Y^{(21)}_{1,3,2}$ of the mixed type

$$\begin{aligned} Y^{(21)}_{1,3,2} = f_1 - Y^{(111)}_{1,2,3} &= \frac{q1}{q+1+q^2} - \frac{(q^3K_4+2q^2K_4+2qK_4-1+K_4)b1}{q^3+2q^2+2q+1} \\ &+ \frac{(K_4q^4+2q^3K_4+2q^2K_4+qK_4+1)b_2}{(q^3+2q^2+2q+1)} \\ &+ \frac{(q^3K_4+2q^2K_4+2qK_4-1+K_4)b_{12}}{(q^3+2q^2+2q+1)} \\ &- \frac{(K_4q^5+K_4q^4+q^3K_4+q^3+q^2K_4+qK_4+q+K_4-1)b_{21}}{(q^3+2q^2+2q+1)q} \\ &- \frac{(K_4q^4+q^3K_4+q^2-qK_4+1-K_4)b_{121}}{(q+1+q^2)q(1+q)}. \end{aligned} \tag{3.14}$$

We define our second Young operator $Y^{(21)}_{1,2,3}$ of the mixed type as the reversion (conjugate) of $Y^{(21)}_{1,3,2}$, that is, $Y^{(21)}_{1,2,3} := Y^{(21)\sim}_{1,3,2}$, and we get

$$\begin{aligned} Y^{(21)}_{1,2,3} &= \frac{q1}{q+1+q^2} + \frac{(q^3K_4-q^2+2q^2K_4+2qK_4+K_4)b_1}{(1+q)(q+1+q^2)} \\ &- \frac{q(q^3K_4+2q^2K_4+2qK_4+q+K_4)b_2}{(1+q)(q+1+q^2)} \\ &- \frac{(q^3K_4+2q^2K_4+2qK_4+q+K_4)b_{12}}{q^3+2q^2+2q+1} \\ &+ \frac{(K_4q^5+K_4q^4+q^3K_4+q^3-q^2+q^2K_4+qK_4-1+K_4)b_{21}}{q(q^3+2q^2+2q+1)} \\ &+ \frac{(K_4q^4+q^3K_4+q^2-qK_4+1-K_4)b_{121}}{(q+1+q^2)q(1+q)}. \end{aligned} \tag{3.15}$$

Furthermore, the idempotent $Y^{(111)}_{1,2,3}$ (resp. $Y^{(3)}_{1,2,3}$) annihilates the idempotent $Y^{(21)}_{1,3,2}$ (resp. $Y^{(21)}_{1,2,3}$) when multiplied from both sides. This reflects the fact that the representation spaces constructed in this way give a direct sum decomposition of the three-dimensional ideal generated by f_1 (resp. $\tilde{f}_1$).

Let's summarize the relationships between the Young operators:

$$f_1 = Y^{(111)}_{1,2,3} + Y^{(21)}_{1,3,2}, \quad Y^{(21)}_{1,2,3} + Y^{(3)}_{1,2,3} = \tilde{f}_1,$$

$$Y^{(111)}_{1,2,3} Y^{(21)}_{1,3,2} = Y^{(21)}_{1,3,2} Y^{(111)}_{1,2,3} = 0, \quad Y^{(21)}_{1,2,3} Y^{(3)}_{1,2,3} = Y^{(3)}_{1,2,3} Y^{(21)}_{1,2,3} = 0, \tag{3.16}$$

$$Y^{(111)}_{1,2,3} = Y^{(3)\,\tilde{}}_{1,2,3}, \quad Y^{(21)}_{1,3,2} = Y^{(21)\,\tilde{}}_{1,2,3}.$$

Thus, we have carefully built in the conjugation of the Young tableaux as the Clifford reversion. It can then be checked with CLIFFORD that the four Young operators displayed in (3.12), (3.13), (3.14), and (3.15) are primitive mutually annihilating idempotents which decompose the unity in the Hecke algebra since $f_1 + \tilde{f}_1 = \mathbf{1}$. It can be easily verified that the Young operators of mixed type decompose into the row-symmetrizer and the colum-antisymmetrizer in accordance to Hamermesh [21] p. 245. Our expressions, however, are different from those in KW.

In order to represent our Young operator $Y^{(21)}_{1,3,2}$ as a product of the row symmetrizer $R(13)$ and the column antisymmetrizer $C(12)$, we use previously defined $f_1 = r_1$ to define $C(12) := f_1$ and compute $R(13)$ from the equation

$$Y^{(21)}_{1,3,2} = R(13) f_1. \tag{3.17}$$

Notice that our $f_1 = r_1$ is a generalization to S_3 of $C(12)$ from S_2 displayed in (2.2). In an effort to be consistent with our previous discussion of $C(12)$ and $R(12)$, which were related by the reversion, we will later define $C(13) := R(13)\tilde{}$ and require that $R(13) + C(13) = R(13) + R(13)\tilde{} = \mathbf{1}$. Thus, when we solve (3.17) for $R(13)$, we get the following solution:

$$\begin{aligned} R(13) = {} & \frac{q\mathbf{1}}{1+q} - \frac{(-q^2 + q^2 P_3 + P_3 q - 1 + P_3) b_1}{(q + 1 + q^2) q} + P_3 b_2 \\ & + \frac{(q^2 P_3 + P_3 q + P_3 - 1) b_{12}}{q(1 + q + q^2)} \\ & - \frac{(q^5 P_3 + q^4 P_3 + q^3 P_3 + q^2 P_3 - q^2 + P_3 q - 1 + P_3) b_{21}}{(1+q) q^2 (1 + q + q^2)} \\ & - \frac{(q^4 P_3 + q^3 P_3 - P_3 q + 1 - P_3) b_{121}}{(1+q) q^2 (1 + q + q^2)} \end{aligned} \tag{3.18}$$

where P_3 is an arbitrary real or complex parameter. It can be also verified with CLIFFORD that $R(13)$ is an idempotent element in the Hecke algebra which is not a feature in KW. It is an interesting fact to note that the free parameter P_3 disappears when the product $R(13) f_1 = Y^{(21)}_{1,3,2}$ is taken: the only free remaining parameter is K_4. As a natural consequence of (3.17) we have

$$Y^{(21)}_{1,2,3} = \tilde{f}_1 R(13)\tilde{}. \tag{3.19}$$

Obviously, we also have the following relations:

$$C(13) := R(13)^{\sim}, \quad C(12) := f_1, \quad R(12) := C(12)^{\sim} = f_1^{\sim}. \tag{3.20}$$

There are no non-trivially intertwining elements in the Hecke algebra which would connect the four Young operators except for the pair $\{Y^{(21)}_{1,2,3}, Y^{(21)}_{1,3,2}\}$. That is, the only intertwiners that connect $Y^{(3)}_{1,2,3}$ with $Y^{(111)}_{1,2,3}$ actually annihilate them, and similarly for the pairs $\{Y^{(3)}_{1,2,3}, Y^{(21)}_{1,2,3}\}$ and $\{Y^{(111)}_{1,2,3}, Y^{(21)}_{1,3,2}\}$. On the other hand, there are many choices for an element T in the Hecke algebra such that

$$TY^{(21)}_{1,2,3} = Y^{(21)}_{1,3,2}T \neq 0. \tag{3.21}$$

In fact, T belongs to a five-parameter family of solutions. By assigning 0 and 1 values to four of those parameters, we have reduced the solutions to a one-parameter family parameterized only by K_4 as follows:

$$\begin{aligned} T = & \frac{(1 - K_4 - q^3 - K_4 q + K_4 q^3 + q^4 K_4)\mathbf{1}}{q(2K_4 - q^2 - q - q^3 + 2q^4 K_4 + 8K_4 q^2 + 6K_4 q + 6K_4 q^3 - 1)} \\ & + \frac{2(-1 - q - q^2 + K_4 + 2K_4 q + K_4 q^3 + 2K_4 q^2)b_1}{q(2K_4 - q^2 - q - q^3 + 2q^4 K_4 + 8K_4 q^2 + 6K_4 q + 6K_4 q^3 - 1)} \\ & - \frac{b_2}{1+q} - \frac{b_{12}}{q(1+q)} + \frac{b_{21}}{1+q} + \frac{b_{121}}{q(1+q)} \end{aligned} \tag{3.22}$$

where $K_4 \neq \dfrac{(q^2+1)}{2(q^3 + 2q^2 + 2q + 1)}$. Thus, the only intertwiners in the Hecke algebra which are not annihilators and which connect the mixed type Young operators constitute, in general, a four-parameter family of intertwiners while T gives a one-parameter family.

We introduce now the Garnir elements $G^{(\lambda_i)}_{i,j}$ (see KW) which will allow us to construct the six-dimensional representation spaces from the Young idempotents. Garnir elements can be seen to act as row or column permutations in Young tableaux, thereby generating non-standard tableaux which correspond to the basis vectors of the representation space. A *Garnir element* $G^{(21)}_{1,1}$ has the following defining properties:

$$Y^{(21)}_{1,2,3}\, G^{(21)}_{1,1} = 0, \tag{3.23}$$

$$G^{(21)}_{1,1}\, Y^{(21)}_{1,2,3} \neq 0. \tag{3.24}$$

Our goal is to use a suitable Garnir element to decompose the three-dimensional one-sided ideals generated by the Young operators of mixed symmetry $Y^{(21)}_{1,3,2}$ and $Y^{(21)}_{1,2,3}$ (see (3.14) and (3.15) respectively) into a direct sum of one-dimensional and two-dimensional vector spaces.

When we first require that a general Hecke element X shown in (3.8) satisfies equation (3.23) when substituted for $G_{1,1}^{(21)}$, we find three different linearly independent solutions which we call XX_1, XX_2, and XX_3. In the Hecke basis they look as follows:

$$XX_1 = (-K_6q + K_2q + K_4)\mathbf{1} + K_2b_1 + \frac{t_1b_2}{q} + K_4b_{12} + K_5b_{21} + K_6b_{121}, \tag{3.25}$$

$$XX_2 = K_1\mathbf{1} + \frac{t_2b_1}{q^2(1+q)} + \frac{t_3b_2}{q^3(1+q)} - \frac{b_{12}}{q(1+q)} + K_5b_{21} + K_6b_{121}, \tag{3.26}$$

$$XX_3 = K_1\mathbf{1} + \frac{t_4b_1}{q(q^3+2q^2+2q+1)} + \frac{t_5b_2}{q^2(q^3+2q^2+2q+1)} - \frac{(q-1)b_{12}}{q^3+2q^2+2q+1} + K_5b_{21} + K_6b_{121}, \tag{3.27}$$

where the polynomial t_1 is parameterized by K_2, K_4, K_5, K_6, and the polynomials t_2, t_3, t_4, t_5 are parameterized by K_1, K_5, K_6.[4]

With CLIFFORD it has been verified that the equation (3.24) is satisfied automatically by each $XX_i, i = 1, 2, 3$ for all possible choices of the parameters. This means that we can use any of the three solutions as the Garnir element. Our choice is

$$G_{1,1}^{(21)} = (-K_6q + K_2q + K_4)\mathbf{1} + K_2b_1 + \frac{t_1b_2}{q} + K_4b_{12} + K_5b_{21} + K_6b_{121}. \tag{3.28}$$

Next we introduce the q-automorphism α_q which replaces the reversion in changing the Garnir elements when acting from the left. This automorphism is, in fact, the inverse of the b_i's generators and their products, the Hecke versors, and it is linearly extended to the entire Hecke algebra by means of the following definition:

$$\alpha_q(b_{i_1}\dots b_{i_s}) = (\frac{-1}{q})^s(b_{i_1}\cdots b_{i_s})^{\tilde{}} = (\frac{-1}{q})^s(\tilde{b_{i_s}}\cdots \tilde{b_{i_1}}) = \alpha_q(b_{i_s})\cdots\alpha_q(b_{i_1}). \tag{3.29}$$

---

[4] $t_1 = K_6q^2+K_4q^2-K_5q-K_4q-K_6q+K_2+K_4$, $t_2 = K_6q^3+K_6q^2+q^2K_1+qK_1+1$, $t_3 = q^5K_6-q^4K_5-q^3-q^3K_5+q^2K_1+K_6q^2+q^2+qK_1-q+1$, $t_4 = q-1+K_6q+2K_6q^2+2K_6q^3+2qK_1+2q^2K_1+K_1+q^3K_1+q^4K_6$, $t_5 = q^6K_6+q^5K_6-q^5K_5+q^4K_6-q^4-2q^4K_5+K_6q^3+q^3K_1+2q^3-2q^3K_5-K_5q^2+K_6q^2+2q^2K_1-2q^2+2qK_1+K_6q+2q+K_1-1$.

For example,

$$\alpha_q(b_1) = \frac{(q-1)\mathbf{1}}{q} + \frac{b_1}{q}, \quad \alpha_q(b_2) = \frac{(q-1)\mathbf{1}}{q} + \frac{b_2}{q},$$
$$\alpha_q(b_1 b_2) = \frac{(1-2q+q^2)\mathbf{1}}{q^2} + \frac{(q-1)b_1}{q^2} + \frac{(q-1)b_2}{q^2} + \frac{b_{21}}{q^2}.$$

Note the fact that

$$\begin{aligned} \alpha_q(b_i)b_i &= \frac{-1}{q}(\tilde{b_i} b_i) = \frac{-1}{q}(((1-q)-b_i)b_i) \\ &= \frac{-1}{q}((1-q)b_i - (1-q)b_i - q) = \mathbf{1}. \end{aligned} \tag{3.30}$$

In fact, α_q acts as the inverse on the generators and, as such, it depends on the presentation. However, while α_q can be extended by the linearity to the entire Hecke algebra, it gives the inverse only of the products of the b_i's (versors) and not of their sums. For example,

$$\alpha_q(\mathbf{1}+b_1) = \frac{(2q-1)\mathbf{1}}{q} + \frac{b_1}{q} \neq$$
$$(\mathbf{1}+b_1)^{-1} = \frac{(-2+q)\mathbf{1}}{2(q-1)} + \frac{b_1}{2(q-1)}, \tag{3.31}$$

$q \neq 0, 1$. One can check that it is not possible to use the reversion to connect the left and the right actions of Garnir elements even if this transformation connects the Young operators of conjugated diagrams. Looking at α_q as a q-inverse, one could try to form a q-Clifford-Lipschitz group [15] in the Hecke algebra by defining:

$$\Gamma_q := \{X \mid \alpha_q(X)X = \mathbf{1}\}. \tag{3.32}$$

When the automorphism α_q is applied to the Garnir element $G_{1,1}^{(21)}$, one gets

$$\begin{aligned} \alpha_q(G_{1,1}^{(21)}) = -\frac{t_6\mathbf{1}}{q^3} + \frac{t_7 b_1}{q^3} - \frac{t_8 b_2}{q^3} + \frac{(-K_6+K_6 q+K_5 q)b_{12}}{q^3} \\ + \frac{(-K_6+K_4 q+K_6 q)b_{21}}{q^3} + \frac{K_6 b_{121}}{q^3}, \end{aligned} \tag{3.33}$$

where t_6, t_7, t_8 are polynomials.[5] With CLIFFORD we have verified that the six elements in the list S below are linearly independent. As such, they

[5] $t_6 = K_2 q - 2K_6 q + K_6 q^3 + K_6 q^2 + K_5 q^2 + K_6 - q^3 K_2 - q^4 K_2 - K_5 q - q^4 K_4$, $t_7 = K_6 + q^2 K_2 - K_5 q + K_5 q^2 - 2K_6 q + K_4 q^2 - K_4 q + K_6 q^2$, $t_8 = -K_2 q + 2K_6 q - K_6 + K_5 q - K_6 q^3 - K_4 q^3$.

provide a basis for the left regular representation of the Hecke algebra:

$$S = [Y^{(3)}_{1,2,3}, Y^{(21)}_{1,2,3}, G^{(21)}_{1,1}Y^{(21)}_{1,2,3}, \alpha_q(G^{(21)}_{1,1})Y^{(21)}_{1,3,2}, Y^{(21)}_{1,3,2}, Y^{(111)}_{1,2,3}]. \qquad (3.34)$$

Recall that the original basis in the Hecke algebra was

$$[\mathbf{1}, b_1, b_2, b_{12}, b_{21}, b_{121}].$$

Each original basis element should be representable in terms of the new basis S. Then we have

$$\mathbf{1} = S_1 + S_2 + S_5 + S_6,$$

$$b_1 = S_1 + \frac{(K_4q^3 + 2K_4q^2 - q^2 + 2K_4q + K_4)S_2}{1+q} + \frac{p_1qS_3}{p_2(1+q)} + \frac{q^2S_4}{p_3} + \frac{q(-K_4q - K_2q + K_6 + K_5)S_5}{p_4} - qS_6,$$

$$b_2 = S_1 - \frac{qp_5S_2}{1+q} - \frac{p_6qS_3}{p_2(1+q)} - \frac{q^3S_4}{p_3} - \frac{p_7S_5}{p_4} - qS_6,$$

$$b_{12} = S_1 + \frac{p_8S_2}{1+q} + \frac{p_9qS_3}{p_2(1+q)} + \frac{(1-q+q^2)q^2S_4}{p_3} - \frac{qp_{10}S_5}{p_4} + q^2S_6,$$

$$b_{21} = S_1 - \frac{qp_5S_2}{1+q} - \frac{p_{11}q^2S_3}{p_2(1+q)} - \frac{q^3S_4}{p_3} - \frac{q^2(-K_4q - K_2q + K_6 + K_5)S_5}{p_4} + q^2S_6,$$

$$b_{121} = S_1 + \frac{p_{13}qS_2}{1+q} + \frac{p_{1,2}q^2S_3}{p_2(1+q)} + \frac{(q-1)q^3S_4}{p_3} - \frac{q^2p_{14}S_5}{p_4} - q^3S_6$$

where to shorten the display we have introduced polynomials p_i, $1 \le i \le 14$ (see the Appendix) and S_i denotes the i-th element of the list S, $i = 1, \dots, 6$. Finally, we can find, as expected, block-structured matrices of the basis elements b_1 and b_2 in the left regular representation (additional polynomials p_i, $15 \le i \le 19$, are also given in the Appendix). They are as follows:

$$M_{b_1} = \begin{pmatrix} 1 & 0 & 0 & 0 & 0 & 0 \\ 0 & \frac{p_{15}}{1+q} & -\frac{p_{16}}{q(1+q)} & 0 & 0 & 0 \\ 0 & \frac{qp_1}{p_{17}} & -\frac{p_{18}}{1+q} & 0 & 0 & 0 \\ 0 & 0 & 0 & -\frac{p_{19}}{p_4} & \frac{q^2}{p_3} & 0 \\ 0 & 0 & 0 & -\frac{p_{20}}{q\,p_4} & \frac{qp_{21}}{p_4} & 0 \\ 0 & 0 & 0 & 0 & 0 & -q \end{pmatrix}, \qquad (3.35)$$

$$M_{b_2} = \begin{pmatrix} 1 & 0 & 0 & 0 & 0 & 0 \\ 0 & -\frac{qp_5}{1+q} & \frac{p_{16}}{1+q} & 0 & 0 & 0 \\ 0 & -\frac{qp_6}{p_{17}} & \frac{p_{21}}{1+q} & 0 & 0 & 0 \\ 0 & 0 & 0 & \frac{qp_{22}}{p_4} & \frac{-q^3}{p_3} & 0 \\ 0 & 0 & 0 & -\frac{p_{23}}{qp_4} & -\frac{p_7}{p_4} & 0 \\ 0 & 0 & 0 & 0 & 0 & -q \end{pmatrix}. \tag{3.36}$$

Matrices M_{b_1} and M_{b_2} satisfy, of course, the same quadratic Hecke relation (1.3) as do b_1 and b_2, and which happens to give the minimum polynomial $p(x) = x^2 - (1-q)x - q = (x-1)(x+q)$ for the basis elements and their matrix representations. The characteristic polynomial for the latter is $c(x) = (x-1)^3(x+q)^3$. The trace of M_{b_1} and M_{b_2} is $3(q-1)$ while their determinants equal $-q^3$.

Finally, we build a left-regular matrix representation of the Young basis elements (3.34) in the Young basis.

$$M_{S_1} = \begin{pmatrix} 1 & 0 & 0 & 0 & 0 & 0 \\ 0 & 0 & 0 & 0 & 0 & 0 \\ 0 & 0 & 0 & 0 & 0 & 0 \\ 0 & 0 & 0 & 0 & 0 & 0 \\ 0 & 0 & 0 & 0 & 0 & 0 \\ 0 & 0 & 0 & 0 & 0 & 0 \end{pmatrix}, \quad M_{S_2} = \begin{pmatrix} 0 & 0 & 0 & 0 & 0 & 0 \\ 0 & 1 & 0 & 0 & 0 & 0 \\ 0 & 0 & 0 & 0 & 0 & 0 \\ 0 & 0 & 0 & 1 & 0 & 0 \\ 0 & 0 & 0 & \frac{p_{24}}{q^2} & 0 & 0 \\ 0 & 0 & 0 & 0 & 0 & 0 \end{pmatrix}, \tag{3.37}$$

$$M_{S_3} = \begin{pmatrix} 0 & 0 & 0 & 0 & 0 & 0 \\ 0 & 0 & 0 & 0 & 0 & 0 \\ 0 & 1 & 0 & 0 & 0 & 0 \\ 0 & 0 & 0 & 0 & 0 & 0 \\ 0 & 0 & 0 & -\frac{p_{25}}{q^3} & 0 & 0 \\ 0 & 0 & 0 & 0 & 0 & 0 \end{pmatrix}, \quad M_{S_4} = \begin{pmatrix} 0 & 0 & 0 & 0 & 0 & 0 \\ 0 & 0 & -\frac{p_{25}}{q^3} & 0 & 0 & 0 \\ 0 & 0 & -\frac{p_{24}}{q^2} & 0 & 0 & 0 \\ 0 & 0 & 0 & -\frac{p_{24}}{q^2} & 1 & 0 \\ 0 & 0 & 0 & 0 & 0 & 0 \\ 0 & 0 & 0 & 0 & 0 & 0 \end{pmatrix}, \tag{3.38}$$

$$M_{S_5} = \begin{pmatrix} 0 & 0 & 0 & 0 & 0 & 0 \\ 0 & 0 & 0 & 0 & 0 & 0 \\ 0 & 0 & 1 & 0 & 0 & 0 \\ 0 & 0 & 0 & 0 & 0 & 0 \\ 0 & 0 & 0 & -\frac{p_{24}}{q^2} & 1 & 0 \\ 0 & 0 & 0 & 0 & 0 & 0 \end{pmatrix}, \quad M_{S_6} = \begin{pmatrix} 0 & 0 & 0 & 0 & 0 & 0 \\ 0 & 0 & 0 & 0 & 0 & 0 \\ 0 & 0 & 0 & 0 & 0 & 0 \\ 0 & 0 & 0 & 0 & 0 & 0 \\ 0 & 0 & 0 & 0 & 0 & 0 \\ 0 & 0 & 0 & 0 & 0 & 1 \end{pmatrix}. \tag{3.39}$$

4 Conclusions

Motivated by the desire to describe symmetries of q-multi-particle systems we have constructed q-symmetrizers and q-antisymmetrizers in the Hecke algebras $H_{\mathbb{F}}(2,q)$ and $H_{\mathbb{F}}(3,q)$ and have related them by the reversion. That is, the Young operators constructed in this paper which correspond to the Young tableaux conjugate to each other in the sense of Macdonald, and which generate representation spaces of dimension greater than one, have been related through the reversion in the Clifford algebra $C\ell_{4,4}$. This feature is not present in King and Wybourne.

In $H_{\mathbb{F}}(2,q)$ we found that the symmetrizer $R(12)$ and its reverse, the antisymmetrizer $C(12)$, were primitive idempotents in the Clifford algebra. We found no non-trivial intertwiners linking these two idempotents.

In $H_{\mathbb{F}}(3,q)$ we first found four mutually annihilating idempotents splitting the unity in the algebra: two without parameters and two parameterized ones. The first two were the Young symmetrizer $Y^{(3)}_{1,2,3}$, defined as in King and Wybourne, and the Young antisymmetrizer $Y^{(111)}_{1,3,2}$, defined in this paper as the reverse of $Y^{(3)}_{1,2,3}$. The two parameterized idempotents were the Young operators of mixed symmetry $Y^{(21)}_{1,2,3}$ and $Y^{(21)}_{1,3,2}$, and they were also related by the reversion. We were able to factor $Y^{(21)}_{1,3,2}$ into the row symmetrizer $R(13)$ and the column antisymmetrizer $C(12)$; that is, we were able to find $R(13)$ as an idempotent element in the Clifford algebra, a feature not found in King and Wybourne. Furthermore, we related the mixed-type Young operators through a five-parameter family of intertwiners.

We have found a Garnir element $G^{(21)}_{1,1}$ (in fact, three distinct families of such elements) which allowed us to further split the representation space of $H_{\mathbb{F}}(3,q)$ from $3 \oplus 3$ to $1 \oplus 2 \oplus 2 \oplus 1$: the one-dimensional spaces being generated by the Young symmetrizer $Y^{(3)}_{1,2,3}$ and the Young antisymmetrizer $Y^{(111)}_{1,3,2}$, while the two-dimensional spaces being generated by $\{Y^{(21)}_{1,2,3}, G^{(21)}_{1,1}Y^{(21)}_{1,2,3}\}$ and $\{\alpha_q(G^{(21)}_{1,1})Y^{(21)}_{1,3,2}, Y^{(21)}_{1,3,2}\}$, respectively. We introduced a Hecke algebra automorphism α_q which acted as the inverse when applied to the Hecke basis elements: when applied to the Garnir element α_q allowed us to generate the representation space of dimension six. The α_q automorphism is expected to be useful in defining a q-Clifford-Lipschitz group in the Hecke algebra [15]. Finally, we computed the matrix representation of the Hecke generators b_1 and b_2 in the Young basis.

In the next step to be considered elsewhere, we intend to extend this approach to $H_{\mathbb{F}}(4,q)$. The connection with the spinor representations of the appropriate Clifford algebras in all three cases needs to be explored. Our approach to the Young operators related to the Young tableaux of various symmetries as Clifford (and Hecke) idempotents has been motivated by the need to describe the q-symmetries of multi-particle states. We have been able to define and construct all needed notations for a classification of tensor spaces w.r.t. the q-symmetry. As mentioned above, the next natural step is to compute multi-particle spinors, that is, spin-tensors, as they are used for mesons or hadrons in QFT. Such spinors are, however, not connected with the spinors of the Clifford algebras used in this paper, but instead they should be connected with spinors of some appropriate sub-Clifford algebras and their tensor product. Our aim to break up a 'container Clifford algebra' into smaller pieces will thus be achieved. We found that the Clifford algebra framework has proven to be most useful for this task.

Appendix

Polynomials below have been introduced as abbreviation to improve readability of the formulas displayed in the main text:

$$
\begin{aligned}
p_1 &= q^4K_4^2 + 3q^3K_4^2 - K_4q^3 + 4q^2K_4^2 - K_4q^2 + 3qK_4^2 - K_4q - q \\
&\quad + K_4^2 - K_4, \\
p_2 &= K_4q^4K_6 + q^4K_4^2 + q^3K_4K_6 - q^3K_4K_2 + K_4q^2 - q^2K_2K_4 + q^2K_4K_6 \\
&\quad + K_6q^2 + q^2K_4K_5 - q^2K_4^2 - K_2q - K_6q + qK_4K_5 - qK_4^2 + K_4qK_6 \\
&\quad - K_5q - qK_2K_4 - K_4q - K_4^2 - K_4K_2 + K_4 + K_2, \\
p_3 &= -K_6q^3 - K_4q^3 + K_6q^2 + K_5q^2 + K_6q + K_5q - K_6 - K_4, \\
p_4 &= -K_6q^2 - K_4q^2 + 2K_6q + K_5q + K_4q - K_6 - K_4, \\
p_5 &= K_4q^3 + 2K_4q^2 + q + 2K_4q + K_4, \\
p_6 &= q^5K_4^2 + 3q^4K_4^2 + 4q^3K_4^2 + K_4q^3 + 3q^2K_4^2 + K_4q^2 + qK_4^2 \\
&\quad + K_4q + K_4 - 1, \\
p_7 &= -K_4q^3 - q^3K_2 + 2K_6q^2 + K_5q^2 + K_4q^2 - 2K_6q - K_5q - K_4q \\
&\quad + K_6 + K_4, \\
p_8 &= q^5K_4 + q^4K_4 + q^3 + K_4q^3 - q^2 + K_4q^2 + K_4q + K_4 - 1, \\
p_9 &= q^6K_4^2 + 2q^5K_4^2 + 2q^4K_4^2 + 2q^4K_4 + 2q^3K_4^2 + K_4q^3 + 2q^2K_4^2 \\
&\quad + q^2 + 2qK_4^2 - K_4q - q + K_4^2 - 2K_4 + 1, \\
p_{10} &= q^3K_2 - K_6q^3 + K_6q^2 - q^2K_2 + K_5q + K_2q - K_5 - K_6, \\
p_{11} &= q^4K_4^2 + 3q^3K_4^2 + 4q^2K_4^2 + K_4q^2 + 3qK_4^2 + K_4^2 - K_4 + 1, \\
p_{12} &= q^5K_4^2 + 2q^4K_4^2 + q^3K_4^2 + 2K_4q^3 - q^2K_4^2 + 2K_4q^2 - 2qK_4^2 \\
&\quad + 2K_4q + q - K_4^2 + 2K_4 - 1, \\
p_{13} &= q^4K_4 + K_4q^3 + q^2 - K_4q - K_4 + 1, \\
p_{14} &= q^2K_2 - K_6q^2 + K_6q - K_2q - K_4 + K_5,
\end{aligned}
$$

$$\begin{aligned}
p_{15} &= K_4q^3 + 2K_4q^2 - q^2 + 2K_4q + K_4,\\
p_{16} &= 3K_4q^4K_6 + 3q^3K_4K_6 + K_4qK_6 + 2q^2K_4K_6 + 2q^5K_4K_6\\
&\quad + 2q^2K_4K_5 + K_6q^4 - 2q^4K_4K_2 + qK_4K_5 + 2q^3K_4K_5 - q^3K_2 - K_5q\\
&\quad - K_6q - K_5q^2 - 2qK_2K_4 + K_2 + q^6K_4K_6 + q^4K_4K_5 + K_4\\
&\quad - 3q^2K_2K_4 + q^4K_4 + K_4q^2 - K_4^2 - 3q^3K_4K_2 - q^5K_4K_2 - 2q^3K_4^2\\
&\quad - q^3K_5 + q^6K_4^2 + q^5K_4^2 - 3q^2K_4^2 - 2qK_4^2 - K_4K_2,\\
p_{17} &= 2K_4q^4K_6 + 2q^3K_4K_6 + K_4qK_6 + 2q^2K_4K_6 + q^5K_4K_6 + 2q^2K_4K_5\\
&\quad - q^4K_4K_2 + qK_4K_5 + q^3K_4K_5 - K_5q - K_6q - K_5q^2 - 2qK_2K_4 + K_2\\
&\quad + q^4K_4^2 + K_4 - 2q^2K_2K_4 + K_6q^3 - q^2K_2 + K_4q^3 - K_4^2 - 2q^3K_4K_2\\
&\quad - q^3K_4^2 + q^5K_4^2 - 2q^2K_4^2 - 2qK_4^2 - K_4K_2,\\
p_{18} &= K_4q^3 + 2K_4q^2 + 2K_4q - 1 + K_4,\\
p_{19} &= -K_4q^3 - K_6q^3 + K_4q^2 + 3K_6q^2 + K_5q^2 - q^2K_2 - 2K_4q - 2K_6q\\
&\quad + K_6 + K_4,\\
p_{20} &= K_4q^4K_6 - 3q^3K_4K_6 + 2K_4qK_6 - 2q^2K_4K_6 - q^5K_4K_6 - q^2K_4K_5\\
&\quad - K_6K_4 + q^4K_4K_2 + qK_4K_5 - q^3K_4K_5 - qK_2K_4 + q^4K_4^2 + q^2K_2K_4\\
&\quad - K_4^2 + 2K_6^2q - K_6^2q^2 + 2q^3K_4K_2 + K_5K_4 + q^5K_4K_2 + K_6q^5K_2\\
&\quad - K_5q^4K_2 - qK_2K_6 + K_6K_5 + q^3K_2^2 - K_5q^2K_2 + K_6K_5q + K_5q^4K_6\\
&\quad + K_5q^3K_6 + 2K_6^2q^4 - K_6^2q^5 - 2q^3K_2K_5 - 2q^4K_2K_6 - 2q^3K_2K_6\\
&\quad + q^4K_2^2,\\
p_{21} &= -K_4q - K_2q + K_6 + K_5,\\
p_{22} &= -q^2K_2 + K_6q^2 - K_6q - K_4q + K_6 + K_4,\\
p_{23} &= K_4q^4K_6 - q^3K_4K_6 + q^5K_4K_6 - q^2K_4K_5 - K_6K_4 - q^4K_4K_2\\
&\quad - qK_4K_5 - q^3K_4K_5 + qK_2K_4 + q^2K_2K_4 - K_6^2 - K_6^2q^3 + K_6^2q^2\\
&\quad - K_5K_4 - q^5K_4K_2 + K_6q^5K_2 + K_5q^4K_2 + qK_2K_6 - K_6K_5 - K_5q^2K_2\\
&\quad + K_6K_5q + 2K_6q^2K_5 + K_5^2q^2 - q^5K_2^2 + K_5^2q - K_5q^4K_6 - K_5q^3K_6\\
&\quad - K_6^2q^4 + 2q^4K_2K_6 - q^4K_2^2 + q^2K_4^2 + qK_4^2,\\
p_{24} &= -q^4K_4^2 + K_6 - q^2K_2 + 2K_6q^2 - q^5K_4K_6 + 2q^3K_4K_5 + qK_4K_5\\
&\quad - q^5K_4^2 - K_6K_4 - K_6q - K_4q^3 + q^2K_4K_6 + K_4 + K_5q^2 - q^2K_4^2\\
&\quad - q^3K_2 + q^3K_4K_6 + 2q^2K_4K_5 + q^4K_4K_5 - K_4q - qK_4^2 - q^3K_4^2 - K_4^2,\\
p_{25} &= -3K_4q^4K_6 + 3q^3K_4K_6 + 3K_4qK_6 - 3q^2K_4K_6 + 3q^5K_4K_6 - K_6K_4\\
&\quad - q^4K_4K_2 + 2qK_4K_5 + 2q^3K_4K_5 + qK_2K_4 - 2q^6K_4K_6 - K_5^2q^4\\
&\quad - K_6^2q^6 - q^3K_5^2 - K_4^3q^8 + q^4K_4^3 + q^5K_4^3 - 2q^4K_4^2 + 2K_6q^5K_5 + qK_4^3\\
&\quad + 2q^2K_4^3 + q^3K_4^3 + K_4q^4K_6^2 + K_4^2q^7K_5 + K_4^2q^7K_2 - 2K_4^2q^8K_6
\end{aligned}$$

$$
\begin{aligned}
&- q^2K_2K_4 + 3q^5K_5K_4K_6 - 2q^5K_2K_4K_5 - q^4K_2K_4K_6 + 3K_5q^3K_4K_6 \\
&- 3K_2q^3K_4K_5 + 2K_5^2q^4K_4 + K_6q^7K_4K_5 + q^7K_2K_4K_6 - K_6^2q^8K_4 \\
&- 3K_2q^4K_4K_5 - K_2q^3K_4K_6 + 4K_5q^4K_4K_6 + K_6^2q^3 - K_4^2 + 2q^5K_4K_5 \\
&+ K_6^2q - 2K_6^2q^2 + q^3K_4K_2 + q^5K_4K_2 + K_5^2q^5K_4 - K_4^2q^6K_6 + 2q^3K_4^2K_2 \\
&+ 2q^4K_4^2K_2 + 2q^5K_4^2K_2 + q^6K_4^2K_2 - 4q^3K_4^2K_6 - 3K_4^2q^4K_6 - K_4^2q^5K_6 \\
&- 3q^4K_4^2K_5 - q^5K_4^2K_5 - q^2K_4^2K_6 + 2q^2K_2K_4^2 - 3q^2K_4^2K_5 - 4q^3K_4^2K_5 \\
&+ qK_2K_4^2 + K_4^3 + q^6K_5K_4K_6 + K_6q^5K_2 + q^3K_4^2 - K_5q^4K_2 + 2qK_2K_6 \\
&+ K_6K_5q - 2K_6q^2K_5 - K_5^2q^2 - K_5q^4K_6 - K_6^2q^4 + 2K_6^2q^5 - 2q^4K_2K_6 \\
&+ q^3K_2K_6 + K_4q^5K_6^2 - q^6K_4^2 + q^5K_4^2 - 2q^2K_5K_4K_2 + q^2K_5K_4K_6 \\
&+ K_6K_4K_2 + K_4^2K_2 + K_4^2K_6 - 2qK_4^2K_5 - qK_4^2K_6 - K_6K_2 - qK_6K_4K_5 \\
&- qK_5K_2K_4 + 2K_5^2K_4q^3 + K_5K_2q - q^6K_5K_2K_4 - K_6K_2q^2 + K_5^2K_4q^2 \\
&- K_6^2K_4q - 2q^2K_4^2 + qK_4^2 - K_4K_2.
\end{aligned}
$$

REFERENCES

[1] R. Abłamowicz, P. Lounesto, On Clifford algebras of a bilinear form with an antisymmetric part, in *Clifford Algebras with Numeric and Symbolic Computations,* Eds. R. Abłamowicz, P. Lounesto, J. M. Parra, Birkhäuser, Boston, 1996, 167–188.

[2] R. Abłamowicz, Clifford algebra computations with Maple, *Clifford (Geometric) Algebras,* Banff, Alberta Canada, 1995, Ed. W. E. Baylis, Birkhäuser, Boston, 1996, 463–501.

[3] R. Abłamowicz, 'CLIFFORD' - Maple V package for Clifford algebra computations, ver. 4, `http://math.tntech.edu/rafal/cliff4/`.

[4] E. Artin, Theory of braids, *Ann. Math.* **48** (1947), 101–126.

[5] C. Chevalley, *The Algebraic Theory of Spinors,* Columbia University Press, New York, 1954.

[6] A. Connes, *Noncommutative Geometry,* Academic Press, San Diego, 1994, Ch. V. 10.

[7] H. S. M. Coxeter, W. O. J. Moser, *Generators and Relations for Discrete Groups, Fourth Edition,* Springer Verlag, Berlin, 1980.

[8] N. Bourbaki, *Algebra 1,* Springer Verlag, Berlin, 1989, Chapters 1–3.

[9] R. Boudet, The Glashow-Salam-Weinberg electroweak theory in the real algebra of space time, in *The Theory of the Electron,* J. Keller, Z. Oziewicz, Eds., Cuautitlan, Mexico, 1995, *Adv. in Appl. Clifford Alg.* **7** (Suppl.) (1997), 321–336.

[10] C. Doran, A. Lasenby, S. Gull, States and operators in spacetime algebra, *Found. Phys.* **23** (9), (1993), 1239. S. Somaroo, A. Lasenby, C. Doran, Geometric algebra and the causal approach to multi-particle quantum mechanics, *J. Math. Phys.* Vol. **40**, No. **7** (July 1999), 3327–3340.

[11] B. Fauser, Hecke algebra representations within Clifford geometric algebras of multivectors, *J. Phys. A: Math. Gen.* **32** (1999), 1919–1936.

[12] B. Fauser, Clifford-algebraische formulierung und regularität der Quantenfeldtheorie, Thesis, Uni. Tübingen, 1996.

[13] B. Fauser, H. Stumpf, Positronium as an example of algebraic composite calculations, in *The Theory of the Electron,* J. Keller, Z. Oziewicz, Eds., Cuautitlan, Mexico, 1995, *Adv. Appl. Clifford Alg.* **7** (Suppl.) (1997), 399–418.

[14] B. Fauser, Clifford geometric parameterization of inequivalent vacua, Submitted to *J. Phys. A.* (hep-th/9710047).

[15] B. Fauser, On the relation of Clifford-Lipschitz groups to q-symmetric groups, *Proc. XXII Int. Colloquium on Group Theoretical Methods in Physics,* Hobart, Tasmania, 1998, Eds. S. P. Corney, R. Delbourgo and P. D. Jarvis, International Press, Cambridge, MA, 1999, 413–417.

[16] B. Fauser, Clifford algebraic remark on the Mandelbrot set of two-component number systems, *Adv. Appl. Clifford Alg.* **6** (1)(1996), 1–26.

[17] B. Fauser, On an easy transition from operator dynamics to generating functionals by Clifford algebras, *J. Math. Phys.* **39** (1998), 4928–4947.

[18] B. Fauser, Vertex normal ordering as a consequence of nonsymmetric bilinear forms in Clifford algebras, *J. Math. Phys.* **37** (1996), 72–83.

[19] W. Fulton, Joe Harris, *Representation Theory,* Springer, New York, 1991.

[20] D. M. Goldschmidt, *Group Characters, Symmetric Functions, and the Hecke Algebra,* University Lecture Series, Berkeley (Spring 1989) Providence, RI: AMS Vol. **4** (1993).

[21] M. Hamermesh, *Group Theory and Its Application to Physical Problems,* Addison-Wesley, London, 1962.

[22] D. Hestenes, *Spacetime Algebra,* Gordon and Beach, 1966.

[23] D. Hestenes, *New Foundation for Classical Mechanics,* Kluwer, Dordrecht, 1986.

[24] H. Hiller, *Geometry of Coxeter Groups,* Pitman Books Ltd., London, 1982.

[25] P. N. Hoffman, J. F. Humphreys, *Projective Representations of the Symmetric Groups,* Oxford University Press, Oxford, 1992.

[26] R. C. King, B. G. Wybourne, Representations and traces of Hecke algebras $H_n(q)$ of type $\mathbf{A}_{n-1}$, *J. Math. Phys.* **33** (1992), 4–14.

[27] I. G. Macdonald, *Symmetric Functions and Hall Polynomials,* Oxford University Press, Oxford, 1979.

[28] S. Majid, *Foundations of Quantum Group Theory,* Cambridge University Press, Cambridge, 1995.

[29] Z. Oziewicz, From Grassmann to Clifford, in *Proceedings Clifford Algebras and Their Application in Mathematical Physics,* Canterbury, UK, J. S. R. Chisholm, A. K. Common, Kluwer, Dordrecht, 1986, 245–256.

[30] Z. Oziewicz, *Clifford Algebra of Multivectors,* in *Theory of the Electron,* Eds. J. Keller and Z. Oziewicz, *Adv. in Appl. Clifford Alg.* **7** (Suppl.) (1997), 467–486.

[31] Z. Oziewicz, Clifford algebra for Hecke braid, in *Clifford Algebras and Spinor Structures,* Special Volume to the Memory of Albert Crumeyrolle, Eds. R. Abłamowicz, P. Lounesto, Kluwer, Dordrecht, 1995, 397–412.

[32] R. Penrose, W. Rindler, Spinors and space-time: two-spinor calculus and relativistic fields, *Cambridge Monographs on Mathematics,* Cambridge University Press, Paperback Reprint Edition, Vol. 1, 1987.

[33] J. Rotman, *Galois Theory,* Springer, New York, 1990.

[34] H. Wenzl, Hecke algebras of type $\mathbf{A}_n$ and subfactors, *Invent. Math.* **92** (1988), 349–383.

Rafał Abłamowicz
Department of Mathematics, Box 5054
Tennessee Technological University
Cookeville, TN 38505, USA
E-mail: rablamowicz@tntech.edu

Bertfried Fauser
Universität Konstanz
Fachbereich Physik, Fach M678
78457 Konstanz, Germany
E-mail: Bertfried.Fauser@uni-konstanz.de

Received: August 12, 1999; Revised: February 2, 2000

On q-Deformations of Clifford Algebras

Gaetano Fiore

ABSTRACT Several Clifford algebras that are covariant under the action of a Lie algebra $\mathbf{g}$ can be deformed in a way consistent with the deformation of $U\mathbf{g}$ into a quantum group (or into a triangular Hopf algebra) $U_q\mathbf{g}$, i.e., so as to remain covariant under the action of $U_q\mathbf{g}$. In this report, after recalling these facts, we review our results regarding the formal realization of the elements of such "q-deformed" Clifford algebras as "functions" (polynomials) in the generators of the undeformed ones, in particular, the intriguing interplay between the original and the q-deformed symmetry. Finally, we briefly illustrate their dramatic consequences on the representation theories of the original and of the q-deformed Clifford algebra and mention how these results could turn out to be useful in quantum physics.
Keywords: q-deformed Clifford algebra, q-deformed anticommutation relations, quantum group, Hopf algebra, enveloping algebra, invariants.

1 Introduction

We first introduce the notions of a q-deformed Clifford algebra and of a deforming map in an explicit way by a simple example.

Consider the Clifford algebra $\mathcal{A}$ generated by $\mathbf{1}, a^{\uparrow}, a^{\downarrow}, a^+_{\uparrow}, a^+_{\downarrow}$, fulfilling the anticommutation relations

$$a^i a^j + a^j a^i = 0, \; a^+_i a^+_j + a^+_j a^+_i = 0, \; a^i a^+_j + a^+_j a^i = \delta^i_j \mathbf{1}, \qquad (1.1)$$

where $i, j = \uparrow, \downarrow$. When equipped with the $*$-structure

$$(a^i)^* = a^+_i, \qquad (1.2)$$

this becomes the familiar algebra of creation and annihilation operators of a fermionic system with two modes (e.g., a spin-up and a spin-down one-particle state). $\mathcal{A}$ is then invariant under the action $\triangleright$ of $su(2)$. This simply means that (1.1) is left invariant by the action of each of the generators X^+, X^-, H of $su(2)$, given by the defining action of $su(2)$ on the

AMS Subject Classification: 15A66, 16W30, 81R50, 17B37.

generators and extended to the whole $\mathcal{A}$ according to linearity and the Leibniz rule

$$x \triangleright (\alpha\alpha') = (x \triangleright \alpha)\alpha' + \alpha(x \triangleright \alpha'). \tag{1.3}$$

We shall not assign the $*$-structure for the moment, so $\mathcal{A}$ will be covariant under the action of $sl(2, \mathbb{C})$.

$\mathcal{A}$ is the simplest algebra one can q-deform. The corresponding q-deformed algebra $\mathcal{A}_q$ ($q \in \mathbb{C} \setminus \{0\}$) is generated by $\mathbf{1}_q, \tilde{A}^{\uparrow}, \tilde{A}^{\downarrow}, \tilde{A}^+_{\uparrow}, \tilde{A}^+_{\downarrow}$, fulfilling the quadratic anticommutation relations

$$\begin{gathered}
\tilde{A}^{\uparrow}\tilde{A}^{\uparrow} = 0 = \tilde{A}^{\downarrow}\tilde{A}^{\downarrow}, \quad \tilde{A}^+_{\uparrow}\tilde{A}^+_{\uparrow} = 0 = \tilde{A}^+_{\downarrow}\tilde{A}^+_{\downarrow} \\
\tilde{A}^{\uparrow}\tilde{A}^{\downarrow} + q\tilde{A}^{\downarrow}\tilde{A}^{\uparrow} = 0, \quad \tilde{A}^+_{\uparrow}\tilde{A}^+_{\downarrow} + q^{-1}\tilde{A}^+_{\downarrow}\tilde{A}^+_{\uparrow} = 0, \\
\tilde{A}^{\uparrow}\tilde{A}^+_{\downarrow} + q^{-1}\tilde{A}^+_{\downarrow}\tilde{A}^{\uparrow} = 0, \quad \tilde{A}^{\downarrow}\tilde{A}^+_{\uparrow} + q^{-1}\tilde{A}^+_{\uparrow}\tilde{A}^{\downarrow} = 0, \\
\tilde{A}^{\uparrow}\tilde{A}^+_{\uparrow} + \tilde{A}^+_{\uparrow}\tilde{A}^{\uparrow} = \mathbf{1}_q + (q^{-1}1)A^+_{\downarrow}\tilde{A}^{\downarrow}, \quad \tilde{A}^{\downarrow}\tilde{A}^+_{\downarrow} + \tilde{A}^+_{\downarrow}\tilde{A}^{\downarrow} = \mathbf{1}_q.
\end{gathered} \tag{1.4}$$

This algebra was first introduced in [14]. Clearly, $\mathcal{A}_q \xrightarrow{q \to 1} \mathcal{A}$ if we identify $\tilde{A}^i \to a^i$, $\tilde{A}^+_i \to a^+_i$ and $\mathbf{1}_q \to \mathbf{1}$ in the limit. Moreover, (1.4), and hence $\mathcal{A}_q$, are covariant under the action $\triangleright_q$ of the quantum group $U_q sl(2)$.

One can easily show that $\mathcal{A}$, $\mathcal{A}_q$ have the same Poincaré series. This means that they have the same dimension (sixteen) as vector spaces, and that the set of ordered monomials in the generators

$$\mathcal{B}_q := \{\mathbf{1}_q, \tilde{A}^{\uparrow}, \tilde{A}^{\downarrow}, \tilde{A}^+_{\uparrow}, \tilde{A}^+_{\downarrow}, \tilde{A}^{\uparrow}\tilde{A}^{\downarrow}, \tilde{A}^+_{\downarrow}\tilde{A}^{\uparrow}, \dots\} \tag{1.5}$$

is a basis of $\mathcal{A}_q$ and becomes a basis $\mathcal{B}$ of $\mathcal{A}$ in the limit $q \to 1$.

What is a q-deformed Clifford algebra like (1.4) good for? Can it be used to describe the same physics as its undeformed counterpart (1.1), e.g., a second-quantized fermionic system or a different (although similar) one? What is the role of the q-deformed symmetries? Answering these questions will require a comparison not only of the algebraic structures of $\mathcal{A}$, $\mathcal{A}_q$, but also of their representations.

Note that the Poincaré series requirement has already an immediate consequence on representation theory. It amounts to saying that the (left or right) regular representation of $\mathcal{A}$ (i.e., the one where the carrier space is the vector space U associated to the algebra $\mathcal{A}$, and the elements of the algebra act on U by (left or right) multiplication) and that of $\mathcal{A}_q$ have the same dimension since the basis $\mathcal{B}$ of U and the basis $\mathcal{B}_q$ of U_q have the same number of elements.

In order to compare the representation theories, it is helpful to first ask a related question: **Can we realize the generators $\tilde{A}^i, \tilde{A}^+_i$ as functions (polynomials) in a^i, a^+_j? And conversely?**

The answer is yes. For instance, an explicit realization of $\tilde{A}^i, \tilde{A}^+_i$, fulfilling (1.4) is through the polynomials

$$
\begin{aligned}
A^+_{\downarrow} := a^+_{\downarrow}, \quad A^+_{\uparrow} := [1+(q^{-1}-1)n^{\downarrow}]a^+_{\uparrow} = q^{-n^{\downarrow}}a^+_{\uparrow}, \\
A^{\downarrow} := a^{\downarrow}, \quad A^{\uparrow} := a^{\uparrow}[1+(q^{-1}-1)n^{\downarrow}] = a^{\uparrow}q^{-n^{\downarrow}},
\end{aligned}
\tag{1.6}
$$

where $n^i := a^i a^+_i$ (with *no* sum over i), $i = \uparrow, \downarrow$. (The last equalities on the right are based on the identity $(n^i)^2 = n^i$). For $q \neq 0$, the transformation (1.6) is clearly invertible; the inverse transformation allows an explicit realization of a^i, a^+_i, fulfilling (1.1) as polynomials $\tilde{a}^i, \tilde{a}^+_i$ in $\tilde{A}^i, \tilde{A}^+_i$:

$$
\begin{aligned}
\tilde{a}^+_{\downarrow} := \tilde{A}^+_{\downarrow}, \quad \tilde{a}^+_{\uparrow} := [1+(q-1)N^{\downarrow}]\tilde{A}^+_{\uparrow} = q^{N^{\downarrow}}\tilde{A}^+_{\uparrow} \\
\tilde{a}^{\downarrow} := \tilde{A}^{\downarrow}, \quad \tilde{a}^{\uparrow} := \tilde{A}^{\uparrow}[1+(q-1)N^{\downarrow}] = \tilde{A}^{\uparrow}q^{N^{\downarrow}},
\end{aligned}
\tag{1.7}
$$

where $N^i := a^i a^+_i$ (with *no* sum over i), $i = \uparrow, \downarrow$. (The last equalities on the right are based on the identity $(N^{\downarrow})^2 = N^{\downarrow}$). In a more abstract language, through the above, one can define an algebra isomorphism $f : \mathcal{A}_q \to \mathcal{A}[[h]]$ ($\mathcal{A}[[h]] \equiv$the algebra of formal power series in $h = q-1$ with coefficients in $\mathcal{A}$), through

$$
\begin{aligned}
f(\tilde{A}^i) = A^i, \quad f(\tilde{A}^+_i) = A^+_i, \\
f(\alpha\beta) = f(\alpha)f(\beta), \quad \forall \alpha, \beta \in \mathcal{A}.
\end{aligned}
\tag{1.8}
$$

Now we illustrate the usefulness of deforming maps to compare the representation theories of $\mathcal{A}$, $\mathcal{A}_q$. We ask: Can f be seen also as an *operator map* intertwining the representations of $\mathcal{A}$ and of $\mathcal{A}_q$? In other words, given a representation of $\mathcal{A}$ (resp. $\mathcal{A}_q$) on a vector space V (resp. V_q), does V (resp. V_q) carry also a representation of $\mathcal{A}_q$ (resp. $\mathcal{A}$)? The answer is clearly yes since A^i, A^+_j (resp. $\tilde{a}^i, \tilde{a}^+_i$), being polynomials in a^i, a^+_i (resp. in $\tilde{A}^i, \tilde{A}^+_i$), result in well-defined operators on V (resp. V_q)[1]. Thus, the classification of the representations of $\mathcal{A}$ (resp. $\mathcal{A}_q$) will determine also the classification of the representations of $\mathcal{A}_q$ (resp. $\mathcal{A}$). From deforming maps, one can also extract more specific informations. For instance, if we endow $\mathcal{A}$ also with the star structure (1.2), the one that is compatible

[1]It is interesting to note that this is *not* true instead for q-deformations of Weyl algebras (the algebras obtained by replacing the anticommutators in relations (1.1) by commutators. The latter are in fact infinite-dimensional as vector spaces, and the corresponding realizations A^i, A^+_i are formal power series (instead of polynomials) in a^i, a^+_i, so strictly speaking do not belong to $\mathcal{A}$ but just to a suitable completion of $\mathcal{A}$. Correspondingly, it is not guaranteed that A^i, A^+_i can be defined as operators on the corresponding vector spaces. In fact, it was e.g., explicitly shown [14] that there are many (inequivalent) irreducible $*$-representations of the simplest $U_q su(N)$ covariant deformed Weyl $*$-algebra $\mathcal{A}_q$, whereas just one of the corresponding undeformed partner $\mathcal{A}$. In [6] we re-read this result by showing that the corresponding objects $\tilde{a}^i, \tilde{a}^+_i \in \mathcal{A}_q$ become ill-defined operators on all but one $*$-representation of $\mathcal{A}_q$. The same might, in principle, occur for Clifford algebras with an infinite number of generators.

with (1.1) and the action of the compact section $Usu(2)$ of $Usl(2)$, the corresponding star structure of $\mathcal{A}_q$, compatible with (1.4) and the action of $U_qsu(2)$, exists only for real q and reads

$$(\tilde{A}^i)^{\star_q} = \tilde{A}_i^+. \tag{1.9}$$

It is easy to see that (1.6) [resp. (1.7)] allows also a realization of $\star_q$ (resp. $\star$) as $\star$ (resp. $\star_q$). Since there is (up to unitary equivalences) a unique $*$-representation of the $*$-algebra $\mathcal{A}$ and a unique $*$-representation of the $*$-algebra $\mathcal{A}_q$ as one can check by direct inspection, we conclude that they correspond to each other in the above identification.

How have we found (1.6)? Does it keep track of the $U_qsu(2)$ symmetry?

In this report, we shall present a systematic approach, based partly on the works [6, 5, 4] (see also 's [7]), to answer the latter questions for arbitrary "q-deformed Clifford algebras". Incidentally, the approach works not only for Clifford but also for q-deformed Weyl algebras.

2 General framework

The general setting is the following. The undeformed Clifford algebra $\mathcal{A}$ is covariant under some Lie algebra $\mathbf{g}$ and the deformed one $\mathcal{A}_q$ under the quantum group [2] (or under a triangular deformation) $U_q\mathbf{g}$. The undeformed algebra $\mathcal{A}$ is generated by $\mathbf{1}, a^i, a_j^+$ fulfilling

$$a^i\, a^j + a^j\, a^i = 0, \quad a_i^+\, a_j^+ + a_j^+\, a_i^+ = 0, \quad a^i\, a_j^+ + a_j^+\, a^i = \delta_j^i\mathbf{1} \tag{2.1}$$

and transforms under the action $\triangleright$ of $\mathbf{g}$ according to some law

$$x \triangleright a_i^+ = \rho(x)_i^j a_j^+, \qquad x \triangleright a^i = \rho(Sx)_j^i a^j; \tag{2.2}$$

here $x \in \mathbf{g}$, $Sx = -x$ and ρ denote some matrix representation of $\mathbf{g}$. Clearly a^i transform under the contragradient representation of the a_i^+ one. The action $\triangleright$ is extended to all of $U\mathbf{g} \times \mathcal{A}$ imposing linearity the Leibniz rule (1.3) and the law

$$(xx') \triangleright \alpha = x \triangleright (x' \triangleright \alpha). \tag{2.3}$$

As a consequence, for $x \in U\mathbf{g}$ we have

$$x \triangleright (\alpha\alpha') \equiv \sum_i (x_{(1)}^i \triangleright \alpha)(x_{(2)}^i \triangleright \alpha'), \tag{2.4}$$

where the coproduct of $U\mathbf{g}$ $\Delta(x) = \sum_i x_{(1)}^i \otimes x_{(2)}^i$ is defined by $\Delta(x) = x \otimes \mathbf{1} + \mathbf{1} \otimes x$ for $x \in \mathbf{g}$ and is extended to all of $U\mathbf{g}$ as an algebra homomorphism. All this is possible because the action of $\mathbf{g}$ is manifestly compatible with the anticommutation relations (2.1). Formulae (2.2), where now we have extended S to the whole $U\mathbf{g}$ as the antipode, give also the standard extension of $\triangleright$ to $x \in U\mathbf{g}$.

The corresponding q-deformed algebra $\mathcal{A}_q$ is generated by $\mathbf{1}_q, \tilde{A}^+_i, \tilde{A}^i$ (the generators are enumerated by the same index), fulfilling deformed anticommutation relations with the a quadratic structure as in (2.1) of the form

$$\begin{gathered} P^{+}_q{}^{hk}_{ij}\tilde{A}^+_h\tilde{A}^+_k = 0, \quad P^{+}_q{}^{ij}_{hk}\tilde{A}^k\tilde{A}^h = 0 \\ \tilde{A}^i\tilde{A}^+_j + P_q{}^{ih}_{jk}\tilde{A}^+_h\tilde{A}^k = \delta^i_j\,\mathbf{1}_q. \end{gathered} \tag{2.5}$$

P^+_q, P_q are matrices with entries in $\mathbb{C}$. P_q is a $U_q\mathbf{g}$-covariant deformation of the ordinary permutator P, defined by $P^{ij}_{hk} = \delta^i_k\delta^j_h$; P^+_q is the $U_q\mathbf{g}$-covariant deformation of the ordinary symmetric projector $P^+ = \frac{1}{2}(1+P)$, and is a projector itself, $(P^+_q)^2 = P^+_q$. Thus, in the limit $q \to 1$, the q-deformed anticommutation relations (2.5) reduce to (2.1). Moreover, we additionally require that $\mathcal{A}$, $\mathcal{A}_q$ have the same Poincaré series. $U_q\mathbf{g}$-covariance means that (2.5) are compatible with the action $\tilde{\triangleright}_q$ of $U_q\mathbf{g}$. The latter is defined on the generators by the law

$$x\,\tilde{\triangleright}_q\,\tilde{A}^+_i = \rho_q{}^j_i(x)\tilde{A}^+_j, \quad x\,\tilde{\triangleright}_q\,\tilde{A}^i = \rho_q{}^i_j(S_qx)\tilde{A}^j \tag{2.6}$$

(here $x \in U_q\mathbf{g}$, S_q is the antipode of $U_q\mathbf{g}$, ρ_q the quantum group deformation of ρ) and is extended on all of $\mathcal{A}_q$ through a modified Leibniz rule, which we shall give below in (3.7). It is exactly the requirement of $U_q\mathbf{g}$-covariance that determines the form of P^+_q, P_q in (2.5).

Sometimes the case that q is a root of unity, any representation ρ of $U\mathbf{g}$ admits a q-deformation into a representation ρ_q of $U_q\mathbf{g}$ (in particular, this implies that ρ, ρ_q have the same dimension), and the corresponding projectors P, P^+ admit $U_q\mathbf{g}$-covariant deformations P^+_q, P_q. However, it is not guaranteed that the corresponding relations (2.5) yield $\mathcal{A}$, $\mathcal{A}_q$ with the same Poincaré series.

This is guaranteed for *arbitrary* ρ only in a less general context, namely if $U_q\mathbf{g}$ is a *triangular* deformation [1] of the Hopf algebra $U\mathbf{g}$ (e.g., a Jordanian [13], or a Reshetikin [16] deformation); then $P_q = \hat{R} := PR$, $P^+_q = \frac{1}{2}(1+\hat{R})$, where $R = (\rho_q \otimes \rho_q)\mathcal{R}$ and $\mathcal{R}$ is a universal object belonging to $U_q\mathbf{g}\otimes\mathbf{U_qg}$ called the universal R-matrix [1, 3]. Triangularity means that $\hat{R}^2 = 1$. A special case with a broad spectrum of potential physical applications is when ρ is a direct sum $\rho = \bigoplus_{\alpha=1}^{M} \rho'$ of M copies of a simpler representation ρ'; if the latter describes the symmetry of some dynamical system $\mathcal{S}'$, ρ should describe the symmetry of the composite dynamical system $\mathcal{S}$ obtained taking M copies of $\mathcal{S}'$. The M copies could correspond to different sites in some d-dimensional lattice, or (if $M = \infty$)

to different space (time) points, respectively, in condensed matter physics or quantum field theory.

If $U_q\mathbf{g}$ is a quantum group in the strict sense, i.e., a *quasitriangular* but not triangular deformation of $U\mathbf{g}$ [2], the Poincaré series condition is fulfilled essentially only if

1. $\mathbf{g} = sl(N)$ [14], $sp(N=2n)$ [4] and $\rho = \rho_N \equiv N$-dimensional defining representation of $\mathbf{g}$ (e.g., the $\mathbf{N}$ of $sl(N)$),
2. $\mathbf{g} = sl(N)$ and $\rho = \bigoplus_{\alpha=1}^{M} \rho_N$ [4] for some integer $M > 1$.

In case 1, $P_q = q^{-1}\hat{R}_N \equiv q^{-1}PR_N$ and the projector P_q^+ is given by

$$P_q^+ = \begin{cases} \frac{1+q\hat{R}_N}{q+q^{-1}} & \text{if } \mathbf{g} = sl(N); \\ \frac{\hat{R}_N^2+(q^{-1-N}+q^{-1})\hat{R}_N+q^{-2-N}1}{(q+q^{-1})(q-q^{-1-N})} & \text{if } \mathbf{g} = sp(N), \end{cases} \tag{2.7}$$

where $R_N = (\rho_{N,q} \otimes \rho_{N,q})\mathcal{R}$ and $\mathcal{R} \in U_q\mathbf{g} \otimes \mathbf{U_q g}$ is the so called universal R-matrix [2] of $U_q\mathbf{g}$. R_N is denoted as the R-matrix of $U_q\mathbf{g}$ in the q-deformed defining representation $\rho_{N,q}$.

In case 2, contrary to the triangular case, it turns out that the resulting commutation relations between the *different* copies automatically order the M copies in a definite way, a phenomenon which we have called a "braided chain" [4]. Consequently, the only physical lattice in which it would be reasonable to arrange the copies would be 1-dimensional. If we use Greek indices $\alpha, \beta, \ldots = 1, \ldots, M$ to enumerate the copies in the prescribed order, then up to some free normalization factors (which we omit for the sake of simplicity) the deformed anticommutation relations (2.5) take the form

$$\tilde{A}^+_{\alpha i}\tilde{A}^+_{\beta j} + q\hat{R}_N{}^{hk}_{ij}\tilde{A}^+_{\beta h}\tilde{A}^+_{\alpha k} = 0, \quad \tilde{A}^{\alpha j}\tilde{A}^{\beta i} + q\hat{R}_N{}^{ij}_{hk}\tilde{A}^{\beta k}\tilde{A}^{\alpha h} = 0 \tag{2.8}$$

and either

$$\tilde{A}^{\alpha i}\tilde{A}^+_{\beta j} + q^{-1}\hat{R}_N{}^{ih}_{jk}\tilde{A}^+_{\beta h}\tilde{A}^{\alpha k} = \delta^i_j\delta^\alpha_\beta \mathbf{1}_q \tag{2.9}$$

or

$$\tilde{A}^{\alpha i}\tilde{A}^+_{\beta j} + \hat{R}_M^{-1}{}^{\alpha\gamma}_{\beta\delta}\hat{R}_N{}^{ih}_{jk}\tilde{A}^+_{\gamma h}\tilde{A}^{\delta k} = \delta^i_j\delta^\alpha_\beta \mathbf{1}_q, \tag{2.10}$$

with $\alpha \leq \beta$ and $\hat{R}_M$ the braid matrix of $sl(M)$. The latter $\mathcal{A}_q$ in fact is covariant not only under $U_q sl(N)$ but also under $U_q(sl(M) \times sl(N))$.

The explicit form of the braid matrix $\hat{R}_N$ of $sl(N)$ is

$$\hat{R}_N = q\sum_{i=1}^{N} e^i_i \otimes e^i_i + \sum_{\substack{i,j=1 \\ i\neq j}}^{N} e^i_j \otimes e^j_i + (q-q^{-1})\sum_{\substack{i,j=1 \\ i<j}}^{N} e^i_i \otimes e^j_j, \tag{2.11}$$

where $q \in \mathbb{C} - 0$ and e^i_j is the matrix with all vanishing elements except a 1 at the i-th row and j-th column.

Our problem can now be formulated more technically as follows: How can we determine all possible deforming maps, i.e., algebra isomorphisms (over $\mathbb{C}[[h]]$) $f : \mathcal{A}_q \to \mathcal{A}[[h]]$, [2] (h:=q-1) for the class of deformed Clifford algebras defined above?

3 Construction procedure

First note that if $\alpha \in \mathcal{A}[[h]]$ is any element of the form $\alpha = \mathbf{1} + O(h)$ and f is a deforming map, one can obtain a new one f_α by the inner automorphism

$$f_\alpha(\cdot) := \alpha f(\cdot)\alpha^{-1}; \tag{3.1}$$

actually the vanishing of the first Hochschild cohomology group [9] of $\mathcal{A}$ implies that *all* deforming maps can be obtained from one in this manner. Therefore, our problem is reduced to finding a particular one that we are going to describe below.

Second, note that given any deforming map f and using $\tilde{\triangleright}_q$ we can draw the solid lines in the diagram

$$\begin{array}{ccccc} U_q\mathbf{g} \times & \mathcal{A}_q & \xrightarrow{\tilde{\triangleright}_q} & \mathcal{A}_q & \\ & \updownarrow \mathrm{id} \times f & & \updownarrow f & \\ U_q\mathbf{g} \times & \mathcal{A}[[h]] & \overset{\triangleright_q}{- - - \rightarrow} & \mathcal{A}[[h]]; & \end{array} \tag{3.2}$$

we define $\triangleright_q$ as the map making the diagram commutative (in other words $\triangleright_q := f \circ \tilde{\triangleright}_q \circ (\mathrm{id} \otimes f^{-1})$, which will realize $\tilde{\triangleright}_q$ on $\mathcal{A}[[h]]$). One can easily realize that $\triangleright_q \neq \triangleright$, since there is no Hopf algebra isomorphism $U_q\mathbf{g} \to \mathbf{Ug}[[\mathbf{h}]]$ [3]. For each f_α in (3.1) one finds correspondingly also a different $\triangleright_q$, in other words by varying α one obtains all pairs $(f, \triangleright_q)$.

Our construction strategy will proceed in the opposite direction: we shall first determine *one* particular $\triangleright_q$ then the corresponding deforming map(s) f. Actually to define the latter, it suffices to find generators $A^i, A^+_j \in \mathcal{A}[[h]]$ fulfilling (2.5) and the analog of (2.6) and apply formula (1.8). We shall first show an Ansatz for A^i, A^+_j which allows us to fulfill at once the transformation law (2.6); the Ansatz is based on the properties of the "Drinfeld's twist" [3]. Then we shall determine in the simplest cases the free parameters appearing in the Ansatz in such a way that the commutation relations (2.5) become fulfilled.

---

[2] We recall that for any algebra B, $B[[h]]$ denotes the ring of formal power series in h with coefficients in B.

Let us summarize the elements of our construction procedure and of the notation we shall adopt:

1. $\mathbf{g}$, a semisimple Lie algebra if the deformation $U_q\mathbf{g}$ we are interested in is triangular, or $sl(N)$, $sp(N)$ if the deformation $U_q\mathbf{g}$ we are interested in is a quantum group. As known, one can associate to $U\mathbf{g}$ a cocommutative Hopf algebra $H \equiv (U\mathbf{g}, \cdot, \Delta, \varepsilon, \mathbf{S})$; $\cdot, \Delta, \varepsilon, S$ denote the product, coproduct, counit, antipode. We shall use Sweedler's notation $\Delta(x) \equiv x_{(1)} \otimes x_{(2)}$: the rhs stands for a sum $\sum_i x^i_{(1)} \otimes x^i_{(2)}$ of different terms, but the symbol $\sum_i$ is dropped. We shall denote by $H_q \equiv (U_q\mathbf{g}, \bullet, \Delta_\mathbf{q}, \varepsilon_\mathbf{q}, \mathbf{S_q}, \mathcal{R})$ the deformation of H we are interested in, respectively a triangular Hopf algebra [1] or a quantum group [2]. $\bullet, \Delta_q, \varepsilon_q, S_q$ denote the deformed product, coproduct, counit, antipode, $\mathcal{R}$ the universal R-matrix. We shall use Sweedler's notation (with barred indices) $\Delta_q(x) \equiv x_{(\bar{1})} \otimes x_{(\bar{2})}$.

2. An algebra isomorphism $\varphi_q : U_q\mathbf{g} \to \mathbf{Ug}[[\mathbf{h}]]$ over $\mathbf{C}[[h]]$ whose existence is proved respectively in [1, 3]: $\varphi_q(x \bullet y) = \varphi_q(x) \cdot \varphi_q(y)$.

3. A corresponding Drinfeld's twist[1, 3], i.e., an element

$$\mathcal{F} \equiv \mathcal{F}^{(1)} \otimes \mathcal{F}^{(2)} = \mathbf{1}^{\otimes^2} + O(h)$$

of $U\mathbf{g}[[\mathbf{h}]] \otimes \mathbf{Ug}[[\mathbf{h}]]$ such that

$$\begin{aligned} (\varepsilon \otimes \mathrm{id})\mathcal{F} = \mathbf{1} = (\mathrm{id} \otimes \varepsilon)\mathcal{F}, \\ \Delta_q(a) = (\varphi_q^{-1} \otimes \varphi_q^{-1})\{\mathcal{F}\Delta[\varphi_q(a)]\mathcal{F}^{-1}\}. \end{aligned} \tag{3.3}$$

The last formula means that, up to the isomorphism φ_q, Δ_q is related to Δ by a similarity transformation.

4. $\gamma' := \mathcal{F}^{(2)} \cdot S\mathcal{F}^{(1)}$ and $\gamma := S\mathcal{F}^{-1(1)} \cdot \mathcal{F}^{-1(2)}$. Up to the isomorphism φ_q, S_q and its inverse are related to S by similarity transformations involving, respectively, γ and γ'.

5. The particular representation ρ_q of $U_q\mathbf{g}$ fulfilling the criteria listed after (2.6) and its classical limit (2.2).

6. The generalized Jordan-Schwinger algebra homomorphism $\sigma : U\mathbf{g}[[\mathbf{h}]] \to \mathcal{A}[[h]]$, defined on the generators by

$$\sigma(\mathbf{1}_{U\mathbf{g}}) = \mathbf{1}, \quad \sigma(x) := \rho(x)^i_j a^+_i a^j \tag{3.4}$$

$x \in \mathbf{g}$, and extended to the whole $U\mathbf{g}[[\mathbf{h}]]$ as an algebra homomorphism, $\sigma(yz) = \sigma(y)\sigma(z)$ and $\sigma(y+z) = \sigma(y) + \sigma(z)$. It is immediate to verify that this extension is possible because $\sigma([x,y]) = [\sigma(x), \sigma(y)]$. In the $su(2)$, σ takes the well-known form

$$\sigma(j_+) = a^+_\uparrow a^\downarrow, \quad \sigma(j_-) = a^+_\downarrow a^\uparrow, \quad \sigma(j_0) = \frac{1}{2}(a^+_\uparrow a^\uparrow - a^+_\downarrow a^\downarrow). \tag{3.5}$$

7. The deformed Jordan-Schwinger algebra homomorphism $\sigma_q : U_q\mathbf{g} \to \mathcal{A}[[\mathbf{h}]]$, defined by $\sigma_q := \sigma \circ \varphi_q$.
8. The $*$-structures $*, *_q, \star, \star_q$ in $H, H_q, \mathcal{A}, \mathcal{A}_q$ if $\mathcal{A}, \mathcal{A}_q$ are $*$-algebras transforming respectively under the Hopf $*$-algebras H, H_q with the compatibility condition

$$(x \triangleright_q a)^{\star_q} = S_q^{-1}(x^{*_q}) \triangleright_q a^{\star_q}. \tag{3.6}$$

As anticipated, our first step is to guess a realization $\triangleright_q$ of $\tilde{\triangleright}_q$ on $\mathcal{A}[[h]]$, instead of $\mathcal{A}_q$. This requires fulfilling

$$(xy) \triangleright_q a = x \triangleright_q (y \triangleright_q a), \quad x \triangleright_q (ab) = (x_{(\bar{1})} \triangleright_q a)(x_{(\bar{2})} \triangleright_q b) \tag{3.7}$$

for any $x, y \in U_q\mathbf{g}$, $a, b \in \mathcal{A}_q$; these are the conditions characterizing a module algebra [also in the undeformed case, see formulae (1.3)]. There is a simple way to find such a realization, namely by setting

$$x \triangleright_q a := \sigma_q(x_{(\bar{1})}) a \sigma_q(S_q x_{(\bar{2})}); \tag{3.8}$$

it is easy to check that (3.7) are indeed fulfilled using the basic axioms characterizing the coproduct, counit, antipode in a generic Hopf algebra. The guess has been suggested by the undeformed case, where the same conditions and realizations are obtained for $U\mathbf{g}$, $\mathcal{A}, \triangleright$ if in the two previous formulae we just erase the suffix $_q$ and replace $\Delta_q(x) \equiv x_{(\bar{1})} \otimes x_{(\bar{2})}$ with the cocommutative coproduct $\Delta(x) \equiv x_{(1)} \otimes x_{(2)}$.

Our second step is to realize elements $A^i, A_j^+ \in \mathcal{A}[[h]]$ that transform under the action $\triangleright_q$ defined by (3.8) as $\tilde{A}^i, \tilde{A}_j^+$ do under $\tilde{\triangleright}_q$ [see (2.6)], namely

$$x \triangleright_q A_i^+ = \rho_q{}_i^j(x) A_j^+, \quad x \triangleright_q A^i = \rho_q{}_j^i(S_q x) A^j. \tag{3.9}$$

Note that a^i, a_j^+ do *not* transform in this way. In [6] we proved that the following objects do:

$$\begin{aligned} A_i^+ &:= u\,\sigma(\mathcal{F}^{(1)}) a_i^+ \sigma(S\mathcal{F}^{(2)}\gamma)\, u^{-1} \\ A^i &:= v\,\sigma(\gamma' S\mathcal{F}^{-1(2)}) a^i \sigma(\mathcal{F}^{-1(1)}) v^{-1}; \end{aligned} \tag{3.10}$$

the result holds for any choice of g-invariant elements $u, v = \mathbf{1} + O(h)$ in $\mathcal{A}[[h]]$, in particular for $u = v = \mathbf{1}$.

The third step is to fix u, v in such a way that the deformed commutation relations (2.5) are fulfilled. One can easily show that the latter may fix at most the product uv^{-1}. In the case that $U_q\mathbf{g}$ is triangular, we showed in [6] that they require $uv^{-1} = 1$. In the case that $U_q\mathbf{g}$ is the quantum group $U_q sl(N)$ and $\rho = \rho_N$, we proved [5] that the deformed commutation relations require

$$uv^{-1} = \frac{\Gamma(n+1)}{\Gamma_{q^2}(n+1)}. \tag{3.11}$$

Here Γ is Euler's γ-function, Γ_{q^2} its q-deformation characterized by $\Gamma_{q^2}(x+1) = (x)_{q^2}\Gamma_{q^2}(x)$, $n := \sum_i a^i a_i^+$. We stress that the above solutions regard the case of ρ being the defining representation ρ_N of $sl(N)$. We have yet no formula yielding the right uv^{-1}, if any, necessary to fulfill the (2.5) in the other cases. However, it is important to recall that [5], in general the (2.5), translate into conditions on uv^{-1} where the Drinfeld's twist $\mathcal{F}$ appears only through the so called "coassociator"

$$\phi := [(\Delta \otimes \mathrm{id})(\mathcal{F}^{-1})](\mathcal{F}^{-1} \otimes \mathbf{1})(\mathbf{1} \otimes \mathcal{F})[(\mathrm{id} \otimes \Delta)(\mathcal{F}). \tag{3.12}$$

ϕ is known, unlike $\mathcal{F}$, for which up to now there is an existence proof. This makes the above conditions explicit and allows us to search uv^{-1} in the general case, if it exists. The explicit expression for ϕ is

$$\phi = \lim_{x_0, y_0 \to 0^+} \left\{ x_0^{-\hbar t_{12}} \vec{P} \exp\left[-\hbar \int_{x_0}^{1-y_0} dx \left(\frac{t_{12}}{x} + \frac{t_{23}}{x-1} \right) \right] y_0^{\hbar t_{23}} \right\}, \tag{3.13}$$

where $t = \Delta(\mathcal{C}) - \mathbf{1} \otimes \mathcal{C} - \mathcal{C} \otimes \mathbf{1}$, $\mathcal{C}$ denoting the quadratic Casimir of $U\mathbf{g}$, $t_{12} = t \otimes \mathbf{1}$, $t_{23} = \mathbf{1} \otimes t$. The symbol $\vec{P}$ means that we must understand a path-ordered integral in the variable x. Note that $\phi = \mathbf{1}^{\otimes^3} + O(h^2)$. (In the triangular case, on the contrary, $\phi \equiv \mathbf{1}^{\otimes^3}$).

Finally, the residual freedom in the choice of u, v is partially fixed if H, H_q, $\mathcal{A}, \mathcal{A}_q$ are matched (Hopf) $*$-algebras, and we make the additional requirement that $\star$ realizes in $\mathcal{A}[[h]]$ the $\star_q$ of $\mathcal{A}_q$. For instance, if $(a^i)^\star = a_i^+$ and $q \in \mathbb{R}^+$, this means

$$(A^i)^\star = A_i^+, \tag{3.14}$$

and is fulfilled if we take $u = v^{-1}$.

In general, one can show that the knowledge of $(\rho \otimes \mathrm{id})\mathcal{F}$ is sufficient to determine the A^i, A_i^+ of formulae (3.10) completely. In the $\mathbf{g} = sl(2)$ case, with ρ being the fundamental representation, $(\rho \otimes \mathrm{id})\mathcal{F}$ is explicitly known [17]. Taking $u = v^{-1}$ one finally finds [6] the result (1.6) we had anticipated.

Above, we have determined in $\mathcal{A}[[h]]$ one particular realization A^i, A_j^+ and $\triangleright_q$ of the generators $\tilde{A}^i, \tilde{A}_j^+$ and of the quantum group action. Its main feature is that the $\mathbf{g}$-invariant ground state $|0\rangle$, as well as, the first excited states $a_i^+|0\rangle$ of the classical Fock space representation, are also respectively $U_q\mathbf{g}$-invariant ground state $|0_q\rangle$ and first excited states $A_i^+|0_q\rangle$ of the deformed Fock space representation.

According to (3.1), all the other realizations are of the form

$$A^{\alpha i} = \alpha A^i \alpha^{-1}, \quad A^+_{\alpha i} = \alpha A_i^+ \alpha^{-1}, \tag{3.15}$$

with $\alpha = \mathbf{1} + O(h) \in \mathcal{A}[[h]]$. They are manifestly covariant under the realization $\triangleright_{h,\alpha}$ of the $U_q\mathbf{g}$-action defined by

$$x \triangleright_{h,\alpha} a := \alpha\, \sigma_q(x_{(\bar{1})})\, a\, \sigma_q(x_{(\bar{2})})\, \alpha^{-1}. \tag{3.16}$$

For these realizations, the deformed ground state in the Fock space representation reads $|0_q\rangle = \alpha|0\rangle$; if $\alpha|0\rangle \neq |0\rangle$ the $\mathbf{g}$ -invariant ground state and first excited states of the classical Fock space representation do not coincide with their deformed counterparts.

4 Ordinary vs. q-deformed invariants

We have introduced two actions on $\mathcal{A}[[h]]$:

$$\triangleright : U\mathbf{g} \times \mathcal{A}[[\mathbf{h}]] \to \mathcal{A}[[\mathbf{h}]], \quad \triangleright_{\mathbf{q}} : \mathbf{U_q g} \times \mathcal{A}[[\mathbf{h}]] \to \mathcal{A}[[\mathbf{h}]]. \tag{4.1}$$

Their respective invariant subalgebras $\mathcal{A}^{inv}[[h]]$, $\mathcal{A}_q^{inv}[[h]]$ are defined by

$$\mathcal{A}_q^{inv}[[h]] := \{I \in \mathcal{A}[[h]] \,|\, x \triangleright_q I = \varepsilon_q(x)I, \,\forall x \in U_q\mathbf{g}\} \tag{4.2}$$

and by the analogous equation where all suffices $_q$ are erased. What is the relation between them? It is easy to prove that [5]

$$\mathcal{A}_q^{inv}[[h]] = \mathcal{A}^{inv}[[h]]. \tag{4.3}$$

In other words, invariants under the $\mathbf{g}$ -action $\triangleright$ are also $U_q\mathbf{g}$-invariants under $\triangleright_q$ and conversely although in general $\mathbf{g}$ -covariant objects (tensors) and $U_q\mathbf{g}$-covariant ones do not coincide in general!

Let us introduce in the vector space $\mathcal{A}^{inv}[[h]] = \mathcal{A}_q^{inv}[[h]]$ bases I^1, $I^2, \dots$ and $I_q^1, I_q^2, \dots$. We realize immediately that we can choose the I^n as homogeneous, normal-ordered polynomials in a^i, a_j^+ and I_q^n as homogeneous, normal-ordered polynomials in A^i, A_j^+, since $\triangleright$ acts linearly without changing the degrees in a^i and a_j^+, and $\triangleright_q$ acts linearly without changing the degrees in A^i and A_j^+. Explicitly,

$$\begin{aligned}
I^1 &= a_i^+ a^i, & I_q^1 &= A_i^+ A^i, \\
I^2 &= d^{ijk} a_i^+ a_j^+ a_k^+, & I_q^2 &= D^{ijk} A_i^+ A_j^+ A_k^+, \\
I^3 &= d'_{kji} a^i a^j a^k, & I_q^3 &= D'_{kji} A^i A^j A^k, \\
I^4 &= \dots, & I_q^4 &= \dots,
\end{aligned} \tag{4.4}$$

where the numerical coefficients $d, d', \dots$ form $\mathbf{g}$ -isotropic tensors and the numerical coefficients D, D' the corresponding $U_q\mathbf{g}$-isotropic tensors. In the quantum group cases considered in Section 2, it is possible to show that $I_q^1 \neq I^1$:

$$I_q^1 = \frac{q^{-2I^1} - 1}{q^{-2} - 1}. \tag{4.5}$$

In general $I_q^n \neq I^n$ although $I_q^n = I^n + O(h)$. The proposition (4.3) implies in particular that

$$I_q^n = g^n(\{I^m\}, h) = k^n(\{a^i, a_j^+\}, h). \tag{4.6}$$

What do the functions g^n, k^n look like?

In [5] we have found universal formulae yielding the k^n's. The latter turn out to be polynomials in a^i, a_i^+ of degree higher than the degree in A^i, A_i^+ (this can be easily worked e.g., for the invariant I_q given in (4.5)), and the degree difference grows very fast with the number of these generators. It is remarkable that in these universal formulae the twist $\mathcal{F}$ appears only through the coassociator ϕ; therefore, all the k^n can be worked out explicitly.

In the case that the Hopf algebra H_q is not a genuine quantum group, but triangular, the coassociator as well as u, v are trivial and one finds $I_q^n = I^n$.

5 Final remarks, outlook and conclusions

We have shown how one can realize a deformed $U_q\mathbf{g}$-covariant Clifford algebra $\mathcal{A}_q$ within the undeformed one $\mathcal{A}[[h]]$. Given a representation (π, V) of $\mathcal{A}$ on a vector space V, does it provide also a representation of $\mathcal{A}_q$? In other words, can one interpret the elements of $\mathcal{A}_q$ as operators acting on V if the elements of $\mathcal{A}$ are? If so, what specific role do the elements A^i, A_i^+ of $\mathcal{A}[[h]]$ play?

Repeating the arguments presented in the introduction for the toy model, one can conclude that the answer to the first question is always positive, at least for finite-dimensional Clifford algebras. In particular, when q is real and the real structure (1.9) is chosen, this allows us to represent the q-deformed Clifford algebra $\mathcal{A}_q$ on the standard Fock space of the original algebra $\mathcal{A}$; in a particle physics interpretation no exotic statistics are then involved but just the ordinary Fermi-Dirac characterizing fermions. Only A^i, A_j^+ do not annihilate/create the undeformed states.

On the other hand, quadratic commutation relations of the type (2.5) mean that A_i^+, A^i can be interpreted as creators and annihilators of some excitations. A glance at (3.10), (3.15) shows that these are not the undeformed excitations, but some 'collective' ones.[3] The last point is: what could the latter be good for. As a Hamiltonian H of the system, we can choose a simple combination of the $U_q\mathbf{g}$-invariants I_q^n of section 4; the Hamiltonian is $U_q\mathbf{g}$-invariant and has a simple polynomial structure in the composite

[3]The idea that deformed excitations should consist of a compound of ordinary ones is not new, both for fermions and for bosons: see for instance [12].

operators A^i, A_j^+. H is also $\mathbf{g}$-invariant but has a higher degree polynomial structure (or more generally a non-polynomial structure if $\mathcal{A}$ has an infinite number of generators) in the undeformed generators a^i, a_j^+. This suggests that the use of the A^i, A_j^+ instead of the a^i, a_j^+ should simplify the resolution of the corresponding dynamics.

The results presented in the previous sections could in principle be applied to models in quantum field theory or condensed matter physics by choosing representations ρ which are the direct sum of many copies of the same fundamental representation ρ_d; this is what we have addressed in [4]. The different copies could correspond respectively to different space(time)-points or crystal sites.

REFERENCES

[1] V.G. Drinfel'd, On constant quasiclassical solutions of the Yang-Baxter quantum equation, *Sov. Math. Dokl.* **28** (1983), 667.

[2] V. G. Drinfeld, Quantum groups, in *Proceedings of the International Congress of Mathematicians,* Berkeley, 1986, Gleason, ed., Providence, (1987), 798. M. Jimbo, *Lett. Math. Phys.* (1985), 10, 63.

[3] V. G. Drinfeld, Quasi Hopf algebrae, *Leningrad Math. J.* **1** (1990), 1419.

[4] G. Fiore, Braided chains of q-deformed Heisenberg algebrae, *J. Phys. A* **31** (1998), 5289.

[5] G. Fiore, Drinfeld's twist and q-deforming maps for Lie group covariant Heisenberg algebras, e-print q-alg/9708017, to appear in *Rev. Math. Phys* **12 1** (January 2000).

[6] G. Fiore, Deforming maps for Lie group covariant creation and annihilation operators, *J. Math. Phys.* (1998), 39, 3437.

[7] G. Fiore, Q-deforming maps for Lie group covariant Heisenberg algebras, in the *Proceedings of the 5th Wigner Symposium,* Vienna, August 1997, ed. L. Kasperkowitz, Embedding q-deformed Heisenberg algebras into undeformed ones, *Rep. Math. Phys.*.

[8] L. D. Faddeev, N. Y. Reshetikhin, and L. A. Takhtajan, Quantization of Lie groups and Lie algebras, *Algebra I Analysis,* **1** (1989), 178; translation: *Leningrad Math. J.* **1** (1990), 193.

[9] M. Gerstenhaber, On the deformation of rings and algebrae, *Ann. Math.* **79** (1964), 59.

[10] T. Hayashi, Q-analogues of Clifford and Weyl algebras-spinor and oscillator representations of quantum enveloping algebras, *Commun. Math. Phys.* **127** (1990), 129.

[11] L. C. Biedenharn, *J. Phys.* **A22 L873**, (1989). A. J. Macfarlane, *J.Phys.* **A22** (1989), 4581. T. Hayashi, *Commun. Math. Phys.* **127** (1990), 129. M. Chaichian and P. Kulish, *Phys. Lett.* **B234** (1990), 72.

[12] S. Murakami, F. Gvhmann, Algebraic solution of the Hubbard Model on the infinite interval, *Nucl. Phys.* **B512** (1998), 637. F. Bonechi, E. Celeghini, R. Giachetti, E. Sorace, M. Tarlini, Quantum Galilei group as symmetry of magnons, *Phys. Rev.* **B46** (1992), 5727.

[13] C. Ohn, A $*$-Product on $SL(2)$ and the corresponding nonstandard quantum-$U(sl(2))$, *Lett. Math Phys.* **166** (1994), 63.

[14] W. Pusz, S. L. Woronowicz, Twisted second quantization, *Reports on Mathematical Physics* (1989), 27, 231.

[15] W. Pusz, Twisted canonical anticommutation relations, *Reports on Mathematical Physics* **27** (1989), 349.

[16] N. Yu. Reshetikhin, Multiparameter quantum groups and twisted quasitriangular Hopf algebrae, *Lett. Math. Phys.* **20** (1990), 331.

[17] T. L. Curtright, G. I. Ghandour, C. K. Zachos, Quantum algebra deforming maps, Clebsch-Gordan coefficients, coproducts, U and R Matrices, *J. Math. Phys.* **32** (1991), 676.

Gaetano Fiore
Dip. di Matematica e Applicazioni, Fac. di Ingegneria
Università di Napoli, V. Claudio 21, 80125 Napoli
I.N.F.N., Sezione di Napoli
Mostra d'Oltremare, Pad. 19, 80125 Napoli
E-mail: gfiore@na.infn.it

Received: October 16, 1999; Revised: November 11, 1999

Dirac Operator, Hopf Algebra of Renormalization, and Structure of Spacetime

Marcos Rosenbaum and J. David Vergara

ABSTRACT We analyze the part that noncommutative geometry could play in determining the structure of spacetime at the Planck length scale and the idea that the geometry of spacetime itself may be dictated by the renormalization processes in quantum field theories. Two recently discovered and intimately related Hopf algebras – the Hopf algebra for the computation of the local index formula of transversally hypoelliptic operators and the algebra of renormalization – lie behind these ideas.
Keywords: Noncommutative geometry, Hopf algebra, quantum field theory, renormalization, Dirac operator, Schwinger-Dyson equations, propagator, rooted trees.

1 Introduction

Gaussian and Riemannian geometric spaces are usually defined as manifolds where the metric is given by the geodesic distance

$$d_\gamma(x,y) = \inf_\gamma\{\text{length of paths } \gamma \text{ from } x \text{ to } y\}. \tag{1.1}$$

However, in order to algebraize the geometry and arrive at a formulation which can be extended to noncommutative spaces, Connes [1] has proposed, as a starting point, the following dualized form of (1.1):

$$d(x,y) = \sup\{|f(x) - f(y)| \mid f \in \mathcal{A}, \|\frac{df}{ds}\| \leq 1\}, \tag{1.2}$$

where $\mathcal{A}$ is the algebra of $C^\infty(M)$ functions over M, and ds is the line element in Riemannian geometry.

To measure distances in a possibly noncommutative space X, equation (1.2) is generalized by first introducing a Fredholm module $(\mathcal{H}, F)$ over the

AMS Subject Classification: 81T15, 81T75, 16W30.

involutive algebra $\mathcal{A}$. Here F is a selfadjoint involutive operator acting on the Hilbert space $\mathcal{H} = L^2(M, S)$ of square integrable sections of the irreducible spinor bundle over M (cf. [2] for an overview of some of the basic concepts associated with noncommutative geometry). The differential calculus is quantized by using the operator quantum theoretic notion for the differential

$$df = [F, f], \tag{1.3}$$

where $f \in \mathcal{A}$. One further specifies a metric structure on X by defining a "unit length" via an operator of the form

$$G = \sum_{\mu=1}^{q} (dx^\mu)^* g_{\mu\nu} (dx^\nu), \tag{1.4}$$

where x^μ are elements of $\mathcal{A}$, $dx = [F, x]$, and $g = g_{\mu\nu}$ $\mu, \nu = 1, ..., q$ is a positive element of the matrix algebra $M_q(\mathcal{A})$. Note that $G \in \mathcal{K}$ (the space of compact operators) and that it is a positive "infinitesimal" by construction. Thus, we can think of its positive square root as the line element of Riemannian geometry, i.e.,

$$G^{\frac{1}{2}} = ds. \tag{1.5}$$

Replacing now the points $x, y \in X$ by the corresponding pure states ϕ, ψ on the $C^\star$-algebra closure of $\mathcal{A}$ and using the evaluation map, by the Gel'fand-Naimark theorem, we get

$$\phi_x(a) = a(x), \quad \psi_y(a) = a(y), \quad \forall a \in \mathcal{A}, \tag{1.6}$$

together with the infinitesimal $da = [F, a]$, and also assuming that G commutes with F (thus avoiding operator ordering ambiguities) so that $dG = 0$, we can rewrite the basic formula (1.2) as

$$d(x, y) = \sup\{|a(x) - a(y)| \mid a \in \mathcal{A}, \ \|[\frac{F}{G^{\frac{1}{2}}}, a]\| \leq 1\}. \tag{1.7}$$

Let

$$D := \frac{F}{G^{\frac{1}{2}}} = F(ds)^{-1} \tag{1.8}$$

and note that this is a selfadjoint operator on $\mathcal{H}$ whose existence assumes that G is nonsingular. We can then reformulate (1.2) as

$$d(x, y) = \sup\{|a(x) - a(y)|; \ a \in \mathcal{A}, \ \|[D, a]\| \leq 1\}. \tag{1.9}$$

Note that by squaring equation (1.8) defining operator D and by making use of the properties of F, we get

$$D^2 = FG^{\frac{1}{2}} FG^{\frac{1}{2}} = F^2 G^{-1} = G^{-1}, \tag{1.10}$$

so

$$|D| = G^{-\frac{1}{2}}, \tag{1.11}$$

or

$$D = F|D|. \tag{1.12}$$

Hence, F is by construction the sign of D and since G is also given in terms of D, the information on the metric structure of our Fredholm module is contained in the selfadjoint unbounded operator D on $\mathcal{H}$. Therefore, it turns out more economical to take as our basic data the triple $(\mathcal{A}, \mathcal{H}, D)$.

Consequently, if (M, g) is a closed n-dimensional Riemannian spin manifold and $D = \gamma^\mu \nabla_\mu$ is the Dirac operator of a Clifford connection, where

$$\nabla_\mu \psi(x) = (\partial_\mu \psi^\alpha(x) + \psi^\beta(x) \Lambda^\alpha{}_{\mu\beta}) \mathbf{l}_\alpha, \tag{1.13}$$

$$\Lambda^\alpha{}_{\mu\beta} = \mathbf{l}^\alpha \blacktriangle \nabla_\mu \mathbf{l}_\beta, \tag{1.14}$$

"$\blacktriangle$" the symplectic spinor product and $\mathbf{l}_\alpha$ is an arbitrary basis for the Dirac spinors $(\psi(x) = \psi^\alpha(x)\mathbf{l}_\alpha)$, then the geodesic distance for the canonical triple $(\mathcal{A}, \mathcal{H}, D)$ is given by

$$d(\phi_x, \phi_y) = \sup_{a \in \mathcal{A}} \{|a(x) - a(y)|;\ \|[D, a]\| \leq 1\}. \tag{1.15}$$

Moreover, the Riemann measure on the structure space of the norm closure $\bar{\mathcal{A}}$ of the commutative C^*-algebra $\mathcal{A}$ is

$$\int_M a = c(n) Tr_\omega (a|D|^{-n}) \qquad \forall a \in \mathcal{A}, \tag{1.16}$$

where

$$c(n) = 2^{(n - [\frac{n}{2}] - 1)} n \pi^{\frac{n}{2}} \Gamma(\frac{n}{2}), \tag{1.17}$$

and the symbol Tr_ω denotes the Dixmier trace [3, 4].

There is an analogue of (1.15) for the natural distance function for a general spectral triple $(\mathcal{A}, \mathcal{H}, D)$ where $\mathcal{A}$ is noncommutative. It is

$$d(\phi, \chi) = \sup_{a \in \mathcal{A}} \{|\phi(a) - \chi(a)|;\ \|[D, a]\| \leq 1\}; \qquad \forall \phi, \chi \in \mathcal{S}(\bar{\mathcal{A}}), \tag{1.18}$$

where $\mathcal{S}(\bar{\mathcal{A}})$ is the space of states on the C^*-algebra $\bar{\mathcal{A}}$.

Because of the interplay of Quantum Mechanics and General Relativity at the Planck length scale, there are good reasons for believing that at such short distances the structure of spacetime may no longer be considered as a 4-dimensional continuum. It then becomes a major issue in the study of the short scale structure of spacetime, in the context of noncommutative geometry, to find the appropriate identification of the C^*-algebra $\mathcal{A}$ that would play the role of the gauge group in the noncommutative scenario.

We have seen that by dualizing Riemannian geometry, we had $ds = G^{\frac{1}{2}} = |D|^{-1}$; so the Dirac operator plays a central role in spectral geometry, and in Riemannian geometry the spin group is essential to define the Dirac operator. It may turn out that the quantum spin group will provide the required generalization to noncommutative geometry of the Dirac operator.

On the other hand, however radical these changes to our conception of the small structure of spacetime may seem, they may still turn out to be insufficient. Perhaps we should think instead that the geometry of spacetime is dictated by Quantum Field Theory. This intriguing idea has recently been pursued by Connes and Kreimer [8] and is based on the fact that, under appropriate boundary conditions, the inverse of the Dirac operator is a Feynman propagator, so D has the dual significance of providing the metric and also the fundamental class for the manifold under consideration. According to Connes and Kreimer [8], this emphasizes the fact that spacetime ought to be regarded as a derived concept whose structure should follow from the properties of Quantum Field Theory. They propose the Schwinger-Dyson equations as the possible starting point to recover the full propagator. A recent remarkable result which appears to give support to the above contention is the observation that two Hopf algebras, one of them discovered by Connes and Moscovici [5] in the context of noncommutative geometry and the other by Kreimer [6] [8] in the context of quantum field theory, are intimately related. In this second algebra, the antipode reproduces the combinatorics of renormalization by producing the precise local counterterms to the divergences in the Feynman diagrams, making them thus finite.

The Connes-Moscovici algebra $\mathcal{H}_T$ and its relation to Kreimer's Hopf Algebra of Renormalization $\mathcal{H}_R$ is the main subject of our discussion. Our presentation of the $\mathcal{H}_T$ algebra (Sections 2 and 3) differs from the one given in the original papers in that we follow an intrinsic, and basically coordinate free approach, as opposed to the local coordinate formulation of Connes and Moscovici, as well as that of Wulkenhaar [10] in his detailed reconstruction of the algebra. In addition to providing a more compact and fairly explicit elegant derivation, we feel our work will help clarify the more essential aspects of these important formalisms for those readers who are familiar with mathematical physics and differential geometric techniques, but who are not experts in noncommutative geometry. We do this without distracting the reader by side calculations with a proliferation of indices. In section 4 we discuss – rather summarily because of space limitations – the $\mathcal{H}_R$ algebra, its presentation in terms of rooted trees, and its relation to $\mathcal{H}_T$. For a more ample presentation of the subject, the reader is referred to the very clear introduction given in [10] and to the much more technical and complete works cited above.

2 Automorphisms on the frame bundle

In what follows, we shall be relying on essential concepts of differential geometry and using standard notation as in [11] or [12]. Thus, let M denote an n-dimensional smooth oriented manifold. An automorphism $\tilde{\psi} : P \to P$ on a principal fiber bundle (PFB) $\pi : P \to M$ is a diffeomorphism such that $\tilde{\psi}(R_g p) = R_g \tilde{\psi}(p)$ (i. e. $\tilde{\psi} \circ R_g = R_g \circ \tilde{\psi}$) $\forall\, g \in G,\ p \in P$. $\tilde{\psi}$ induces a well defined diffeomorphism $\psi : M \to M$ given by $\psi(\pi(p)) = \pi\tilde{\psi}(p)$.

Lemma 2.1. *If ω is a connection 1-form, then $\tilde{\psi}^*\omega$ is a connection 1-form.*

Proof. Let $A \in \mathcal{G}$ and let A^* be the fundamental field on P. Then

$$\begin{aligned}(\tilde{\psi}^*\omega)(A^*_p) &= \omega_{\tilde{\psi}(p)}(\tilde{\psi}_* A^*_p) = \omega_{\tilde{\psi}(p)}\left(\frac{d}{dt}\tilde{\psi}(p \exp tA)|_{t=0}\right) \\ &= \omega_{\tilde{\psi}(p)}\left(\frac{d}{dt}\tilde{\psi}(p) \exp tA|_{t=0}\right) = \omega_{\tilde{\psi}(p)}(A^*_{\tilde{\psi}(p)}) = A. \end{aligned} \tag{2.1}$$

Since $R_g \circ \tilde{\psi} = \tilde{\psi} \circ R_g$, we have

$$\begin{aligned} R^*_g \tilde{\psi}^*\omega &= (\tilde{\psi} \circ R_g)^*\omega = (R_g \circ \tilde{\psi})^*\omega \\ &= \tilde{\psi}^* R^*_g \omega = \tilde{\psi}^* a\delta_{g-1}\omega = a\delta_{g-1}\tilde{\psi}^*\omega. \end{aligned} \tag{2.2}$$

Let X_i be horizontal vector fields associated to ω, and let $\tilde{X}_i$ be new vector fields defined by

$$\tilde{X}_i = \tilde{\psi}_*^{-1} X_i. \tag{2.3}$$

Then

$$0 = \omega(X_i) = \omega(\tilde{\psi}_*\tilde{X}_i) = (\tilde{\psi}^*\omega)(\tilde{X}_i), \tag{2.4}$$

i.e., $\tilde{X}_i$ is horizontal with respect to $\tilde{\psi}^*\omega$. □

Now let our PFB be a frame bundle $\pi : L(M) \to M$ with group $GL(n)$. The canonical 1-form on $L(M)$ is the $\mathbb{R}^n$-valued form $\varphi \in \Lambda^1(L(M))$ defined by $\varphi(W_u) = u^{-1}(\pi_* W_u)$ for $W_u \in T_u L(M)$ and $u : \mathbb{R}^n \to T_x M$, with $\pi(u) = x$, so u is a frame $u \in L(M)_x$. Furthermore, since in this case

$$\tilde{\psi} = \psi_{\pi(u)*} \circ u \ : \ \mathbb{R}^n \to T_{\psi(\pi(u))}M, \tag{2.5}$$

we obtain

$$\begin{aligned}(\tilde{\psi}^*\varphi)_p(W_p) &= \varphi_{\tilde{\psi}(p)}(\tilde{\psi}_* W_p) - (\psi_* \circ u)^{-1}\pi_*\tilde{\psi}_* W_p \\ &= u^{-1}(\psi_*^{-1}\psi_* X_x) = u^{-1}(\pi_* W_p) = \varphi_p(W_p), \end{aligned} \tag{2.6}$$

with $p = u$. Hence the canonical 1-form is invariant under diffeomorphisms.

3 The Hopf algebra of Connes and Moscovici

Consider now a basis of horizontal and vertical vector fields $\{X_i\}, \{Y_k^{j*}\}$, respectively. Here Y_k^j is a basis of the Lie algebra $\mathfrak{gl}(n)$, so the connection ω can be written as $\omega = \omega_j^k Y_k^j$, i.e., ω is a matrix valued 1-form, and Y_k^j is a matrix with zeros everywhere except for a 1 in the k-th row and j-th column. Y_k^{j*} is a basis of fundamental vectors tangent to the fiber. Let Γ be the pseudogroup of local diffeomorphisms on M which preserve orientation. Consider further the semidirect product algebra

$$\mathcal{A} = C_c^\infty(F^+) \rtimes \Gamma, \tag{3.1}$$

where $C_c^\infty(F^+)$ is of compact support on the frame bundle F^+. As a set this algebra $\mathcal{A}$ can be regarded as the tensor product of $C_c^\infty(F^+)$ with Γ, and is generated by monomials of the form fU_ψ^* with $f \in C_c^\infty(\mathrm{Dom}(\tilde{\psi}))$, $\psi \in \Gamma$. Here U^* denotes the right action of the pseudogroup, following the notation in [5]. As an algebra, $\mathcal{A}$ has the multiplication rule

$$f_1 U_{\psi_1}^* f_2 U_{\psi_2}^* := f_1(f_2 \circ \tilde{\psi}_1) U_{\psi_2\psi_1}^*, \tag{3.2}$$

where

$$f_1(f_2 \circ \tilde{\psi}_1)(p) = f_1(p) f_2(\tilde{\psi}_1(p)) \in \mathbb{R}(\text{ or } \mathbb{C}) \tag{3.3}$$

and

$$U_{\psi_1}^* U_{\psi_2}^* = U_{\psi_2\psi_1}^*. \tag{3.4}$$

Note that

$$(f_1 U_{\psi_1}^* f_2 U_{\psi_2}^*) f_3 U_{\psi_3}^* = ((f_1(f_2 \circ \tilde{\psi}_1))(f_3 \circ (\tilde{\psi}_2 \tilde{\psi}_1)) U_{\psi_3\psi_2\psi_1}^*, \tag{3.5}$$

and

$$f_1 U_{\psi_1}^* (f_2 U_{\psi_2}^* f_3 U_{\psi_3}^*) = f_1((f_2(f_3 \circ \tilde{\psi}_2)) \circ \tilde{\psi}_1) U_{\psi_3\psi_2\psi_1}^*. \tag{3.6}$$

Comparing equations (3.5) and (3.6), we get the associative rule of $\mathcal{A}$:

$$(f_1(f_2 \circ \tilde{\psi}_1))(f_3 \circ (\tilde{\psi}_2 \tilde{\psi}_1))(p) = f_1((f_2(f_3 \circ \tilde{\psi}_2)) \circ \tilde{\psi}_1)(p). \tag{3.7}$$

The action of X_i and Y_k^{j*} on the monomials generating the algebra is defined by

$$X_i(fU_\psi^*) = X_i(f) U_\psi^*, \tag{3.8}$$

$$Y_k^{j*}(fU_\psi^*) = Y_k^{j*}(f) U_\psi^*. \tag{3.9}$$

For any vector field W on F^+, we have

$$\begin{aligned} W_p(f_1 U^*_{\psi_1} f_2 U^*_{\psi_2}) &= W_p(f_1(f_2 \circ \tilde{\psi}_1)) U^*_{\psi_2\psi_1} \\ &= \left[W_p(f_1)(f_2 \circ \tilde{\psi}_1)(p) + f_1(p) W_p(f_2 \circ \tilde{\psi}_1) \right] U^*_{\psi_2\psi_1} \\ &= W_p(f_1) U^*_{\psi_1} f_2(p) U^*_{\psi_2} + f_1(p) W_p(f_2 \circ \tilde{\psi}_1) U^*_{\psi_2\psi_1}, \quad (3.10) \end{aligned}$$

where in the last step we have used (3.2). Moreover, from the definition of differential of a mapping $\tilde{\psi}_{1*} W_p(f_2) := W_p(f_2 \circ \tilde{\psi}_1)$, we get

$$\begin{aligned} W_p(f_1 U^*_{\psi_1} f_2 U^*_{\psi_2}) &= W_p(f_1) U^*_{\psi_1} f_2 U^*_{\psi_2} + f_1(p) \tilde{\psi}_{1*} W_p(f_2) U^*_{\psi_2\psi_1} \\ &= W_p(f_1) U^*_{\psi_1} f_2 U^*_{\psi_2} + \\ &\qquad f_1(p) U^*_{\psi_1} U^*_{\psi_1^{-1}} \tilde{\psi}_{1*} W_p(f_2) U^*_{\psi_2\psi_1}. \quad (3.11) \end{aligned}$$

Writing the last term in (3.11) as

$$\begin{aligned} f_1(p) U^*_{\psi_1} U^*_{\psi_1^{-1}} \tilde{\psi}_{1*} W_p(f_2) U^*_{\psi_2\psi_1} &= f_1(p) U^*_{\psi_1} U^*_{\psi_1^{-1}} ((\tilde{\psi}_{1*} W) f_2)_{\tilde{\psi}_1(p)} U^*_{\psi_2\psi_1} \\ &= f_1(p) U^*_{\psi_1} ((\tilde{\psi}_{1*} W) f_2)_p U^*_{\psi_2} \\ &= f_1(p) U^*_{\psi_1} \tilde{\psi}_{1*} W_{\tilde{\psi}_1^{-1}} f_2(p) U^*_{\psi_2}, \quad (3.12) \end{aligned}$$

we arrive at

$$\begin{aligned} W_p(f_1 U^*_{\psi_1} f_2 U^*_{\psi_2}) &= W_p(f_1) U^*_{\psi_1} f_2(p) U^*_{\psi_2} + \\ &\qquad f_1 U^*_{\psi_1} (\tilde{\psi}_{1*} W_{\tilde{\psi}^{-1}}) f_2(p) U^*_{\psi_2}. \quad (3.13) \end{aligned}$$

Applying (3.13) to the fundamental vertical fields Y_k^{j*} gives

$$Y_k^{j*}(ab)|_p = Y_k^{j*}(a) b|_p + a_p \left(\tilde{\psi}_* Y^{j*}_{k\,\tilde{\psi}^{-1}} \right) (b)|_p\,; \quad (3.14)$$

and since

$$\tilde{\psi}_* Y^{j*}_{k\,\tilde{\psi}^{-1}(p)} = Y^{j*}_{k\,\tilde{\psi}(\tilde{\psi}^{-1})}|_p = Y_k^{j*}|_p, \quad (3.15)$$

we obtain

$$Y_k^{j*}(ab)|_p = Y_k^{j*}(a) b|_p + a Y_k^{j*}(b)|_p. \quad (3.16)$$

Consider now horizontal fields in (3.13). We have

$$\begin{aligned} X_i(f_1 U^*_{\psi_1} f_2 U^*_{\psi_2})|_p &= X_i(f_1 U^*_{\psi_1}) f_2 U^*_{\psi_2}|_p \\ &\qquad + f_1 U^*_{\psi_1} (\tilde{\psi}_{1*} (X_i)_{\tilde{\psi}_1^{-1}}) f_2 U^*_{\psi_2}|_p, \quad (3.17) \end{aligned}$$

and recalling that (see (2.3))

$$\tilde{X}_i|_p = \tilde{\psi}_{1*}(X_i|_{\tilde{\psi}_1^{-1}(p)}) = (\tilde{\psi}_{1*} X_i)_p, \quad (3.18)$$

where $\tilde{X}_i|_p$ is horizontal to $((\tilde{\psi}_1^{-1})^*\omega)_p$, we can write

$$\tilde{X}_i|_p = \tilde{\psi}_{1*}X_i|_{\tilde{\psi}_1^{-1}(p)} = X_i|_p + (\omega_j^k(\tilde{X}_i)Y_k^{j^*})_p. \tag{3.19}$$

The validity of this last expression can be established by first acting with the connection 1-form on both sides of (3.19) to obtain

$$\begin{aligned}\omega_p(\tilde{X}_{ip}) &= \omega_p(X_{ip}) + \omega_j^k(\tilde{X}_{ip})\omega_m^l(Y_k^{j^*})Y_l^m \\ &= 0 + \omega_j^k(\tilde{X}_{ip})\delta_k^l\delta_m^j Y_l^m \\ &= \omega_j^k(\tilde{X}_i)|_p Y_k^j,\end{aligned} \tag{3.20}$$

which is an identity. Also, since

$$\begin{aligned}\varphi_p(\tilde{X}_{ip}) &= \varphi_p(\tilde{\psi}_*X_i) = (\tilde{\psi}^*\varphi)_{\tilde{\psi}^{-1}(p)}(X_i) = \varphi_{\tilde{\psi}^{-1}(p)}(X_i) \\ &= (\psi_*^{-1}\circ u)^{-1}(\pi_*X_i|_{\tilde{\psi}^{-1}(p)}) = u^{-1}\circ\psi_*\pi_*X_i|_{\psi^{-1}(p)} \\ &= u^{-1}(\psi_*\psi_*^{-1}(u(e_i)) = e_i = \varphi_p(X_{ip}),\end{aligned} \tag{3.21}$$

the X_{ip} is the horizontal component of $\tilde{X}_{ip}$. Thus, the r.h.s. of (3.19) is a unique splitting of $\tilde{X}_i|_p$. Furthermore, if we denote

$$\omega_k^j(\tilde{X}_i)|_p := \hat{\gamma}_{ki}^j|_p^{(\psi)}, \tag{3.22}$$

then (3.17) becomes

$$\begin{aligned}X_i(f_1U_{\psi_1}^* f_2U_{\psi_2}^*)|_p = X_i(f_1U_{\psi_1}^*)f_2U_{\psi_2}^*|_p + f_1U_{\psi_1}^* X_i(f_2U_{\psi_2}^*)|_p \\ + \hat{\gamma}_{ki}^j|_{\tilde{\psi}_1(p)}^{(\psi_1)} f_1U_{\psi_1}^* Y_j^{k^*}(f_2U_{\psi_2}^*)|_p.\end{aligned} \tag{3.23}$$

Defining now the operator

$$\delta_{ki}^j(fU_\psi^*) := (\hat{\gamma}_{ki}^j|_{\tilde{\psi}(p)}^{(\psi)} f)U_\psi^*, \tag{3.24}$$

we arrive at

$$X_i(ab)|_p = X_i(a)b|_p + aX_i(b)|_p + \delta_{ki}^j(a)Y_j^{k^*}(b)|_p. \tag{3.25}$$

We next show that δ_{ki}^j is a derivation operator. To this end, note that by (3.19)

$$(\tilde{\psi}_2\circ\tilde{\psi}_1)_*(X_i)_{\tilde{\psi}_1^{-1}\circ\tilde{\psi}_2^{-1}(p)} = X_i|_p + \hat{\gamma}_{ki}^j|_p^{(\psi_2\psi_1)}Y_j^{k^*}|_p. \tag{3.26}$$

On the other hand,

$$\begin{aligned}(\tilde{\psi}_2\circ\tilde{\psi}_1)_*(X_i)_{\tilde{\psi}_1^{-1}\circ\tilde{\psi}_2^{-1}(p)} &= \tilde{\psi}_{2*}(\tilde{\psi}_{1*}X_{i(\tilde{\psi}_1^{-1}\circ\tilde{\psi}_2^{-1}(p))}) \\ &= \tilde{\psi}_{2*}(X_{i\tilde{\psi}_2^{-1}(p)} + \hat{\gamma}_{ki}^j|_{\tilde{\psi}_2^{-1}(p)}^{(\psi_1)} Y_j^{k^*}{}_{\tilde{\psi}_2^{-1}(p)}) \\ &= X_{ip} + \hat{\gamma}_{ki}^j|_p^{\psi_2}Y_j^{k^*} + \hat{\gamma}_{ki}^j|_{\tilde{\psi}_2^{-1}(p)}^{(\psi_1)} Y_j^{k^*}|_p.\end{aligned} \tag{3.27}$$

Hence,

$$\hat{\gamma}^j_{ki}|^{(\psi_2\psi_1)}_p = \hat{\gamma}^j_{ki}|^{(\psi_2)}_p + \hat{\gamma}^j_{ki}|^{(\psi_1)}_{\tilde{\psi}_2^{-1}(p)}. \tag{3.28}$$

Moreover, from (3.24) and (3.28)

$$\begin{aligned}\delta^j_{ki}(ab)|_{(p)} &= \big(\hat{\gamma}^j_{ki}|^{\psi_2\psi_1}_{\tilde{\psi}_2\circ\tilde{\psi}_1}(f_1f_2\circ\tilde{\psi}_1)\big)U^*_{\psi_2\psi_1}|_p \\ &= \big(\hat{\gamma}^j_{ki}|^{(\psi_1)}_{\tilde{\psi}_1(p)}(f_1f_2\circ\tilde{\psi}_1)\big)U^*_{\psi_2\psi_1}|_p \\ &\quad + \big(\hat{\gamma}^j_{ki}|^{(\psi_2)}_{\tilde{\psi}_2\circ\tilde{\psi}_1(p)}(f_1f_2\circ\tilde{\psi}_1)\big)U^*_{\psi_2\psi_1}|_p \\ &= \big(\hat{\gamma}^j_{ki}|^{(\psi_1)}_{\tilde{\psi}_1(p)}f_1\big)U^*_{\psi_1}f_2U^*_{\psi_2}|_p + f_1U^*_{\psi_1}|_p\big(\hat{\gamma}^j_{ki}|^{(\psi_2)}_{\tilde{\psi}_2(p)}f_2\big)U^*_{\psi_2}|_p,\end{aligned} \tag{3.29}$$

which shows that δ^j_{ki} indeed satisfies the Leibniz rule

$$\delta^j_{ki}(ab) = (\delta^j_{ki}a)b + a(\delta^j_{ki}b). \tag{3.30}$$

3.1 Commutation relations

We shall now investigate the commutation relations of the vector fields X_i, Y^{k*}_j, and δ^j_{ki}. Recall first that (cf. [12])

$$[Y^{j\,*}_k, Y^{l\,*}_m] = [Y^j_k, Y^l_m]^*\,; \tag{3.31}$$

and also note that, since $(Y^j_k)^\alpha_\beta = \delta^\alpha_k\delta^j_\beta$,

$$[Y^j_k, Y^l_m]^\alpha_\gamma = \delta^j_m\delta^\alpha_k\delta^l_\gamma - \delta^l_k\delta^\alpha_m\delta^j_\gamma = \delta^j_m(Y^l_k)^\alpha_\gamma - \delta^l_k(Y^j_m)^\alpha_\gamma. \tag{3.32}$$

Consequently,

$$[Y^{j\,*}_k, Y^{l\,*}_m] = \delta^j_m Y^{l*}_k - \delta^l_k Y^{j*}_m. \tag{3.33}$$

Moreover, since

$$\begin{aligned}\varphi(R_{A*}X_{iu}) &= \varphi(X_{i(uA)}) = (uA)^{-1}(\pi_*X_{iuA}) \\ &= A^{-1}\circ u^{-1}(\pi_*X_i) = A^{-1}\varphi(X_{iu}),\end{aligned} \tag{3.34}$$

we have

$$R_{A*}X_i = u(A^{-1}e_i), \tag{3.35}$$

after making use of the notation $\pi_*X_i = u(e_i)$. Thus,

$$\begin{aligned}[Y^{j\,*}_k, u(e_i)] = \mathcal{L}_{Y^{j*}_k}(u(e_i)) &= \lim_{t\to 0}\frac{1}{t}[u(e_i) - R_{\exp(tY^j_k)*}u(e_i)] \\ &= u\lim_{t\to 0}[e_i - \exp(-tY^j_k)e_i] \\ &= u(Y^j_k e_i) = \delta^j_i u(e_k),\end{aligned} \tag{3.36}$$

from where it clearly follows that

$$[Y_k^{j\,*}, X_i] = \delta_i^j X_k. \tag{3.37}$$

As the next step in deriving the commutation relations for the vector fields X_i, Y_j^{k*}, and δ_{ki}^j, we calculate $[Y_m^{l\,*}, \delta_{ki}^j]$. This follows by first acting on (3.25) with $Y_m^{l\,*}$ on the left and using (3.16). We get

$$\begin{aligned}(Y_m^{l\,*} X_i)(ab) = (Y_m^{l\,*} X_i(a))b + X_i(a) Y_m^{l\,*}(b) + (Y_m^{l\,*} a) X_i(b) +& \\ a Y_m^{l\,*}(X_i(b)) + (Y_m^{l\,*} \delta_{ki}^j)(a) Y_j^{k*}(b) + \delta_{ki}^j(a) Y_j^{k*} Y_m^{l\,*} b.&\end{aligned} \tag{3.38}$$

Now acting on (3.16) with X_i from the left, we obtain

$$\begin{aligned}(X_i Y_m^{l\,*})(ab) = (X_i Y_m^{l\,*} a)b + (Y_m^{l\,*} a)(X_i b) + \delta_{ki}^j (Y_m^{l\,*} a) Y_j^{k*}(b) +& \\ (X_i a) Y_m^{l\,*} b + a(X_i Y_m^{l\,*} b) + \delta_{ki}^j(a) Y_j^{k*} Y_m^{l\,*} b.&\end{aligned} \tag{3.39}$$

Subtracting (3.39) from (3.38) results in

$$\begin{aligned}[Y_m^{l\,*}, X_i](ab) = ([Y_m^{l\,*}, X_i]a)b + a[Y_m^{l\,*}, X_i]b& \\ + ([Y_m^{l\,*}, \delta_{ki}^j]a) Y_j^{k*} b + \delta_{ki}^j(a)[Y_m^{l\,*}, Y_j^{k*}]b.&\end{aligned} \tag{3.40}$$

Using now (3.33), (3.37), (3.25) and combining terms, we get

$$\delta_i^l \delta_{pm}^n(a) Y_n^{p*}(b) - \delta_{ki}^l Y_m^{k*}(b) + \delta_{mi}^j(a) Y_j^{l*}(b) = ([Y_m^{l\,*}, \delta_{ki}^j]a) Y_j^{k*} b, \tag{3.41}$$

which we can rewrite as

$$((\delta_i^l \delta_{km}^j - \delta_m^j \delta_{ki}^l + \delta_k^l \delta_{mi}^j)a) Y_j^{k*} b = ([Y_m^{l*}, \delta_{ki}^j]a) Y_j^{k*} b. \tag{3.42}$$

Consequently,

$$[Y_m^{l\,*}, \delta_{ki}^j] = \delta_i^l \delta_{km}^j - \delta_m^j \delta_{ki}^l + \delta_k^l \delta_{mi}^j. \tag{3.43}$$

We also need the commutator $[X_i, X_j]$. This can be derived by recalling that for a Levi-Civita connection the torsion Θ vanishes, so

$$\begin{aligned}0 = \Theta(X_i, X_j) &= d\varphi(X_i, X_j) \\ &= X_i[\varphi(X_j)] - X_j[\varphi(X_i)] - \varphi([X_i, X_j]) \\ &= -\varphi([X_i, X_j]).\end{aligned} \tag{3.44}$$

Thus, for a Levi-Civita connection

$$[X_i, X_j]^H = 0. \tag{3.45}$$

The above implies that we can write

$$[X_i, X_j] = \rho_{lij}^m Y_m^{l\,*}. \tag{3.46}$$

Using now this expression in the definition of the curvature Ω, we get

$$\begin{aligned}\Omega(X_i, X_j) &= d\omega(X_i, X_j)\\ &= X_i(\omega(X_j)) - X_j(\omega(X_i)) - \omega([X_i, X_j])\\ &= -\omega([X_i, X_j]) = -\rho^m_{lij} Y^l_m. \end{aligned} \tag{3.47}$$

However,

$$\begin{aligned}R(\bar{X}_i, \bar{X}_j)\bar{X}_k = u(\Omega(X_i, X_j)(u^{-1}\bar{X}_k)) &= -\rho^m_{lij} u(Y^l_m \varphi(X_k))\\ &= -\rho^m_{lij} u(\delta^l_k e_m), \end{aligned} \tag{3.48}$$

so

$$R^m_{\ kij} \bar{X}_m = -\rho^m_{kij} \bar{X}_m, \tag{3.49}$$

and, therefore, the commutator for two horizontal vector fields is

$$[X_i, X_j] = -R^m_{\ kij} Y^{k*}_m. \tag{3.50}$$

If we impose the additional restriction of a flat connection (as shown in [5]) this restriction can be bypassed by replacing $\mathcal{A}$ by a Morita strong equivalent algebra), then clearly

$$[X_i, X_j] = 0. \tag{3.51}$$

Finally, let us define

$$\delta^i_{jk,l} := [X_l, \delta^i_{jk}], \tag{3.52}$$

and

$$\delta^i_{jk,l_1,\ldots,l_n} := [X_{l_n}, \ldots, [X_{l_1}, \delta^i_{jk}]]. \tag{3.53}$$

Since $\hat{\gamma}^j_{ki}$ in δ^j_{ik} is just a multiplicative factor, all the δ's given by (3.53) commute with each other. Hence, the linear space generated by X_i, $Y^{k^*}_j$, $\delta^i_{jk,l_1,\ldots,l_n}$ forms a Lie algebra.

3.2 The Hopf algebra $\mathcal{H}_T$

We now have all the ingredients needed to construct the Hopf algebra of Connes-Moscovici. Thus, first construct the enveloping algebra $\mathcal{H}_T$ as the unital algebra of polynomials in X_i, $Y^{k^*}_j$, $\delta^i_{jk,l_1,\ldots,l_n}$. By the Poincaré-Birkhoff-Witt theorem, a basis for $\mathcal{H}_T$ is

$$\{X_{i_1} \ldots X_{i_k} Y^{k_1*}_{m_1} \ldots Y^{k_j*}_{m_j} \delta^{i_1}_{j_1 p_1} \ldots \delta^{i_s}_{j_s p_s} \delta^{i_1}_{j_1 p_1, l_1} \ldots \delta^{i_r}_{j_r p_r, l_r} \ldots\}. \tag{3.54}$$

Observe now that elements in the linear space $\mathcal{B}$, generated by $\mathbf{1}, X_i, Y_j^{k*}$ and δ^i_{jk} through the multiplication in the algebra $\mathcal{A}$, together with (3.25), (3.16), and (3.30), induce the additional coproduct structure

$$\begin{aligned}\Delta(X_i) &= X_i \otimes \mathbf{1} + \mathbf{1} \otimes X_i + \delta^j_{ki} \otimes Y_j^{k*},\\ \Delta(Y_j^{k*}) &= Y_j^{k*} \otimes \mathbf{1} + \mathbf{1} \otimes Y_j^{k*},\\ \Delta(\delta^j_{ki}) &= \delta^j_{ki} \otimes \mathbf{1} + \mathbf{1} \otimes \delta^j_{ki},\\ \Delta(\mathbf{1}) &= \mathbf{1} \otimes \mathbf{1}.\end{aligned} \tag{3.55}$$

While the inclusion map, $\eta : \mathbb{C} \to \mathcal{A}$ induces the evaluation (counit) map ϵ given by

$$\epsilon(\mathbf{1}) = 1, \quad \epsilon(X_i) = \epsilon(Y_j^{k*}) = \epsilon(\delta^i_{jk}) = 0. \tag{3.56}$$

Moreover, for $\mathcal{B}$ the associativity of $\mathcal{A}$ implies the coassociativity axiom for Δ expressed by

$$(\Delta \otimes id) \circ \Delta = (id \otimes \Delta) \circ \Delta, \tag{3.57}$$

and also the counit ϵ satisfies the counit axiom expressed by

$$(id \otimes \epsilon) \circ \Delta = (\epsilon \otimes id) \circ \Delta = id. \tag{3.58}$$

In this way, $\mathcal{B}$ becomes a coalgebra.

Next we extend the coproduct (3.55) and counit (3.56) to elements in $\mathcal{H}_T$ by means of the connection axiom

$$\begin{aligned}\Delta(h_1 h_2) = \Delta(h_1)\Delta(h_2) = (h_1^1 \otimes h_1^2)(h_2^1 \otimes h_2^2) = h_1^1 h_2^1 \otimes h_1^2 h_2^2,\\ \epsilon(h_1 h_2) = \epsilon(h_1)\epsilon(h_2),\end{aligned} \tag{3.59}$$

for $h_1, h_2 \in \mathcal{H}_T$. Note, in particular, that from (3.59) we obtain

$$\begin{aligned}\Delta(\delta^i_{jk,l}) &= \Delta([X_l, \delta^i_{jk}]) = [\Delta X_l, \Delta\delta^i_{jk}]\\ &= [X_l \otimes \mathbf{1} + \mathbf{1} \otimes X_l + \delta^m_{nl} \otimes Y_m^{n*}, \delta^i_{jk} \otimes \mathbf{1} + \mathbf{1} \otimes \delta^i_{jk}]\\ &= \delta^i_{jk,l} \otimes \mathbf{1} + \mathbf{1} \otimes \delta^i_{jk,l} + \delta^m_{nl} \otimes [Y_m^{n*}, \delta^i_{jk}].\end{aligned} \tag{3.60}$$

Similarly, for the case of two derivatives, we get

$$\begin{aligned}
\Delta(\delta^i_{jk,lm}) &= \Delta[X_m, \delta^i_{jk,l}] = [\Delta X_m, \Delta\delta^i_{jk,l}] \\
&= [X_m \otimes \mathbf{1} + \mathbf{1} \otimes X_m + \delta^p_{qm} \otimes Y^{q*}_p, \delta^i_{jk,l} \otimes \mathbf{1} + \\
&\quad \mathbf{1} \otimes \delta^i_{jk,l} + \delta^r_{sl} \otimes [Y^{s*}_r, \delta^i_{jk}]] \\
&= \delta^i_{jk,lm} \otimes \mathbf{1} + \mathbf{1} \otimes \delta^i_{jk,lm} + \delta^r_{sl,m} \otimes [Y^{s*}_r, \delta^i_{jk}] + \\
&\quad \delta^r_{sl} \otimes [X_m, [Y^{s*}_r, \delta^i_{jk}]] + \delta^p_{qm} \otimes [Y^{q*}_p, \delta^i_{jk,l}] + \\
&\quad \delta^p_{qm}\delta^r_{sl} \otimes [Y^{q*}_p, [Y^{s*}_r, \delta^i_{jk}]] \\
&= \delta^i_{jk,lm} \otimes + \mathbf{1} \otimes \delta^i_{jk,lm} + \\
&\quad [X_m \otimes \mathbf{1} + \mathbf{1} \otimes X_m + \delta^p_{qm} \otimes Y^{q*}_p, \delta^r_{sl} \otimes [Y^{s*}_r, \delta^i_{jk}]] + \\
&\quad \delta^p_{qm} \otimes [Y^{q*}_p, \delta^i_{jk,l}].
\end{aligned} \tag{3.61}$$

Iterating these last two results, one can verify that

$$\Delta\delta^i_{jk,l_1\dots,l_n} = \delta^i_{jk,l_1,\dots,l_n} \otimes \mathbf{1} + \mathbf{1} \otimes \delta^i_{jk,l_1,\dots,l_n} + \mathcal{R}_{l_1,\dots,l_n}, \tag{3.62}$$

where

$$\begin{aligned}
\mathcal{R}_{l_1,\dots,l_n} = [X_{l_n} \otimes \mathbf{1} + \mathbf{1} \otimes X_{l_n} + \delta^{p_n}_{q_n l_n} \otimes Y^{q_n *}_{p_n}, \mathcal{R}_{l_1,\dots,l_{n-1}}] \\
+ \delta^{p_n}_{q_n l_n} \otimes [Y^{q_n *}_{p_n}, \delta^i_{jk,l_1,\dots,l_n}].
\end{aligned} \tag{3.63}$$

Using the above results, it is easily seen that (3.57) and (3.58) are verified for all $\mathcal{H}_T$. Therefore, $\mathcal{H}_T$ is a bialgebra.

Finally, to make $\mathcal{H}_T$ into a Hopf algebra we need to shown that there exists an antipode S on $\mathcal{H}_T$ that satisfies the antipode axioms

$$\begin{aligned}
S(h_1 h_2) &= S(h_2)S(h_1), \\
\mathbf{1}\epsilon &= m \circ (S \otimes \mathrm{id}) \circ \Delta, \\
\mathbf{1}\epsilon &= m \circ (\mathrm{id} \otimes S) \circ \Delta.
\end{aligned} \tag{3.64}$$

It is easy to verify that

$$\begin{aligned}
S(\mathbf{1}) &:= \mathbf{1}, \\
S(Y^{k*}_l) &:= -Y^{k*}_l, \\
S(\delta^i_{jk}) &:= -\delta^i_{jk}, \\
S(X_i) &:= -X_i + \delta^l_{ki} Y^{k*}_l
\end{aligned} \tag{3.65}$$

satisfy the axioms (3.64) for elements in the vector space $\mathcal{B}$.

To obtain $S(\delta^i_{jk,l})$, we make use of the antihomomorphism given by the first equation in (3.64). We thus get

$$\begin{aligned} S(\delta^i_{jk,l}) &= S[X_l, \delta^i_{jk}] = [S(\delta^i_{jk}), S(X_l)] \\ &= [-\delta^i_{jk}, -X_l + \delta^p_{ql} Y^{q*}_p] = -\delta^i_{jk,l} + [\delta^p_{ql} Y^{q*}_p, \delta^i_{jk}] \\ &= -\delta^i_{jk,l} + \delta^p_{ql}[Y^{q*}_p, \delta^i_{jk}] \\ &= -\delta^i_{jk,l} + \delta^p_{jl}\delta^i_{pk} + \delta^p_{kl}\delta^i_{jp} - \delta^i_{ql}\delta^q_{jk}. \end{aligned} \tag{3.66}$$

Note that

$$\begin{aligned} m \circ (S \otimes \mathrm{id}) \circ \Delta(\delta^i_{jk,l}) &= m \circ (S \otimes \mathrm{id})\big(\delta^i_{jk,l} \otimes \mathbf{1} + \mathbf{1} \otimes \delta^i_{jk,l} \\ &\qquad + \delta^m_{nl} \otimes [Y^{n*}_m, \delta^i_{jk}]\big) \\ &= S(\delta^i_{jk,l}) + \delta^i_{jk,l} + S(\delta^m_{nl})[Y^{n*}_m, \delta^i_{jk}] \\ &= -\delta^i_{jk,l} + \delta^p_{ql}[Y^{q*}_p, \delta^i_{jk}] + \delta^i_{jk,l} \\ &\qquad - \delta^m_{nl}[Y^{n*}_m, \delta^i_{jk}] \\ &= 0 = \epsilon(\delta^i_{jk,l}), \end{aligned} \tag{3.67}$$

so $\delta^i_{jk,l}$ satisfies the second axiom in (3.64) and, similarly, can be shown for the third axiom.

In order to arrive at a general expression for $\delta^i_{jk,l_1,\ldots,l_n}$, observe that from (3.62) we get

$$\begin{aligned} 0 &= \epsilon(\delta^i_{jk,l_1,\ldots,l_n}) \\ &= m \circ (S \otimes \mathrm{id}) \circ \Delta(\delta^i_{jk,l_1,\ldots,l_n}) \\ &= m \circ (S \otimes \mathrm{id})(\delta^i_{jk,l_1,\ldots,l_n} \otimes \mathbf{1} + \mathbf{1} \otimes \delta^i_{jk,l_1,\ldots,l_n} + \mathcal{R}_{l_1,\ldots,l_n}) \\ &= S(\delta^i_{jk,l_1,\ldots,l_n}) + \delta^i_{jk,l_1,\ldots,l_n} + m \circ (S \otimes \mathrm{id})(\mathcal{R}_{l_1,\ldots l_n}). \end{aligned} \tag{3.68}$$

This last equation implies that

$$S(\delta^i_{jk,l_1,\ldots,l_n}) = -\delta^i_{jk,l_1,\ldots,l_n} - m \circ (S \otimes \mathrm{id})(\mathcal{R}_{l_1,\ldots l_n}). \tag{3.69}$$

With these results it is now straightforward to verify that the whole enveloping algebra $\mathcal{H}_T$ satisfies the antipode axioms (3.64) and that it is, consequently, a Hopf algebra.

4 The Hopf algebra $\mathcal{H}_R$ of renormalization

Even though renormalization has been quite successful in curing the problems of infinities in perturbative quantum field theories of submicroscopic particle physics above the Planck length, there is a common feeling among many physicists that this procedure appears as an *ad hoc* artifact which exhibits the present lack of understanding of the theory. The intricacy of the calculations, the lack of an obvious mathematical structure, and the seemingly "miraculous" properties of the renormalization group have made this formalism practically inaccessible to mathematicians. These perceptions, however, may be changed dramatically due to the discovery by Kreimer [6] that the antipode in a very specific Hopf algebra of decorated rooted trees lies behind the mechanism of the actual computations of radiative corrections. The essential idea behind this approach resides in the observation that Feynman diagrams can be grouped into classes, where each class has similar subgraphs. Each divergent subgraph conforms a forest, and the forest formula provides a renormalization procedure for each subgraph. Thus, a given subgraph in a Feynman graph Γ can be replaced by an expression of the form

$$R[Z[\gamma_i]\Gamma/\gamma_i], \tag{4.1}$$

where γ_i is a non-overlapping subdivergence, Γ/γ_i is the expression resulting form shrinking to a point the γ_i in Γ, and the symbol R denotes a particular renormalization scheme. A Feynman graph Γ is then a sum of expressions of the above form.

In other words, renormalization is an operation which replaces a graph Γ by products of graphs, and then all possible subgraphs are summed over. More specifically, the forest formula constructs an expression $\bar{\Gamma}$ which has all its subdivergences renormalized. From this quantity, one obtains a counterterm expression $Z[\bar{\Gamma}]$ (which is at most a polynomial in the external parameters) such that $\bar{\Gamma} + Z[\bar{\Gamma}]$ is finite. Both Γ and its corresponding counterterms have the form of sums of (4.1). Moreover, the finiteness of $\bar{\Gamma}+Z[\bar{\Gamma}]$ makes it possible to establish an equivalence relationship between the various schemes of renormalization such that

$$A \sim B \Leftrightarrow \lim_{\hbar\to 0}(\lim_{\varepsilon\to 0}[A-B]) = 0, \tag{4.2}$$

where ε is the regularization parameter.

The fact that overlapping subintegrations in the bare Green functions of quantum field theory can be resolved into disjoint or nested types makes it possible to describe a given Feynman graph and its divergent subgraphs by means of an algebra of rooted trees. A rooted tree is assigned to each graph, and each vertex in the tree corresponds to a divergent subgraph. Further, recall first that a *rooted tree* t is a connected and simply connected set of oriented edges and vertices such that there is a distinguished vertex with no incoming edge. This vertex is called the *root* of t. Two vertices are connected by an edge, and the *fertility* of a given vertex is the number of edges going out from it. One can introduce a Hopf algebra on such rooted trees by using the possibility of cutting the trees in pieces. The structure of this algebra [6, 7] is then given by

$$\begin{aligned}
&\epsilon(e) = 1, \quad \epsilon(X) = 0, \text{ for any } X \neq e,\\
&\epsilon(t_1 \ldots t_n) = \epsilon(t_1) \ldots \epsilon(t_n),\\
&\Delta\,(e) = e \otimes e,\\
&\Delta\,(t_1 \ldots t_n) = \Delta(t_1) \cdots \Delta\,(t_n),\\
&\Delta\,(t) = t \otimes e + e \otimes t + \sum_{\text{adm. cuts } C \text{ of } t} P^C(t) \otimes R^C(t),\\
&S(e) = e,\\
&S(t) = -t - \sum_{\text{adm. cuts } C \text{ of } t} S[P^C(t)] R^C(t).
\end{aligned} \tag{4.3}$$

Here e denotes the unit element in the algebra, and $t_1 \ldots t_n$ is a monomial of n rooted trees. C is an admissible cut which maps a tree to a monomial in trees; R^C is the resulting distinguished tree in the monomial which contains the root, and P^C is the monomial resulting from the $n-1$ other factors. Extending the definitions of C, P, and R in the natural way to monomials of trees, one can confirm that the algebra of rooted trees satisfies the coassociativity of the coproduct as well as the counit and antipode axioms, so it is indeed a Hopf algebra which is usually denoted by $\mathcal{H}_R$. The relation between the algebras $\mathcal{H}_R$ and $\mathcal{H}_T$ can now be established by identifying the $\delta^i_{jk,l_1 \ldots, l_n}$, defined in the previous section, with the sum of all possible trees with $(n+1)$-decorated vertices, as may be readily verified by the respective actions of the coproduct operations in (3.62) and (4.3).

Moreover, we can now see how the quantum field renormalization scheme arises from $\mathcal{H}_R$. To this end, first note that the terms resulting from the action of the antipode (cf. (4.3)) on the algebra of rooted trees can be alternatively viewed as a series of full cuts C_f (where a full cut is defined by $P^{C_f}(t) = t$ and $R^{C_f}(t) = e$) which, in turn, induce a natural bracket structure on trees. In addition, because any admissible cut in a rooted tree corresponds to a divergent subintegration in a Feynman diagram, and vice versa, the bracket structure induced by the antipode allows us to generate local counterterms, so for equivalent (modulo finite terms) renormalization maps R, such that $R[t] \sim t$, the antipode $S[R(t)]$ coincides with the counterterm $Z[t]$ for a given Feynman graph. One then obtains a renormalized Green function according to the following scheme:

Let t be the elements in the algebra of rooted trees that are annihilated by the counit ϵ and which correspond to all possible divergent Feynman diagrams. Then the coproduct Δ produces all the terms necessary to compensate subdivergences, and the antipode axiom

$$m[(S \otimes id)\Delta(t)] \sim \epsilon(t) = 0 \tag{4.4}$$

will yield a renormalized Feynman diagram.

For a more detailed analysis of these procedures, we refer the reader to the original works of Connes and Kreimer cited above.

5 The algebra $\mathcal{H}_R$ and the Schwinger-Dyson equations

We mentioned in the previous section that overlapping divergences can be avoided by using Green functions instead of individual Feynman diagrams, and that this, in turn, made it possible to relate the Hopf algebra $\mathcal{H}_R$ of rooted trees with a renormalization scheme. Moreover, since the resolution of overlapping divergences into disjoint or nested types is also encoded in the Schwinger-Dyson (SD) equations which satisfy these Green functions, it follows that some relation between $\mathcal{H}_R$ and the SD equations must exist.

Indeed, the SD equations can be obtained in the functional integral representation by using the simple observation that the integral of a total derivative vanishes, so

$$\int \mathcal{D}\phi \frac{\delta}{\delta\phi}(\ldots) \equiv 0. \tag{5.1}$$

In particular, let $W(J)$ be the generating functional for a field ϕ

$$W(J) = \int \mathcal{D}\phi \exp\left(iS(\phi) + i\int d^4xJ\phi\right); \tag{5.2}$$

then, using the above observation, we have

$$0 = \int \mathcal{D}\phi \frac{\delta}{\delta\phi} \exp\left(iS(\phi) + i\int d^4xJ\phi\right), \tag{5.3}$$

where $S(\phi)$ is the classical action and J the source of the field. More explicitly, from (5.3) we get

$$\int \mathcal{D}\phi \left[J(x) + \frac{\delta S}{\delta\phi(x)}\right] \exp\left(iS(\phi) + i\int d^4xJ\phi\right) = 0, \tag{5.4}$$

which can be rewritten as

$$\left[\frac{\delta S}{\delta\phi(x)}\left(-i\frac{\delta}{\delta J}\right) + J(x)\right] W(J) = 0. \tag{5.5}$$

This is then the Schwinger-Dyson equation of motion of field theory, which is independent of perturbation theory. From this functional equation, it is possible to obtain an infinite system of coupled integral equations relating the Green functions of the theory [13]. Thus, for example in quantum electrodynamics (QED), the above equation has the form

$$\left[\frac{\delta S}{\delta A^\mu(x)}\left(-i\frac{\delta}{\delta J}, -i\frac{\delta}{\delta\bar{\eta}}, i\frac{\delta}{\delta\eta}\right) + J_\mu(x)\right] W(J,\eta,\bar{\eta}) = 0, \tag{5.6}$$

with J_μ the source of the electromagnetic potential, and $\eta(x)$ and $\bar{\eta}(x)$ are sources of the electron-positron field. We also have the equivalent equations for the fields ψ and $\bar{\psi}$. Introducing now the generating functional of the one-particle irreducible Green function $\Gamma(A_\mu, \bar{\psi}, \psi)$:

$$\Gamma(A_\mu, \bar{\psi}, \psi) := -i \ln W(J, \eta, \bar{\eta}) - \int d^4x (J \cdot A + \bar{\psi}\eta + \bar{\eta}\psi) \tag{5.7}$$

and using (5.6) we get integral equations for the propagator, the vertex, and the kernel functions. In explicit form, the vertex SD equation is (Ward identity)

$$e_0 \Lambda_\mu(y; z_1, z_2) = \frac{\delta}{\delta A^\mu(y)} \frac{\delta^2 \Gamma}{\delta \bar{\psi}(z_1) \delta \psi(z_2)} \bigg|_{A_\mu = \psi = \bar{\psi} = 0} . \tag{5.8}$$

This expression, together with the normalization condition which to lowest order in perturbation theory is given by

$$\Lambda_\mu(y; z_1, z_2) = \gamma_\mu \delta^4(y - z_1) \delta^4(y - z_2) + \ldots, \tag{5.9}$$

allows us to rewrite the SD equation for the fermion-photon vertex in the momentum space and in renormalized form as [14]

$$\Lambda(q) = Z\gamma + \int d^4k \; \Lambda(k) P(k)^2 K(k, q), \tag{5.10}$$

where $P(k)$ is the propagator and $K(k,q)$ is the four legs Green function. In this way, the perturbative iterations of this expression have the same form as those that are generated from $\mathcal{H}_R$, according to (4.4). Consequently, $\mathcal{H}_R$ contains essentially all the perturbative structure of quantum field theory.

Notice, however, that (5.6) contains, in general, higher order functional derivatives than those in (5.8). One could then ask if this additional information is included in noncommutative geometry. We speculate that the answer is affirmative since the full information, equivalent to (5.1) in the case of noncommutative geometry, is encoded in cyclic cohomology [1]. So we could hope that all the information, both perturbative and nonperturbative of field theory, is in fact included in noncommutative geometry, with $\mathcal{H}_R$ being only the perturbative part.

An additional aspect that might further probe the depth of the relation between noncommutative geometry and quantum field theory arises when we recall that the basic ingredient of noncommutative geometry is the spectral triple $(\mathcal{A}, \mathcal{H}, D)$. For the case of the canonical triple, we observed in the introduction that the Riemannian distance is determined by the inverse of the Dirac operator, i.e., $ds = |D|^{-1}$. But the inverse of the Dirac operator is the bare propagator P, which in renormalized quantum field theory is invariably dressed up. On the other hand, the SD equation (5.5) gives us a functional equation for the generating functional $W(J)$, from which we can get, *in principle*, this fully dressed propagator. This in turn means, as also commented at the beginning of this paper, that the geometry of the spacetime is determined by the quantum field theory structure. An interesting possibility within this program would be to try to formulate (5.5) in the context of noncommutative geometry, thus giving a complete meaning to (5.5), and not just the formal definition that it has in quantum field theory.

Acknowledgments

The authors are grateful for partial support for this work provided by grant DGAPA-UNAM-IN106897.

REFERENCES

[1] A. Connes, *Noncommutative Geometry,* Academic Press, Inc., San Diego, 1994.

[2] M. Rosenbaum, The short scale structure of space-time and the Dirac operator, *Proceedings of the International Conference on Clifford Algebras and Their Applications in Mathematical Physics,* to appear in *Int. J. of Theor. Phys.*

[3] J. Dixmier, *Les C^*-Algèbres et Leurs Représentations,* Gauthier-Villars, 1964.

[4] J. Dixmier, *Existence de Traces non Normals,* C.R. Acad. Sci. Paris, Ser. A-B, **262 A1107–A1108** (1966).

[5] A. Connes and H. Moscovici, Hopf algebras, cyclic cohomology and the transverse index theorem, *Commun. Math. Phys.* **198** (1998), 198–246.

[6] D. Kreimer, On the Hopf algebra structure of perturbative quantum field theories, *Adv. Theor. Math. Phys.* **2** (1998), 303–334.

[7] A. Connes and D. Kreimer, Hopf algebras, renormalization and noncommutative geometry, *Commun. Math. Phys.* **199** (1998), 203–242.

[8] A. Connes and D. Kreimer, Lessons from quantum field theory, *Lett. Math. Phys.* **48** (1999), 85–96.

[9] D. Kreimer, Chen's iterated integral represents the operator product expansion, *Adv. Theor. Math. Phys.* **3.3** (1999); hep-th/9901099.

[10] R. Wulkenhaar, On the Connes-Moscovici Hopf algebra associated to the diffeomorphisms of a manifold, math-ph/9904009.

[11] S. Kobayashi and K. Nomizu, *Foundations of Differential Geometry,* Interscience Publishers, New York, 1963.

[12] D. Bleecker, *Gauge Theory and Variational Principles,* Addison-Wesley Publishing Company, Reading, 1981.

[13] C. Itzykson and J. B. Zuber, *Quantum Field Theory,* McGraw-Hill, New York, 1980.

[14] J. D. Bjorken and S. D. Drell, *Relativistic Quantum Fields,* McGraw-Hill, New York, 1965.

Marcos Rosenbaum
Instituto de Ciencias Nucleares
Universidad Nacional Autónoma de Mexico, A. Postal 70-543
Mexico D.F., Mexico
E-mail: mrosen@nuclecu.unam.mx

J. David Vergara
Instituto de Ciencias Nucleares
Universidad Nacional Autónoma de Mexico, A. Postal 70-543
Mexico D.F., Mexico
E-mail: vergara@nuclecu.unam.mx

Received: September 28, 1999; Revised: January 13, 2000

Non-commutative Spaces for Graded Quantum Groups and Graded Clifford Algebras

Michaela Vancliff

ABSTRACT We propose that the notion of "quantum space" from Artin, Tate and Van den Bergh's non-commutative algebraic geometry be considered the "non-commutative space" of a quantum group.
Keywords: Regular algebra, quadratic algebra, point module, line module, point scheme, line scheme, quantum group, Clifford algebra, Poisson algebra, symplectic leaves.

1 Introduction

Analysis of solutions to the quantum Yang-Baxter equation from the quantum inverse scattering method is best approached via the irreducible finite-dimensional representations of the coordinate ring of a quantum group ([11, 12, 13, 35]). For this reason and many others, it is now a well-established fact that quantum groups are an important algebraic tool in mathematical physics.

In the spirit of classical physics, a quantum group is commonly viewed as a "deformation" of an algebra of functions of some variety or topological group, and so the quantum group is considered to be an algebra of non-commuting functions acting on some "non-commutative space" ([8]). However, there is still much debate over whether or not such a non-commutative space exists although it is believed that if it does, then it should be useful in determining properties of the quantum group and its representation theory. The objective of this paper is to demonstrate that a good candidate for this non-commutative space is the notion of "quantum space" that arises in the theory of non-commutative algebraic geometry à la M. Artin, J. Tate and M. Van den Bergh ([1, 2, 3, 4]).

Roughly speaking, Artin, Tate and Van den Bergh use the category of graded modules of a non commutative algebra as the space in which to

Supported in part by NSF grant DMS-9996056.
AMS Subject Classification: 16W50, 14A22, 17B37, 81R50.

do geometry. The first main success in applying this theory to quantum groups was seen in the analysis by S. P. Smith et al., in their classification of the finite-dimensional, irreducible representations of the Sklyanin algebra on four generators ([21, 32, 33, 31]). It should be emphasized that many physicists and mathematicians had attacked this problem since 1981, but without success until Smith et al., succeeded in 1991. Smith et al., used properties of the quantum space of the Sklyanin algebra in order to determine structural properties of the algebra. An attractive feature of this geometric theory is that it recovers commutative algebro-geometric results in addition to being applicable to non-commutative algebras.

In Sections 2-4, some examples from quantum groups will be discussed via these geometric ideas, namely, the Sklyanin algebra, the coordinate ring of quantum matrices and graded Clifford algebras. In Section 5, the theory of non-commutative algebraic geometry will be outlined in general with recent developments also discussed.

Surprisingly, within the context of Poisson algebras and quadratic quantum groups, the points of the quantum space of the quantum group are related to the projective zero-dimensional symplectic leaves of the Poisson manifold (Theorem 6).

Recently, the theory has begun to grow independently of deformational issues, with a shift in emphasis towards analyzing non-commutative projective schemes in their own right ([6, 14, 34, 42, 43]). For example, in [42], Van den Bergh developed a theory of non-commutative blowing up at a point on a non-commutative projective surface in order to classify the finite-dimensional irreducible representations of the family of three-dimensional regular algebras in [2, 3]. Thus, an intersection theory and a notion of isomorphism between non-commutative geometric objects is needed in order to compare geometric objects in one "quantum space" with those in another, and this is the current focus of P. Jorgensen, S. P. Smith and J. Zhang in [14, 34].

Throughout the paper, k denotes an algebraically closed field such that $\text{char}(k) \neq 2$ and $\mathcal{V}(X)$ denotes the locus of common zeros of the set X.

2 The Sklyanin algebra

In the early 1980s, using Baxter's elliptic solutions to the Yang-Baxter equation, Sklyanin constructed a family of graded algebras on four generators defined in terms of elliptic functions ([29]). In [10], these algebras (now called Sklyanin algebras) were shown to depend on an elliptic curve in $\mathbb{P}^3$ and an automorphism. In this section, we discuss the Sklyanin algebra on three generators and the Sklyanin algebra on four generators.

It should be noted that methods which had been effective for enveloping algebras, group algebras of polycyclic-by-finite groups and polynomial-identity rings were ineffective when applied to the Sklyanin algebras, mainly because they have nothing like a Poincaré-Birkhoff-Witt basis.

2.1 The Sklyanin algebra on three generators

Let $a, b, c \in k^\times$ such that $(3abc)^3 \neq (a^3 + b^3 + c^3)^3$. The k-algebra A on generators x, y, and z, with defining relations

$$\begin{aligned} cx^2 + bzy + ayz &= 0, \\ cy^2 + bxz + azx &= 0, \\ cz^2 + byx + axy &= 0, \end{aligned}$$

is the Sklyanin algebra on three generators ([2, 3]). Consider these relations as bihomogeneous $(1,1)$-forms on $\mathbb{P}^2 \times \mathbb{P}^2$, and let Γ denote their locus of common zeros in $\mathbb{P}^2 \times \mathbb{P}^2$. It is shown in [3] that Γ is the graph of an automorphism σ of an elliptic curve E in $\mathbb{P}^2$, where

$$E = \mathcal{V}(\,(a^3 + b^3 + c^3)xyz - abc(x^3 + y^3 + z^3)\,).$$

The following is proved in [3, 4].

A *twisted homogeneous coordinate ring* $B(E)$ of E is obtained by using the automorphism σ to deform the multiplication in the commutative homogeneous coordinate ring of E. We refer the reader to [3, 5] for details. This process produces a non-commutative algebra $B(E)$ whose algebraic properties are entwined in the geometric properties of E. The algebra $B(E)$ is a quotient of A via $B(E) = A/\langle g \rangle$ where $g \in A$ is homogeneous of degree three. As such, the algebraic properties of A are encoded in E and σ, and this is demonstrated by the fact that g belongs to the centre of A. Indeed, A is a finite module over its centre if and only if σ has finite order.

The geometry determined by E and σ were important tools in establishing that A is a Noetherian domain with Hilbert series the same as that of the polynomial ring on three generators.

The above discussion suggests that the elliptic curve E, together with its automorphism σ, should play the role of a non-commutative space for A. In fact, Section 4 will demonstrate that sometimes the locus of zeros of the defining relations is too small to be tractable. The notion of "quantum space" (to be discussed in Section 5) for A includes E and σ plus some other geometric data in such a way that the points of E may be viewed as points of the quantum space.

2.2 The Sklyanin algebra on four generators

The Sklyanin algebra on four generators is the k-algebra A on generators x_0, x_1, x_2 and x_3, with six defining relations

$$\begin{aligned} x_0x_i - x_ix_0 &= \alpha_i(x_jx_k + x_kx_j), \\ x_0x_i + x_ix_0 &= x_jx_k - x_kx_j \end{aligned}$$

where (i, j, k) is a cyclic permutation of $(1, 2, 3)$, and $(\alpha_1, \alpha_2, \alpha_3) \in k^3$ where $\alpha_i \neq 0, \pm 1$ for all i and $\alpha_1 + \alpha_2 + \alpha_3 + \alpha_1\alpha_2\alpha_3 = 0$ ([29]). The reader is referred to [21, 32, 33] for details of the following.

Consider these relations as bihomogeneous $(1,1)$-forms on $\mathbb{P}^3 \times \mathbb{P}^3$. Their locus of common zeros $\Gamma \subset \mathbb{P}^3 \times \mathbb{P}^3$ is the graph of an automorphism σ of a subscheme $E \cup S \subset \mathbb{P}^3$ where E is an elliptic curve and S consists of the four points

$$e_1 = (1, 0, 0, 0),\ e_2 = (0, 1, 0, 0),\ e_3 = (0, 0, 1, 0) \text{ and } e_4 = (0, 0, 0, 1)$$

in $\mathbb{P}^3$. As in the three-generator case, one may form the twisted homogeneous coordinate ring $B(E)$ of E by using the automorphism σ to deform the multiplication of the commutative homogeneous coordinate ring of E. As before, the algebra $B(E)$ is a quotient of A via $B(E) = A/\langle g_1, g_2\rangle$ where $g_1, g_2 \in A$ are homogeneous of degree two. In fact, $kg_1 \oplus kg_2$ belongs to the centre of A and generates the centre of A if σ has infinite order. Moreover, A is a finite module over its centre if and only if σ has finite order. As before, the geometry of Γ establishes that A is a Noetherian domain with Hilbert series the same as that of the polynomial ring on four generators.

The Abelian group structure of the elliptic curve E and the position of the e_i in $\mathbb{P}^3$ with respect to E were critical components in conclusively determining the finite-dimensional irreducible representations of A ([31, 33]).

The notion of "quantum space" (to be discussed in Section 5) for A includes $E \cup S$ and σ plus some other geometric data in such a way that the points of $E \cup S$ may be viewed as points of the quantum space.

3 The coordinate ring of quantum matrices

Let M_n denote the ring of $n \times n$ matrices over k and, for $q \in k^\times$, let $\mathcal{O}_q(M_n)$ denote the coordinate ring of quantum $n \times n$ matrices over k [9]. The algebra $\mathcal{O}_q(M_n)$ is the k-algebra on n^2 generators $\{x_{ij} : 1 \leq i, j \leq n\}$, with defining relations determined by the requirement that whenever $r < s$ and $l < m$ there exists a k-algebra isomorphism

$$k_q\begin{bmatrix} a & b \\ c & d \end{bmatrix} \xrightarrow{\sim} k_q\begin{bmatrix} x_{rl} & x_{rm} \\ x_{sl} & x_{sm} \end{bmatrix},$$

where the k-algebra $k_q[a, b, c, d] = k_q\begin{bmatrix} a & b \\ c & d \end{bmatrix}$ has the six defining relations:

$$\begin{array}{lll} ab = qba, & bd = qdb, & bc = cb, \\ cd = qdc, & ac = qca, & ad - da = (q - q^{-1})bc. \end{array}$$

The results stated in this section are given in [38].

We first focus on the case $n = 2$. As in Section 2.2, consider the relations of $\mathcal{O}_q(M_2)$ as bihomogeneous $(1,1)$-forms on $\mathbb{P}^3 \times \mathbb{P}^3$. If $q^2 \neq 1$, then the locus of common zeros in $\mathbb{P}^3 \times \mathbb{P}^3$ of the relations is the graph of an automorphism σ of a subscheme $Q \cup L$ of $\mathbb{P}^3$ where $Q = \mathcal{V}(ad - bc)$ is a nonsingular quadric in $\mathbb{P}^3$ corresponding to the matrices with zero determinant, and $L = \mathcal{V}(b, c)$ is a line in $\mathbb{P}^3$ which meets Q at two distinct points and corresponds to the diagonal matrices.

There is a unique nonzero homogeneous element (up to nonzero scalar multiples) of degree two in $\mathcal{O}_q(M_2)$ which vanishes on the graph of $\sigma|_Q$, but not on the graph of σ. This distinguished element Ω is the famous quantum determinant of $\mathcal{O}_q(M_2)$ which is central in $\mathcal{O}_q(M_2)$. In fact, $\mathcal{O}_q(M_2)/\langle\Omega\rangle$ is the twisted homogeneous coordinate ring of Q determined by $\sigma|_Q$. Moreover, the twisted homogeneous coordinate ring of the line L determined by $\sigma|_L$ is given by $\mathcal{O}_q(M_2)/\langle b, c\rangle$. In other words, it is as if the geometry is determining the quantum determinant and the normal elements b and c in $\mathcal{O}_q(M_2)$.

Since $\mathcal{O}_q(M_2)$ is a finite module over its centre if and only if q has finite order, it follows from [38] that $\mathcal{O}_q(M_2)$ is a finite module over its centre if and only if σ has finite order.

The case $n > 2$ generalizes that of $n = 2$ as follows. For all $n \geq 2$ and for all $q \in k^\times$, the locus of common zeros of the defining relations of $\mathcal{O}_q(M_n)$ is the graph of an automorphism of a subscheme of $\mathbb{P}(M_n)$. If $q^2 \neq 1$, this subscheme is independent of q and is the nondisjoint union of $\binom{n}{2}^2$ copies of $Q \cup L$ and $\binom{n}{d}^2$ copies of $\mathbb{P}^{d-1}$ for all $d = 1, \ldots, n$. As in the case $n = 2$, the quantum determinant of $\mathcal{O}_q(M_n)$ may be read off from this subscheme and its automorphism and so can the normal elements of degree one in $\mathcal{O}_q(M_n)$. The notion of "quantum space" (to be discussed in Section 5) for $\mathcal{O}_q(M_n)$ includes this subscheme together with its automorphism, plus some other geometric data, in such a way that the points of this subscheme may be viewed as points of the quantum space.

4 Graded Clifford algebras

Let $R = k[y_1, \ldots, y_n]$ denote the commutative polynomial ring on n variables, and let $Y = (Y_{ij}) \in M_n(R)$ denote a symmetric matrix whose entries are homogeneous linear polynomials in the variables y_i.

Definition 1. *In the terminology of [41] and [17, §4], the graded Clifford algebra $A = A(Y)$ over R associated to Y is the k-algebra on generators $x_1, \ldots, x_n, y_1, \ldots, y_n$ with defining relations $x_i x_j + x_j x_i = Y_{ij}$ for all i, j, and y_i central for all i. The algebra A is graded by taking $\deg(x_i) = 1$ and $\deg(y_i) = 2$ for all i.*

Writing $Y = Y_1 y_1 + \cdots + Y_n y_n$ where the $Y_i \in M_n(k)$ are symmetric matrices, we may associate to Y an n-dimensional linear system

$$\mathcal{Q} = kQ_1 + \cdots + kQ_n$$

of quadrics $Q_1, \ldots, Q_n \subset \mathbb{P}^{n-1}$ by taking each Q_i to be the quadric in $\mathbb{P}^{n-1}$ corresponding to Y_i, that is, $Q_i = \{y \in \mathbb{P}^{n-1} : y^T Y_i y = 0\}$. A base point of $\mathcal{Q}$ is a common point of intersection of all the Q_i.

If $\mathcal{Q}$ has no base points, then A is generated by the x_i only and is a finite module over R and, hence, Noetherian. In this case, by [7, §3], A is the enveloping algebra of a Lie superalgebra and so has $n(n-1)/2$ defining relations, and they are of the form $\sum_{ij} \alpha_{ijm}(x_i x_j + x_j x_i) = 0$ where the $\alpha_{ijm} \in k$, and A has Hilbert series the same as that of the polynomial ring on n variables. Henceforth, we assume that $\mathcal{Q}$ has no base points.

We write $A = k\langle x_1, \ldots, x_n \rangle / \langle W \rangle$ where W is the span of the defining relations of A. As in the preceding sections, consider the relations as bihomogeneous $(1,1)$-forms on $\mathbb{P}^{n-1} \times \mathbb{P}^{n-1}$ and let Γ denote their locus of common zeros. Let $\pi_i : \Gamma \to \mathbb{P}^{n-1}$ denote the ith projection map. Since the defining relations of A have the aforementioned description, $\pi_1(\Gamma) = \pi_2(\Gamma)$. Owing to the symmetry of the defining relations, if $(a, b) \in \Gamma$, then so is (b, a); in other words, Γ is the graph of an automorphism $\sigma : \pi_1(\Gamma) \to \pi_1(\Gamma)$ where σ has order two.

In this situation, a point $(a, b) \in \Gamma$ if and only if $\sum_{ij} \alpha_{ijm}(a_i b_j + a_j b_i) = 0$ for all $m \leq n(n-1)/2$. In other words, $(a, b) \in \Gamma$ if and only if the symmetric matrix $ab^T + ba^T = (a_i b_j + a_j b_i)$ is a zero of $\sum_{ij} \alpha_{ijm} X_{ij}$ for all $m \leq n(n-1)/2$ where X_{ij} denotes the ijth coordinate function. Following [40], we consider symmetric $n \times n$ matrices as $\mathbb{P}^N$, where $N+1 = n(n+1)/2$, and define a map $\phi : \mathbb{P}^{n-1} \times \mathbb{P}^{n-1} \to \mathbb{P}^N$ by $\phi(u, v) = uv^T + vu^T$ (which is a matrix of rank ≤ 2). Since k is algebraically closed with $\operatorname{char}(k) \neq 2$ and since $\mathrm{GL}_n(k)$ acts transitively on matrices of the same rank, the image of ϕ consists of all symmetric $n \times n$ matrices X such that $\operatorname{rank}(X) \leq 2$.

Depending on the defining relations, it is possible for Γ to be finite. In fact, by [40, Theorem 1.7], the cardinality of Γ is $2r_2 + r_1 \in \mathbb{N} \cup \{0, \infty\}$ where r_j denotes the number of matrices in $\mathbb{P}(\sum_{i=1}^n kY_i)$ which have rank j. If Γ is finite, then $r_1 \in \{0, 1\}$. The surprising fact is that it is possible for Γ to consist of only one point (counted with multiplicity twenty) ([40]). It is perhaps even more surprising that even Γ consists of only one point, the space W consists of all those bihomogeneous $(1,1)$-forms which vanish on Γ ([26]). It follows that properties of the algebra are entwined in properties of Γ. In spite of this, one finds that algebraic properties of the algebra do not leap to the reader's eye from Γ as they seem to be doing in the preceding examples. This suggests that perhaps there is more geometric data that one may associate to A which, although completely determined by Γ, would be more insightful. The notion of "quantum space" (to be discussed in Section 5) for A includes Γ plus some other geometric data in such a way that the points of Γ may be viewed as points of the quantum space.

5 Non-commutative algebraic geometry

In the previous sections, the algebras were quadratic, and the locus Γ of common zeros of each algebra's defining relations was considered. In this section the notion of quantum space will be defined, and the scheme Γ will constitute (some of the) points in the quantum space. Roughly speaking, the quantum space of an algebra is a (quotient) category of graded modules, and the scheme Γ will constitute (some of the) points in of the algebra, in which certain modules play the role of points, certain modules play the role of lines, and so forth. In addition, there are incidence relations between the modules which determine whether or not a point lies on a line, a line lies on a plane, etc. Within this category there are certain modules which are parameterized by projective schemes and may be determined via computation. Two such schemes are the so-called *point scheme* and *line scheme* of [28, 39].

Until Section 5.3, unless otherwise stated, A denotes a finitely generated, connected, positively graded k-algebra generated by homogeneous elements of degree one, and A_i denotes the homogeneous degree-i elements of A. Any A-module will be either a left A-module or a right A-module.

5.1 The point scheme

Definition 2. [3] *A point module of A is a graded, cyclic A-module which is generated by elements of degree zero and which has the Hilbert series $(1-t)^{-1}$. A truncated point module of length r is a graded, cyclic A-module which is generated by elements of degree zero and which has Hilbert series $1+t+\cdots+t^{r-1}$.*

In particular, if M is a point module, then $M=\oplus_{i=0}^{\infty}M_i$ where $\dim_k(M_i)=1$ for all i.

If A is commutative and generated by $n+1$ elements, then it is the homogeneous coordinate ring of a subscheme S of $\mathbb{P}^n$. In this case, any truncated point module of length three may be extended to a point module, and the set of point modules of A is in one-to-one correspondence with the points of S. In fact, if $M=AM_0$ is a point module, then $\mathcal{V}(\{a\in A_1 : aM_0=0\})$ is a point of S and $M\cong A/A(\{a\in A_1 : aM_0=0\})$. Conversely, if $p=\mathcal{V}(a_1,\dots,a_n)\in S$, then $A/(Aa_1+\cdots+Aa_n)$ is a point module.

Suppose A is non-commutative. If M is a truncated (left) point module of length three, then $\mathcal{V}(\{a\in A_1 : aM_0=0\})$ is a point of $\mathbb{P}^n$ and, moreover, $A/A(\{a\in A_1 : aM_0=0\})\twoheadrightarrow M$.

Theorem 1. [3] *There is a scheme which represents the functor of truncated point modules of length three.* □

If A has the property that every truncated point module of length three may be extended to a point module, then this result says that the point modules are parameterized by a scheme (called the *point scheme* in [39]). Algebras of global dimension three which are "regular", according to the following definition, were considered in [2, 3, 4] and have this property.

Definition 3.

1. [2, Page 171] *The algebra A is called regular if the global (homological) dimension of A (gldim(A)) is finite, A has polynomial growth, and A is Gorenstein in the sense that* $Ext^q_A(k, A) = \delta^q_n k$ *where* $n = gldim(A)$.

2. [20, §2] *The algebra A is called Auslander-regular if gldim(A) is finite, and if, for every finitely generated A-module M and for every* $i \geq 0$ *and for every A-submodule N of* $Ext^i_A(M, A)$, *we have* $j(N) \geq i$, *where*
$$j(N) = \inf\{j : \mathrm{Ext}^j_A(N, A) \neq 0\}.$$

3. [20, §5] *The algebra A is said to satisfy the Cohen-Macaulay property if* $GKdim(A) = j(M) + GKdim(M)$ *for all nonzero finitely generated A-modules M where GKdim denotes Gelfand-Kirillov dimension.*

In [20], it was shown that any Auslander-regular algebra which has polynomial growth is regular. The algebras considered in Sections 2-4 of this article are Noetherian, Auslander-regular, have polynomial growth and satisfy the Cohen-Macaulay property ([2, 3, 4, 21, 32, 37, 38, 40]). If the global dimension is three, then regularity is sufficient to ensure that the zero locus of the defining relations is the graph of an automorphism, and in this case the point scheme is the zero locus of the defining relations ([3]).

The following result was proved in [39] with an additional geometric hypothesis which was subsequently removed in [27].

Theorem 2. [27] *Suppose that A is a quadratic, connected, finitely generated, Noetherian k-algebra, which is Auslander-regular of global dimension four, and satisfies the Cohen-Macaulay property. If the Hilbert series of A is the same as that of the polynomial ring on four variables, then the zero locus* Γ *of the defining relations of A is the graph of an automorphism of a subscheme of* $\mathbb{P}^3$. *In particular, any truncated point module of length three may be extended to a point module, and* Γ *represents the functor of point modules, so it is the point scheme.* □

These results on regularity suggest that Noetherian (Auslander-)regular algebras of polynomial growth which satisfy the Cohen-Macaulay property are non-commutative analogues of the polynomial ring. Moreover, they suggest that to understand such algebras, one should first classify which subschemes and automorphisms arise as the zero locus of defining relations.

The regular algebras of global dimension three were classified and fully analyzed in [2, 3, 4]. However the regular algebras of global dimension four are not yet classified as this case appears to be less accessible; for instance, such an algebra's defining relations need not be determined by their zero locus Γ even if Γ is the point scheme ([27, 39, 44]). Instead, attention has shifted to *generic* regular algebras of global dimension four, in particular those which are quadratic on four generators with six defining relations. It is now well-known that the relations of a quadratic algebra on four generators with six *generic* defining relations have zero locus consisting of twenty points. This motivates the classification of regular algebras of global dimension four which have a finite point scheme. Indeed, the following result demonstrates that even if a quadratic algebra is not regular, the defining relations may be recovered from their zero locus, providing the latter is finite.

Theorem 3. [28] *Suppose A is a quadratic, connected, finitely-generated k-algebra on four generators with six defining relations. If the zero locus Γ of the defining relations of A is finite, then the span of the defining relations consists of those $(1,1)$-forms which vanish on Γ.* □

This result motivates classifying the geometric data which arises as the zero locus Γ of the defining relations of a quadratic algebra. Recall from Section 4 that it is possible to construct graded Clifford algebras which have one point of multiplicity twenty as the zero locus of the defining relations. In that case, the point is the point scheme. It follows from Theorem 3 that the defining relations of such a Clifford algebra are completely determined by that one point (and automorphism) (see [26] for additional details).

5.2 The line scheme

It was remarked in Section 4 that sometimes the point scheme is too small to give much insight into the structure of the algebra even if the point scheme determines the defining relations.

Definition 4. [3] *A line module of A is a graded, cyclic A-module which is generated by elements of degree zero and which has the Hilbert series $(1-t)^{-2}$. A truncated line module of length r is a graded, cyclic A-module which is generated by elements of degree zero and which has Hilbert series $1+2t+\cdots+rt^{r-1}$.*

In particular, if M is a line module, then $M = \oplus_{i=0}^{\infty} M_i$ where $\dim_k(M_i) = i+1$ for all i.

If A is commutative with $n+1$ generators, then any truncated line module of length three may be extended to a line module, and the set of line modules of A is in one-to-one correspondence with the lines on the scheme S for which A is the homogeneous coordinate ring. Mimicking the discussion for point modules, if $M = AM_0$ is a line module, then $\mathcal{V}(\{a \in A_1 : aM_0 = 0\})$ is a line on S and $M \cong A/A(\{a \in A_1 : aM_0 = 0\})$. Conversely, if $\ell = \mathcal{V}(a_1, \ldots, a_{n-1})$ is a line on S, then $A/(Aa_1 + \cdots + Aa_{n-1})$ is a line module.

Suppose A is non-commutative. If M is a truncated (left) line module of length three, then $\mathcal{V}(\{a \in A_1 : aM_0 = 0\})$ is a line in $\mathbb{P}^n$ and, moreover, $A/A(\{a \in A_1 : aM_0 = 0\}) \twoheadrightarrow M$.

Theorem 4. [28] *There is a scheme which represents the functor of truncated line modules of length three.* □

If A is regular of global dimension three or if, instead, A is quadratic, Noetherian, Auslander-regular of global dimension four, satisfies the Cohen-Macaulay property and has Hilbert series the same as that of the polynomial ring on four variables, then every truncated line module of length three may be extended to a line module ([3, 21]). In this case, Theorem 4 says that the line modules are parameterized by a scheme, the *line scheme*. It is straightforward to show that under these conditions, the line scheme is at least one-dimensional (see [40]). Moreover, if the line scheme has minimal dimension, then it determines the defining relations ([28]).

Even if the line scheme is well-defined, it might not be straightforward to compute. The reader is referred to [28] for methods on computing the line scheme and for results on the line scheme of graded Clifford algebras which have a singleton point scheme.

5.3 Quantum spaces

Let $M(p)$ denote a point module corresponding to a point p, and let $M(\ell)$ denote a line module corresponding to a line ℓ. If $M(\ell) \twoheadrightarrow M(p)$, then $p \in \ell$. The converse holds if A is Auslander-regular of global dimension three or four. Thus, one may do geometry with the graded modules, where the incidence relations between the modules play the role of containment of the physical objects.

Of course, one may define higher dimensional linear modules, and truncated d-linear modules, in analogy with point and line modules. It is proved in [28] that there is a scheme which represents the functor of truncated d-linear modules.

In the commutative setting, this idea of doing geometry with modules in place of geometric objects was considered in the 1950s by J.-P. Serre as follows. Let B be a finitely generated, commutative, positively graded, connected k-algebra generated by B_1, and let gr-B denote the category of finitely generated, graded B-modules.

Theorem 5. [25] *The category of coherent sheaves on* Proj B *is equivalent to the quotient category* (gr-B)$/\mathcal{T}$, *where* $\mathcal{T}$ *denotes the full subcategory of* gr-B *of modules of finite length.*

Moreover, Proj B may be recovered from the quotient category (gr-B)$/\mathcal{T}$ as follows. The shift $M[m]$ of a module M, defined by $M[m]_n = M_{m+n}$, corresponds to tensor product by the polarizing invertible sheaf $\mathcal{L} = \mathcal{O}_{\mathrm{Proj}B}(1)$ of linear forms on Proj B, that is, $\mathcal{M}[1] = \mathcal{M} \otimes \mathcal{L}$ where $\mathcal{M}$

is the coherent sheaf corresponding to M. It follows that this shift operation defines an autoequivalence on the quotient category (gr-B)/$\mathcal{T}$ so that the graded algebra $\bigoplus_{n=0}^{\infty} H^0(\text{Proj } B, \mathcal{L}^{\otimes n})$ is isomorphic to B in sufficiently high degree.

Inspired by Serre's theorem, M. Artin extended these ideas in [1] to the non-commutative setting as follows. Let A be a Noetherian, positively graded, connected k-algebra generated by A_1, and let gr-A denote the category of finitely generated, graded A-modules.

Definition 5. [1] *Define* Proj A *to be the triple* $((\text{gr-A})/\mathcal{T}, \mathcal{O}, \sigma)$ *where* $\mathcal{T}$ *denotes the subcategory of gr-A of torsion modules,* $\mathcal{O}$ *denotes an object of (gr-A)/*$\mathcal{T}$ *which is represented by the left module A, and* σ *is the operation* $\mathcal{M} \to \mathcal{M}[1]$ *on (gr-A)/*$\mathcal{T}$ *induced by the shift of degree on an A-module. A quantum* $\mathbb{P}^2$*, or quantum projective plane, is* Proj A *where A is a regular algebra of global dimension three.*

Given a quantum group A which satisfies the hypotheses of Definition 5, one may analyze Proj A and consider it the non-commutative space of the quantum group. The very nature of the quantum space of a quantum group is to produce certain graded modules which are tools for finding and classifying other modules. For instance, the finite-dimensional irreducible representations of the Sklyanin algebra on four generators are quotients of line modules ([33]). In general, a quantum (projective) space should be Proj A for some A, but it is unclear if regularity of A alone is sufficient to guarantee that A may be sensibly viewed as a deformation of a polynomial ring. As such, there is still debate over the definition of a quantum $\mathbb{P}^3$ and of a quantum projective space. However, the quantum groups in Sections 2-4 satisfy all the homological properties discussed herein, and so their associated category Proj are quantum spaces.

It should be emphasized that the point scheme and line scheme are subschemes of Proj A. In fact, d-linear modules play the role of d-dimensional linear objects. However, some non-commutative algebras have other graded modules which also play the role of d-dimensional linear objects, and yet such modules are not equivalent to d-linear modules for any d. Although there is no obvious $\mathbb{P}^n$ in which these objects may be visualized (unlike linear modules), the presence of such modules means that even more geometric structure may be associated to A.

We remark that even if the algebra A is not regular, Proj A might still encode properties of A. This was demonstrated in [36, Chapter 2] where a quadratic algebra $A_{(4)}$ was produced from a procedure which is the quantum analogue of a classical construction as follows. Let $U_q(\mathfrak{sl}_2)$ denote the quantum universal enveloping algebra of $\mathfrak{sl}_2$ ([16, 30]). It is proved in [22, 23, 24] that if q is not a root of unity, then for each $n \in \mathbb{N}$ there are four non-isomorphic simple n-dimensional $U_q(\mathfrak{sl}_2)$-modules depending on a fourth root of unity ω and that every finite-dimensional $U_q(\mathfrak{sl}_2)$-module is semisimple. Let V_n denote either an n-dimensional simple $U_q(\mathfrak{sl}_2)$-module corresponding to "$\omega = 1$", or an n-dimensional simple $U(\mathfrak{sl}_2)$-module. Then $V_n \otimes V_n = W_1 \oplus W_3 \oplus \cdots \oplus W_{2n-1}$ where W_i is a simple i-dimensional module of $U_q(\mathfrak{sl}_2)$ (respectively, $U(\mathfrak{sl}_2)$). Let $A_{(n)} := T(V_n)/\langle U_n \rangle$ where $T(V_n)$ denotes the tensor algebra on V_n and U_n denotes an $n(n-1)/2$-dimensional submodule of $V_n \otimes V_n$; so, for example, $A_{(3)} = T(V_3)/\langle W_3 \rangle$ and $A_{(4)} = T(V_4)/\langle W_1 \oplus W_5 \rangle$. Classically, $A_{(n)} \cong S(V_n)$, the symmetric algebra on V_n; in particular, the classical $A_{(n)}$ is Noetherian and regular of Gelfand-Kirillov dimension n.

In the quantum case, $A_{(3)} \cong \mathcal{O}_q(\mathfrak{so}k^3)$, the coordinate ring of quantum Euclidean three-dimensional space. The quantum $A_{(4)}$ is not so straightforward to understand. In particular, in [36], the quantum $A_{(4)}$ is shown not to be regular nor a domain. Nevertheless, its point scheme is well-defined and is the graph of an automorphism σ of a twisted cubic curve C in $\mathbb{P}^3$, even though the point scheme is not the locus of zeros of the defining relations. The structure of $A_{(4)}$ is determined by a nilpotent ideal N which is generated by the nonzero degree-two elements of $A_{(4)}$ which vanish on the graph of σ. The quotient algebra $A_{(4)}/N$ is a twisted homogeneous coordinate ring of C. Using this geometric data, it is straightforward to show that $A_{(4)}$ is Noetherian of Gelfand-Kirillov dimension two (not four).

6 Quantum spaces via Poisson geometry

The setting in this section is that of a non-commutative quadratic deformation A of the polynomial ring which induces a Poisson bracket on the polynomial ring. Our main result, Theorem 6, states (essentially) that the point scheme is contained in the scheme of projective zero-dimensional symplectic leaves.

Example 1. *If $q \in \mathbb{C}$ is generic, then $\mathcal{O}_q(M_2)$ from Section 3 induces a Poisson bracket on the polynomial ring $\mathcal{O}(M_2)$ on four variables via*

$$\{b,c\} = 0,\ \{a,d\} = 2bc,\ \{a,b\} = ab,\ \{a,c\} = ac,\ \{b,d\} = bd,\ \{c,d\} = cd,$$

and similarly for $\mathcal{O}_q(M_n)$ ([38]). Since the bracket is homogeneous, it lifts to $(\mathcal{O}(M_n)[z^{-1}])_0$, the degree-zero part of the localization $\mathcal{O}(M_n)[z^{-1}]$, for all nonzero, homogeneous $z \in \mathcal{O}(M_n)$. In this way, $\mathbb{P}(M_n)$ is equipped with a Poisson structure, and it is proved in [38] that if $q^2 \neq 1$, then the scheme of zero-dimensional symplectic leaves is precisely the first projection of the point scheme of $\mathcal{O}_q(M_n)$ to $\mathbb{P}(M_n)$.

There are many other quantum groups where the point scheme gives precisely the scheme of zero-dimensional symplectic leaves, such as the Sklyanin algebra on three generators, the coordinate ring $\mathcal{O}_q(\mathfrak{so}\mathbb{C}^3)$ of quantum Euclidean three-dimensional space, and the coordinate ring $\mathcal{O}_q(\mathfrak{sp}\mathbb{C}^{2n})$ of quantum symplectic $2n$-dimensional space, to name a few. However, the following example points out that equality need not hold in general.

Example 2. *Let $\mathbb{C}[h]$ denote the polynomial ring in an indeterminate h, and let A denote the $\mathbb{C}[h]$-algebra on generators x, y and z with defining relations*

$$xy = (1+2h)yx, \quad yz = (1-h)zy \quad \text{and} \quad zx = (1-h)xz$$

so that $A(0) = A/\langle h\rangle$ is the polynomial ring on three variables. Moreover, $A(q) = A/\langle h-q\rangle$ is an iterated Ore extension for all $q \in \mathbb{C}$ where $(1+2q)(1-q) \neq 0$. A Poisson bracket is induced by A on $A(0)$ and is given by

$$\{x,y\} = 2xy, \quad \{y,z\} = -yz, \quad \{z,x\} = -xz$$

(see page 316). In this case, the scheme consisting of the projective zero-dimensional symplectic leaves is $\mathbb{P}^3$, but for all $q \in \mathbb{C}^\times$ such that $(1+2q)(1-q) \neq 0$, the projection of the point scheme of $A(q)$ to its first component in $\mathbb{P}^3$ is $\mathcal{V}(xyz)$. So, for this example, the point scheme encodes the normal elements x, y and z of $A(q)$ and the Poisson ideals $\langle x\rangle$, $\langle y\rangle$ and $\langle z\rangle$ of $A(0)$ via the components of the point scheme whereas the scheme of zero-dimensional symplectic leaves loses this information.

Our goal is to prove that in general the point scheme is contained in the scheme of zero-dimensional symplectic leaves.

Let $\mathbb{C}'$ denote a subset of $\mathbb{C}$ containing zero whose complement is a finite set. Let h be an indeterminate, and let $\mathbb{C}(h)$ denote the ring of rational functions on $\mathbb{C}$, and let R denote $\mathbb{C}[h][\Pi_r(h-r)^{-1}] \subset \mathbb{C}(h)$ where r runs through a subset of $\mathbb{C}\setminus\mathbb{C}'$; in particular, R is a ring of rational functions on $\mathbb{C}$ which are defined on $\mathbb{C}'$ (or possibly more than $\mathbb{C}'$). For each $q \in \mathbb{C}'$, let $\mathfrak{m}_q$ denote the maximal ideal $R(h-q)$. Fix a finite-dimensional $\mathbb{C}$-vector space V and an R-submodule $W \subset R\otimes_{\mathbb{C}} V \otimes_{\mathbb{C}} V$. Let $A = (R\otimes_{\mathbb{C}} T(V))/\langle W\rangle$ where $T(V)$ denotes the tensor algebra on V.

Definition 6. *We define the family $\{A(q)\}_{q\in\mathbb{C}'}$ of quadratic $\mathbb{C}$-algebras parameterized by $\mathbb{C}'$ to be $A(q) := R/\mathfrak{m}_q \otimes_R A$ for all $q \in \mathbb{C}'$.*

If A is a flat R-module, then the family $\{A(q)\}$ is said to be a flat family, and $A(q)$ is said to be a flat deformation of $A(0)$.

Lemma 1. [36, Chapter 4] *If the family $\{A(q)\}_{q\in\mathbb{C}'}$ is flat, then the Hilbert series of $A(q)$ is independent of q for all $q \in \mathbb{C}'$.* □

For each $q \in \mathbb{C}'$ there is a $\mathbb{C}$-algebra homomorphism $T(V) \twoheadrightarrow A(q)$ given by the composition $T(V) \longrightarrow R\otimes_{\mathbb{C}} T(V) \longrightarrow A \longrightarrow A(q)$ where the first map is $x \mapsto 1\otimes x$. For all $q \in \mathbb{C}'$, let W_q denote the kernel of the $\mathbb{C}$-algebra homomorphism $T(V)_2 \twoheadrightarrow A(q)_2$ so that $A(q) = T(V)/\langle W_q\rangle$.

Lemma 2. [36, Chapter 4] *If A is a flat R-module, then W is a free R-module, and $W_q \cong R/\mathfrak{m}_q \otimes_R W$ for every $q \in \mathbb{C}'$.* □

Henceforth, we assume that A is a flat R-module and that $A(0)$ is the polynomial ring $S(V)$. In particular, the first condition implies that A is torsion-free (since R is a principal ideal domain) and that Lemmas 1 and 2 hold. The second condition implies that $A(0) = R \otimes_{\mathbb{C}} S(V)/\langle \mathfrak{m}_0 \rangle$ and that $W_0 = \mathbb{C} \otimes \Lambda^2(V)$ where $\Lambda^2(V)$ denotes the second exterior power of V. Hence, if x, $y \in V$, then there exists $r_{xy} \in W$ of the form

$$r_{xy} = 1 \otimes x \otimes y - 1 \otimes y \otimes x - hb_h(x, y)$$

for some $b_h(x, y) \in R \otimes_{\mathbb{C}} V \otimes_{\mathbb{C}} V$. Since A is torsion-free over R, it follows that $r_{xx} = 0$.

Lemma 3. *The module W is generated by the elements r_{xy}.*

Proof. For a contradiction, suppose there exists $w \in W \setminus \sum Rr_{xy}$. Since $A(0) \cong S(V)$, the image of w in W_0 is zero. Hence, there exists w' in $R \otimes V \otimes V$ such that $w = hw'$. However, A is torsion-free so $w' \in W$. It follows that $w' \in W \setminus Rr_{xy}$. Repeating this argument, we find $w \in \bigcap(\mathfrak{m}_0{}^n W)$. By Lemma 2, W is free, so $\bigcap(\mathfrak{m}_0{}^n W) = (\bigcap \mathfrak{m}_0{}^n)W$. However R is a Noetherian domain, so Krull's intersection theorem ensures that $\bigcap \mathfrak{m}_0{}^n = 0$. Hence $w = 0$, which is a contradiction. □

Let f, $g \in A(0)$, and let $\tilde{f}$, $\tilde{g}$ denote preimages of f, g in A respectively. Then $\tilde{f}\tilde{g} - \tilde{g}\tilde{f} \in Ah$. Since A is torsion-free, we may define a Poisson bracket $\{\ ,\ \}$ on $A(0)$ by [8] to be

$$\{f, g\} = h^{-1}(\tilde{f}\tilde{g} - \tilde{g}\tilde{f}) \quad \text{modulo } Ah,$$

which is independent of the choice of preimages of f and g. In particular, if $x, y \in V$, then $\{x, y\} = b_0(x, y)$.

Since A is graded and since the Poisson bracket is homogeneous of degree zero, the bracket extends to $A(0)[z^{-1}]$ whenever z is a nonzero homogeneous element of $A(0)$, and it then restricts to the degree zero part $(A(0)[z^{-1}])_0$. By covering the projective space $\mathbb{P}(V^*)$ with affine open sets $\{p \in \mathbb{P}(V^*) : z(p) \neq 0\}$, where z is any homogeneous element of $A(0)$, the Poisson bracket on all such $(A(0)[z^{-1}])_0$ induces a Poisson structure on $\mathbb{P}(V^*)$. The maximal connected components of $\mathbb{P}(V^*)$ on which the Poisson structure is nondegenerate are symplectic manifolds, and $\mathbb{P}(V^*)$ is the disjoint union of these symplectic manifolds, which are called *symplectic leaves* (see [15]).

Proposition 1. *The projective zero-dimensional symplectic leaves are given by the zero locus of the polynomials $[x, y, z] := \{x, y\}z + \{y, z\}x + \{z, x\}y$ for all x, y, $z \in A(0)_1$.*

Proof. The proof given in [38] for $\mathcal{O}_q(M_n)$ applies. □

Since A is quadratic, the polynomials $[x, y, z]$ have degree three.

The assumption that A is a flat R-module implies that $W_q \hookrightarrow V \otimes V$. Let $\Gamma_q \subset \mathbb{P}(V^*) \times \mathbb{P}(V^*)$ denote the zero locus of W_q, and let $\pi_i \colon \Gamma_q \to \mathbb{P}(V^*)$ denote the projection onto the ith component for $i = 1, 2$.

Theorem 6. *Suppose that for each $q \in \mathbb{C}'$, Γ_q is the graph of an automorphism. If $\pi_1(\Gamma_q)$ is independent of q for all nonzero $q \in \mathbb{C}'$, then it is contained in the scheme consisting of the projective zero-dimensional symplectic leaves.*

Proof. Throughout the proof, we assume $q \in \mathbb{C}'$.

The homogeneous coordinate ring of Γ_q is obtained by bilinearizing the defining relations of $A(q)$ as follows. If $x, y \in V$, write the relation $xy - yx - qb_q(x, y)$ as $x_1y_2 - y_1x_2 - qb_q(x, y)_{12}$ which we view as a relation ρ_{xy} for the commutative $\mathbb{C}$-algebra generated by $\{x_1, x_2, y_1, y_2 : x, y \in V\}$. The commutative algebra B_q obtained this way, whose defining relations are the ρ_{xy}, for all $x, y \in V$, is the homogeneous coordinate ring of Γ_q (see [3, §3] and [19]). If $x, y, z \in V$, then, in B_q, we have

$$qx_1b_q(y, z)_{12} = x_1(y_1z_2 - z_1y_2) = q(y_1b_q(x, z)_{12} - z_1b_q(x, y)_{12})$$

from which it follows that if $q \neq 0$, then

$$z_1b_q(x, y)_{12} + x_1b_q(y, z)_{12} + y_1b_q(z, x)_{12} \qquad (*)$$

is zero in B_q. Since Γ_q is the graph of an automorphism σ_q, it follows that $X_2 = \sigma_q(X_1)$ for all $X \in V$. Substitution into $(*)$ yields a polynomial p_q in the variables X_1 where X runs through a basis for V. That is, if $q \neq 0$, then p_q is a relation for the homogeneous coordinate ring of $\pi_1(\Gamma_q)$. However, since, by hypothesis, the $\pi_1(\Gamma_q)$ are equal if $q \neq 0$, it follows that $p_q = f_qP$ where $f_q \in (R/\mathfrak{m}_q) \setminus \{0\}$ and P is a (possibly zero) polynomial in the variables X_1 where X runs through a basis for V.

On the other hand, since σ_0 is the identity, it follows that substituting $q = 0$ and $X_2 = X_1$ for all $X \in V$ into $(*)$ yields

$$z_1b_0(x_1, y_1) + x_1b_0(y_1, z_1) + y_1b_0(z_1, x_1) = p_0 = f_0P$$

where $f_0 \in R/\mathfrak{m}_0$. Hence, $p_0 = f_0f_q^{-1}p_q$ for all $q \neq 0$. However, the polynomials $[x, y, z]$ which define the zero-dimensional symplectic leaves are the polynomials p_0. It follows that all the polynomials $[x, y, z]$, for all $x, y, z \in V$, belong to the ideal defining $\pi_1(\Gamma_q)$ for all $q \neq 0$ which completes the proof. □

In view of this section and Sections 2-5, the notion of quantum space seems to be a good candidate for the non-commutative space of a quantum group. This is further supported by other examples which have been analyzed in the literature ([18, 19, 37, 39]). Example 2, Theorem 6, and the striking resemblance between the polynomials defining the projective zero-dimensional symplectic leaves and the quantum Yang-Baxter equation give further credibility to the quantum space of a quantum group being the quantum group's non-commutative space.

Acknowledgment

The author would like to thank S. P. Smith for suggesting that the point scheme of a quadratic quantum group be compared with the scheme of projective zero-dimensional symplectic leaves of the Poisson structure induced on the polynomial ring.

References

[1] M. Artin, Geometry of quantum planes, Azumaya algebras, actions and modules, D. Haile and J. Osterburg, eds., *Contemp. Math.* **124** (1992), 1–15.

[2] M. Artin and W. Schelter, Graded algebras of global dimension 3, *Adv. Math.* **66** (1987), 171–216.

[3] M. Artin, J. Tate, and M. Van den Bergh, Some algebras associated to automorphisms of elliptic curves, *The Grothendieck Festschrift* **1**, P. Cartier, et al., eds., Birkhäuser, 1990, 33–85.

[4] M. Artin, J. Tate, and M. Van den Bergh, Modules over regular algebras of dimension 3, *Invent. Math.* **106** (1991), 335–388.

[5] M. Artin and M. Van den Bergh, Twisted homogeneous coordinate rings, *J. Alg.* **188** (1990), 249–271.

[6] M. Artin and J. Zhang, Noncommutative projective schemes, *Adv. Math.* **109** No. **2** (1994), 228–287.

[7] M. Aubry and J. -M. Lemaire, Zero divisors in enveloping algebras of graded Lie algebras, *J. Pure and App. Algebra* **38** (1985), 159–166.

[8] V. G. Drinfel'd, Quantum groups, *Proc. Int. Cong. Math., Berkeley* **1** (1986), 798–820.

[9] L. D. Faddeev, N. Yu. Reshetikhin, and L. A. Takhtadzhyan, Quantization of Lie groups and Lie algebras, *Leningrad Math. J.* **1** No. **1** (1990), 193–225.

[10] B. L. Feigin and A. B. Odesskii, Elliptic sklyanin algebras, *Func. Anal. Appl.* **23** (1989), 45–54.

[11] M. Jimbo, A q-difference analogue of $U(\mathfrak{g})$ and the Yang-Baxter equation, *Lett. Mat. Phys.* **10** (1985), 63–69.

[12] M. Jimbo, Quantum R matrix for the generalized Toda system, *Comm. Math. Phys.* **102** (1986), 537–547.

[13] M. Jimbo, A q-analogue of $U(\mathfrak{gl}(n+1))$, Hecke algebra and the Yang-Baxter equation, *Lett. Mat. Phys.* **11** (1986), 247–252.

[14] P. Jorgensen and S. P. Smith, Intersection theory for the blowup of a noncommutative surface, in preparation, 1999.

[15] A. A. Kirillov, Local Lie algebras, *Uspekhi Mat. Nauk* **31** No. **4** (1976), 57–76; English transl., in *Russian Math. Surveys* **31** (1976).

[16] P. P. Kulish and N. Yu. Reshetikhin, The quantum linear problem for the Sine-Gordon equation and higher representations, *Zap. Nauchn. Sem. L.O.M.I.(Russian)* **101** (1981), 101–110.

[17] L. Le Bruyn, Central singularities of quantum spaces, *J. Alg.* **177** (1995), 142–153.

[18] L. Le Bruyn and S. P. Smith, Homogenized $\mathfrak{sl}(2)$, *Proc. Amer. Math. Soc.* **118** (1993), 725–730.

[19] L. Le Bruyn and M. Van den Bergh, On quantum spaces of Lie algebras, *Proc. Amer. Math. Soc.* **119** No. **2** (1993), 407–414.

[20] T. Levasseur, Some properties of non-commutative regular graded rings, *Glasgow Math. J.* **34** (1992), 277–300.

[21] T. Levasseur and S. P. Smith, Modules over the 4-dimensional Sklyanin algebra, *Bull. Soc. Math. de France* **121** (1993), 35–90.

[22] G. Lusztig, Quantum deformations of certain simple modules over enveloping algebras, *Adv. Math.* **70** (1988), 237–249.

[23] M. Rosso, Representations irreducibles de dimension finie du q-analogue de L'algebre enveloppante d'une algebre de Lie semisimple, *C. R. Acad. Sci. Paris* **305** (1987), 587–590.

[24] M. Rosso, Finite dimensional representations of the quantum analog of the enveloping algebra of a complex semisimple Lie algebra, *Comm. Math. Phys.* **117** (1988), 581–593.

[25] J. -P. Serre, Faisceaux algébriques cohérents, *Ann. of Math.* **61** (1955), 197–278.

[26] B. Shelton and M. Vancliff, Some quantum $\mathbb{P}^3$s with one point, *Comm. Alg.* **27** No. **3** (1999), 1429–1443.

[27] B. Shelton and M. Vancliff, Embedding a quantum rank three quadric in a quantum $\mathbb{P}^3$, *Comm. Alg.* **27** No. **6** (1999), 2877–2904.

[28] B. Shelton and M. Vancliff, Schemes of line modules, preprint 1999.

[29] E. K. Sklyanin, Some algebraic structures connected to the Yang-Baxter equation, *Func. Anal. Appl.* **16** No. **4** (1982), 27–34.

[30] S. P. Smith, Quantum groups: an introduction and survey for ring theorists, *Noncommutative Rings,* S. Montgomery and L. Small, eds., Springer-Verlag, 1992.

[31] S. P. Smith, The 4-dimensional Sklyanin algebra at points of finite order, preprint, 1999.

[32] S. P. Smith and J. T. Stafford, Regularity of the four dimensional Sklyanin algebra, *Compositio Math.* **83** No. **3** (1992), 259–289.

[33] S. P. Smith and J. M. Staniszkis, Irreducible representations of the 4-dimensional Sklyanin algebra at points of infinite order, *J. Alg.* **160** No. **1** (1993), 57–86.

[34] S. P. Smith and J. Zhang, Curves on quasi-schemes, *Algebras and Representation Theory* **1** No. **4** (1998), 311–351.

[35] T. Tanisaki, Finite dimensional representations of quantum groups, *Osaka J. Math.* **28** No. **1** (1991), 37–53.

[36] M. Vancliff, *The Non-commutative Algebraic Geometry of Some Quadratic Algebras,* Ph.D. Thesis, University of Washington, August 1993.

[37] M. Vancliff, Quadratic algebras associated with the union of a quadric and a line in $\mathbb{P}^3$, *J. Alg.* **165** No. **1** (1994), 63–90.

[38] M. Vancliff, The defining relations of quantum $n \times n$ matrices, *J. London Math. Soc.* **52** No. **2** (1995), 255–262.

[39] M. Vancliff and K. Van Rompay, Embedding a quantum nonsingular quadric in a quantum $\mathbb{P}^3$, *J. Alg.* **195** No. **1** (1997), 93–129.

[40] M. Vancliff, K. Van Rompay, and L. Willaert, Some quantum $\mathbb{P}^3$s with finitely many points, *Comm. Alg.* **26** No. **4** (1998), 1193–1208.

[41] M. Van den Bergh, An example with 20 points, Notes, 1988.

[42] M. Van den Bergh, Blowing up of non-commutative smooth surfaces, *Memoirs Amer. Math. Soc.*, to appear.

[43] M. Van den Bergh, Abstract blowing down, *Proc. Amer. Math. Soc.* **128** No. **2** (2000), 375–381.

[44] K. Van Rompay, Segre product of Artin-Schelter regular algebras of dimension 2 and embeddings in quantum $\mathbb{P}^3$'s, *J. Alg.* **180** No. **2** (1996), 483–512.

Michaela Vancliff
Department of Mathematics, Box 19408
University of Texas at Arlington
Arlington, TX 76019-0408
E-mail: vancliff@uta.edu

Received: September 28, 1999; Revised: February 20, 2000

5.

MATHEMATICS: STRUCTURES

Clifford Algebras and the Construction of the Basic Spinor and Semi-Spinor Modules

Johan Gijsbertus Frederik Belinfante

ABSTRACT Tensor powers of the defining module of a complex orthogonal Lie algebra can be used to construct all of the basic modules except for the spinor and semispinor modules. Thus, for the constructive representation theory of these simple Lie algebras, one must supplement the methods of tensor algebra with those of spinor algebra. For the simple complex Lie algebras of type B_l, there is a single basic spinor module, while for the simple complex Lie algebras of type D_l, there are two basic semispinor modules. The classical constructions of these spinor and semispinor modules by factoring the second Clifford algebra will be reviewed.
Keywords: Clifford algebra, spinor module.

1 Introduction

Spinors were first discovered by Élie Cartan in 1913 as a byproduct of his classification of all simple Lie groups [5]. Later, these spinor representations became famous in the special case of the Lie algebra $A_1 \approx B_1 \approx C_1$ through their role in the Pauli theory of electron spin [15]. For a more thorough account of the fascinating history of our subject, one may consult van der Waerden's book [16].

Through the classic work of Cartan, Freudenthal, Dynkin, Chevalley and others, it is known that the representation theory of the classical simple Lie algebras over the complex number field can, in principle, be reduced to algorithms involving only integers [9]. From the Cartan matrix, or the equivalent Dynkin diagram, one can set up a standard basis for the Lie algebra. Each irreducible representation is characterized by its highest weight,

---

AMS Subject Classification: 15A66, 17B10.

which can be thought of as an assignment of non-negative integers to the nodes of the Dynkin diagram. The action of the Lie algebra on the irreducible module can be deduced from a study of the action of lowering elements on an extreme vector.

In practice, much of the work involved in the construction process can be reduced by using tensor products. First, the construction of the irreducible representations can be reduced via the process of Cartan composition to that of constructing the basic irreducible modules, corresponding to highest weight assignments in which exactly one node of the Dynkin diagram is assigned the integer 1, and the others are assigned 0. Using tensor powers, the construction of these basic modules can be further reduced to the problem of constructing the basic modules corresponding to the ends of the Dynkin diagram.

In this paper, we consider only the case of the orthogonal Lie algebras. Let $M = \mathbb{C}^n$ be an n-dimensional complex vector space equipped with a nonsingular symmetric bilinear form (x, y). This vector space M may be regarded as a module over the special orthogonal group $SO(n, \mathbb{C})$ or over the corresponding complex Lie algebra. The complex orthogonal Lie algebra $L = so(n, \mathbb{C})$ is defined as the complex vector space of all linear operators a on M which are antisymmetric with respect to the bilinear form, that is, which satisfy $(x, ay) = -(ax, y)$ for all $x, y \in \mathbb{C}^n$.

The structure of the Lie algebra $so(n, \mathbb{C})$ differs for even and odd dimension n, and there are some exceptions for small values of n. The number of nodes in the Dynkin diagram is the rank l of the Lie algebra. (The rank is the dimension of a Cartan subalgebra of the Lie algebra.) If n is an odd integer, $n = 2l + 1$ with $l \geq 1$, then L is a simple Lie algebra called B_l; and if n is an even integer, $n = 2l$ with $l \geq 3$, then L is a simple Lie algebra called D_l. The Lie algebra $so(2, \mathbb{C})$ is Abelian, and $so(4, \mathbb{C})$ is semisimple, being the direct sum of two ideals each isomorphic to $so(3, \mathbb{C})$. For a simple Lie algebra, the Dynkin diagram is connected. The Dynkin diagram for B_1 has only one node, but it has two ends for B_l when $l \geq 2$. The Dynkin diagram for D_3 has only two ends, but it has three ends for D_l when $l \geq 4$. For $n = 5$ and $n \geq 7$, the defining module M of the orthogonal algebra $so(n, \mathbb{C})$ is a basic irreducible module corresponding to one of the ends of the Dynkin diagram. Using tensor methods, the defining module can be used to construct all of the basic modules except the spinor and semispinor modules, which correspond to the other end(s) of the Dynkin diagram. For the constructive representation theory of the orthogonal Lie algebras, we must, therefore, supplement the methods of tensor analysis with those of spinor analysis [3].

In this paper, we review the use of the Clifford algebra of the defining module M of an orthogonal Lie algebra L to construct the basic spinor and semispinor modules. For the simple Lie algebras of type $B_l = so(2l+1, \mathbb{C})$, there is a single basic spinor module, while for the simple Lie algebras of type $D_l = so(2l, \mathbb{C})$, there are two basic semispinor modules. The Dynkin

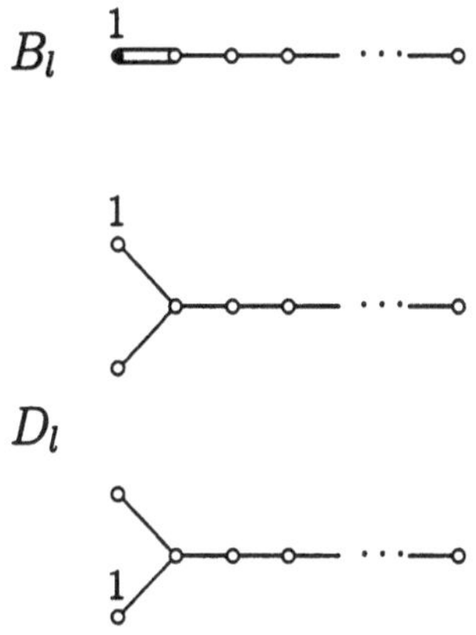

FIGURE 1. Dynkin diagrams for the basic spinor modules over B_l and semi-spinor modules over D_l.

diagrams of the basic spinor and semispinor modules are shown in Figure 1.

With the advent of symbolic computer algebra systems, it has become much more convenient to actually carry out these constructions. The importance of doing so to avoid conceptual errors has been forcefully emphasized by Pertti Lounesto [14]. We found it convenient to use a Mathematica™ program for Clifford algebra computations to check the results reported here.

2 Commutators and anticommutators

We denote the commutator of two elements x and y in an associative algebra A by

$$[x, y] = xy - yx.$$

This operation is antisymmetric and bilinear and satisfies the rule

$$[x, yz] = [x, y]\, z + y\, [x, z],$$

which resembles the Leibniz rule for differentiating a product. If we interchange y and z and subtract, we obtain the Jacobi identity

$$[x, [y, z]] - [y, [x, z]] = [[x, y], z].$$

Thus, the vector space A is a Lie algebra under the operation of commutation. We will also find it convenient to introduce the anticommutator

$$\{x, y\} = xy + yx,$$

which also satisfies various interesting identities, among which are the identity

$$[x, \{y, z\}] = \{[x, y], z\} + \{y, [x, z]\},$$

resembling the Leibniz rule, and another special identity

$$\{x,\{y,z\}\}-\{y,\{x,z\}\}=[[x,y],z],$$

which resembles the Jacobi identity. We shall have occasion to refer to this special identity in the sequel.

3 Clifford algebras

Clifford algebras were originally discovered as generalizations of Hamilton's quaternions [10]. The Clifford algebra, related to the Lorentz group $SO(3,1;\mathbb{R})$, is the Dirac gamma algebra, which is widely used in relativistic quantum mechanics [11] [15]. Since Clifford algebras are determined abstractly by their generators and relations, we may construct them explicitly via tensor algebras [8]. Let K be the ideal in the tensor algebra $T(M)$ over the module M generated by all elements of the form

$$z\otimes z-(z,z)\,1,$$

where $z\in M$. The quotient algebra

$$C\ell(M)=T(M)/K$$

is an associative algebra with unity called the first Clifford algebra of the module M with respect to the bilinear form (x,y). Since the composition of the canonical mapping of $T(M)$ onto $C\ell(M)$ with the inclusion mapping of M into $T(M)$ yields a one-to-one mapping, we may identify M with its image in $C\ell(M)$. The Clifford algebra $C\ell(M)$ obtained in this way satisfies the following universality property. If A is an associative algebra with unity, then any linear mapping $\gamma: M\to A$ which satisfies

$$\gamma(z)^2=(z,z)\,1$$

can be extended to a unique homomorphism from $C\ell(M)$ into A.

To study the properties of the Clifford algebra, it is convenient to describe the algebra in terms of generators and relations. Denoting the product in $C\ell(M)$ of elements x and y in M by xy, we have the Jordan anticommutation relations

$$\{x,y\}=2(x,y)\,1$$

for all $x,y\in M$. The generators and relations for a Clifford algebra take the especially simple form

$$\{o_i,o_j\}=2\delta_{ij}1$$

if we introduce an orthonormal basis $o_1,\ldots,o_n$ in M. We can then show that the dimension of the first Clifford algebra is given by $\dim C\ell(M)=2^n$ since ${o_1}^{r_1}\cdots{o_n}^{r_n}$, where $r_i=0,1$, is a basis for the Clifford algebra.

4 The second Clifford algebra

The Jordan relations are invariant under replacement of all vectors in M by their negatives. Consequently, the mapping $z \mapsto -z$ in M induces an involutive automorphism of the Clifford algebra $C\ell(M)$. Denoting the image of an element $t \in C\ell(M)$ under this automorphism by t^*, we have $t^{**} = t$, and $z^* = -z$ for all $z \in M$. The Clifford algebra $C\ell(M)$ can be split up as the direct sum of two subspaces corresponding to the eigenvalues ± 1 of this automorphism. The part $C\ell^+(M)$ corresponding to the eigenvalue $+1$ is a subalgebra of $C\ell(M)$ called the second Clifford algebra [1].

We can clarify the structures of the first and second Clifford algebras somewhat as follows. For each vector $z \in M$, we introduce a linear operator $\alpha(z)$ on $C\ell(M)$ by

$$\alpha(z)t = zt + t^* z.$$

By direct computation, one verifies that these operators anticommute, that is, they satisfy $\alpha(z)\alpha(z') = -\alpha(z')\alpha(z)$ for all $z, z' \in M$. If we let M_k be the subspace of $C\ell(M)$ spanned by all elements of the form $\alpha(z_1)\cdots\alpha(z_k)\,1$, then

$$C\ell(M) = M_0 \oplus M_1 \oplus \cdots \oplus M_n.$$

In particular, we note that

$$\alpha(z_1)\,1 = 2z_1, \quad \alpha(z_1)\alpha(z_2)\,1 = 2[z_1, z_2],$$

and, hence,

$$M_0 = \mathbb{C},\ M_1 = M \quad \text{and} \quad M_2 = [M, M].$$

Since the operators $\alpha(z_i)$ anticommute, the expression $\alpha(z_1)\cdots\alpha(z_k)\,1$ is completely antisymmetric in the vectors $z_1, \ldots, z_k$.

The mapping $z_1 \wedge \cdots \wedge z_k \mapsto \alpha(z_0)\cdots\alpha(z_k)\,1$ establishes a vector space isomorphism between the exterior power $\bigwedge^k M$ and the subspace M_k; and, hence,

$$t = \alpha(z_1)\cdots\alpha(z_k)\,1$$

satisfies $t^* = (-1)^k t$, it should be apparent that the second Clifford algebra is given by

$$C\ell^+(M) = M_0 \oplus M_2 \oplus M_4 \oplus \cdots.$$

In other words, the second Clifford algebra is spanned by the even elements of the first Clifford algebra, and it follows that it is a subalgebra with $\dim C\ell^+(M) = 2^{n-1}$.

For example, the first Clifford algebra for the complex Lie algebra $A_1 \approx B_1 \approx C_1$ corresponding to the ordinary rotation group $SO(3,\mathbb{R})$ is generated by three orthonormal vectors o_1, o_2 and o_3 satisfying the Jordan relations. In this case, the first Clifford algebra $C\ell(M)$ is an eight-dimensional algebra spanned by the elements 1, $o_1, o_2, o_3, o_1o_2, o_2o_3, o_1o_3$ and $o_1o_2o_3$. The second Clifford algebra $C\ell^+(M)$ is the four-dimensional subalgebra of $C\ell(M)$ spanned by 1, $i = o_2o_3$, $j = o_1o_3$ and $k = o_1o_2$. Since i, j and k satisfy $i^2 = j^2 = k^2 = -1$, and $ij = -ji = k$, $jk = -kj = i$, and $ki = -ik = j$, we may identify $C\ell^+(M)$ as the complex quaternion algebra. Unlike the real quaternion algebra, this complex quaternion algebra is not a division algebra. Indeed, the complex quaternion algebra contains nilpotent elements such as $i + \sqrt{-1}\, j$.

5 Embedding the orthogonal Lie algebra

Any associative algebra A may be regarded as a Lie algebra A_L under commutation. In the case of the Clifford algebra, the orthogonal Lie algebra L itself may be regarded as a subalgebra of the Lie algebra $C\ell(M)_L$. To see this, we argue as follows. For all x, y and z in M we may obtain the relation

$$[[x, y], z] = 4\,(y, z)\,x - 4\,(x, z)\,y$$

from the Jordan relation by using the previously cited special identity connecting commutators and anticommutators. An immediate corollary of this formula is that $[M_2, M] \subset M$, and, hence, we also find

$$[M_2, M_2] = [M_2, [M, M]] \subset [[M_2, M], M] \subset [M, M] = M_2$$

by using the Jacobi identity. Thus, M_2 is a subalgebra of the Lie algebra $C\ell(M)_L$, and we shall argue that $M_2 \approx L$. For each $a \in M_2$, define a linear operator $f(a)$ on M by $f(a)x = [a, x]$. These linear operators are antisymmetric with respect to the inner product on M,

$$(f(a)x, y) = -(x, f(a)y)$$

for all $a \in M_2$ and all $x, y \in M$. To prove this, one can use the Jordan relations and the identity

$$[a, \{x, y\}] = \{[a, x], y\} + \{x, [a, y]\}.$$

The left side is zero because $\{x, y\} = 2(x, y)$ is a scalar, and, therefore, commutes with a. Since $f(a)$ is an antisymmetric linear operator, $f(a) \in L$. One may show that $f : M_2 \to L$ is a one-to-one Lie algebra homomorphism, from which it follows that $\dim f[M_2] = \dim M_2 = \binom{n}{2}$. Since the orthogonal Lie algebra L has the same dimension, $\dim L = \binom{n}{2}$, the mapping f is onto, establishing the isomorphism of M_2 with L.

For the simplest case of the Lie algebra $L = B_1 = so(3, \mathbb{C})$, the elements i, j, k form a basis satisfying the commutation relations

$$[i, j] = 2k, \quad [j, k] = 2i, \quad [k, i] = 2j.$$

In applications to physics, it is customary to use a different basis J_x, J_y, J_z satisfying the commutation relations

$$[J_x, J_y] = \sqrt{-1}\, J_z, \quad [J_y, J_z] = \sqrt{-1}\, J_x, \quad [J_z, J_x] = \sqrt{-1}\, J_y.$$

To translate the notation used here, one simply needs to make the identifications

$$J_x = \frac{1}{2}\sqrt{-1}i, \quad J_y = \frac{1}{2}\sqrt{-1}j, \quad J_z = \frac{1}{2}\sqrt{-1}k.$$

For Lie algebraic calculations, a more convenient basis is the Chevalley basis

$$\begin{aligned} h &= 2J_z = \sqrt{-1}\, k, \\ e &= J_x + \sqrt{-1}\, J_y = \frac{1}{2}(j - \sqrt{-1}i), \\ f &= J_x - \sqrt{-1}\, J_y = \frac{1}{2}(j + \sqrt{-1}i). \end{aligned}$$

The factor 2 in the definition of h is included convert the half-integer spins used in physics to integers.

6 Constructing the spinor and semispinor modules

A module over a Lie algebra L consists of a vector space M and a bilinear mapping $\mu : L \times M \to M$, such that

$$\mu([l_1, l_2], m) = \mu(l_1, \mu(l_2, m)) - \mu(l_2, \mu(l_1, m))$$

for all $l_1, l_2 \in L$ and all $m \in M$. One such action is the trivial one with $\mu = 0$. When a Lie algebra L is a subalgebra of the Lie algebra A_L of an associative algebra A, there are, in general, besides the trivial action, always three other natural module actions of L on A. One of these actions is by commutation. In this case

$$\mu(l, a) = [l, a] = la - al$$

for all $l \in L$ and all $a \in A$. The fact that this is a module action follows from the Jacobi identity for commutators. A second natural module action is obtained by using left-multiplication; in this case

$$\mu(l, a) = la,$$

for all $l \in L$ and all $a \in A$. That this is a module action amounts to the statement

$$[l_1, l_2]\, a = l_1(l_2 a) - l_2(l_1 a),$$

which follows from the associative law in A and the fact that the Lie bracket is the commutator. A third natural action of L on A is by negative right-multiplication

$$\mu(l, a) = -al.$$

These natural module actions work for any Lie subalgebra of an associative algebra; for particular algebras, of course, there may be other possibilities in addition to these generic ones.

In the case of the Clifford algebra $C\ell(M)$ and the Lie subalgebra $L = [M, M]$, the module that we get by using the commutation action is a direct sum of the fundamental module M and various other submodules that one can obtain from the fundamental module M by using tensor algebra. To obtain the spinor and semispinor modules from the Clifford algebra, we should instead use the left-multiplication action of L on $C\ell(M)$ or, better yet, the left-multiplication action on the second Clifford algebra $C\ell^+(M)$. One could also use the negative right-multiplication action, but the results would not be essentially different.

For the actual construction of the spinor and semispinor modules, we must distinguish the case $n = 2l + 1$ with $L = B_l$ from the case $n = 2l$ with $L = D_l$.

For odd n, the first Clifford algebra $C\ell(M)$ is a semisimple associative algebra, isomorphic to the direct sum of two copies of $C\ell^+(M)$. The second Clifford algebra $C\ell^+(M)$ for odd n is a simple associative algebra, isomorphic to the algebra $\mathrm{lin}(N)$ of linear operators on some vector space N. This vector space N of dimension 2^l will turn out to be the basic spinor module over B_l. We may write $C\ell^+(M) \approx \mathrm{lin}(N) \approx N \otimes N^*$, and since the basic spinor module N also happens to be self-dual, this simplifies to $C\ell^+(M) \approx N \otimes N$. Thus, for the case n odd, the second Clifford algebra $C\ell^+(M)$ is isomorphic, as a module over L, to the tensor square $N \otimes N$ of the basic spinor module N.

For even n, the first Clifford algebra $C\ell(M)$ itself is simple, and thus is isomorphic to the algebra of linear operators on some vector space of dimension 2^l. This vector space may be identified with the direct sum $N_1 \dot{+} N_2$ of the two basic semispinor modules N_1 and N_2 over D_l, each having dimension 2^{l-1}. While the semispinor modules themselves are self-dual only for even l, and each is the dual of the other for odd l, their direct sum $N_1 \dot{+} N_2$ is self-dual in either case. Thus, for even n we may write the first Clifford algebra as

$$C\ell(M) \approx \mathrm{lin}(N) \approx N \otimes N,$$

where $N = N_1 \dot{+} N_2$. The second Clifford algebra $C\ell^+(M)$ for even n is not simple, but only semisimple, being the direct sum of two simple ideals, namely,

$$C\ell^+(M) \approx \mathrm{lin}(N_1) \dot{+} \mathrm{lin}(N_2).$$

We shall give a detailed explanation of these constructions only for the case that of odd $n = 2l+1$. The basic 2^l-dimensional spinor module N appears, roughly speaking, as the square root of the 2^{l+1}-dimensional second Clifford algebra $C\ell^+(M) \approx N \otimes N$.

In order to extract this square root explicitly, it is convenient to introduce a special basis for the orthogonal Lie algebra $L = B_l$. We begin by introducing a set of $2l$ elements

$$u_i = \frac{1}{2}\left(\sqrt{-1}\, o_{2i} o_{2l+1} + o_{2i-1} o_{2l+1}\right)$$

and

$$v_i = \frac{1}{2}\left(\sqrt{-1}\, o_{2i} o_{2l+1} - o_{2i-1} o_{2l+1}\right),$$

where $i = 1, \ldots, l$, which satisfy the Jordan-Wigner canonical anticommutation relations [12]

$$\{u_i, v_j\} = \delta_{ij}, \quad \{u_i, u_j\} = \{v_i, v_j\} = 0.$$

Since these elements are formally analogous to the annihilation and creation operators used in quantum field theory, we refer to the u_i as lowering elements and the v_j as raising elements. These raising and lowering elements, together with their commutators, generate the Lie algebra $M_2 \approx L = B_l$. Specifically, the l commutators $[v_i, u_i]$ form a basis for the Cartan subalgebra H of this Lie algebra. The $2l$ elements u_i and v_j are root vectors, as are the $l(l-1)$ commutators $[u_i, u_j]$ and $[v_i, v_j]$, where $i \neq j$, and the $l(l-1)$ commutators $[u_i, v_j]$, where $i \neq j$.

Under multiplication, these raising and lowering elements generate the whole second Clifford algebra $C\ell^+(M)$. A basis for $C\ell^+(M)$ consists of the set of elements

$$u_{i_1} \cdots u_{i_r} v_{j_1} \cdots v_{j_s},$$

where $i_1 < \cdots < i_r$ and $j_1 < \cdots < j_s$. The second Clifford algebra then factorizes as a product $C\ell^+(M) = UV = VU$, where U and V are the 2^l-dimensional subalgebras generated by the u's and v's, respectively, together with 1.

The first step toward analyzing the structure of $C\ell^+(M)$, considered as a module over the Lie algebra $L \approx M_2$ under the left-multiplication action, is to identify all the extreme vectors $x \in C\ell^+(M)$. These are the vectors which are annihilated by the root vectors corresponding to the positive roots of the Lie algebra. These vectors must therefore satisfy $u_i x = 0$ for all $i = 1, \ldots, l$, and it can be shown that for the module $C\ell^+(M)$ these conditions are also sufficient for x to be extreme. The extreme vectors form the 2^l-dimensional subspace $X = (v_1 \cdots v_l)\, U$.

From each extreme vector $x \in X$, we obtain a spinor module $U x$ by applying the lowering subalgebra U. Since $V v_1 \cdots v_l = v_1 \cdots v_l = v_1 \cdots v_l V$, the extreme subspace X can also be described as the right ideal of the second Clifford algebra generated by the element $v_1 \cdots v_l$. Each of the subspaces $U\,x$ is a 2^l-dimensional left ideal of the second Clifford algebra $C\ell^+(M)$. Since $M_2\,U\,x \subset U\,x$, we may regard each $U\,x$ as a 2^l-dimensional module over $L \approx M_2$. These submodules are all isomorphic to the particular spinor module $N = U\,v_1 \cdots v_l$.

7 Examples: B_1 and B_2

The construction of the spinor module for the simplest case, corresponding to the Lie algebra $A_1 \approx B_1 \approx C_1$, is somewhat atypical in that the defining module M is not a basic module; the irreducible module M has Dynkin index 2. For each integer $n \geq 0$ there is an $n+1$-dimensional irreducible module over B_1 with Dynkin index n. In physics, $\frac{n}{2}$ is called the spin. We avoid the square roots encountered in angular momentum theory by introducing Chevalley bases for the irreducible modules over B_1 [2]. One can chose a basis for this module with respect to which the Chevalley basis elements e, f, h are represented by integer matrices:

$$e \mapsto \begin{pmatrix} 0 & n & 0 & \cdots & 0 & 0 \\ 0 & 0 & n-1 & \cdots & 0 & 0 \\ 0 & 0 & 0 & \cdots & 0 & 0 \\ \vdots & \vdots & \vdots & & \vdots & \vdots \\ 0 & 0 & 0 & \cdots & 0 & 1 \\ 0 & 0 & 0 & \cdots & 0 & 0 \end{pmatrix}, \quad f \mapsto \begin{pmatrix} 0 & 0 & 0 & \cdots & 0 & 0 \\ 1 & 0 & 0 & \cdots & 0 & 0 \\ 0 & 2 & 0 & \cdots & 0 & 0 \\ \vdots & \vdots & \vdots & & \vdots & \vdots \\ 0 & 0 & 0 & \cdots & 0 & 0 \\ 0 & 0 & 0 & \cdots & n & 0 \end{pmatrix},$$

$$h \mapsto \begin{pmatrix} n & 0 & \cdots & 0 & 0 \\ 0 & n-2 & \cdots & 0 & 0 \\ \vdots & \vdots & & \vdots & \vdots \\ 0 & 0 & \cdots & -n+2 & 0 \\ 0 & 0 & \cdots & 0 & -n \end{pmatrix}.$$

Starting from the defining 3-dimensional module M, the methods of tensor analysis can be used to construct all the irreducible modules over B_1 whose Dynkin indices are even, but not those whose Dynkin indices are odd. The second Clifford algebra $C\ell^+(M)$, which in this case is the complex quaternion algebra, may be generated from the elements

$$u = J_x - \sqrt{-1}\,J_y = \frac{1}{2}(\jmath + \sqrt{-1}\imath),$$
$$v = J_x + \sqrt{-1}\,J_y = \frac{1}{2}(\jmath - \sqrt{-1}\imath),$$

which belong to $M_2 \approx L$. The Chevalley basis for B_1 can be identified with the vectors $e = v$, $f = u$ and $h = [e, f]$ in M_2. The second Clifford algebra $C\ell^+(M)$, spanned by $1, u, v$ and uv, factorizes as UV, where the lowering subalgebra U is the 2-dimensional subalgebra spanned by 1 and u. The raising subalgebra V is spanned by 1 and v. The extreme subspace $X = vU$ is spanned by the vectors v and vu. Each vector $x \in X$ generates a two-dimensional irreducible spinor module $U\,x$. The vector x is a weight vector $hx = x$. Hence the highest weight of the module $U\,x$ has Dynkin index 1. With respect to the basis $b_1 = x$, $b_2 = u\,x$, the Chevalley basis elements e, f, h for the Lie algebra B_1 are represented by the following integer matrices:

$$e \mapsto \begin{pmatrix} 0 & 1 \\ 0 & 0 \end{pmatrix}, \quad f \mapsto \begin{pmatrix} 0 & 0 \\ 1 & 0 \end{pmatrix}, \quad h \mapsto \begin{pmatrix} 1 & 0 \\ 0 & -1 \end{pmatrix}.$$

The extreme vector $x = v$ generates the spinor module $N = Uv$, spanned by v and uv. The extreme vector vu generates another submodule Uvu, with basis vu and $u = uvu$, which is isomorphic to N. The factorization of the second Clifford algebra, $UX = UvU = NU = C\ell^+(M) = M_0 + M_2 \approx \mathbb{C} \dot{+} L$, for the Lie algebra $A_1 \approx B_1 \approx C_1$, resembles the Clebsch-Gordan series $\mathbf{2} \otimes \mathbf{2} \approx \mathbf{1} \dot{+} \mathbf{3}$.

The orthogonal Lie algebra B_1 is isomorphic to the symplectic Lie algebra C_1 consisting of all linear operators on a 2-dimensional complex vector space that are antisymmetric with respect to a nonsingular antisymmetric bilinear form. This complex vector space may be identified with the spinor module for B_2. The representation matrices for e, f, h, above, are antisymmetric with respect to the antisymmetric bilinear form given by the matrix

$$\begin{pmatrix} 0 & -1 \\ 1 & 0 \end{pmatrix}.$$

For the Lie algebra $L = B_2 = so(5, \mathbb{C})$, the 5-dimensional defining module M is a basic irreducible module with Dynkin indices (0,1). The basic spinor module with Dynkin indices (1,0) does not arise as a submodule of any tensor power of M, but it can be constructed using the Clifford algebra $C\ell^+(M)$. The elements u_1, u_2 and v_1, v_2 generate $C\ell^+(M)$ under multiplication. Under commutation these elements generate the Lie algebra $M_2 \approx L$.

A Chevalley basis for the 10-dimensional Lie algebra $M_2 \approx L = B_2$ can be constructed by using the elements

$$e_1 = v_2, \quad f_1 = u_2, \quad h_1 = [e_1, f_1],$$

$$e_2 = \frac{1}{2}[u_2, v_1], \quad f_2 = \frac{1}{2}[u_1, v_2], \quad h_2 = [e_2, f_2]$$

and their commutators, $[e_1, e_2]$, $[e_1, [e_1, e_2]]$, $[f_1, f_2]$, and $[f_1, [f_1, f_2]]$. (See Figure 2.)

We consider the 16-dimensional algebra $A = C\ell^+(M)$ as a module over the Lie subalgebra $M_2 \approx L = B_2$ under the left multiplication action.

To reduce the module A as a direct sum of irreducible submodules, we begin by identifying the subspace $X \subset A$ of all extreme vectors $x \in A$ satisfying $e_1 x = e_2 x = 0$. Every such vector can be written as $x = v_1 v_2 a$, where $a \in A$ is arbitrary; that is, the extreme subspace is the right ideal $X = v_1 v_2 A$ of the associative algebra A. Writing $A = VU$ and using the canonical anticommutation relations satisfied by the raising and lowering elements, one can write $X = v_1 v_2 U$, from which one can deduce that the extreme subspace X has dimension 4. The module A is the direct sum of four isomorphic spinor submodules $Ux_1, \ldots, Ux_4$, where $x_1, \ldots, x_4$ is any basis for X.

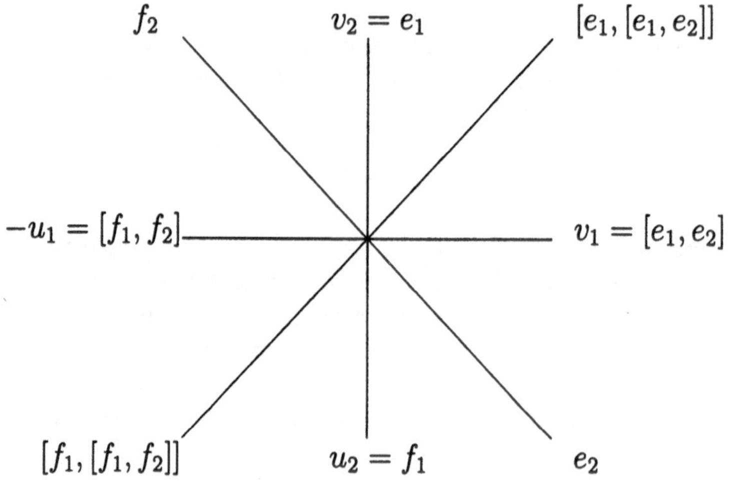

FIGURE 2. Root vectors of a Chevalley basis for B_2

Each element $x \in X$ generates an irreducible four-dimensional spinor submodule Ux. Each element $x \in X$ satisfies $h_1 x = x$ and $h_2 x = 0$, so the highest weight of the submodule $U x$ has Dynkin indices (1,0). With respect to the basis $b_1 = x$, $b_2 = f_1 b_1$, $b_3 = f_2 b_2$, $b_4 = f_1 b_3$, the Chevalley basis elements of the Lie algebra are represented by the following integer matrices:

$$e_1 \mapsto \begin{pmatrix} 0 & 1 & 0 & 0 \\ 0 & 0 & 0 & 0 \\ 0 & 0 & 0 & 1 \\ 0 & 0 & 0 & 0 \end{pmatrix}, \quad e_2 \mapsto \begin{pmatrix} 0 & 0 & 0 & 0 \\ 0 & 0 & 1 & 0 \\ 0 & 0 & 0 & 0 \\ 0 & 0 & 0 & 0 \end{pmatrix},$$

$$f_1 \mapsto \begin{pmatrix} 0 & 0 & 0 & 0 \\ 1 & 0 & 0 & 0 \\ 0 & 0 & 0 & 0 \\ 0 & 0 & 1 & 0 \end{pmatrix}, \quad f_2 \mapsto \begin{pmatrix} 0 & 0 & 0 & 0 \\ 0 & 0 & 0 & 0 \\ 0 & 1 & 0 & 0 \\ 0 & 0 & 0 & 0 \end{pmatrix},$$

$$h_1 \mapsto \begin{pmatrix} 1 & 0 & 0 & 0 \\ 0 & -1 & 0 & 0 \\ 0 & 0 & 1 & 0 \\ 0 & 0 & 0 & -1 \end{pmatrix}, \quad h_2 \mapsto \begin{pmatrix} 0 & 0 & 0 & 0 \\ 0 & 1 & 0 & 0 \\ 0 & 0 & -1 & 0 \\ 0 & 0 & 0 & 0 \end{pmatrix}.$$

The orthogonal Lie algebra B_2 is isomorphic to the symplectic Lie algebra C_2 consisting of all linear operators on a 4-dimensional complex vector space that are antisymmetric with respect to a nonsingular antisymmetric bilinear form. This vector space we identify with the spinor module for B_2. The representation matrices found above are antisymmetric with respect to the bilinear form defined by

$$\begin{pmatrix} 0 & 0 & 0 & -1 \\ 0 & 0 & 1 & 0 \\ 0 & -1 & 0 & 0 \\ 1 & 0 & 0 & 0 \end{pmatrix}.$$

8 Spin groups

We have concentrated in our paper on applications of Clifford algebras to representations of orthogonal Lie algebras. A closely related, but technically more complicated subject, is the study of the Lie groups associated with these Lie algebras. Here too one can apply Clifford algebra ideas [13, 4].

The simply connected Lie groups corresponding to the compact real forms of B_l and D_l are not the orthogonal groups $SO(n,\mathbb{R})$, but their covering groups Spin(n). To obtain these spin groups from the Clifford algebra, we consider the group Γ of all invertible elements $u \in C\ell(M)$ such that $uzu^{-1} \in M$ for all $z \in M$. The spin group Spin(n) is the identity component of the subgroup of the group Γ consisting of all elements whose left regular representation restricted to any ideal of $C\ell(M)$ has determinant one [6], [7].

A Appendix

```
(* :Title: CliffAlg.m -- a simple Clifford algebra package *)
(* :Author: Johan G. F. Belinfante                         *)
(* :Context: CliffAlg`                                     *)
(* :Summary:
   Basic Mathematica tools for Clifford algebra calculations. *)
(* :Package Version: 1999 July 25 *)
(* :Mathematica Version: 3.0      *)
(* :Keywords: Clifford algebra    *)
(* :Sources:
   Johan G. F. Belinfante and Bernard Kolman,
   A Survey of Lie Groups and Lie Algebras with Applications and
```

```
    Computational Methods, 2nd ed., Society for Industrial and
    Applied Mathematics, Philadelphia, 1989 *)
(* :Warnings: <description of global effects, incompatibilities>
    The Mathematica system functions Dot and Expand are redefined
    to allow Dot to be used as multiplication in a Clifford algebra.
    The functions scalarQ and vectorQ are not protected. *)

(* :Limitations:
   <special cases not handled>
   Only Clifford algebras over the complex numbers are considered.
   The inner product is assumed to be nonsingular.  An orthonormal
   basis o[1], ... , o[n] is used for the vector space on which
   the Clifford algebra is built.

   <known problems>
   Some calculations could be speeded up by using memofunctions.
   Symbolic numbers like E and Pi may not be treated as scalars
   unless one makes explicit declarations for them.
   We purposely left the scalarQ function unprotected to
   facilitate adding such declarations if one needs them. *)

(* :Discussion: <description of algorithm, information for experts>
    See text of Ixtapa paper. *)
(* :Requirements: No other packages are needed. *)

(* :Examples:
    <sample input that demonstrates the features of this package>
    Sample notebooks were prepared for the Ixtapa paper. *)
(*  set up the package context, including public imports *)

BeginPackage["CliffAlg`"]
(* usage messages for the exported functions and the context *)

CliffAlg::usage = "CliffAlg.m is a package for Clifford Algebras"
brace::usage = "brace[x,y] is the anticommutator of x and y"
bracket::usage = "bracket[x,y] is the commutator of x and y"

CliffordBasis::usage = "CliffordBasis[n] produces a list of the
 2^n elements of a standard basis for the Clifford algebra on
 complex n-space"

ComponentExpand::usage = "ComponentExpand[c,n] produces a generic
 element of the Clifford algebra Cl(n) expanded in the standard
 basis with scalar coefficients called c[i,j,...]"

Components::usage = "Components[x,n] produces a list of the
 components of an element x of the Clifford algebra Cl[n]"

Dot::usage = "Dot[x,y,...] adapts Mathematica's built in Dot to be
 used for multiplication in a Clifford algebra"

DotPower::usage = "DotPower[x,n] is the n-th power of x in the
 Clifford algebra"
```

```
Expand::usage = "Expand[x] is an extension of Mathematica's
 builtin Expand[x]"

inner::usage = "inner[x,y] is the inner product of vectors x and y"

o::usage ="o[i,j,...] are basis elements for the Clifford algebra"

scalarQ::usage = "scalarQ[x] tests whether x is a scalar"

Star::usage = "Star denotes the involution of the Clifford algebra
 which sends vectors to their negatives.  Star[x] is the result of
 applying Star to x."

vectorQ::usage = "vectorQ[x] tests whether x is a vector"

Begin["'Private'"]    (* begin the private context *)

(* definition of auxiliary functions and local variables *)

outcomes[n_]:=Distribute[Table[{True,False},{n}],List]
PowerSetRange[n_]:=Union[Map[Map[First,Position[#,False]]&,
  outcomes[n]]]
PowerSet[x_List]:=Map[Part[x,#]&, PowerSetRange[Length[x]]]

(* definitions of the exported functions *)

(* rules for scalars *)
scalarQ[c_?NumberQ] := True
scalarQ[a_ + b_] := scalarQ[a] && scalarQ[b]
scalarQ[a_ b_] := scalarQ[a] && scalarQ[b]
scalarQ[Power[c_,n_]] := scalarQ[c]

(* vectorQ *)
vectorQ[o[i__]] := Equal[Length[List[i]],1]
vectorQ[a_?scalarQ x_?vectorQ] := True
vectorQ[x_?vectorQ + y_?vectorQ] := True

(* new rules for System function Expand *)
Unprotect[Expand]
Expand[x_ + y_] := Expand[x] + Expand[y]
Expand[x_ y_] := Module[{times = Unique[]},
    Distribute[times[Expand[x],Expand[y]], Plus] /. times -> Times]
Expand[x_ . y_] := Module[{dot = Unique[]},
    Distribute[dot[Expand[x],Expand[y]], Plus] /. dot -> Dot]
Protect[Expand]      (* restore protection to Expand *)

(* new rules for the System function Dot *)
Unprotect[Dot]
o[i__] . o[j__] := o[i,j]
a_?scalarQ . x_ := a x
x_ . a_?scalarQ := a x
(a_?scalarQ x_) . y_ := a (x . y)
```

```
x_ . (b_?scalarQ y_) := b (x . y)
(a_?scalarQ x_) . (b_?scalarQ y_) := a b (x . y)
Protect[Dot]          (* restore protection of system symbols *)

(* Clifford algebra rules for o *)
o[a___,i_,j_,b___] := -o[a,j,i,b] /; i > j
o[a___,i_,i_,b___] := o[a,b]
o[] := 1

(* DotPower rules *)
DotPower[x_,0] := 1
DotPower[x_,n_Integer] := x . DotPower[x,n-1] /; n > 0

(* rules for the involution Star *)
Star[Star[x_]] := x
Star[c_] := c /; scalarQ[c]
Star[a_ + b_] := Star[a] + Star[b]
Star[a_ b_] := Star[a] Star[b]
Star[a_ . b_] := Star[a] . Star[b]
Star[o[i__]] := (-1)^Length[List[i]] o[i]

(* bracket and brace *)
bracket[x_,y_] := (x . y - y . x) // Expand
brace[x_,y_] := (x . y + y . x) // Expand

(* inner product *)
inner[x_?vectorQ,y_?vectorQ] := (1/2) brace[x,y]

(* bases and components *)
ComponentList[c_,n_] := Map[Apply[c,#]&,PowerSetRange[n]]
CliffordBasis[n_] := ComponentList[o,n]
ComponentExpand[c_,n_] := Module[{}, scalarQ[c[___]]^:=True;
                Expand[ComponentList[c,n] . CliffordBasis[n]]]
ConstantTerm[x_] := Expand[x] /. o[__] :> 0
Components[x_,n_] := Join[{ConstantTerm[x]},
                     Map[Coefficients[Expand[x],#]&,
                     Drop[CliffordBasis[n],1]]]

End[ ]          (* end the private context *)

(* protect exported symbols other than scalarQ and vectorQ *)
Protect[ brace, bracket, CliffordBasis, ComponentExpand,
         Components, DotPower, inner, o, Star ]

EndPackage[ ]  (* end the package context *)
```

REFERENCES

[1] Emil Artin, *Geometric Algebra,* Interscience Tracts in Pure and Applied Mathematics, Vol. **3**, Interscience, New York, 1957.

[2] Johan Gijsbertus Frederik Belinfante, Integer Clebsch-Gordan coefficients for Lie algebra representations, in *Computers in Nonassociative Rings and*

Algebras, Robert E. Beck and Bernard Kolman, eds., Academic Press, New York, 1977, 209–234.

[3] Johan Gijsbertus Frederik Belinfante and Bernard Kolman, *A Survey of Lie Groups and Lie Algebras, with Applications and Computational Methods, Second Edition,* Society for Industrial and Applied Mathematics, Philadelphia, 1989.

[4] Richard Brauer and Hermann Weyl, Spinors in n dimensions, *American Journal of Math.* **57** (1935), 425–449.

[5] Élie Joseph Cartan, Projective groups which do not leave any flat manifold invariant, *Bull. Soc. Math. France* **41** (1913), 53–96.

[6] Claude Chevalley, *Theory of Lie Groups,* Volume 1, Princeton University Press, Princeton, 1946.

[7] Claude Chevalley, *The Algebraic Theory of Spinors,* Columbia Univ. Press, New York, 1954.

[8] Claude Chevalley, *The Construction and Study of Certain Important Algebras,* Mathematical Society of Japan, Tokyo, 1955.

[9] Claude Chevalley, On certain simple groups, *Tôhoku Math. J.* **7** (2) (1955), 14–66.

[10] William Kingdon Clifford, Applications of Grassmann's extensive algebra, *American Journal of Math.* **1** (1978), 350–358.

[11] R. H. Good, Properties of the Dirac matrices, *Reviews of Modern Physics* **27** (1955), 187–211.

[12] Ernst Pascual Jordan and Eugene Paul Wigner, On the Pauli equivalence exclusion principle, *Zeitschrift für Physik* [in German] **47** (1928), 631–651.

[13] Rudolf Lipschitz, *Investigations on the Sums of Squares,* Max Cohen and Sohn, Bonn, 1884 [in German].

[14] Pertti Lounesto, Counter-examples in Clifford algebras, *Advances in Applied Clifford Algebras* **6** (1996), 69–104.

[15] Wolfgang Pauli, Jr., Mathematical contributions to the theory of Dirac matrices, *Ann. Inst. H. Poincaré* **6** (1936), 109–136.

[16] Bartel Leenert van der Waerden, *A History of Algebra,* Springer-Verlag, Berlin, 1985.

Johan Gijsbertus Frederik Belinfante
Georgia Institute of Technology
Atlanta, GA 30332–0160 (U.S.A.)
E-mail: belinfan@math.gatech.edu

Received: August 11, 1999; Revised: October 20, 1999

On the Decomposition of Clifford Algebras of Arbitrary Bilinear Form

Bertfried Fauser and Rafał Abłamowicz

ABSTRACT Clifford algebras are naturally associated with quadratic forms. These algebras are $\mathbb{Z}_2$-graded by construction. However, only a $\mathbb{Z}_n$-gradation induced by a choice of a basis, or even better, by a Chevalley vector space isomorphism $C\ell(V) \leftrightarrow \bigwedge V$ and an ordering, guarantees a multi-vector decomposition into scalars, vectors, tensors, and so on, mandatory in physics. We show that the Chevalley isomorphism theorem cannot be generalized to algebras if the $\mathbb{Z}_n$-grading or other structures are added, e.g., a linear form. We work with pairs consisting of a Clifford algebra and a linear form or a $\mathbb{Z}_n$-grading which we now call *Clifford algebras of multi-vectors* or *quantum Clifford algebras*. These quantum Clifford algebras are in fact Clifford algebras of a bilinear form in a functorial way. It turns out that in this sense, all multi-vector Clifford algebras of the same quadratic but different bilinear forms are non-isomorphic. The usefulness of such algebras in quantum field theory and superconductivity was shown elsewhere. Allowing for arbitrary bilinear forms however spoils their diagonalizability which has a considerable effect on the tensor decomposition of the Clifford algebras governed by the periodicity theorems, including the Atiyah-Bott-Shapiro mod 8 periodicity. We consider real algebras $C\ell_{p,q}$ which can be decomposed in the symmetric case into a tensor product $C\ell_{p-1,q-1} \otimes C\ell_{1,1}$. The general case used in quantum field theory lacks this feature. Theories with non-symmetric bilinear forms are however needed in the analysis of multi-particle states in interacting theories. A connection to q-deformed structures through nontrivial vacuum states in quantum theories is outlined.

Keywords: Clifford algebras of multi-vectors, Clifford map, quantum Clifford algebras, periodicity theorems, index theorems, spinors, spin-tensors, Chevalley map, quadratic forms, bilinear forms, deformed tensor products, multi-particle geometric algebra, multi-particle states, compositeness, inequivalent vacua.

AMS Subject Classifications: 15A66, 17B37, 81R25, 81R50.

1 Why study Clifford algebras of an arbitrary bilinear form?

1.1 Notation, basics and naming

Notation

To fix our notation, we want to give some preliminary material. If nothing is said about the *ring* linear spaces or algebras are build over, we denote it by $\mathbf{R}$ and assume usually that it is unital, commutative and not of characteristic 2. In some cases we specialize our base ring to the field of real or complex numbers denoted as $\mathbb{R}$ and $\mathbb{C}$. A *quadratic form* is a map $Q : V \mapsto \mathbf{R}$ with the following properties $(\alpha \in \mathbf{R},\ \mathbf{x}, \mathbf{y} \in V)$

$$\begin{aligned} &i)\quad Q(\alpha \mathbf{x}) = \alpha^2 Q(\mathbf{x}), \\ &ii)\quad 2g(\mathbf{x}, \mathbf{y}) = Q(\mathbf{x}+\mathbf{y}) - Q(\mathbf{x}) - Q(\mathbf{y}), \end{aligned} \tag{1.1}$$

where $g(\mathbf{x}, \mathbf{y})$ is bilinear and necessarily symmetric. $g(\mathbf{x}, \mathbf{y})$ is called *polar bilinear form* of Q. Transposition is defined as $g(\mathbf{x}, \mathbf{y})^T = g(\mathbf{y}, \mathbf{x})$. Quadratic forms over the reals can always be diagonalized by a *choice* of a basis. That is, in every equivalence class of a representation there is a diagonal representative.

We consider a *quadratic space* $\mathcal{H} = (V, Q)$ as a pair of a linear space V – over the ring $\mathbf{R}$ – and a quadratic form Q. This is extended to a *reflexive space* $\mathcal{H}' = (V, B)$ viewed as a pair of a linear space V and an arbitrary non-degenerate bilinear form $B = g + A$, where $g = g^T$ and $A = -A^T$ are the symmetric and antisymmetric parts respectively. g is connected to a certain Q.

We denote the finite additive group of n elements under addition modulo n as $\mathbb{Z}_n$. This should not be confused with the ring $\mathbb{Z}_n$ also denoted the same way.

Algebras or modules can be graded by an Abelian group. If the linear space W – not the same as V – of an algebra can be divided into a direct sum $W = W_0 + W_1 + \ldots + W_{n-1}$; and if the algebra product maps these spaces in a compatible way one onto another, see examples, so that the index labels behave like an Abelian group, one refers to a *grading* [8].

Example 1: $W = W_0 + W_1$ and $W_0 W_0 \subseteq W_0$, $W_0 W_1 \simeq W_1 W_0 \subseteq W_1$ and $W_1 W_1 \subseteq W_0$. The indices are added modulo 2 and form a group $\mathbb{Z}_2$. If $W = W_0 + W_1 + \ldots + W_{n-1}$ one has, e.g., $W_i W_j \subseteq W_{i+j \bmod n}$ which is a $\mathbb{Z}_n$-grading. If n goes to infinity – that is, we have no modulo relation left – one speaks about a $\mathbb{Z}_\infty$-grading.

In the case of $\mathbb{Z}_n$-grading, elements of W_m are called m-vectors or homogenous multi-vectors. The elements of $W_0 \simeq \mathbf{R}$ are also called *scalars*, and the elements of W_1 are *vectors*. When the $\mathbb{Z}_2$-grading is considered, one speaks about even and odd elements collected in W_0 and W_1, respectively.

However, observe that the Clifford product is *not* graded in this way since with $V \simeq W_1$ and $\mathbf{R} \simeq W_0$ one has $V \times V = \mathbf{R} + W_2$ which is not group-like. Only the even/odd grading, sometimes called parity grading, is preserved, $C\ell_+ C\ell_+ \subseteq C\ell_+$, $C\ell_+ C\ell_- \simeq C\ell_- C\ell_+ \subseteq C\ell_-$ and $C\ell_- C\ell_- \subseteq C\ell_+$. Hence, $C\ell$ is $\mathbb{Z}_2$ graded and $C\ell \simeq C\ell_+ + C\ell_- \simeq W_0 + W_1$.

Clifford algebras are displayed as follows: $C\ell(B, V)$ is a *quantum Clifford algebra*, $C\ell(Q, V)$ is a basis-free Clifford algebra, $C\ell(g, V)$ is a Clifford algebra with a choice of a basis, $C\ell_{p,q}$ and $C\ell_n$ are real and complex Clifford algebras of symmetric bilinear forms with signature p, q or of complex dimension n, respectively.

Basic constructions of Clifford algebras

Constructions of Clifford algebras can be found at various places in the literature. We give only notation and refer the reader to these publications [6, 8, 10, 12, 14, 35, 49, 61].

Functorial: The main advantage of the tensor algebra method is its formal strength. Existence and uniqueness theorems are most easily obtained in this language. Mathematicians derive almost all algebras from the tensor algebra – the real mother of algebras – by a process called factorization. If one singles out a two-sided ideal $\mathcal{I}$ of the tensor algebra, one can calculate "modulo" this ideal. That is all elements in the ideal are collected to form a class called "zero" $[0] \simeq \mathcal{I}$. Every element is contained in an equivalence class due to this construction. Denote the tensor algebra as $T(V) = \mathbf{R} \oplus V \oplus \ldots \otimes_n V \oplus \ldots$ and let $\mathbf{x}, \mathbf{y}, \ldots \in V$ and $L, M, \ldots \in T(V)$. This algebra is by construction naturally $\mathbb{Z}_\infty$-graded for any dimension of V.

In the case of Clifford algebras, one selects an ideal of the form

$$\mathcal{I}_{C\ell} = \{X \mid X = L \otimes (\mathbf{x} \otimes \mathbf{x} - Q(\mathbf{x})\mathbf{1}) \otimes M\} \tag{1.2}$$

which implements essentially the "square law" of Clifford algebras. Note that elements of different tensor grades – scalar and grade two – are identified. Hence this ideal is not grade-preserving and the factor algebra, the Clifford algebra, cannot be $\mathbb{Z}_\infty$-graded (finiteness of $C\ell(V)$), and not even multi-vector or $\mathbb{Z}_n$-graded with $n = \dim V$ because all indices are now mod 2. However, the ideal $\mathcal{I}_{C\ell}$ is $\mathbb{Z}_2$-graded, that is, it preserves the evenness and the oddness of the tensor elements. One defines now the Clifford algebra as

$$C\ell(Q, V) := T(V) \bmod \mathcal{I}_{C\ell}. \tag{1.3}$$

It is clear from the construction that a Clifford algebra is unital and associative, a heritage from the tensor algebra.

Generators and relations: Physicists and most people working in Clifford analysis prefer another construction of Clifford algebras by generators and relations [17]. One chooses a set of *generators* $\mathbf{e}_i$, images of some arbitrary basis elements $\mathbf{x}_i$ of V under the usual Clifford map $\gamma : V \mapsto C\ell(V)$ in the Clifford algebra $C\ell(V)$, and asserts the validity, in the case of $\mathbf{R} = \mathbb{R}$ or $\mathbb{C}$, of the normalized, "square law":

$$\mathbf{e}_i^2 = \pm\mathbf{1}. \tag{1.4}$$

Using the linearity, that is polarizing these equations by $\mathbf{e}_i \mapsto \mathbf{e}_i + \mathbf{e}_j$, one obtains the usual set of relations which have to be used to "canonify" the algebraic expressions

$$\mathbf{e}_i\mathbf{e}_j + \mathbf{e}_j\mathbf{e}_i = 2g(\mathbf{e}_i, \mathbf{e}_j)\mathbf{1} = 2g_{ij}\mathbf{1}. \tag{1.5}$$

Let $\mathbf{Alg}(\mathbf{e}_i)$ be the algebra freely generated by the $\mathbf{e}_i$ vector elements. This is $\mathbf{Alg}(\mathbf{e}_i) \cong T(V)$ with V the linear span of the $\mathbf{e}_i$. The definition of the Clifford algebra reads

$$C\ell(g_{ij}, V) \simeq \mathbf{Alg}(\mathbf{e}_i) \ \text{mod} \ (\, \mathbf{e}_i\mathbf{e}_j = 2g_{ij}\mathbf{1} - \mathbf{e}_j\mathbf{e}_i \, ; \quad i > j\,). \tag{1.6}$$

While the – image of the – numbers of the base field are called *scalars*, the $\mathbf{e}_i$ and their linear combinations are called *vectors*. The entire algebra is constructed by multiplying and linear-combining the generators $\mathbf{e}_i$ modulo the relation (1.5). This "modulo relation" is in fact nothing but a "cancelation law" which provides one with a *unique* representative of the class of tensor elements. A basis of the linear space underlying the Clifford algebra is given by *reduced monomials* in the generators, where a certain ordering has to be chosen in the index set, e.g., ascending indices or antisymmetry [62]. A monomial build out of n generators and the linear span of such monomials is called a homogenous n-vector. Thereby a unique $\mathbb{Z}_n$-grading is introduced by the choice of a basis and an ordering, as will be discussed below.

This method has the advantage of being plain in construction, easy to remember, and powerful in computational means.

Naming

A very important and delicate point in mathematics and physics is the appropriate naming of objects and structures. Since we deal with a very well-known structure, but want to highlight special novel features, we have to give distinguishing names to different albeit well-known objects, which otherwise could not be properly addressed. This section shall establish such a coherent naming, at least for this article.

Clifford algebra is often denoted, following Clifford himself and Hestenes, as "Geometric Algebra", GA, or "Clifford Geometric Algebra", CGA, or

"Clifford Grassmann Geometric Algebra", CGGA [59]. Having the advantage of being descriptive, this notation has, however, also a peculiar tendency to call upon connotations and intuitions which might *not in all cases* be appropriate. Even at this stage, one has to distinguish "Metric Geometric Algebra", MGA, and "Projective Geometric Algebra", PGA, which relies on the identification of the homogenous multi-vector objects and geometrical entities [43]. In the former case, "vectors" are identified with "places" – position vectors – of pseudo-Euclidean or unitary spaces while in the second case "vectors" are identified with "points" of a projective space.

Both variants, metric or projective, use unquestionably the artificial multi-vector structure introduced by the mere notation of a basis and foreign to Clifford algebras to assert "ontological" statements such as: "**x** is a place in Euclidean space" or "**x** is a point in a projective space".

Both of these interpretations have one thing in common, namely, they assert an *object character* to the Clifford elements themselves. We will coin for this case the term "**Classical Clifford Algebra**".

To our current experience, the Wick isomorphism, developed below, guarantees that such interpretation of Clifford algebras is *independent* from the chosen $\mathbb{Z}_n$-grading. That is, we make the following *conjecture:* if the Clifford elements themselves are "ontologically" interpreted as "place" or "point", then all $\mathbb{Z}_n$-gradings are equivalent through the Wick isomorphism.

We turn to the second aspect. In [57] Oziewicz introduced the term "Clifford algebras of multi-vectors" to highlight the fact that he considered different $\mathbb{Z}_n$-gradings or, equivalently, different multi-vector structures. However, Clifford algebras have in nearly every case been used as multi-vector Clifford algebras since mathematicians and physicists want to consider the n-vectors or multi-vectors for different purposes.

Following the introduction of Clifford algebras of arbitrary bilinear forms, implicitly in [12] and explicitly in [1, 24, 25, 26, 27, 28, 29, 31, 32, 50, 56], situations have occurred for good physical reasons where different $\mathbb{Z}_n$-gradings have led to different physical outcomes. In those situations a theory of gradings is mandatory.

A new point is the *operational approach* to Clifford elements. If one considers a Clifford number to be an operator, it has to act on another object, a "state vector". This "*quantum point of view*" moves also the ontological assertions into the states. Their interpretation however is difficult.

Moreover, one has to deal with representation theory which was not necessary in the "classical" Clifford algebraic approach – in both senses of classical, i.e., also as opposed to quantum, here. Adopting Wigner's definition of a particle as an irreducible representation – of the Poincaré group – one has to seek irreducible representations of Clifford algebras. It is a well-known fact that these representations are faithfully realized in spinor spaces. It is exactly at this place where it will be shown in this article that one obtains different $\mathbb{Z}_n$-gradings or different multi-vector structures leading to different results. In fact we are able to find *irreducible* spinor spaces of dimension 8 in $C\ell_{2,2}(B,V)$, where 2, 2 denotes the signature of the symmetric part g of B, and not of dimension 4 as predicted by the "classical" Clifford algebra theory.

For the case of Clifford algebras of multi-vectors, we coin the term **Quantum Clifford Algebra.**[1,2] It is clear to us that we risk creating a confusion with this term, which looks like a q-deformed version of an ordinary Clifford algebra, while also in our case the common "square law" is fully valid! However, this link is not wrong! As we show elsewhere in this volume [4], one is able to find Hecke algebras and q-symmetry *within* the structure of the quantum Clifford algebra. It is also in accord with the attempt of Fiore [33] to describe q-deformed algebras in terms of undeformed generators. This is just a reverse of our argument. However, the characteristic point in our consideration is that we dismiss the classical ontological interpretation in favor of an operational interpretation. Thereby it is necessary to study states which are now $\mathbb{Z}_n$-grade dependent. Our approach should be contrasted by the recent developments excellently described in [15, 52]. A different treatment of Clifford algebras in connection with Hecke algebras was given in [58].

As a last point, we emphasize that indecomposable spinor representations of unconventionally large dimensions are expected to be spinors of bound systems, see [27]. Hence, studying decomposability is the first step towards an algebraic theory of compositeness, including stability of bound states.

1.2 Why study $C\ell(B,V)$ and not $C\ell(Q,V)$? – Physics

Clifford algebras play without any doubt a predominant role in physics and mathematics. This fact was clearly addressed and put forward by D. Hestenes [39, 40, 41, 42]. Based on this solid ground, we give an analysis of Clifford algebras of an arbitrary bilinear form which exhibit novel features especially regarding their representation theory. The most distinguishing fact between our approach and usual treatments of Clifford algebras, e.g., [6, 10, 14, 49, 61], is that we seriously consider how the $\mathbb{Z}_n$-grading is introduced in Clifford algebras. This is most important since Clifford algebras are *only* $\mathbb{Z}_2$-graded by their natural – functorial – construction. The introduction of a further finer grading does therefore put new assumptions into the theory. One might therefore ask if theses additional structures are important or even necessary in physics and mathematics.

Indeed, after examining various cases we notice that *every* application of Clifford algebras which is computational – not only functorial – deals in fact

[1] This is close to Saller's notion of a "quantum algebra" which denotes however a special choice of grading [63].

[2] Classical Clifford algebras emerge as a particular case of quantum Clifford algebras.

with the so-called Clifford algebras of multi-vectors [57] or *quantum Clifford algebras*. However, the additional $\mathbb{Z}_n$-grading, even if mathematically and physically necessary for applications, is usually introduced without any ado. Looking at literature we can however find lots of places where $\mathbb{Z}_n$-graded Clifford algebras are not only appropriate but needed. This is in general evident in every quantum mechanical setup.

If one analyzes functional hierarchy equations of quantum field theory (QFT), one is able to translate these functionals with a help of Clifford algebras. Such attempts have already been made by Caianiello [11]. He noticed that at least two types of orderings are needed in QFT, namely the time-ordering and normal-ordering. Since one has – at least – two possibilities to decompose Clifford algebras into basis monomials, he introduces Clifford and Grassmann bases. A basis of a Clifford algebra is usually given by monomials with totally ordered index sets. If one has a finite number of "vector" elements $\mathbf{e}_i$, one can, by using the anti-commutation relations (1.5) of the Clifford algebra, introduce the following bases

$$\begin{aligned} &i) \quad \{\mathbf{1}; \mathbf{e}_1, \ldots, \mathbf{e}_n; \mathbf{e}_1\mathbf{e}_2, \ldots; \mathbf{e}_{i_1}\mathbf{e}_{i_2}\mathbf{e}_{i_3\,(i_1<i_2<i_3)}, \ldots\} \\ &ii) \quad \{\mathbf{1}; \mathbf{e}_1, \ldots, \mathbf{e}_n; \mathbf{e}_{[1}\mathbf{e}_{2]}, \ldots; \mathbf{e}_{[i_1}\mathbf{e}_{i_2}\mathbf{e}_{i_3]}, \ldots\}. \end{aligned} \tag{1.7}$$

We used the $[\ldots]$ bracket to indicate antisymmetrization in the index set. An ordering of index sets is inevitable since the $\mathbf{e}_i\mathbf{e}_j$ and $\mathbf{e}_j\mathbf{e}_i$ monomials are not algebraically independent due to the anti-commutation relations $\mathbf{e}_i\mathbf{e}_j = -\mathbf{e}_j\mathbf{e}_i + g_{ij}\mathbf{1}$. Caianiello identifies then the two above choices with time- and normal-ordering. However, already at this point it is questionable why one uses "lexicographical" ordering "$<$" and not, e.g., the "anti-lexicographical" ordering "$>$" or an ordering which results from a permutation of the index set.

A detailed study shows that fermionic QFT needs antisymmetric index sets and that there are infinitely many such choices [25, 31]. Using this fact we have been able to show that singularities, which arise usually due to the reordering procedures such as the normal-ordering, are no longer present in such algebras [32]. Studying the transition from operator dynamics to functional hierarchies, the so-called Schwinger-Dyson-Freese hierarchies, in [25, 31] it turned out that the multi-vector structure, or, equivalently a uniquely chosen $\mathbb{Z}_n$-grading, was a *necessary input* to QFT.

Multi-particle systems provide a further place where a careful study of gradings will be of great importance. It is a well-known fact that one has the Clebsch-Gordan decomposition of two spin-$\frac{1}{2}$ particles as follows [34, 38]:

$$\frac{\mathbf{1}}{\mathbf{2}} \otimes \frac{\mathbf{1}}{\mathbf{2}} = \mathbf{0} \oplus \mathbf{1}. \tag{1.8}$$

However, since this is an identity, it can be used either from left to right to form bosonic spin 0 and spin 1 "composites" *or* from right to left! There is no way – besides the experience – to distinguish if such a system is composed, that is, dynamically stable or not, see [26]. From a mathematical point of view one cannot distinguish n free particles from an n-particle bound system by means of algebraic considerations. This is seen clearly in the decomposition theorems for Clifford algebras where larger Clifford algebras are decomposed into smaller blocks of Clifford tensor factors. This cannot be true for bound objects which lose their physical character when being decomposed. An electron and proton system is quite different from a hydrogen atom. In this work we will see, that one can indeed find such *indecomposable* states in quantum Clifford algebras.

This raises a question how to distinguish such situations. One knows from QFT that interacting systems have to be described in non-Fock states and that there are infinitely many such representations [36]. It is, thus, necessary to introduce the concept of *inequivalent states* in finite dimensional systems [27, 45, 46]. Such states are necessarily non-Fock states since Fock states belong to systems of non-interacting particles. This is the so-called *free case* which is however very useful in perturbation theory. The present paper supports the situation found in [27].

Closely related to these inequivalent states are *condensation phenomena*. As it was shown in [27], one can algebraically determine boundedness using an appropriate $\mathbb{Z}_n$-grading. Furthermore, it was shown that the dynamics *determines* correct grading. In BCS theory of superconductivity the fact that bound states can or cannot be build was shown to imply a gap-equation [27] which governs the phase transition.

A further point related to $\mathbb{Z}_n$-graded Clifford algebras is q-quantization. This can be seen when studying physical systems as in [30] and when adopting a more mathematical point of view as in [24, 28]. In these articles a detailed example was worked out to show how q-symmetry and Hecke algebras can be described within quantum Clifford algebras [4]. It is quite clear that this structure should play a major role in the discussion of the Yang-Baxter equation, knot theory, link invariants and in other related fields which are crucial for the physics of integrable systems in statistical physics.

However, the most important implication from these various applications is that the q-symmetry and more general deformations are *symmetries of composites*.This was already addressed in [30] and more recently in [24]. Also the present work provides full support for this interpretation, as the talk of G. Fiore at this conference. Providing as much evidence as possible to this fact was a major motivation for the present work.

1.3 Why study $C\ell(B,V)$ and not $C\ell(Q,V)$? – Mathematics

There are also arguments of purely mathematical character which force us to consider quantum Clifford algebras.

If we look at the construction of Clifford algebras by means of the tensor algebra, we notice that $C\ell$ is a functor. To every quadratic space

$\mathcal{H} = (V, Q)$, a pair of a linear space V over a ring $\mathbf{R}$ and a quadratic form Q, there is a uniquely connected Clifford algebra $C\ell(Q, V)$. That is, one can introduce the algebra structure without any further input or choices, so to say for free. One may further note that if the characteristic of the ring $\mathbf{R}$ is not 2, then there is a one-to-one correspondence between quadratic forms and classes of symmetric matrices [64]. In other words, every symmetric matrix is a representation of a quadratic form in a special basis. Over the reals (complex numbers) the classes of quadratic forms can be labeled by dimension n and signature s (dimension n only, no signature in $\mathbb{C}$). Equivalently one can use the numbers p, q of positive and negative eigenvalues of the quadratic form. This leads to a classification (naming) of real (and complex) Clifford algebras. One writes $C\ell(Q, V) \simeq C\ell_{p,q}$ ($C\ell_n$) where $\dim V = n = p + q$ and Q has signature $s = p - q$. The remarkable fact is that the "square law" for vectors $Q(\mathbf{v}) \equiv \mathbf{v}^2 = \alpha \mathbf{1} \in C\ell(Q, V)$ ($\alpha \in \mathbb{R}$ or $\alpha \in \mathbb{C}$) is a diagonal map determining only the symmetric part of the map $Q(\mathbf{V}) \mapsto \mathbb{R}$. Following Clifford one should note that the product operation can be seen as acting on the second factor $2 \times x$ as a doubling of x; that is, $2\times$ is a doubling operator or endomorphism acting on the space of the second factor. In this sense any "Clifford number" induces an endomorphism on the graded space W underlying the algebra, and it is questionable why one should use only diagonal maps and their symmetric polarizations. Furthermore, note that one has

$$\text{quadratic forms} \simeq \text{bilinear forms} \mod \text{alternating forms}. \tag{1.9}$$

The dualization $V \mapsto V^* \simeq \text{lin-Hom}(V, \mathbb{R})$ is performed by an arbitrary (non-degenerate) bilinear form. Endomorphisms have in general the following form

$$\text{End}(V) \simeq V \otimes V^*, \tag{1.10}$$

so why do we restrict ourselves to the symmetric case? If we consider a pair (V, B) of a space V and an arbitrary bilinear form B, can we construct functorially an algebra like the Clifford algebra for the pair $\mathcal{H} = (V, Q)$?

It can be easily checked that if one insists on the validity of the "square law" $\mathbf{v}^2 = \alpha\mathbf{1}$, the *anti-commutation relations* of the resulting algebra are the same as for usual Clifford algebras while the *commutation relations* – and thus the meaning of ordering and grade – is changed. Let $B = g + A$, $A^T = -A$, $g^T = g$. We denote $B(\mathbf{x}, \mathbf{y}) = \mathbf{x} \lrcorner_B \mathbf{y}$, $A(\mathbf{x}, \mathbf{y}) = \mathbf{x} \lrcorner_A \mathbf{y}$ and $g(\mathbf{x}, \mathbf{y}) = \mathbf{x} \lrcorner_g \mathbf{y}$ (the latter also denoted by Hestenes and Sobczyk as $\mathbf{x} \cdot \mathbf{y}$).[3] Then, the B-dependent Clifford product $\underset{B}{\mathbf{xy}}$ of two 1-vectors $\mathbf{x}$ and $\mathbf{y}$ in $C\ell(B, V)$ can be decomposed in *different ways* into scalar and

[3]The symbols $\lrcorner_B$, $\lrcorner_A$ and $\lrcorner_g$ denote the left contraction in $C\ell(B, V)$ with respect to B, A and g, respectively.

bi-vector parts as follows:

$$\underset{B}{\mathbf{x}\mathbf{y}} = \mathbf{x} \underset{g}{\rfloor} \mathbf{y} + \mathbf{x} \dot{\wedge} \mathbf{y} \quad \text{(Hestenes, common case, } A = 0\text{)}$$

$$\underset{B}{\mathbf{x}\mathbf{y}} = \mathbf{x} \underset{B}{\rfloor} \mathbf{y} + \mathbf{x} \wedge \mathbf{y} \quad \text{(Oziewicz, Lounesto, Abłamowicz, Fauser),} \tag{1.11}$$

where $\mathbf{x} \dot{\wedge} \mathbf{y} = \mathbf{x} \wedge \mathbf{y} + A(\mathbf{x}, \mathbf{y}) = \mathbf{x} \wedge \mathbf{y} + \mathbf{x} \rfloor_A \mathbf{y}$. Of course, for any 1-vector $\mathbf{x}$ and any element u in $C\ell(B, V)$ we have

$$\underset{B}{\mathbf{x}u} = \mathbf{x} \underset{B}{\rfloor} u + \mathbf{x} \wedge u = \mathbf{x} \underset{g}{\rfloor} u + \mathbf{x} \underset{A}{\rfloor} u + \mathbf{x} \wedge u = \mathbf{x} \underset{g}{\rfloor} u + \mathbf{x} \dot{\wedge} u. \tag{1.12}$$

Notice that the element $\mathbf{x} \dot{\wedge} u = \mathbf{x} \rfloor_A u + \mathbf{x} \wedge u$ is not even a homogenous multi-vector in $\bigwedge V$. We have, thus, established that the multi-vector structure is uniquely connected with the antisymmetric part A of the bilinear form, see also [1, 29, 31].

This has an immediate consequence: in some cases one finds bi-vector elements which satisfy minimal polynomial equations of the Hecke type [24, 28]. This feature is treated extensively elsewhere in this Volume [4].

Some mathematical formalisms, not treated here, are closely connected to this structure. One is the structure theory of Clifford algebras over arbitrary rings [37] where a classification is still lacking. Connected to these questions is the arithmetic theory of Arf invariants and the Brauer-Wall groups.

Much more surprising is the fact that, due to central extensions, the ungraded bi-vector Lie algebras turn into Kac-Moody and Virasoro algebras [55] and, as it is also shown in [4], to some q-deformed algebras.

Since Clifford algebras naturally contain reflections, automorphisms generated by non-isotropic vectors, we expect to find infinite dimensional Coxeter groups [17, 44], affine Weyl groups, etc., connected to $\mathbb{Z}_n$-graded or quantum Clifford algebras.

Involutions connected to special elements, norms and traces [37] are also affected by different gradings. This has considerable effects. One important point is that the Cauchy-Riemann differential equations are altered which makes probably the concept of monogeneity [49] grade dependent. However, this is speculative.

2 Chevalley's approach to Clifford algebras

2.1 Confusion with Chevalley's approach

Chevalley's book *The Algebraic Theory of Spinors* [12] seems to have been badly accepted by working mathematicians and physicists despite its frequent citation. Albert Crumeyrolle stated the following in [14], p. xi:

> *In spite of its depth and rigor, Chevalley's book proved too abstract for most physicists and the notions explained in it have not been applied much until recently, which is a pity.*

The more compact and readable book *The Study of Certain Important Algebras* [13] seems to be little known. However, one can find in many physical writings, e.g., Berezin [7] very analogous structures, without mentioning the much more complete work of Chevalley.

When looking for the most general construction of Clifford algebras over arbitrary rings including the case where the characteristic of $\mathbf{R}$ is 2, Chevalley constructed the so-called *Clifford map.* This map is an injection of the linear space V into the algebra $C\ell(V)$ which establishes the "square law". This construction emphasizes the operator character of Clifford algebras and establishes a connection between the spaces underlying the $\mathbb{Z}_n$-graded Grassmann algebra and the thereby constructed Clifford algebra, the Chevalley isomorphism. For our purpose it is important that *only* Chevalley's construction allows a non-symmetric bilinear form in constructing Clifford algebras. However, this fact is not explicit in Chevalley's writings, but it is clearly emphasized in [56].

Ironically, a careful analysis of Lounesto shows that even Crumeyrolle made a mistake in describing the Chevalley isomorphism connecting Grassmann and Clifford algebra spaces. In [51] Lounesto points out that Crumeyrolle rejects the Chevalley isomorphism for *any* characteristic. This seems to be implied by Crumeyrolle's frequent questioning, also in previous Clifford conferences of this series: "What is a bi-vector?" [54]. However, an isomorphism can be uniquely given if the characteristic of $\mathbf{R}$ is not 2, see [50, 51]. On the other hand, Lounesto points out that Lawson and Michelsohn [48] postulate such an isomorphism which is wrong in the exceptional case of characteristic 2. One should note in this context that their point of view is taken by almost all working mathematicians and physicists.

At this point we submit that we insist on Chevalley's construction even in the case of characteristic not 2. Lounesto claims that in these cases $C\ell(B,V)$ is isomorphic to $C\ell(Q,V)$ with Q the quadratic form associated to B. In fact, this is true for the Clifford algebraic structure and was proved in [1] up to the dimension 9 of V. However, this, the so-called *Wick isomorphism* between $C\ell(B,V)$ and $C\ell(Q,V)$, has to be rejected when the $\mathbb{Z}_n$-grading is considered, or, in other words, the multi-vector structure. Hence, we reject Lounesto's judgment that it is worth studying $C\ell(B,V)$ only in characteristic 2 for the reason of carefully treating the involved $\mathbb{Z}_n$-grading or multi-vector structure. This is one of the main points of our analysis.

2.2 Chevalley's construction of $C\ell(B,V)$

A detailed and mathematical rigorous development of quantum Clifford algebras $C\ell(B,V)$ can be found in [24, 31]. We will develop only the notation and point out some peculiar features insofar as they appear in the present study, see also [1, 29].

The main feature of the Chevalley approach is that Clifford algebras are constructed as special – satisfying the "square law" – endomorphism algebras on the linear space of a Grassmann algebra. In this way the Grassmann algebra, which is naturally $\mathbb{Z}_n$-graded, induces via the Chevalley isomorphism a grading or multi-vector structure in the Clifford algebra. This grading is however not preserved by the Clifford product which renders the Clifford algebra as a deformation of the Grassmann algebra.

To proceed along this line we construct the Grassmann algebra as a factor algebra of the tensor algebra. Let

$$\mathcal{I}_G := \{X \mid X = L \otimes (\mathbf{x} \otimes \mathbf{x}) \otimes M\} \tag{2.1}$$

with notation as in (1.2) and define

$$\bigwedge V := T(V) \bmod \mathcal{I}_G, \quad \pi : T(V) \mapsto \bigwedge V. \tag{2.2}$$

The projected tensor product $\pi(\otimes) \mapsto \wedge$ is denoted as wedge or outer product. The induced grading is

$$\bigwedge V = \mathbf{R} \oplus V \wedge V \oplus \ldots \oplus \wedge^n V \oplus \ldots \quad . \tag{2.3}$$

As the next step, we consider reflexive duals of the linear space V. Define

$$V^* := \text{lin-Hom}(V, \mathbf{R}) \tag{2.4}$$

where $\dim V^* = \dim V$ (reflexivity). Using the action of the dual elements on V we define the (left) contraction $\rfloor_B$ as

$$i_{\mathbf{x}}(\mathbf{y}) = \mathbf{x} \underset{B}{\rfloor} \mathbf{y} = B(\mathbf{x}, \mathbf{y}). \tag{2.5}$$

Note that $i_{\mathbf{x}} \in V^*$ is the dualized element $\mathbf{x}$ and that here a *certain* duality map is employed. If this is the usual duality map $i_{\mathbf{e}_i}(\mathbf{e}_j) = \delta_{ij}$, one denotes this as Euclidean dual isomorphism and writes the map as $\star$ [63]. The notation $\mathbf{x} \rfloor_B \mathbf{y}$ and much more $B(\mathbf{x}, \mathbf{y})$ is very peculiar since we have

$$\underset{B}{\rfloor} : V \times V \mapsto V, \quad B : V \times V \mapsto V. \tag{2.6}$$

Hence, $\rfloor_B$ and B are in $\text{lin-Hom}(V \times V, \mathbf{R}) \simeq V^* \times V^*$. In this notation a dual isomorphism is *implicitly* involved since we consider really maps of the form

$$< . \mid . >: V^* \times V \mapsto \mathbf{R} \tag{2.7}$$

which might be called *a dual product* or *a pairing* [8, 63].

Having defined the action of V^* on V, we lift this action to the entire Grassmann algebras $\bigwedge V$ and $\bigwedge V^*$. For $\mathbf{x}, \mathbf{y} \in V$, and $u, v, w \in \bigwedge V$ we have

$$\begin{aligned} &i) \quad \mathbf{x} \underset{B}{\lrcorner} \mathbf{y} = B(\mathbf{x}, \mathbf{y}), \\ &ii) \quad \mathbf{x} \underset{B}{\lrcorner} (u \wedge v) = (\mathbf{x} \underset{B}{\lrcorner} u) \wedge v + \hat{u} \wedge (\mathbf{x} \underset{B}{\lrcorner} v), \\ &iii) \quad (u \wedge v) \underset{B}{\lrcorner} w = u \underset{B}{\lrcorner} (v \underset{B}{\lrcorner} w), \end{aligned} \tag{2.8}$$

where $\hat{}$ is the involutive map – grade involution – $\hat{} : V \mapsto -V$ lifted to $\bigwedge V$. The Clifford algebra $C\ell(B, V)$ is then constructed in the following way. Define an operator $L_{\mathbf{x}}^{\pm} : \bigwedge V \mapsto \bigwedge V$ for any $\mathbf{x} \in V$ as

$$(L_{\mathbf{x}}^{\pm})^2 := \mathbf{x} \underset{B}{\lrcorner} \cdot \, \pm \, \mathbf{x} \wedge \cdot \tag{2.9}$$

and observe that this is a Clifford map [12, 24, 31]

$$(L_{\mathbf{x}}^{\pm})^2 = \pm Q(\mathbf{x}) \mathbf{1}, \tag{2.10}$$

where $Q(\mathbf{x}) = B(\mathbf{x}, \mathbf{x})$. This is nothing else as again the "square law", and one proceeds as in the case of generators and relations. Chevalley has thus established that

$$C\ell(B, V) \subset \operatorname{End}(\bigwedge V). \tag{2.11}$$

This inclusion is strict.

3 Wick isomorphism and $\mathbb{Z}_n$-grading

3.1 Wick isomorphism

In this section we will prove the following

Main Theorem. *$C\ell(B, V) \cong C\ell(Q, V)$ as $\mathbb{Z}_2$-graded Clifford algebras.*

This isomorphism, denoted below by ϕ, is the *Wick isomorphism* since it is the well-known normal-ordering transformation of the quantum field theory [21, 31, 66]. This was not noticed for a long time, which is another "missed opportunity" [22].

Proof. The proof proceeds in various steps, numbered by letters a, b, c, etc. After defining the outer exponential, we prove the following important formulas:

$$\begin{aligned} &i) \quad e_{\wedge}^{-F} \wedge e_{\wedge}^{F} = \mathbf{1}, \\ &ii) \quad e_{\wedge}^{-F} \wedge \mathbf{x} \wedge e_{\wedge}^{F} \wedge u = \mathbf{x} \wedge u, \\ &iii) \quad e_{\wedge}^{-F} \wedge (\mathbf{x} \underset{g}{\lrcorner} (e_{\wedge}^{F} \wedge u)) = \mathbf{x} \underset{g}{\lrcorner} u + (\mathbf{x} \underset{g}{\lrcorner} F) \wedge u, \end{aligned} \tag{3.1}$$

and finally we show that the Wick isomorphism ϕ is given as

$$\begin{aligned} C\ell(B,V) &= \phi^{-1}(C\ell(g,V)) \\ &= e_{\wedge}^{-F} \wedge C\ell(Q,V) \wedge e_{\wedge}^{F} \qquad (3.2) \\ &\cong (\, C\ell(g,V), <.>_r^A \,) \qquad (3.3) \end{aligned}$$

where $<.>_r^A$ denotes the A-dependent $\mathbb{Z}_n$-grading.

That is, the isomorphism is given by the following transformation of *vector* variables which is then algebraically lifted to the entire algebra:

$$\begin{aligned} &\mathbf{x} \underset{g}{\rfloor} \cdot \to \mathbf{x} \underset{B}{\rfloor} \cdot = \mathbf{x} \underset{g}{\rfloor} \cdot + (\mathbf{x} \underset{g}{\rfloor} F) \wedge \cdot \\ &\mathbf{x} \wedge \cdot \to \mathbf{x} \wedge \cdot \qquad (3.4) \end{aligned}$$

a) According to Hestenes and Sobczyk [40] it is possible to express every antisymmetric bilinear form in the following way:

$$A(\mathbf{x},\mathbf{y}) := F \underset{g}{\rfloor} (\mathbf{x} \wedge \mathbf{y}) \qquad (3.5)$$

where F is an appropriately chosen bi-vector. F can be split in a non-unique way into decomposable parts $F_i = \mathbf{a}_i \wedge \mathbf{b}_i$, $F = \sum F_i$. We define the outer exponential of this bi-vector as $(\wedge^0 F = \mathbf{1})$

$$e_{\wedge}^{F} := \sum \frac{1}{n!} \wedge^n F = \mathbf{1} + F + \frac{1}{2} F \wedge F + \ldots + \frac{1}{n!} \wedge^n F + \ldots . \qquad (3.6)$$

This series is finite when the dimension of V is finite since in that case there exists a term of the highest grade.

b) Substitute the series expansion (3.6) into (3.1-i) and note that after applying the Cauchy product formula for sums we have

$$e_{\wedge}^{-F} \wedge e_{\wedge}^{F} = \sum_{r=0}^{\infty} \left(\sum_{l=0}^{r} (-1)^l \binom{r}{l} \right) \frac{1}{r!} \wedge^r F. \qquad (3.7)$$

The alternating sum of the binomial coefficient is zero except in the case $r = 0$ when we obtain $\mathbf{1}$, which proves formula (3.1-i).

c) To prove (3.1-iii) one needs the commutativity of $\mathbf{x} \rfloor_g F_i$ with F_j. If the contraction is zero, it commutes trivially; if not, the contraction is a vector $\mathbf{y}$. From $\mathbf{y} \wedge F = F \wedge \mathbf{y}$ for every bi-vector, we have that $\mathbf{x} \rfloor_g F_i$ commutes with any F_j and thus with F. This allows us to write

$$\mathbf{x} \underset{g}{\rfloor} (\wedge^n F) = n(\mathbf{x} \underset{g}{\rfloor} F) \wedge (\wedge^{(n-1)} F). \qquad (3.8)$$

Once more using $\wedge^0 F = \mathbf{1}$, the Leibniz rule, and the fact that $\hat{F} = F$, we obtain

$$\mathbf{x} \underset{g}{\rfloor} (e_{\wedge}^{F} \wedge u) = e_{\wedge}^{F} \wedge (\mathbf{x} \underset{g}{\rfloor} u + (\mathbf{x} \underset{g}{\rfloor} F) \wedge u), \qquad (3.9)$$

which proves (3.1-iii).

d) Since any vector $\mathbf{y}$ commutes under the wedge with any bi-vector F, the case (3.1-ii) reduces to **b)**.

e) The Wick isomorphism is now given as $C\ell(B,V) = \phi^{-1}(C\ell(g,V)) = e_\wedge^{-F} \wedge C\ell(Q,V) \wedge e_\wedge^{F}$. The same transformation can be achieved by decomposing every Clifford "operator" into vectorial parts and then into contraction and wedge parts w.r.t. $(g,\wedge)$ and then performing the substitution laws given in (3.4) and a final renaming of the contractions; see [31] for an application in quantum field theory.

Note that since the wedges are *not* altered and the new contractions are given by $\mathbf{x} \,\lrcorner_B \cdot \equiv \mathrm{d}_\mathbf{x}(\cdot) := \mathbf{x} \,\lrcorner_g \cdot + (\mathbf{x} \,\lrcorner_g F)\wedge\cdot$, this transformation does mix grades, but it respects the parity. It is thus a $\mathbb{Z}_2$-graded isomorphism.

□

An equivalent proof was delivered in [65] without using (explicitly) Clifford algebras but index doubling – see below. The Wick isomorphism was called "nonperturbative normal-ordering".

3.2 $C\ell(B,V) \leftrightarrow C\ell(Q,V)$ – Isomorphic yet different?

We have already discussed that many researchers reject the idea that $C\ell(B,V)$ is of any use because of the Wick isomorphism. However, as our proof has shown this isomorphism is only $\mathbb{Z}_2$-graded. Indeed it was not the mathematical opportunity, but a necessity in modeling quantum physical multi-particle systems and quantum field theory, which forced us to investigate quantum Clifford algebras [25, 27, 31, 32].

Decomposing B into g, A as in (1.12) and noting that in our case, of characteristic not 2 one has $Q(\mathbf{x}) = g(\mathbf{x},\mathbf{x})$, one concludes that $C\ell(Q,V)$ is exactly the equivalence class of $C\ell(B,V) \simeq C\ell(g+A,V)$ with A varying arbitrarily:

$$C\ell(Q,V) = [C\ell(g+A,V)]\,. \tag{3.10}$$

In other words, one does not have a single Clifford algebra $C\ell(Q,V)$ but an entire class of equivalent – under the $\mathbb{Z}_2$-graded Wick isomorphism – Clifford algebras $C\ell(B,V)$. This can be written as

$$C\ell(Q,V) \simeq C\ell(g+A,V) \bmod A \tag{3.11}$$

which induces a unique projection from the class of quantum Clifford algebras onto the classical Clifford algebra. Such a projection π can be defined as

$$\begin{aligned} &i) \quad \pi : T(V) \mapsto C\ell(B,V) \\ &ii) \quad <.>_r^A := \pi(\otimes^r V). \end{aligned} \tag{3.12}$$

This is once more a sort of "cancellation law". The important fact is that *only* those properties belong to $C\ell(Q,V)$ which do *not* depend on the particular choice of a representant parameterized by A. Physically speaking, only those properties belong to $C\ell(Q,V)$ which are homogenous over the entire equivalence class.

As we will show now, especially the multi-vector $\mathbb{Z}_n$-grading is *not* of this simple type. Recall that it is possible to decompose the Clifford product in various ways as in (1.11) and (1.12). Hence we obtain a relation between the $\wedge$- and the $\dot\wedge$-grading as

$$\mathbf{x} \,\dot\wedge\, \mathbf{y} = A(\mathbf{x},\mathbf{y}) + \mathbf{x} \wedge \mathbf{y} \tag{3.13}$$

which shows that a $\dot\wedge$-bi-vector is an inhomogeneous $\wedge$-multi-vector and vice versa. Since the antisymmetric part can be absorbed in the wedge product, using the Wick isomorphism, we can give the grading explicitly by writing

$$< . >_r^A = < . >_r^{\dot\wedge} \tag{3.14}$$

with respect to the dotted wedge $\dot\wedge$ *within* the undeformed algebra $C\ell(Q,V)$, see also Fiore's contribution. This gives us a second characterization of $C\ell(B,V)$, namely

$$C\ell(B,V) \simeq (C\ell(Q,V), < . >_r^A) \,. \tag{3.15}$$

That is, $C\ell(B,V)$ can be seen as a pair of a classical $\mathbb{Z}_2$-graded Clifford algebra $C\ell(Q,V)$ and a unique multi-vector structure given by the projectors $< . >_r^A$. As a main result, we have that these algebras are *not* isomorphic under the Wick isomorphism

$$C\ell(g+A_1,V) \underset{\text{Wick}}{\not\simeq} C\ell(g+A_2,V) \quad \text{iff} \quad A_1 \neq A_2 \,. \tag{3.16}$$

4 Periodicity theorems

Our theory will have an impact on all famous periodicity theorems of Clifford algebras, especially on the Atiyah-Bott-Shapiro mod 8 index theorem [5]. But to be as concrete and explicit as possible, we restrict ourself to the case $C\ell_{p,q} \simeq C\ell_{p-1,q-1} \otimes C\ell_{1,1}$, where one needs obviously $p \geq 1$, $q \geq 1$. Periodicity theorems can be found, for example, in [6, 10, 47, 53, 61].

We need some further notation. Let $V_{p,q} = (g_{p,q}, V)$ be a quadratic space, where $g = \mathrm{diag}(1, \ldots, 1, -1, \ldots, -1)$ with p plus signs and q minus signs, and let V be a linear space of dimension $p+q$. According to the Witt theorem [67] one can split off a quadratic space of the hyperbolic type $M_{1,1}$. This split is orthogonal with respect to g:

$$V_{p,q} = N_{p-1,q-1} \perp_g M_{1,1} \,. \tag{4.1}$$

If one applies the Clifford map $\gamma : V_{p,q} \mapsto C\ell_{p,q}$ and defines its natural restrictions $\gamma' : N_{p-1,q-1} \mapsto C\ell_{p-1,q-1}$, $\gamma'' : M_{1,1} \mapsto C\ell_{1,1}$, one obtains the following

Periodicity Theorem. $C\ell_{p,q} \simeq C\ell_{p-1,q-1} \otimes C\ell_{1,1}$.

While in this special case the tensor product may be ungraded, in general the tensor product in such decompositions may be graded or not, see [10, 47, 53].

Using the obvious notation $C\ell(V_{p,q}) = C\ell_{p,q}(Q)$ and introducing the restrictions of the Wick isomorphism $\phi^{-1}|_N$ and $\phi^{-1}|_M$, (here $N = N_{p-1,q-1}$ and $M = M_{1,1}$), we can calculate the decomposition of $C\ell_{p,q}(B)$. However, if there are terms in the bi-vector F which connect spaces N and M, that is, if $F = \sum F_i$ and if there exists $F_s = \mathbf{a}_s \wedge \mathbf{b}_s$ with $\mathbf{a}_s \in N$, $\mathbf{b}_s \in M$, this part of the construction belongs *neither* to the restriction $\phi^{-1}|_N$ *nor* to $\phi^{-1}|_M$. We have *either* **no tensor decomposition** *or* a **deformed tensor product**. Expressed in formulas we get

$$\begin{aligned} C\ell_{p,q} &= \phi^{-1}(C\ell_{p,q}(Q)) \\ &= \phi^{-1}\left[C\ell_{p-1,q-1}(Q|_N) \otimes C\ell_{1,1}(Q|_M)\right] \\ &= C\ell_{p-1,q-1}(B|_N)(\phi^{-1}\otimes)C\ell_{1,1}(B|_M) \\ &= C\ell_{p-1,q-1}(B|_N) \otimes_{\phi^{-1}} C\ell_{1,1}(B|_M). \end{aligned} \tag{4.2}$$

Remark. *The deformed tensor product $\otimes_{\phi^{-1}}$ is not braided by construction since we have no restrictions on ϕ^{-1}. But one is able to find, e.g., Hecke elements, etc., necessary for a common q-deformation or, more generally, a braiding.*

As the main result of our investigation we have shown that quantum Clifford algebras *do not* come in general with periodicity theorems as, e.g., the famous Atiyah-Bott-Shapiro mod 8 index theorem. This has *enormous* impact on quantum manifold theory and the topological structure of such spaces as well as on their analytical properties. However, we have constructed a deformed – not necessarily braided – tensor product $\otimes_{\phi^{-1}}$ which gives a decomposition at the cost of losing (anti)-commutativity. To fully support this view and convince also those readers who might consider our reasoning too abstract and only formal in nature, we proceed to provide some examples.

5 Examples

In this section we consider three examples, each of them pointing out a peculiar feature of quantum Clifford algebras and $\mathbb{Z}_n$-gradings. Two of these examples have been found by using CLIFFORD, a Maple V Rel. 5 package for quantum Clifford algebras [2, 3]. While the second example is generic, the third one was taken from [27] and provides an example of a physical theory which benefits extraordinarily from using quantum Clifford algebras.

5.1 Example 1

This example shows that even in classical Clifford algebras one does not have a unique access to the *objects* of the graded space. Consider the well-known Dirac γ matrices which generate the Dirac-Clifford algebra $C\ell_{1,3}$ and satisfy $\gamma_i\gamma_j + \gamma_j\gamma_i = 2\eta_{ij}\mathbf{1}$ with the Minkowski metric $\eta_{ij} = \mathrm{diag}(1,-1,-1,-1)$. The linear span of the γ-matrices (generators) contains 1-vectors $\mathbf{x} = \sum x^i\gamma_i$. Define $\gamma_5 = \gamma_0\gamma_1\gamma_2\gamma_3$ and note that $\gamma_5^2 = -\mathbf{1}$. If we define *new generators* $\alpha_i := \gamma_i\gamma_5$ which are 3-vectors(!), it is easily checked that they nevertheless fulfill $\alpha_i\alpha_j + \alpha_j\alpha_i = 2\eta_{ij}\mathbf{1}$. They might be called *vectors* on an equal right.

Define the map $\underline{\gamma_5} : C\ell_{1,3} \mapsto C\ell_{1,3},\ \mathbf{x} \mapsto \mathbf{x}' := \mathbf{x}\gamma_5$, lifted to $C\ell_{1,3}$. We have thus defined two *different* Clifford maps $\gamma : V \mapsto C\ell_{1,3}$ and $\gamma' : V \mapsto C\ell_{1,3}$ with $\gamma' := \underline{\gamma_5} \circ \gamma$. That is one can't know for sure which elements are "vectors" even in this case.

We emphasized earlier that we did not expect the interpretation and the mathematical aspects of classical Clifford algebras to change in such a transformation. However, see [16] for a far more elaborate application of a similar situation where *both* gradings are used.

5.2 Example 2

In this example we examine the split case $C\ell_{2,2} \simeq C\ell_{1,1} \otimes C\ell_{1,1}$ and show the existence and irreducibility of an 8-dimensional representation not known in the classical representation theory of Clifford algebras.

We start with $C\ell_{1,1}(B)$ where B is given as

$$B := \begin{pmatrix} 1 & a \\ 0 & -1 \end{pmatrix}. \tag{5.1}$$

If a is zero, we have two choices for an idempotent element generating a spinor space

$$\mathbf{f}_{11}^{-} := \frac{1}{2}(\mathbf{1} + \mathbf{e}_1), \quad \mathbf{f}_{11}^{+} := \frac{1}{2}(\mathbf{1} + \mathbf{e}_1 \wedge \mathbf{e}_2). \tag{5.2}$$

A spinor basis can be found in both cases by left multiplying by $\mathbf{e}_2$ which yields $\mathcal{S}^{\pm} = < \mathbf{f}_{11}^{\pm}, \mathbf{e}_2\mathbf{f}_{11}^{\pm} >$. The spinor spaces $\mathcal{S}^{\pm}$ are 2-dimensional, and the Clifford elements are represented as 2×2 matrices. If a is not zero, an analogous construction runs through.

Now let us put together two such algebras, as shown in [53], generated by $C\ell_{1,1} = < \mathbf{e}_1, \mathbf{e}_2 >$ and $C\ell_{1,1} = < \mathbf{e}_3, \mathbf{e}_4 >$. The bilinear form B which reduces in both cases to the above setting *and* which contains connecting elements is

$$B := \begin{pmatrix} 1 & a & n_{11} & n_{12} \\ 0 & -1 & n_{21} & n_{22} \\ 0 & 0 & 1 & a \\ 0 & 0 & 0 & -1 \end{pmatrix} . \tag{5.3}$$

We expect the n_{ij} parameters to govern the deformation of the tensor product in the decomposition theorem.

Searching with CLIFFORD for idempotents in this general case yields the following fact. Let λ be a fixed parameter. Among six choices for an idempotent $\mathbf{f}$, we found

$$\mathbf{f} := \frac{1}{2}(\mathbf{1} + X_1) = \frac{1}{4}(2 + \lambda a)\mathbf{1} + \frac{1}{4}\sqrt{4 - \lambda^2 a^2 - 4\lambda^2}\,\mathbf{e}_1 + \frac{1}{2}\lambda\,\mathbf{e}_1 \wedge \mathbf{e}_2,$$

where X_1 is one of six different, non-trivial, and general elements X in $C\ell(B,V)$ satisfying $X^2 = \mathbf{1}$. This is an *indecomposable idempotent* which therefore generates an *irreducible* 8 *dimensional representation* since the regular representation of $C\ell(B,V)$ is of dimension 16. This fact depends on the appearance of the non-zero n_{ij} parameters. It was proved by brute force that none of the remaining five non-trivial elements $X_i, i = 2, \ldots, 6$, and squaring to $\mathbf{1}$ commuted with X_1. Thus, the search showed that there is no second Clifford element $X_2 \neq X_1$ which would square to $\mathbf{1}$ and which would commute with X_1. Such an element would be necessary to decompose $\mathbf{f}$ into a product $\mathbf{f} = \prod_i \frac{1}{2}(1 + X_i)$ where $X_i X_j = X_j X_i$ and $X_i^2 = \mathbf{1}$. Since this type of reasoning can be used to classify Clifford algebras [19], we have found a way to classify quantum Clifford algebras.

This type of an indecomposable exotic representation will occur in the next example of a physical model and is thereby not academic.

5.3 Example 3

Index doubling

For a simple treatment with a computer algebra, using CLIFFORD package, and for physical reasons not discussed here, see [25, 27, 31], we introduce an index doubling which provides us with a possibility to map the contraction and the wedge onto a new Clifford product in the larger algebra. The benefits of such a treatment are: the associativity of the mapped products, only one algebra product needed during calculations, etc.

Define the self-dual (reflexive) space $\mathbf{V} = V \oplus V^*$ and introduce generators $\mathbf{e}_i$ which span V and V^*

$$V = < \mathbf{e}_1, \ldots, \mathbf{e}_n >, \qquad V^* = < \mathbf{e}_{n+1}, \ldots, \mathbf{e}_{2n} > . \tag{5.4}$$

In this transition we require that the elements $\mathbf{e}_i$ from V generate a Grassmann sub-algebra and the $\mathbf{e}_{n+1}, \ldots, \mathbf{e}_{2n} \in V^*$ are duals which act via the contraction on V. This gives the following conditions on the form $\mathbf{B}:\mathbf{V}\times\mathbf{V}\mapsto\mathbf{R}$:

$$\begin{aligned} &i) \quad \mathbf{e}_i^2 = \mathbf{e}_i\wedge\mathbf{e}_i\wedge\cdot = 0 \\ &ii) \quad \mathbf{e}_{n+i}^2 = \mathbf{e}_{n+1}\,\underset{\mathbf{B}}{\rfloor}\,\mathbf{e}_{n+i}\,\underset{\mathbf{B}}{\rfloor}\,\cdot = (\mathbf{e}_{n+i}\wedge\mathbf{e}_{n+i})\,\underset{\mathbf{B}}{\rfloor}\,\cdot = 0\,. \end{aligned} \tag{5.5}$$

Thus, with respect to the basis $<\mathbf{e}_1,\ldots,\mathbf{e}_n,\mathbf{e}_{n+1},\cdots,\mathbf{e}_{2n}>$, $\mathbf{B}$ has the following matrix:

$$\mathbf{B} := \begin{pmatrix} 0 & g \\ g^T & 0 \end{pmatrix} + A = g + A, \tag{5.6}$$

where, with an abuse of notation, the symmetric part of $\mathbf{B}$ is again denoted by g. Note that we have introduced here a further freedom since A may be non-trivial also in the V-V and V^*-V^* sectors. This fact has certain physical consequences which were discussed in [27]. The $\mathbf{e}_i$'s from V can be identified with Schwinger sources of quantum field theory [25, 31].

The $U(2)$-model

We simply report here the result from [27] and strongly encourage the reader to consult this work since we quote here only a part of that work which shows the indecomposability of quantum Clifford algebra representations and the therefrom following physical consequences.

Define $C\ell(B,V)\simeq C\ell_{2,2}(B)$ by specifying $\mathbf{V}=<\mathbf{e}_i>=<\mathbf{a}_1^\dagger,\mathbf{a}_2^\dagger,\mathbf{a}_3,\mathbf{a}_4>$ and

$$\mathbf{B} := \frac{1}{2}\begin{pmatrix} 0 & I_2 \\ I_2 & 0 \end{pmatrix} + A, \tag{5.7}$$

where I_2 is the 2×2 unit matrix and A is an arbitrary but fixed 4×4 antisymmetric matrix with respect to the $\mathbf{e}_i$ or $\mathbf{a}_i$ basis. Note, furthermore, that the $\mathbf{a}_i$ and $\mathbf{a}_i^\dagger$ fulfill the canonical anti-commutation relations, CAR, of a quantum system: $\{\mathbf{a}_i,\mathbf{a}_j^\dagger\}_+ = \delta_{ij}$. Define, furthermore, Clifford elements $N, S_i \in \mathbb{R}\oplus\mathbf{V}\wedge\mathbf{V}$, $i\in\{1,2,3\}$ such that the following relations hold:

$$\begin{aligned} [N,a_i]_- &= -a_i, \quad [N,a_i^\dagger]_- = +a_i^\dagger, \quad N^\dagger = N, \\ [S_k,a_i]_- &= \sigma_{ij}a_j, \quad \text{h.c.}, \quad k\in\{1,2,3\}, \\ [S_k,N]_- = 0, \quad [S_k,S_l]_- &= i\epsilon_{klm}\,S_m, \quad S_k^\dagger = S_k\,, \end{aligned} \tag{5.8}$$

where $\dagger$ is the anti-involutive map (includes a product reversion) interchanging $\mathbf{a}_i\leftrightarrow\mathbf{a}_i^\dagger$. This is the $U(2)$ algebra if $A\equiv 0$.

Define a "vacuum", for a discussion see [27], simply by defining the expectation function –linear functional– as the projector onto the scalar part $<\,.\,>_0^A$ which depends now explicitly on A. In a physicist's notation $<0\,|\,\hat{\mathcal{H}}\,|\,0>\simeq\ <\mathcal{H}>_0^A$ for any operator $\hat{\mathcal{H}}$ resp. Clifford element $\mathcal{H}$.

An algebraic analysis which coincides in the positive definite case with C^*-algebraic results shows that this linear functional called "vacuum" can be uniquely decomposed in certain extremal, that is indecomposable, states. Denoting these states as spinor like $\mathcal{S}_1, \mathcal{S}_2$ and exotic $\mathcal{E}$ we obtain the following identity:

$$< . >_0^A = \lambda_1 < . >^{\mathcal{S}_1} + \lambda_2 < . >^{\mathcal{S}_2} + \lambda_3 < . >^{\mathcal{E}}, \quad \sum \lambda_i = 1. \tag{5.9}$$

Since the regular representation of $C\ell_{2,2}(B)$ is 16 dimensional and we find dim $\mathcal{S}_1$ = dim $\mathcal{S}_2 = 4$, dim $\mathcal{E} = 8$ this is a direct sum decomposition into irreducible representations. The "classical" case would have led to four representations of the spinor type each 4 dimensional. The indecomposable exotic representation obtained from $< . >^{\mathcal{E}}$ is therefore new, and it is a direct outcome of the structure of the quantum Clifford algebra, see previous example. This representation decomposes into two spinor like parts if A vanishes identically $A \equiv 0$.

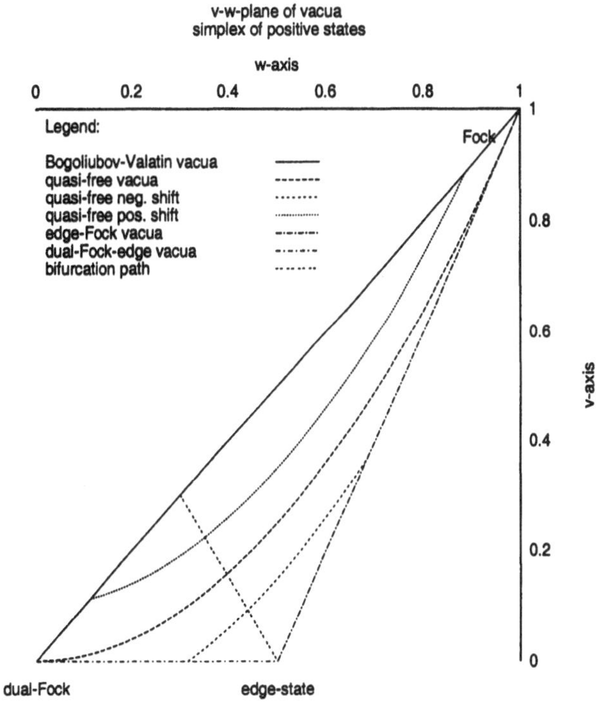

FIGURE 1. $C\ell(B)$-deformation of $U(2)$ algebra

In [27] we obtained a v-w-plane of vacua while implementing the $\sum \lambda_i = 1$ condition and renaming of variables into v, w. There it was shown, see Figure 1, that we find free systems of Fock and dual-Fock type which constitute the spinor representations $\mathcal{S}_1$, $\mathcal{S}_2$ and that the line connecting them contains Bogoliubov-transformed ground states of BCS-superconductivity. Quasi free – that is, correlation free – states are on the displayed parabola. In the exotic state one finds spin 1 and spin 0 components which are beyond Bogoliubov transformations. Every choice of A fixes *exactly* one particular state in the v-w-plane. Hence, we have solved the problem of finding an algebraic condition on which side of the Clebsch-Gordan identity $\mathbf{\frac{1}{2}} \otimes \mathbf{\frac{1}{2}} = \mathbf{0} \oplus \mathbf{1}$ our algebraic system *has to be* treated.

Our model, even if only marginally discussed, shows all features we want to see in the composite and multi-particle theory. Moreover, exotic representations which describe "bound objects" not capable of a decomposition are *beyond* the treatment in [20] which mimics in Clifford algebraic terms the usual tensor method which generically bears this problem. In this context we refer to the interesting work of Daviau [18] on de Broglie's spin fusion theory [9] and to the joint works with Stumpf and Dehnen [23, 26] which are connected with algebraic composite theories.

Acknowledgment

The first author (BF) acknowledges a travel grant of the DFG and a critical reading of the manuscript by Th. Konrad.

References

[1] R. Abłamowicz, P. Lounesto, On Clifford algebras of a bilinear form with an antisymmetric part, in *Clifford Algebras with Numeric and Symbolic Computations,* Eds. R. Abłamowicz, P. Lounesto, J. M. Parra, Birkhäuser, Boston, 1996, 167–188.

[2] R. Abłamowicz, *Clifford Algebra Computations with Maple,* Proc. Clifford (Geometric) Algebras, Banff, Alberta Canada, 1995, Ed. W. E. Baylis, Birkhäuser, Boston, 1996, 463–501.

[3] R. Abłamowicz, 'CLIFFORD' - Maple V package for Clifford algebra computations, ver. **4**, `http://math.tntech.edu/rafal/cliff4/`.

[4] R. Abłamowicz, B. Fauser, *Hecke algebra representations in ideals generated by q-Young Clifford idempotents,* in this volume and math.QA/9908062.

[5] M. F. Atiyah, R. Bott, A. Shapiro, Clifford modules, *Topology* **3** (Suppl. 1) (1964), 3–38.

[6] I. M. Benn, R. W. Tucker, *Introduction to Spinors and Geometry with Applications in Physics,* Adam Hilger, Bristol, 1987.

[7] F. A. Berezin, *The Method of Second Quantization,* Academic Press, London, 1966.

[8] N. Bourbaki, *Algebra 1,* Springer Verlag, Berlin, 1989, Chapters 1–3.

[9] L. de Broglie, *La méchanique du photon, Une nouvelle théorie de la Lumière* tome 1, La Lumière dans le vide, Hermann Paris, 1940, tome 2, Les interactions entre les photons et la metière, Hermann, Paris, 1942.

[10] P. Budinich, A. Trautmann, *The Spinorial Chessboard,* Trieste Notes in Physics, Springer, 1988.

[11] E. R. Caianiello, *Combinatorics and Renormalization in Quantum Field Theory,* W. A. Benjamin, Inc., London, 1973.

[12] C. Chevalley, *The Algebraic Theory of Spinors,* Columbia University Press, New York, 1954.

[13] C. Chevalley, *The Construction and Study of Certain Important Algebras,* The Mathematical Society of Japan 1955, in Collected Works, Vol. 2, Springer, Berlin, 1997.

[14] A. Crumeyrolle, *Orthogonal and Symplectic Clifford Algebras, Spinor Structures,* Mathematics and Its Applications, Kluwer, Dordrecht, 1990.

[15] A. Connes, *Noncommutative Geometry,* Academic Press, San Diego, 1994, Ch. V. 10.

[16] O. Conradt, The principle of duality in Clifford algebra and projective geometry, this volume.

[17] H. S. M. Coxeter, W. O. J. Moser, *Generators and Relations for Discrete Groups, Fourth Edition,* Springer Verlag, Berlin, 1980.

[18] C. Daviau, Application à la théorie de la lumiére de Louis de Broglie d'une réécriture de l'équation de Dirac, *Ann. de la Fond. Louis de Broglie,* Vol. **23** (3–4) (1998), 121–127.

[19] A. Dimakis, Geometrische Behandlung von Cliffordalgebren und spinoren mit Anwendungen auf die Dynamik von Spinteilchen, Thesis, Univ. Göttingen 1983. A new representation of Clifford algebras, *J. Phys.* **A22** (1989), 3171.

[20] C. Doran, A. Lasenby, S. Gull, States and operators in spacetime algebra, *Found. Phys.* **23** (9), (1993), 1239. S. Somaroo, A. Lasenby, C. Doran, Geometric algebra and the causal approach to multiparticle quantum mechanics, *J. Math. Phys.* Vol. **40**, No. **7** (July 1999), 3327–3340.

[21] F. J. Dyson, The S-matrix in electrodynamics, *Phys. Rev.* **75** (1949), 1736–1735.

[22] F. J. Dyson, Missed opportunities, *Bull. Amer. Math. Soc.* **78** (1972), 635–652.

[23] B. Fauser, H. Dehnen, Isospin from spin by compositeness, in Proceedings of the International Workshop *Lorentz Group, CPT and Neutrinos,* V. Dvoeglazov, ed., Zacatecas, Mexico, 1999.

[24] B. Fauser, Hecke algebra representations within Clifford geometric algebras of multivectors, *J. Phys. A: Math. Gen.* **32** (1999), 1919–1936.

[25] B. Fauser, *Clifford-algebraische Formulierung und Regularität der Quantenfeldtheorie,* Thesis, Uni. Tübingen, 1996.

[26] B. Fauser, H. Stumpf, Positronium as an example of algebraic composite calculations, in Proceedings of *The Theory of the Electron,* J. Keller, Z. Oziewicz, eds., Cuautitlan, Mexico, 1995, *Adv. Appl. Clifford Alg.* **7** (Suppl.) (1997), 399–418.

[27] B. Fauser, Clifford geometric parameterization of inequivalent vacua, Submitted to *J. Phys. A.* (hep-th/9710047).

[28] B. Fauser, On the relation of Clifford-Lipschitz groups to q-symmetric groups, Group 22: Proc. XXII *Int. Colloquium on Group Theoretical Methods in Physics,* Hobart, Tasmania, 1998, eds., S. P. Corney, R. Delbourgo and P. D. Jarvis, International Press, Cambridge, 1999, 413–417.

[29] B. Fauser, Clifford algebraic remark on the Mandelbrot set of two-component number systems, *Adv. Appl. Clifford Alg.* **6** (1) (1996), 1–26.

[30] B. Fauser, Vertex functions and generalized normal-ordering by triple systems in non-linear spinor field models, preprint hep-th/9611069.

[31] B. Fauser, On an easy transition from operator dynamics to generating functionals by Clifford algebras, *J. Math. Phys.* **39** (1998), 4928–4947.

[32] B. Fauser, Vertex normal ordering as a consequence of nonsymmetric bilinear forms in Clifford algebras, *J. Math. Phys.* **37** (1996), 72–83.

[33] G. Fiore, On q-deformations of Clifford algebras, this volume.

[34] W. Fulton, J. Harris, *Representation Theory,* Springer, New York, 1991.

[35] W. H. Greub, *Multilinear Algebra,* Springer, Berlin, 1967.

[36] R. Haag, On quantum field theories, *Mat. Fys. Medd. Dann. Vid. Selsk.* **29**, No. **12** (1955). *Local Quantum Physics* Springer, Berlin, 1992.

[37] A. J. Hahn, *Quadratic Algebras, Clifford Algebras, and Arithmetic Witt Groups,* Springer, New York, 1994.

[38] M. Hamermesh, *Group Theory and Its Application to Physical Problems,* Addison-Wesley, London, 1962.

[39] D. Hestenes, *Spacetime Algebra,* Gordon and Beach, 1966.

[40] D. Hestenes, G. Sobczyk, *Clifford Algebra to Geometric Calculus,* Reidel, Dordrecht, 1984.

[41] D. Hestenes, A unified language for mathematics and physics, in Proceedings of *Clifford Algebras and Their Application in Mathematical Physics,* Canterbury, 1985, J. S. R. Chisholm, A. K. Common, eds., Kluwer, Dordrecht, 1986.

[42] D. Hestenes, *New Foundation for Classical Mechanics,* Kluwer, Dordrecht, 1986.

[43] D. Hestenes, R. Ziegler, Projective geometry with Clifford algebra, *Acta Appl. Math.* **23** (1991), 25–64.

[44] H. Hiller, *Geometry of Coxeter Groups,* Pitman Books Ltd., London, 1982.

[45] R. Kerschner, *Effektive Dynamik zusammengesetzter Quantenfelder,* Thesis, Univ. Tübingen, 1994, Chapter 4.

[46] R. Kerschner, On quantum field theories with finitely many degrees of freedom, preprint, hep-th/9505063.

[47] T. Y. Lam, *The Algebraic Theory of Quadratic Forms,* The Benjamin/Cummings Publishing Company, Reading, 1973, [Theorem 2.6 and Corollary 2.7, p. 113].

[48] H. B. Lawson, M.-L. Michelsohn, *Spin Geometry,* Universidade Federal de Ceará, Brazil, 1983, Princeton Univ. Press, Princeton, 1989.

[49] P. Lounesto, *Clifford Algebras and Spinors,* Cambridge University Press, Cambridge, 1997.

[50] P. Lounesto appendix in M. Riesz, *Clifford Numbers and Spinors,* Lecture Series No. 38, The Institute of Fluid Dynamics and Applied Mathematics, University of Maryland (1958), 1993 Facsimile, Kluwer, Dordrecht, 1993.

[51] P. Lounesto, Counterexamples in Clifford algebras with CLICAL, in *Clifford Algebras with Numeric and Symbolic Computations,* R. Abłamowicz, P. Lounesto, J. M. Parra, eds., Birkhäuser, Boston 1996.

[52] S. Majid, *Foundations of Quantum Group Theory,* Cambridge University Press, Cambridge, 1995.

[53] J. Maks, *Modulo (1,1) Periodicity of Clifford Algebras and the Generalized (anti-) Möbius Transformations,* Thesis, TU Delft, 1989.

[54] A. Micali, Albert Crumeyrolle, La démarche algébrique d'un géometre, p. xi in *Clifford Algebras and Spinor Structures, A Special Volume Dedicated to the Memory of Albert Crumeyrolle (1919–1992),* R. Abłamowicz, P. Lounesto, eds., Kluwer, Dordrecht 1995 ix–xiv. P. Lounesto, Crumeyrolle's strange question: "What is a bi-vector?", Preprint 1993.

[55] R. V. Moody, A. Pianzola, *Lie Algebras with Triangular Decompositions,* Wiley-Interscience Publ., New York, 1995.

[56] Z. Oziewicz, From Grassmann to Clifford, in Proceedings *Clifford Algebras and Their Application in Mathematical Physics,* Canterbury, 1985, UK, J.S.R. Chisholm, A.K. Common, eds., Kluwer, Dordrecht, 1986, 245–256.

[57] Z. Oziewicz, Clifford algebra of multivectors, in Proceedings of the International Conference on the Theory of the Electron, Cuautitlan, Mexico, 1995. Eds. J. Keller and Z. Oziewicz, *Adv. in Appl. Clifford Alg.* **7** (Suppl.) (1997), 467–486.

[58] Z. Oziewicz, Clifford algebra for Hecke braid, in *Clifford Algebras and Spinor Structures, Special Volume Dedicated to the Memory of Albert Crumeyrolle,* eds. R. Abłamowicz, P. Lounesto, Kluwer, Dordrecht, 1995, 397–412.

[59] J. M. Parra Serra, In what sense is the Dirac-Hestenes electron a representation of the Poincaré group?, in Proceedings of the International Workshop *Lorentz Group, CPT and Neutrinos,* V. Dvoeglazov, ed., Zacatecas, 1999, Mexico.

[60] R. Penrose, W. Rindler, *Spinors and Space-time: Two-spinor Calculus and Relativistic Fields,* Cambridge Monographs on Mathematics, Cambridge University Press, Paperback Reprint Edition, Vol. 1, June, 1987.

[61] I. Porteous, *Topological Geometry,* Van Nostrand, New York, 1969.

[62] M. Riesz, *Clifford Numbers and Spinors,* Lecture Series No. 38, The Institute of Fluid Dynamics and Applied Mathematics, University of Maryland 1958, Facsimile by Kluwer, 1993.

[63] H. Saller, Quantum algebras I, *Nuovo Cim. 108B* (6) (1993), 603–630. Quantum algebras II, *Nuovo Cim.* **109B**, (3) (1994), 255–280.

[64] G. Scheja, U. Storch, *Lehrbuch der Algebra, Teil 2,* Teubner, Stuttgart, 1988.

[65] H. Stumpf, Th. Borne, *Composite Particle Dynamics in Quantum Field Theory,* Vieweg, Braunschweig, 1994.

[66] G. C. Wick, The evaluation of the collision matrix, *Phys. Rev.* **80** (1950), 268–272.

[67] E. Witt, Theorie der quadratischen formen in beliebigen Körpern, *J. Reine. Angew. Math.* **176** (1937), 31–44.

Bertfried Fauser
Universität Konstanz
Fachbereich Physik, Fach M678
D-78457 Konstanz
E-mail: Bertfried.Fauser@uni-konstanz.de

Rafał Abłamowicz
Department of Mathematics Box 5054
Tennessee Technological University
Cookeville, TN 38505, USA
E-mail: rablamowicz@tntech.edu

Received: October 1, 1999; Revised: February 12, 2000

Covariant Derivatives on Minkowski Manifolds

Virginia V. Fernández, Antonio M. Moya, and Waldyr A. Rodrigues, Jr.

ABSTRACT We present a general theory of covariant derivative operators (linear connections) on a Minkowski manifold (represented as an affine space $(M, \mathcal{M}^*)$ using the powerful multiform calculus. When a gauge metric extensor G (generated by a gauge distortion extensor h) is introduced in the Minkowski manifold, we get a theory that permits the introduction of general Riemann-Cartan-Weyl geometries. The concept of gauge covariant derivatives is introduced as the key notion necessary to generate linear connections that are compatible with G, thus, permitting the construction of Riemann-Cartan geometries. Many results of genuine mathematical interest are obtained. Moreover, such results are fundamental for building a consistent formulation of a theory of the gravitational field in flat spacetime. Some important examples of applications of our theory are worked in details.
Keywords: Covariant derivatives, Clifford calculus, extensor calculus, Minkowski manifolds.

1 Introduction

In this paper we study the theory of covariant derivative operators on the Minkowski manifold using the multiform calculus (Hestenes and Sobczyk, 1984: Moya, Fernández, and Rodrigues, 2000, Moya, 1999). Before making use of this formalism, we recall the main ideas and problems discussed in this paper in the language of ordinary tensor calculus.[1] Let (M, η, τ_η) be the Minkowski manifold and $(M, \eta, \tau_\eta, D^\eta)$ Minkowski spacetime (Wu and Sachs, 1993, Rodrigues and Rosa, 1989).[2] The quadruple $(M, \eta, \tau_\eta, \nabla)$, where ∇ is an arbitrary covariant derivative operator (not necessarily compatible with η) and such that its torsion and curvature tensors are non-

AMS Subject Classification: 15A66, 53C20, 53C50.

[1] We hope that this will encourage physicists to read the paper.

[2] M is a 4-dimensional manifold oriented by τ_η (the volume element 4-form) and time oriented, which is diffeomorphic to R^4, $\eta \in \sec T_2^0(M)$ is a Lorentzian flat metric and D^η is the Levi-Civita connection of η.

zero, will be called a *Riemann-Cartan-Weyl (RCW) spacetime.*[3] We can introduce into the Minkowski manifold an infinite number of nondegenerate symmetric tensors $G \in \sec T_2^0(M)$ (not necessarily of Lorentzian signature). A quadruple (M, G, τ_G, ∇), where ∇ is an arbitrary covariant derivative operator (not necessarily compatible with G) and such that its torsion and curvature are nonzero, will be called an *RCW space.* Such a quadruple will be called a spacetime only if G has Lorentzian signature, in which case it will be denoted by $\mathbf{g}$, and will be called a gauge metric field. Now, given an invertible mapping $\mathbf{h} : \sec TM \to \sec TM$, called a gauge distortion field, it induces on the Minkowski manifold a nondegenerate symmetric tensor G, by

$$G(\mathbf{u}, \mathbf{v}) = \eta(\mathbf{h}(\mathbf{u}), \mathbf{h}(\mathbf{v})), \ \forall \mathbf{u}, \mathbf{v} \in \sec TM. \tag{1.1}$$

A *nontrivial* mathematical question is the following: given an arbitrary RCW space (or spacetime) (M, G, τ_G, ∇), where G is generated by $\mathbf{h}$, construct a covariant derivative operator from ∇ and $\mathbf{h}$, denoted by $\hat{\nabla}$, such that it is compatible with G, i.e., $\hat{\nabla} G = 0$. In this paper a solution for this problem is found by introducing the concept of gauge covariant derivatives.

Now, from the physical point of view, our interest in this problem comes from the so-called flat spacetime formulations of gravitational theory (Logunov and Mestvirishvilli, 1989; Rodrigues and Souza, 1993; Pommaret, 1994), where the gravitational field is supposed to be a field in the Faraday sense and not a manifestation of a geometry, as in Einstein's general relativity. Recently, (Doran, Lasenby and Gull (DLG), 1998) developed a theory of this kind (using the multivector calculus (Hestenes and Sobczyk, 1984)) which they called the gauge theory of gravitation.[4] They believe to have produced a theory that is a generalization of the flat spacetime formulation of Einstein's general relativity. One of the main ingredients of their theory is the introduction of a certain covariant derivative operator that they called also a gauge covariant derivative. However, it happens that what DLG called a gauge covariant derivative is one like ∇ and not one as $\hat{\nabla}$ (this statement is proved below). It follows that in their theory[5] ∇ is not g-compatible[6], contrary to their statements (see, in particular Appendix C of (DLG,1993)), and as a consequence their theory has the same deficiency regarding physical interpretation as the Weyl theory (Weyl, 1918, 1922,

---

[3]More details and precise definitions using the multiform calculus are given in Section 5.

[4]Applications of his theory appear in several papers, e.g., (Challinor et al., 1997; Dabrowski et al., 1999; Doran et al., 1999).

[5]g is a $(1,1)$- extensor field representing a metric $\mathbf{g} \in \sec T_2^0(M)$. The definition of (p,q)-extensor fields is given in definition (1.8).

[6]A fact expressed by $\nabla_{\bar{a}}\, g \neq 0$ in the formalism of the present paper (and by an analogous equation in the formalism of (DLG), see Section 4.

Adler, Bazin and Schiffer, 1965). We are not going to discuss gravitational theories in this paper. We only observe that a gauge theory of gravitational field in flat spacetime, i.e., on Minkowski manifold, incorporating the compatibility of the gauge covariant derivative with g (the gauge metric field) and using the multiform calculus, is given in (Fernández, Moya and Rodrigues, 2000a).

Before proceeding, we would like to observe that one of the authors of the present paper already studied the geometry of Riemann-Cartan-Weyl spaces in (Rodrigues and Souza, 1993; Souza and Rodrigues, 1994; Rodrigues et al., 1995) using the Clifford bundle of differential forms. However, that formalism in the general case cannot be considered complete. Indeed, the covariant derivative operators acting on the sections of the Clifford bundle used there are the ones coming from covariant derivative operator defined in the tensor bundle. As it is well-known, a covariant derivative operator defined in the tensor bundle in general does not define a covariant derivative operator in the Clifford bundle of differential forms, unless it is compatible with the metric field used in the definition of the Clifford bundle. Thus, in the approach of (Rodrigues and Souza, 1993; Souza and Rodrigues, 1994; Rodrigues et al., 1995) only Riemann-Cartan geometries can be properly studied, the RCW cases being treated in a correct, but artificial way.[7] To overcome this problem it is necessary to define the concept of a general covariant derivative operator on the sections of the Clifford bundle. However, this enterprise is more easily done by developing the multiform calculus in a special way that enhances its power (Moya, Fernández, and Rodrigues, 2000, Moya 1999) and formulating a general theory of covariant derivative operators on Minkowski manifolds.[8] This can be done once we represent the Minkowski manifold as an *affine space*, with vector space $\mathcal{M}^*$.

Due to limitation of space, in this paper we study only the action of covariant derivative operators on form fields and on $(1,1)$-extensor fields (which are the objects that represent the tensor fields $\mathbf{h}$ and $\mathbf{g}$ in our formalism). A study of the action of covariant derivative operators on general multiform and extensor fields, including a generalization of Cartan differential operator acting on exform fields (i.e., completely antisymmetric extensor fields) and Lie derivatives of multiform and extensor fields, is given in (Fernández, Moya and Rodrigues, 2000b). Despite these limitations the results to be presented are really worth studying and we hope that our readers will enjoy them. Finally, it is necessary to recall that the developments that follow can be only understood by readers who are famil-

[7]This does not means that a gauge theory of the gravitational field in the sense of (Fernández, Moya and Rodrigues, 2000b) cannot be done in the Clifford bundle formalism.

[8]The case of arbitrary manifolds is studied in (Fernández, Moya, and Rodrigues, 2000).

iar with Clifford algebras as presented, e.g., in (Lounesto, 1997)[9] and the multivector calculus as presented in (Hestenes and Sobczyk, 1984: Moya, Fernández, and Rodrigues, 2000; Moya, 1999).

1.1 Some preliminaries and notations

Given a *global coordinate system* over M, say $M \ni x \leftrightarrow x^\mu(x) \in R$ $(\mu = 0,1,2,3)$ associated to a inertial reference frame (Rodrigues and Rosa, 1989) at $x \in M$. $\langle \frac{\partial}{\partial x^\mu}\big|_x \rangle$ and $\langle dx^\mu|_x \rangle$ are the *natural basis* for the tangent vector space T_xM and the tangent covector space T_x^*M.

We have

$$\eta = \eta_{\mu\nu} dx^\mu \otimes dx^\nu,$$
$$\eta_{\mu\nu} = \eta(\frac{\partial}{\partial x^\mu}, \frac{\partial}{\partial x^\nu}) = \text{diag}(1,-1,-1,-1). \tag{1.2}$$

Definition 1. *$T_xM \ni \mathbf{v}_x$ is said to be equipolent to $\mathbf{v}_{x'} \in T_{x'}M$ (written $\mathbf{v}_x = \mathbf{v}_{x'}$) if and only if*

$$\eta_{(x)}(\frac{\partial}{\partial x^\mu}\bigg|_x, \mathbf{v}_x) = \eta_{(x')}(\frac{\partial}{\partial x^\mu}\bigg|_{x'}, \mathbf{v}_{x'}),\ (\mu = 0,1,2,3). \tag{1.3}$$

Note that $\frac{\partial}{\partial x^\beta}\bigg|_x = \frac{\partial}{\partial x^\beta}\bigg|_{x'}$ $(\beta = 0,1,2,3)$.

Definition 2. *The set of equivalent classes of tangent vectors over the tangent bundle*

$$\mathcal{M} = \{C_{\mathbf{v}_x} \mid \textit{for all } x \in M\}, \tag{1.4}$$

has a natural structure of vector space; it is called Minkowski vector space.

Note that $\left\langle C_{\frac{\partial}{\partial x^\mu}\big|_x} \right\rangle$ is a natural basis for $\mathcal{M}$ $(\dim \mathcal{M} = 4)$. With the notations: $\vec{v} \equiv C_{\mathbf{v}_x}$ and $\vec{e}_\mu \equiv C_{\frac{\partial}{\partial x^\mu}\big|_x}$, we can write $\vec{v} = v^\mu \vec{e}_\mu$.

Definition 3. *The 2-tensor over $\mathcal{M}$,*

$$\eta : \mathcal{M} \times \mathcal{M} \to R, \tag{1.5}$$

such that for each $\vec{v} = C_{\mathbf{v}_x}$ and $\vec{w} = C_{\mathbf{w}_x} \in \mathcal{M} : \eta(\vec{v},\ \vec{w}) = \eta_{(x)}(\mathbf{v}_x,\ \mathbf{w}_x)$, for all $x \in M$, is called Minkowski metric tensor.

---

[9]It is particularly important to emphasize that in our approach (contrary to the approach in (Hestenes and Sobczyk, 1984), it is necessary to distinguish between contractions and internal products, as done in (Lounesto, 1997; Moya, Fernández and Rodrigues, 2000a; Moya, 1999).

Note that, for each pair of basis vectors $\vec{e}_\mu \equiv C \left.\frac{\partial}{\partial x^\mu}\right|_x$ and $\vec{e}_\nu \equiv C \left.\frac{\partial}{\partial x^\nu}\right|_x$, it holds that

$$\eta(\vec{e}_\mu, \vec{e}_\nu) = \text{diag}(1, -1, -1, -1). \tag{1.6}$$

Definition 4. *The dual basis of* $\langle \vec{e}_\mu \rangle$ *will be symbolized by* $\langle \gamma^\mu \rangle$, *i.e.,* $\gamma^\mu \in \mathcal{M}^* \equiv \Lambda^1(\mathcal{M})$ *and* $\gamma^\mu(\vec{e}_\nu) = \delta^\mu_\nu$.

To continue, we observe the existence of a fundamental isomorphism between $\mathcal{M}$ and $\Lambda^1(\mathcal{M})$ given by

$$\mathcal{M} \ni \vec{a} \leftrightarrow a \in \Lambda^1(\mathcal{M}), \tag{1.7}$$

such that if $\vec{a} = a^\mu \vec{e}_\mu$, then $a = \eta_{\mu\nu} a^\mu \gamma^\nu$ and if $a = a_\mu \gamma^\mu$, then $\vec{a} = \eta^{\mu\nu} a_\mu \vec{e}_\nu$, where $\eta_{\mu\nu} = \eta(\vec{e}_\mu, \vec{e}_\nu)$, $\eta^{\mu\nu} = \eta_{\mu\nu}$.

Remark 1. *To each basis vector* $\vec{e}_\mu$ *correspond a basis form* $\gamma_\mu = \eta_{\mu\nu}\gamma^\nu$.

Definition 5. *A scalar product of forms can be defined by*

$$\Lambda^1(\mathcal{M}) \times \Lambda^1(\mathcal{M}) \ni (a, b) \mapsto a \cdot b \in R, \tag{1.8}$$

such that if $\vec{a} \leftrightarrow a$ *and* $\vec{b} \leftrightarrow b$ *then* $a \cdot b = \eta(\vec{a}, \vec{b})$.

Remark 2. $\gamma_\mu \cdot \gamma_\nu = \eta_{\mu\nu}$, $\gamma^\mu \cdot \gamma_\nu = \delta^\mu_\nu$ *(*$\langle \gamma_\mu \rangle$ *is called the reciprocal basis of* $\langle \gamma^\mu \rangle$*) and* $\gamma^\mu \cdot \gamma^\nu = \eta^{\mu\nu}$. *Thus,* η *admits the expansions* $\eta = \eta_{\mu\nu}\gamma^\mu \otimes \gamma^\nu = \eta^{\mu\nu}\gamma_\mu \otimes \gamma_\nu$.

Remark 3. *The oriented affine space* $(M, \mathcal{M}^*)$ *(oriented by* $\gamma^5 = \gamma^0 \wedge \gamma^1 \wedge \gamma^2 \wedge \gamma^3$*) is a representation of the Minkowski manifold.*

Remark 4. $(M, \mathcal{M}^*)$ *equipped with the scalar product given by (1.8) is a representation of Minkowski spacetime.*

Definition 6. *Let* $\langle x^\mu \rangle$ *be a global affine coordinate system for* $(M, \mathcal{M}^*)$, *relative to an arbitrary point* $o \in M$. *A position form associated to* $x \in M$, *is the form over* $\mathcal{M}$ *(designed by the same letter), given by the correspondence*

$$M \ni x \leftrightarrows x = x^\mu \gamma_\mu \in \Lambda^1(\mathcal{M}). \tag{1.9}$$

Remark 5. *We denote by* $C\ell(M) \approx C\ell_{1,3} \approx \mathbb{H}(2)$ *the spacetime algebra, i.e., the Clifford algebra (Lounesto, 1997) of* $\mathcal{M}^*$ *equipped with the scalar product defined by (1.8).*

Remark 6. *As a vector space over the reals, we have* $C\ell(\mathcal{M}) = \sum_{p=0}^{4} \Lambda^p(\mathcal{M})$.

Definition 7. *A smooth multiform field* A *on Minkowski spacetime is a multiform valued function of position form*

$$\Lambda^1(\mathcal{M}) \ni x \mapsto A(x) \in \Lambda(\mathcal{M}). \tag{1.10}$$

Definition 8. *Let* $0 \leq p, q \leq 4$. *A* (p,q)*-extensor* t *is a linear mapping*

$$t : \Lambda^q(\mathcal{M}) \to \Lambda^p(\mathcal{M}). \tag{1.11}$$

Remark 7. *The set of all* (p,q)*-extensors will be denoted from now on by* $\mathrm{ext}(\Lambda^p(\mathcal{M}), \Lambda^q(\mathcal{M}))$.

Definition 9. *A smooth* (p,q)*-extensor field* t *on Minkowski spacetime is a differentiable* (p,q)*-extensor valued function of position form*

$$\Lambda^1(\mathcal{M}) \ni x \mapsto t_x \in \mathrm{ext}(\Lambda^p(\mathcal{M}), \Lambda^q(\mathcal{M})). \tag{1.12}$$

Definition 10. *The* a*-directional derivative (*a *is an arbitrary form) of a smooth multiform field* X*, denoted as* $a \cdot \partial X$*, is defined by*

$$a \cdot \partial X = \lim_{\lambda \to 0} \frac{X(x + \lambda a) - X(x)}{\lambda} = \frac{d}{d\lambda} X(x + \lambda a)\bigg|_{\lambda=0}. \tag{1.13}$$

Remark 8. *The* γ_μ*-directional derivative* $\gamma_\mu \cdot \partial X$ *coincides with the coordinate derivative* $\frac{\partial X}{\partial x^\mu}$. *For short, we will use the notation* $\partial_\mu \equiv \gamma_\mu \cdot \partial$.

Definition 11. *The gradient, divergence and curl of a smooth multiform field* X*, respectively denoted by* ∂X, $\partial \lrcorner X$ *and* $\partial \wedge X$*, are defined by*

$$\partial X = \gamma^\mu (\partial_\mu X), \tag{1.14}$$

$$\partial \lrcorner X = \gamma^\mu \lrcorner (\partial_\mu X), \tag{1.15}$$

$$\partial \wedge X = \gamma^\mu \wedge (\partial_\mu X). \tag{1.16}$$

Remark 9. *For any* X*, it holds* $\partial X = \partial \lrcorner X + \partial \wedge X$.

2 Covariant derivative of form fields

Definition 12. *Let* a *be any form. A covariant derivative operator* ∇ *(or connection) acting on the set of smooth form fields on Minkowski manifold, modeled by* $\mathcal{M}^*$*, is the mapping*

$$\nabla_a : \mathrm{hom}[\Lambda^1(\mathcal{M}), \Lambda^1(\mathcal{M})] \to \mathrm{hom}[\Lambda^1(\mathcal{M}), \Lambda^1(\mathcal{M})], \tag{2.1}$$

satisfying the following axioms:

(i) For all scalars α, α' *and forms* a, a' *it holds* $\nabla_{\alpha a + \alpha' a'} b = \alpha \nabla_a b + \alpha' \nabla_{a'} b$ $(b \in \mathrm{hom}[\Lambda^1(\mathcal{M}), \Lambda^1(\mathcal{M})])$,

(ii) For all smooth scalar fields f, f' *and form fields* b, b' *it holds that* $\nabla_a (fb + f'b') = (a \cdot \partial f) b + f \nabla_a b + (a \cdot \partial f') b' + f' \nabla_a b' (a \in \Lambda^1(\mathcal{M}))$.

$\nabla_a b$ *is called the directional covariant derivative of the form field* b.

Remark 10. *The definition above is not empty. For example, the ordinary directional derivative* $a \cdot \partial b$ *is a well-defined directional covariant derivative. If* h *is a smooth* $(1,1)$*-extensor field which has inverse* h^{-1}*, then* $h^{-1}(a \cdot \partial h(b))$ *is also a well-defined directional covariant derivative.*

2.1 Connection extensor fields

We will show that, associated to ∇_a, there exist just two fundamental extensor fields, say $b \mapsto \gamma_a(b)$ and $a \mapsto \Omega(a)$, the first one being of type $(1,1)$ and the second one of type $(1,2)$. They are smooth extensor fields over the Minkowski spacetime.

Proposition 1. *There exists a unique smooth* $(1,1)$*-extensor field* $b \mapsto \gamma_a(b)$,

$$\gamma_a(b) = b \cdot \partial_n \nabla_a n, \tag{2.2}$$

such that for any smooth form field b, *it holds that*

$$\nabla_a b = a \cdot \partial b + \gamma_a(b). \tag{2.3}$$

Proof. Let b be an arbitrary form field; by using axiom (ii) into definition (2.1), we have

$$\begin{aligned}\nabla_a b = \nabla_a(b \cdot \gamma^\nu \gamma_\nu) &= (a \cdot \partial b \cdot \gamma^\nu)\gamma_\nu + b^\nu \nabla_a \gamma_\nu \\ &= a \cdot \partial b + b \cdot \gamma^\nu \nabla_a \gamma_\nu.\end{aligned} \tag{2.4}$$

This shows that there exists a $(1,1)$-extensor field, defined as

$$n \mapsto \gamma_a(n) = n \cdot \gamma^\nu \nabla_a \gamma_\nu, \tag{2.5}$$

(note the linearity with respect to $n \in \Lambda^1(\mathcal{M})$) such that for any form field b, it holds that

$$\nabla_a b = a \cdot \partial b + \gamma_a(b). \tag{2.6}$$

Now, by applying the *directional derivative operator* $b \cdot \partial_n$ (b is an arbitrary form) on $\nabla_a n = a \cdot \partial n + \gamma_a(n)$, we get

$$b \cdot \partial_n \nabla_a n = b \cdot \partial_n a \cdot \partial n + b \cdot \partial_n \gamma_a(n) = 0 + \gamma_a(b). \tag{2.7}$$

Thus, the $(1,1)$-extensor field γ_a is given by

$$b \mapsto \gamma_a(b) = b \cdot \partial_n \nabla_a n.$$

□

Remark 11. γ_a *is indeed a smooth* $(1,1)$*-extensor field over Minkowski spacetime, associated to the differential operator* ∇_a. *It is convenient to use the short notations* $\nabla_a b = a \cdot \partial b + \gamma_a(b)$ *and* $\gamma_a(b) = b \cdot \partial_n \nabla_a n$ *for* $\nabla_a b(x) = a \cdot \partial b(x) + \gamma_a|_x(b(x))$ *and* $\gamma_a|_x(b) = b \cdot \partial_{n(x)} \nabla_a n(x)$.

Definition 13. *The* $(1,1)$*-extensor field* γ_a *will be called first connection extensor field associated to* ∇_a.

Proposition 2. *There exists a unique smooth* $(1,2)$*-extensor field* $a \mapsto \Omega(a)$,

$$\Omega(a) = -\frac{1}{2}\partial_n \wedge \nabla_a n \tag{2.8}$$

such that the skew-symmetric part *of* γ_a *(i.e.,* $\gamma_{a-} = \frac{1}{2}(\gamma_a - \gamma_a^\dagger)$ *and* $\gamma_a^\dagger$ *is the adjoint*[10]*) of* γ_a *can be factorized by*

$$\gamma_{a-}(b) = \Omega(a) \times b, \tag{2.9}$$

for any form field b.

Proof. Recall that for any $(1,1)$-extensor, say $v \mapsto t(v)$, the *skew-symmetric part of* t (i.e., $t_-(v) = \frac{1}{2}[t(v) - t^\dagger(v)]$) can be factorized as $t_-(v) = \frac{1}{2}\text{bif}(t) \times v$, where $\text{bif}(t) = -\partial_n \wedge t(n)$ is the so-called *biform of* t.

Thus, for the skew-symmetric part of γ_a, taking into account (2.2), we have

$$\begin{aligned}\gamma_{a-}(b) = \frac{1}{2}\text{bif}(\gamma_a) \times b &= (-\frac{1}{2}\partial_n \wedge \gamma_a(n)) \times b \\ &= (-\frac{1}{2}\partial_n \wedge \nabla_a n) \times b.\end{aligned} \tag{2.10}$$

This implies the existence of a $(1,2)$-extensor field, defined as

$$a \mapsto \Omega(a) = -\frac{1}{2}\partial_n \wedge \nabla_a n, \tag{2.11}$$

(note the linearity with respect to $a \in \Lambda^1(\mathcal{M})$), such that for any form field b, it holds that

$$\gamma_{a-}(b) = \Omega(a) \times b.$$

□

Remark 12. Ω *is indeed a smooth* $(1,2)$*-extensor field over the Minkowski manifold, associated to* ∇_a. *Convenient short notations for*

$$\Omega_x(a) = -\frac{1}{2}\partial_{n(x)} \wedge \nabla_a n(x) \quad \textit{and} \quad \gamma_{a-}\big|_x(b) = \Omega_x(a) \times b$$

are $\Omega(a) = -\frac{1}{2}\partial_n \wedge \nabla_a n$ *and* $\gamma_{a-}(b) = \Omega(a) \times b$.

Definition 14. *The* $(1,2)$*-extensor field* Ω *will be called second connection extensor field associated to* ∇_a.

---

[10]Let m be and arbitrary (p,q)-extensor and let $A \in \Lambda^q(\mathcal{M})$, $B \in \Lambda^p(\mathcal{M})$. The adjoint of m, denoted $m^\dagger$, is the (q,p)-extensor such that $m^\dagger(A) \cdot B = A \cdot m(B)$.

3 Associated covariant derivatives

Definition 15. *Associated to a given directional covariant derivative operator ∇_a, we introduce two other directional covariant derivative operators ∇_a^- and ∇_a^0, by*

$$\nabla_a^- b = a \cdot \partial b - \gamma_a^\dagger(b), \tag{3.1}$$

$$\nabla_a^0 b = \frac{1}{2}(\nabla_a b + \nabla_a^- b). \tag{3.2}$$

Remark 13. *∇_a^- and ∇_a^0 are in fact operators acting on the set of smooth form fields satisfying the axiomatic of definition (2.1).*

Proposition 3. *For any smooth form field b, it holds that*

$$\nabla_a^0 b = a \cdot \partial b + \gamma_{a-}(b) = a \cdot \partial b + \Omega(a) \times b. \tag{3.3}$$

Proof. By using (2.3), definition (3.1) and (2.9), we obtain the required result. □

Proposition 4. *For any smooth form field b, it holds that*

$$\nabla_a b = \nabla_a^0 b + \gamma_{a+}(b), \tag{3.4}$$

$$\nabla_a^- b = \nabla_a^0 b - \gamma_{a+}(b). \tag{3.5}$$

Proof. The proof of formulas (3.4) and (3.5) follows directly from (2.3), definition (3.1) and (3.3). □

Proposition 5. *For any smooth form fields b, c it holds that*

$$a \cdot \partial(b \cdot c) = (\nabla_a b) \cdot c + b \cdot (\nabla_a^- c), \tag{3.6}$$

$$a \cdot \partial(b \cdot c) = (\nabla_a^0 b) \cdot c + b \cdot (\nabla_a^0 c). \tag{3.7}$$

Proof. In order to prove the identity (3.6) we must use (2.3) and definition (3.1). The proof of the identity (3.7) is left to the reader. □

4 Covariant derivative of (1,1)-extensor fields

The differential operator ∇_a acting on smooth form fields can be extended in order to act on smooth multiform fields and on extensor fields. The general theory concerning these extensions is given in (Fernández, Moya and Rodrigues, 2000b).[11] Here we need only the action of ∇_a on $(1,1)$-extensor fields.

[11] In (Fernández, Moya and Rodrigues, 2000b) we introduce also generalizations of the concepts of Cartan's differential and Lie derivative operators which act on the set of completely skew-symmetric extensor fields (the so-called exform fields).

Definition 16. *If t is a smooth $(1,1)$-extensor field, then $\nabla_a t$ is another smooth $(1,1)$-extensor field such that, for any smooth form field b, it holds that*

$$(\nabla_a t)(b) = \nabla_a t(b) - t(\nabla_a^- b). \tag{4.1}$$

Proposition 6. *The extended covariant derivative $t \mapsto \nabla_a t$ satisfies the following fundamental properties:*

For all scalars α, α' and forms a, a' and for any smooth form field b, it holds that

$$(\nabla_{\alpha a + \alpha' a'} t)(b) = \alpha(\nabla_a t)(b) + \alpha'(\nabla_{a'} t)(b), \tag{4.2}$$

For all smooth scalar fields f, f' and for any form a and form fields b, b' it holds that

$$(\nabla_a t)(fb + f'b') = f(\nabla_a t)(b) + f'(\nabla_a t)(b'). \tag{4.3}$$

Proof. It follows from the definition (2.1) and the linearity properties of the smooth $(1,1)$-extensor fields. $\qquad\square$

Proposition 7. *For any smooth form field b, it holds that*

$$(\nabla_a t)(b) = a \cdot \partial t(b) - t(\nabla_a^- b) - \partial_n(t(b) \cdot \nabla_a^- n). \tag{4.4}$$

Proof. Let b be an arbitrary smooth form field. Taking into account (2.3) and definitions (4.1) and (3.1), we have

$$
\begin{aligned}
(\nabla_a t)(b) &= a \cdot \partial t(b) + \gamma_a t(b) - t(\nabla_a^- b) \\
&= a \cdot \partial t(b) - t(\nabla_a^- b) - \partial_n(\gamma_a t(b) \cdot n) \\
&= a \cdot \partial t(b) - t(\nabla_a^- b) - \partial_n(t(b) \cdot (a \cdot \partial n)) + \partial_n(t(b) \cdot \gamma_a^\dagger(n)), \\
&= a \cdot \partial t(b) - t(\nabla_a^- b) - \partial_n(t(b) \cdot \nabla_a^- n).
\end{aligned}
$$
$\qquad\square$

Proposition 8. *For all smooth form fields b, c, it holds that*

$$(\nabla_a t)(b) \cdot c = a \cdot \partial(t(b) \cdot c) - t(\nabla_a^- b) \cdot c - t(b) \cdot (\nabla_a^- c). \tag{4.5}$$

Proof. Take two arbitrary smooth form fields b, c. Using (4.4), (2.2) and (2.3) we get

$$
\begin{aligned}
(\nabla_a t)(b) \cdot c &= (a \cdot \partial t(b)) \cdot c - t(\nabla_a^- b) \cdot c - \partial_n(t(b) \cdot \nabla_a^- n) \cdot c \\
&= (a \cdot \partial t(b)) \cdot c + t(b) \cdot (a \cdot \partial c) - t(\nabla_a^- b) \cdot c - t(b) \cdot (a \cdot \partial c) - \\
&\quad c \cdot \partial_n(t(b) \cdot \nabla_a^- n) \\
&= a \cdot \partial(t(b) \cdot c) - t(\nabla_a^- b) \cdot c - t(b) \cdot (a \cdot \partial c) + t(b) \cdot \gamma_a^\dagger(c), \\
&= a \cdot \partial(t(b) \cdot c) - t(\nabla_a^- b) \cdot c - t(b) \cdot \nabla_a^- c.
\end{aligned}
$$
$\qquad\square$

Proposition 9. *For all smooth* $(1,1)$*-extensor field* t*, it holds that*

$$\nabla_a t^\dagger = (\nabla_a t)^\dagger. \tag{4.6}$$

Proof. Let b, c be two arbitrary smooth form fields. Using (4.5) and the fundamental scalar product property in the adjoint $t^\dagger$ of the extensor t, we have

$$\begin{aligned}(\nabla_a t^\dagger)(b) \cdot c &= a \cdot \partial(t^\dagger(b) \cdot c) - t^\dagger(\nabla_a^- b) \cdot c - t^\dagger(b) \cdot \nabla_a^- c \\ &= a \cdot \partial(b \cdot t(c)) - \nabla_a^- b \cdot t(c) - b \cdot t(\nabla_a^- c) \\ &= a \cdot \partial(t(c) \cdot b) - t(\nabla_a^- c) \cdot b - t(c) \cdot \nabla_a^- b \\ &= (\nabla_a t)(c) \cdot b = c \cdot (\nabla_a t)^\dagger(b).\end{aligned}$$

This implies that $(\nabla_a t^\dagger)(b) = (\nabla_a t)^\dagger(b)$, that is, $\nabla_a t^\dagger = (\nabla_a t)^\dagger$. □

Proposition 10. *Let* t *be a smooth* $(1,1)$*-extensor field. Then*

$$(\nabla_a^- t)(b) = \nabla_a^- t(b) - t(\nabla_a b). \tag{4.7}$$

Proof. The proof is trivial and left to the reader. □

Finally, we will present two very important properties: one for an *identity extensor field* i_d and another for the so-called *gauge metric extensor field* g. In the general gauge theory of gravitation, $g \equiv h^\dagger h$, where h is a smooth $(1,1)$-extensor field which has inverse h^{-1} (h is the so-called *gauge distortion extensor field*).

Proposition 11. *If* γ_{a+} *is the symmetric part of* γ_a *(*γ_a *is the first connection extensor field associated to any covariant derivative* ∇_a*), then*

$$\nabla_a^- i_d = -2\gamma_{a+}. \tag{4.8}$$

Proof. From (4.1) we can write

$$(\nabla_a^- i_d)(b) = \nabla_a^- i_d(b) - i_d(\nabla_a b) = \nabla_a^-(b) - \nabla_a b. \tag{4.9}$$

Now, using (3.1) and (2.3), we get

$$(\nabla_a^- i_d)(b) = a \cdot \partial b - \gamma_a^\dagger(b) - a \cdot \partial b - \gamma_a(b) = -2\gamma_{a+}(b).$$

□

Definition 17. *A connection* ∇_a *is said to be* i_d *compatible if and only if* $\nabla_a^- i_d = 0$.

Corollary 1. ∇_a *is* i_d*-compatible if and only if* $\gamma_{a+} = 0$ *(i.e.,* $\gamma_a = -\gamma_a^\dagger$*,* γ_a *is skew-symmetric).*

Remark 14. *It is important to have in mind that* i_d *is the* $(1,1)$*-extensor field that represents the Minkowski metric tensor* $\eta \in \sec T_0^2(M)$.

Note that given any covariant derivative $b \mapsto \nabla_a b$, it is possible (and convenient) to introduce a well-defined covariant derivative by $b \mapsto \widehat{\nabla}_a b = h^{-1}(\nabla_a h(b))$. In the general gauge theory of gravitation, $\widehat{\nabla}_a$ is the so-called *h-gauge of* ∇_a.

Proposition 12. *If γ_{a+} is the symmetric part of γ_a (γ_a is the first connection extensor field associated to ∇_a), then*

$$\widehat{\nabla}_a^- g = -2h^\dagger \gamma_{a+} h. \tag{4.10}$$

Proof. Let b, c be smooth form fields; by (4.5) and (2.3) we have

$$\begin{aligned}(\widehat{\nabla}_a^- g)(b) \cdot c &= a \cdot \partial(g(b) \cdot c) - g(\widehat{\nabla}_a b) \cdot c - g(b) \cdot (\widehat{\nabla}_a c) \\ &= a \cdot \partial(h(b) \cdot h(c)) - \nabla_a h(b) \cdot h(c) - h(b) \cdot \nabla_a h(c) \\ &= -\gamma_a h(b) \cdot h(c) - h(b) \cdot \gamma_a h(c) \\ &= -(\gamma_a h(b) + \gamma_a^\dagger h(b)) \cdot h(c) \\ &= -2h^\dagger \gamma_{a+} h(b) \cdot c,\end{aligned} \tag{4.11}$$

that is, $\widehat{\nabla}_a^- g = -2h^\dagger \gamma_{a+} h$. □

Definition 18. *An arbitrary connection ∇_a is said to be g-compatible if and only if $\nabla_a^- g = 0$.*

Corollary 2. *$\widehat{\nabla}_a$ is g-compatible if and only if $\gamma_{a+} = 0$ (i.e., $\gamma_a = -\gamma_a^\dagger$, γ_a is skew-symmetric).*

Remark 15. *Taking into account the first corollary above, the second corollary above set: $\widehat{\nabla}_a$ is g-compatible if and only if ∇_a is i_d-compatible.*

In (*DLG*, 1998) the authors introduce a covariant derivative $b \mapsto \mathcal{D}_a b = a \cdot \partial b + \Omega(a) \times b$. According to our theory of covariant derivation, $\mathcal{D}_a$ would be the most general covariant derivative with the property of being i_d-compatible (i.e., $\mathcal{D}_a^- i_d = 0$). However, it is not g-compatible as claimed by the mentioned authors.

Observe that (4.5) implies a logical equivalence between the property $\mathcal{D}_a^- i_d = 0$ and the following property $a \cdot \partial(b \cdot c) = (\mathcal{D}_a b) \cdot c + b \cdot (\mathcal{D}_a c)$, where a, b, c are smooth form fields. Hence, by putting $a = \partial_\mu x$, $b = g_\nu$ and $c = g_\lambda$ into the last identity and taking into account the definitions and properties used in (*DLG*,1998), it is not difficult to get the differential equation $\partial_\mu g_{\nu\lambda} = \Gamma^\alpha_{\mu\nu} g_{\alpha\lambda} + \Gamma^\alpha_{\mu\lambda} g_{\alpha\nu}$, where $\Gamma^\alpha_{\mu\nu} = (\mathcal{D}_{\partial_\mu x} g_\nu) \cdot g^\alpha$. That one is only a *coordinate expression for the i_d-compatibility of $\mathcal{D}_a$*, $\Gamma^\alpha_{\mu\nu}$ are *hybrid connection coefficients* among the natural basis and gauge basis.

A correct g-compatible gauge theory of gravitation should be formulated by taking into account the corollary from proposition (4.10). Such a theory is presented in (Fernández, Moya and Rodrigues, 2000a).

5 Structural extensor fields

5.1 Nonmetricity extensor field

Definition 19. *Given a symmetric* $(1,1)$ *extensor field* g $(g = g^{\dagger})$ *and an arbitrary covariant derivative operator* ∇ *on the Minkowski manifold. The smooth biextensor field, say* $(a,b) \mapsto A(a,b)$ *given by*

$$A(a,b) = \nabla_a^{-} g(b), \tag{5.1}$$

for any smooth form fields a, b *is called nonmetricity of* g *relative to* ∇.

Definition 20. ∇ *is said to be* g*-compatible if and only if* $A(a,b) = 0$ *for any form fields* a, b.

5.2 Torsion extensor field

We show now that there exists a well-defined smooth *bi-exform field* (i.e., a skew-symmetric bi-extensor field), associated to γ_a, that measures the so-called *torsion* of ∇_a. After that, we introduce the *torsion extensor field.*

Proposition 13. *There exists a unique smooth bi-exform field* $(a,b) \mapsto \tau(a,b)$,

$$\tau(a,b) = \gamma_a(b) - \gamma_b(a), \tag{5.2}$$

such that for all smooth form fields a, b *it holds that*

$$\nabla_a b - \nabla_b a = [a,b] + \tau(a,b). \tag{5.3}$$

Proof. Let a, b be two smooth form fields. Using (2.3), we have

$$\begin{aligned} \nabla_a b - \nabla_b a &= a \cdot \partial b + \gamma_a(b) - b \cdot \partial a - \gamma_b(a) \\ &= [a,b] + \gamma_a(b) - \gamma_b(a), \end{aligned} \tag{5.4}$$

where $[a,b] = a \cdot \partial b - b \cdot \partial a$ is the so-called *Lie bracket* of the form fields a, b. Equation (5.4) implies the existence of a bi-exform field, defined as

$$(a,b) \mapsto \tau(a,b) = \gamma_a(b) - \gamma_b(a), \tag{5.5}$$

(note the linearity with respect to $a, b \in \Lambda^1(\mathcal{M})$ and the skew-symmetry under interchange of their variables), such that

$$\nabla_a b - \nabla_b a = [a,b] + \tau(a,b)$$

for all smooth form fields a, b. □

Remark 16. τ *is indeed a smooth bi-exform field associated to* ∇_a. *From (2.2) it follows that* $\tau(a,b) = b \cdot \partial_n \nabla_a n - a \cdot \partial_n \nabla_b n$.

Definition 21. *The bi-exform field τ will be called torsion bi-exform field.*

Proposition 14. *For any form fields a, b, it holds that*

$$\tau(a,b) = \gamma_{a+}(b) - \gamma_{b+}(a) + \Omega(a) \times b - \Omega(b) \times a. \tag{5.6}$$

Proof. It is enough to use the decomposition of γ_a into symmetric and skew-symmetric parts and the factorization (2.9) into the definition (5.2). □

Remark 17. *In (5.6), the bi-exform field $\gamma_{a+}(b) - \gamma_{b+}(a)$ comes from the symmetric part of γ_a, and the bi-exform field $\Omega(a) \times b - \Omega(b) \times a$ comes from the skew-symmetric part of γ_a.*

Definition 22. *The smooth $(1,2)$-extensor field, say $n \mapsto T(n)$, such that for each $x \in \Lambda^1(\mathcal{M})$ its adjoint is the smooth $(2,1)$-extensor field given by*

$$\Lambda^2(\mathcal{M}) \ni B \mapsto T_x^\dagger(B) = \frac{1}{2!} B \cdot (\partial_a \wedge \partial_b)\tau_x(a,b) \in \Lambda^1(\mathcal{M}), \tag{5.7}$$

will be called torsion extensor field.

Remark 18. *We usually employ the short notation*

$$T^\dagger(B) = \frac{1}{2} B \cdot (\partial_a \wedge \partial_b)\tau(a,b).$$

Proposition 15. *For any forms a, b, it holds that*

$$T^\dagger(a \wedge b) = \tau(a,b). \tag{5.8}$$

Proof. Taking into account the skew-symmetry of τ, we have

$$\begin{aligned} T^\dagger(a \wedge b) &= \frac{1}{2}(a \wedge b) \cdot (\partial_p \wedge \partial_q)\tau(p,q) = \frac{1}{2}(a \cdot \partial_p b \cdot \partial_q - a \cdot \partial_q b \cdot \partial_p)\tau(p,q) \\ &= \frac{1}{2}[a \cdot \partial_p b \cdot \partial_q \tau(p,q) - a \cdot \partial_q b \cdot \partial_p \tau(p,q)] = \frac{1}{2}[\tau(a,b) - \tau(b,a)], \\ &= \tau(a,b). \end{aligned}$$

□

5.3 Curvature extensor field

We show now the existence of two well-defined operators, associated to γ_a, involved in the concept of *curvature* of the covariant derivative operator ∇. After that, we introduce the *curvature extensor field.*

Proposition 16. *There exists a unique operator acting on the set of smooth vector fields, say $(a,b,c) \mapsto \widehat{\omega}_1(a,b,c)$,*

$$\widehat{\omega}_1(a,b,c) = c \cdot \partial_n(a \cdot \partial\gamma_b(n) - b \cdot \partial\gamma_a(n) + [\gamma_a, \gamma_b](n)), \tag{5.9}$$

such that for all smooth form fields a, b, c, it holds that

$$[\nabla_a, \nabla_b]c = [a \cdot \partial, b \cdot \partial]c + \widehat{\omega}_1(a,b,c). \tag{5.10}$$

Proof. Let a, b, c be three smooth form fields; by using (2.3) we have

$$\begin{aligned}\nabla_a(\nabla_b c) &= a \cdot \partial b \cdot \partial c + a \cdot \partial \gamma_b(c) + \gamma_a(b \cdot \partial c) + \gamma_a \gamma_b(c).\\ &= b \cdot \partial a \cdot \partial c + b \cdot \partial \gamma_a(c) + \gamma_b(a \cdot \partial c) + \gamma_b \gamma_a(c). \qquad (5.11)\end{aligned}$$

Subtracting we get

$$\begin{aligned}[\nabla_a, \nabla_b]c = [a \cdot \partial, b \cdot \partial]c &+ a \cdot \partial \gamma_b(c) - \gamma_b(a \cdot \partial c)\\ &- b \cdot \partial \gamma_a(c) + \gamma_a(b \cdot \partial c) + [\gamma_a, \gamma_b](c). \qquad (5.12)\end{aligned}$$

Now, using the formula $b \cdot \partial_n a \cdot \partial t(n) = a \cdot \partial t(b) - t(a \cdot \partial b)$ which is valid for any smooth $(1,1)$-extensor field, where a is a form and b, n are smooth form fields, we obtain

$$\begin{aligned}[\nabla_a, \nabla_b]c = [a \cdot \partial, b \cdot \partial]c &+ c \cdot \partial_n(a \cdot \partial \gamma_b(n)\\ &- b \cdot \partial \gamma_a(n) + [\gamma_a, \gamma_b](n)). \qquad (5.13)\end{aligned}$$

Equation (5.13) implies the existence of an operator,

$$\underbrace{\hom[\Lambda^1(\mathcal{M}), \Lambda^1(\mathcal{M})]}_{\text{3-copies}} \ni (a, b, c) \mapsto \widehat{\omega}_1(a, b, c) \in \hom[\Lambda^1(\mathcal{M}), \Lambda^1(\mathcal{M})],$$

defined by

$$\widehat{\omega}_1(a, b, c) = c \cdot \partial_n(a \cdot \partial \gamma_b(n) - b \cdot \partial \gamma_a(n) + [\gamma_a, \gamma_b](n)), \qquad (5.14)$$

such that

$$[\nabla_a, \nabla_b]c = [a \cdot \partial, b \cdot \partial]c + \widehat{\omega}_1(a, b, c).$$

for all smooth form fields a, b, c. □

Remark 19. *$\widehat{\omega}_1$ is an operator associated to the directional covariant derivative operator ∇_a, which is linear with respect to its third variable and skew-symmetric under interchange of first and second variables. In terms of ∇_a, it is obvious that $\widehat{\omega}_1(a, b, c) = c \cdot \partial_n[\nabla_a, \nabla_b]n$.*

Definition 23. *The operator $\widehat{\omega}_1$ will be called first curvature operator.*

Besides $\widehat{\omega}_1$, we introduce another operator related to curvature.

Definition 24. *The second curvature operator $\widehat{\omega}_2$ acting on the set of smooth form fields is given by*

$$\underbrace{\hom[\Lambda^1(\mathcal{M}), \Lambda^1(\mathcal{M})]}_{\text{2 copies}} \ni (a, b) \mapsto \widehat{\omega}_2(a, b) \in \hom[\Lambda^1(\mathcal{M}), \Lambda^2(\mathcal{M})],$$

such that

$$\begin{aligned}\widehat{\omega}_2(a, b) &= -\frac{1}{2}\partial_c \wedge \widehat{\omega}_1(a, b, c)\\ &= -\frac{1}{2}\partial_n \wedge (a \cdot \partial \gamma_b(n) - b \cdot \partial \gamma_a(n) + [\gamma_a, \gamma_b](n)). \qquad (5.15)\end{aligned}$$

Remark 20. *The operator $\widehat{\omega}_2$ is skew-symmetric under interchange of their variables. In terms of ∇_a, we have $\widehat{\omega}_2(a,b) = -\frac{1}{2}\partial_n \wedge [\nabla_a, \nabla_b]n$.*

Proposition 17. *For any smooth vector form fields a, b, it holds that*

$$\widehat{\omega}_2(a,b) = \frac{1}{2}\operatorname{bif}([\gamma_{a+}, \gamma_{b+}]) + a \cdot \partial\Omega(b) - b \cdot \partial\Omega(a) + \Omega(a) \times \Omega(b). \tag{5.16}$$

Proof. By straightforward calculation we have

$$\begin{aligned}
\widehat{\omega}_2(a,b) &= -\frac{1}{2}\partial_n \wedge (a \cdot \partial\gamma_b(n) - b \cdot \partial\gamma_a(n) + [\gamma_a, \gamma_b](n)) \\
&= a \cdot \partial(-\frac{1}{2}\partial_n \wedge \gamma_b(n)) - b \cdot \partial(-\frac{1}{2}\partial_n \wedge \gamma_a(n)) - \frac{1}{2}\partial_n \wedge [\gamma_a, \gamma_b](n) \\
&= a \cdot \partial\Omega(b) - b \cdot \partial\Omega(a) - \\
&\quad \frac{1}{2}\partial_n \wedge ([\gamma_{a+}, \gamma_{b+}] + [\gamma_{a+}, \gamma_{b-}] + [\gamma_{a-}, \gamma_{b+}] + [\gamma_{a-}, \gamma_{b-}])(n) \\
&= a \cdot \partial\Omega(b) - b \cdot \partial\Omega(a) \\
&\quad - \frac{1}{2}\partial_n \wedge [\gamma_{a+}, \gamma_{b+}](n) - \frac{1}{2}\partial_n \wedge [\gamma_{a+}, \gamma_{b-}](n) \\
&\quad - \frac{1}{2}\partial_n \wedge [\gamma_{a-}, \gamma_{b+}](n) - \frac{1}{2}\partial_n \wedge [\gamma_{a-}, \gamma_{b-}](n).
\end{aligned} \tag{5.17}$$

Now, using (2.8), (2.9) and Jacobi's identity

$$A \times (B \times C) = (A \times B) \times C + B \times (A \times C),$$

we get

$$\begin{aligned}
\widehat{\omega}_2(a,b) = a \cdot \partial\Omega(b) - b \cdot \partial\Omega(a) + \frac{1}{2}\operatorname{bif}([\gamma_{a+}, \gamma_{b+}]) \\
+ \frac{1}{2}\operatorname{bif}([\gamma_{a+}, \gamma_{b-}]) + \frac{1}{2}\operatorname{bif}([\gamma_{a-}, \gamma_{b+}]) \\
- \frac{1}{2}\partial_n \wedge ([\Omega(a) \times \Omega(b)] \times n).
\end{aligned} \tag{5.18}$$

Since $[\gamma_{a+}, \gamma_{b+}]$ and $[\gamma_{a+}, \gamma_{b-}]$ are symmetric extensors, their biforms vanish. Using the formula $\partial_n \wedge (B \times n) = -2B$, where B is a biform and n is a form, we get

$$\widehat{\omega}_2(a,b) = \frac{1}{2}\operatorname{bif}([\gamma_{a+}, \gamma_{b+}]) + a \cdot \partial\Omega(b) - b \cdot \partial\Omega(a) + \Omega(a) \times \Omega(b).$$

□

Remark 21. *In (5.16), the bi-exform field $\frac{1}{2}\operatorname{bif}([\gamma_{a+}, \gamma_{b+}])$ comes from the symmetric part of γ_a, and the operator*

$$(a,b) \mapsto a \cdot \partial\Omega(b) - b \cdot \partial\Omega(a) + \Omega(a) \times \Omega(b)$$

comes from the skew-symmetric part of γ_a.

Definition 25. *The smooth* $(2,2)$*-extensor field, say* $B \mapsto R(B)$*, for each* $x \in \Lambda^1(\mathcal{M})$ *:*

$$\Lambda^2(\mathcal{M}) \ni B \mapsto R_x(B) = \frac{1}{2!}B \cdot (\partial_{a(x)} \wedge \partial_{b(x)})\widehat{\omega}_2(a,b)(x) \in \Lambda^2(\mathcal{M}) \quad (5.19)$$

will be called a curvature extensor field.

Remark 22. *We usually employ the short notation*

$$R(B) = \frac{1}{2}B \cdot (\partial_a \wedge \partial_b)\widehat{\omega}_2(a,b).$$

Proposition 18. *For all smooth form fields* a, b*, it holds that*

$$R(a \wedge b) = \widehat{\omega}_2(a,b) - \Omega([a,b]). \quad (5.20)$$

Proof. Due to the skew-symmetry of $\widehat{\omega}_2(a,b)$, we have

$$\begin{aligned} R(a \wedge b) &= \frac{1}{2}(a \wedge b) \cdot (\partial_p \wedge \partial_q)\widehat{\omega}_2(a,b) \\ &= \frac{1}{2}[a \cdot \partial_p b \cdot \partial_q \widehat{\omega}_2(p,q) - a \cdot \partial_q b \cdot \partial_p \widehat{\omega}_2(p,q)] \\ &= \frac{1}{2}[a \cdot \partial_p b \cdot \partial_q \widehat{\omega}_2(p,q) - a \cdot \partial_p b \cdot \partial_q \widehat{\omega}_2(q,p)] \\ &= \frac{1}{2}[a \cdot \partial_p b \cdot \partial_q \widehat{\omega}_2(p,q) + a \cdot \partial_p b \cdot \partial_q \widehat{\omega}_2(p,q)], \\ &= a \cdot \partial_p b \cdot \partial_q \widehat{\omega}_2(p,q). \end{aligned} \quad (5.21)$$

Taking into account (5.16) and using the general formula for smooth $(1,k)$-extensor fields $b \cdot \partial_n a \cdot \partial t(n) = a \cdot \partial t(b) - t(a \cdot \partial b)$, where a is a form and b, n are smooth form fields, we obtain

$$\begin{aligned} R(a \wedge b) &= a \cdot \partial_p b \cdot \partial_q (\frac{1}{2}\operatorname{bif}([\gamma_{p+}, \gamma_{q+}]) + p \cdot \partial\Omega(q) - q \cdot \partial\Omega(p) \\ &\quad + \Omega(p) \times \Omega(q)) \\ &= \frac{1}{2}\operatorname{bif}([\gamma_{a+}, \gamma_{b+}]) + a \cdot \partial\Omega(b) - \Omega(a \cdot \partial b) - b \cdot \partial\Omega(a) + \Omega(b \cdot \partial a) \\ &\quad + \Omega(a) \times \Omega(b), \\ &= \widehat{\omega}_2(a,b) - \Omega([a,b]). \end{aligned}$$

□

6 Examples

6.1 Levi-Civita covariant derivative

We suppose the existence of a smooth $(1,1)$-extensor field on the Minkowski manifold (modeled by $(M, \mathcal{M}^*)$), say $n \mapsto g(n)$, which is symmetric (i.e., $g = g^\dagger$) and has inverse. It will be called *gauge metric extensor field.*

We introduce two fundamental operators acting on the set of smooth form fields on M :

(a) The *Christoffel operator of the first kind,*

$$\underbrace{\hom[\Lambda^1(\mathcal{M}),\Lambda^1(\mathcal{M})]}_{\text{3 copies}} \ni (a,b,c) \mapsto [a,b,c] \in \hom[\Lambda^1(\mathcal{M}),R], \tag{6.1}$$

where

$$2[a,b,c] = a\cdot\partial(g(b)\cdot c) + b\cdot\partial(g(c)\cdot a) - c\cdot\partial(g(a)\cdot b) \\ + g(c)\cdot[a,b] + g(b)\cdot[c,a] - g(a)\cdot[b,c], \tag{6.2}$$

and where $[a,b]$ is the *Lie bracket* of the form fields a,b.

(b) The *Christoffel operator of the second kind,*

$$\underbrace{\hom[\Lambda^1(\mathcal{M}),\Lambda^1(\mathcal{M})]}_{\text{3 copies}} \ni (a,b,c) \mapsto \left\{ \begin{matrix} c \\ a,b \end{matrix} \right\} \in \hom[\Lambda^1(\mathcal{M}),R] \tag{6.3}$$

such that

$$\left\{ \begin{matrix} c \\ a,b \end{matrix} \right\} = [a,b,g^{-1}(c)]. \tag{6.4}$$

The Christoffel operator of the first kind has several useful properties, namely, for any form fields a,a',b,c and any scalar field f, we have

$$\begin{aligned} [a+a',b,c] &= [a,b,c]+[a',b,c] \\ [fa,b,c] &= f[a,b,c]. \end{aligned} \tag{6.5}$$

$$\begin{aligned} [a,b+b',c] &= [a,b,c]+[a,b',c] \\ [a,fb,c] &= f[a,b,c]+(a\cdot\partial f)g(b)\cdot c. \end{aligned} \tag{6.6}$$

$$\begin{aligned} [a,b,c+c'] &= [a,b,c]+[a,b,c'] \\ [a,b,fc] &= f[a,b,c]. \end{aligned} \tag{6.7}$$

Now, we can define the Levi-Civita covariant derivative D as follows:

$$\hom[\Lambda^1(\mathcal{M}),\Lambda^1(\mathcal{M})] \ni b \mapsto D_a b \in \hom[\Lambda^1(\mathcal{M}),\Lambda^1(\mathcal{M})], \tag{6.8}$$

such that if $a \in \hom[\Lambda^1(\mathcal{M}),\Lambda^1(\mathcal{M})]$, then

$$D_a b \equiv \partial_n \left\{ \begin{matrix} n \\ a,b \end{matrix} \right\} = \partial_n[a,b,g^{-1}(n)].$$

Observe that for any smooth form fields a,b,c we have

$$(D_a b)\cdot c = \left\{ \begin{matrix} c \\ a,b \end{matrix} \right\}. \tag{6.9}$$

D_a is a well-defined directional covariant derivative. In fact, the linearity with respect to the direction a is consequence of (6.5). Also, if f is a scalar field and b, b' are form fields. Using (6.7), we have

$$\begin{aligned} D_a(b+b') &= \partial_n[a, b+b', g^{-1}(n)] \\ &= \partial_n[a, b, g^{-1}(n)] + \partial_n[a, b', g^{-1}(n)] \\ &= D_a b + D_a b'. \end{aligned} \tag{6.10}$$

$$\begin{aligned} D_a(fb) &= \partial_n[a, fb, g^{-1}(n)] \\ &= f\partial_n[a, b, g^{-1}(n)] + \partial_n(a \cdot \partial f)(g(b) \cdot g^{-1}(n)) \\ &= (a \cdot \partial f)b + f D_a b. \end{aligned} \tag{6.11}$$

Thus, the axiom (ii) from definition (2.1) is satisfied.

Theorem 1. *Let a, b, c be arbitrary smooth form fields. Then*

$$g(D_a b) \cdot c + g(b) \cdot D_a c = a \cdot \partial(g(b) \cdot c). \tag{6.12}$$

Proof.

$$\begin{aligned} g(D_a b) \cdot c + g(b) \cdot D_a c &= \partial_n[a, b, g^{-1}(n)] \cdot g(c) + g(b) \cdot \partial_n[a, c, g^{-1}(n)] \\ &= [a, b, g^{-1}g(c)] + [a, c, g^{-1}g(b)] \\ &= a \cdot \partial(g(b) \cdot c). \end{aligned}$$

We have used definition (6.8), and in the last line we have used an important identity for the Christoffel operator of the first kind, namely,

$$[a, b, c] + [a, c, b] = a \cdot \partial(g(b) \cdot c).$$

□

Remark 23. *The above theorem is known as Ricci theorem for the Levi-Civita derivative.*

Corollary 3. *Let a, b be arbitrary smooth form fields,*

$$D_a^- b = g(D_a g^{-1}(b)). \tag{6.13}$$

Proof. Once again, let a, b, c be arbitrary smooth form fields. Using the Ricci theorem (6.12), we have

$$\begin{aligned} a \cdot \partial(b \cdot c) &= a \cdot \partial(g(b) \cdot g^{-1}(c)) \\ &= g(D_a b) \cdot g^{-1}(c) + g(b) \cdot D_a g^{-1}(c) \\ &= (D_a b) \cdot c + b \cdot g(D_a g^{-1}(c)). \end{aligned} \tag{6.14}$$

Comparing (6.14) with the identity (3.6), it follows that

$$D_a^- b = g(D_a g^{-1}(b)).$$

□

Proposition 19. *g is compatible with D, i.e., for any form field a the extension of the Levi-Civita derivative acting on the smooth $(1,1)$-extensor field g (the gauge metric extensor field) vanishes,*

$$D_a^- g = 0. \tag{6.15}$$

Proof.

$$(D_a^- g)(b) = D_a^- g(b) - g(D_a b) = g(D_a g^{-1} g(b)) - g(D_a b) = 0.$$

□

Finally, we will see that Levi-Civita derivative is torsion less.

In fact, according to (5.3), we can write

$$\begin{aligned} \tau(a,b) &= D_a b - D_b a - [a,b] \\ &= \partial_n (D_a b \cdot n - D_b a \cdot n) - [a,b] \\ &= \partial_n (\left\{ \begin{matrix} n \\ a,b \end{matrix} \right\} - \left\{ \begin{matrix} n \\ a,b \end{matrix} \right\} - [a,b] \\ &= \partial_n ([a,b,g^{-1}(n)] - [b,a,g^{-1}(n)]) - [a,b], \end{aligned} \tag{6.16}$$

and using the remarkable identity $[a,b,c] - [b,a,c] = g(c) \cdot [a,b]$ yields

$$\tau(a,b) = \partial_n g g^{-1}(n) \cdot [a,b] - [a,b] = 0, \tag{6.17}$$

that is, $T^\dagger(a \wedge b) = 0$.

6.2 Hestenes covariant derivative

Take a smooth, even multiform field over the Minkowski manifold modeled by $(M, \mathcal{M}^*)$, say $\Lambda^1(\mathcal{M}) \ni x \mapsto R(x) \in \Lambda^+(\mathcal{M})$, such that $R(x)\widetilde{R}(x) = 1$. It is called a *Lorentz rotator* since for any $x \in \forall\Lambda^1(\mathcal{M}), R(x) \in \mathbf{Spin}^+(1,3)$ $\approx \mathrm{SL}(2,\mathbb{C})$, which is the universal covering group of $\mathfrak{L}_+^\uparrow$, the restricted orthochronous Lorentz group (Lounesto, 1997).

Definition 26. *A Lorentz extensor field l_x is a smooth $(1,1)$-extensor field over M such that for each $x \in \Lambda^1(\mathcal{M})$:*

$$\begin{aligned} \Lambda^1(\mathcal{M}) &\ni n \mapsto l_x(n) \in \Lambda^1(\mathcal{M}), \\ l_x(n) &= R(x) n \widetilde{R}(x). \end{aligned} \tag{6.18}$$

Some important properties for Lorentz extensor field $n \mapsto l(n) = Rn\widetilde{R}$, are: For any multiform field X,the extension[12] of l, denoted $\underline{l}$ is such that $\underline{l}(X) = RX\widetilde{R}$. Also $\det(l) = 1$, $l^\dagger(n) = \widetilde{R}nR$ (where $l^\dagger$ is the adjoint of l), and $l^{-1} = l^\dagger$.

[12]The extension of a $(1,1)$-extensor t is the general extensor $\underline{t}$ defined by the properties: $\underline{t}(\alpha) = \alpha$, $\alpha \in R$ and $\underline{t}(a_1 \wedge \ldots \wedge a_k) = t(a_1) \wedge \ldots \wedge t(a_k)$, $a_1, \ldots, a_k \in \Lambda^1(\mathcal{M})$.

Definition 27. *Let $(\langle\varepsilon_\mu\rangle, \langle\varepsilon^\mu\rangle)$ be any pair of reciprocal frames (i.e., $\varepsilon_\mu \cdot \varepsilon^\nu = \delta_\mu^\nu$), then the pair of reciprocal frames $(\langle l(\varepsilon_\mu)\rangle, \langle l^\star(\varepsilon^\mu)\rangle)$, where $l^\star \equiv (l^{-1})^\dagger = (l^\dagger)^{-1}$ (in this case, it holds that $l^\star = l$), are called the l-gauge frames of $(\langle\varepsilon_\mu\rangle, \langle\varepsilon^\mu\rangle)$.*

The existence of a multiform field R, with the properties $R(x) \in \Lambda^+(\mathcal{M})$ (i.e., $R(x) = \overline{R}(x)$) and $R(x)\widetilde{R}(x) = 1$, implies the existence of a smooth $(1,2)$-extensor field. For each $x \in \Lambda^1(\mathcal{M})$,

$$\Lambda^1(\mathcal{M}) \ni a \mapsto \omega_x(a) \in \Lambda^2(\mathcal{M}) \tag{6.19}$$

such that $\omega_x(a) = (-2a \cdot \partial R(x))\widetilde{R}(x)$.

Observe that the canonical biforms $\omega(\gamma_\mu) = (-2\partial_\mu R)\widetilde{R}$ are the so-called Darboux biform fields. This suggests calling ω the *Darboux extensor field* (Rodrigues et al., 1996).

We prove now two important properties involving the Darboux extensor field.

Proposition 20. *Let $(\langle l(\gamma_\mu)\rangle, \langle l(\gamma^\mu)\rangle)$ be the l-gauge frames of the fundamental frames $(\langle\gamma_\mu\rangle, \langle\gamma^\mu\rangle)$. Then*

$$a \cdot \partial l(\gamma_\mu) = l(\gamma_\mu) \times \omega(a), \quad a \cdot \partial l(\gamma^\mu) = l(\gamma^\mu) \times \omega(a). \tag{6.20}$$

Proof. Using the known property $R\widetilde{R} = \widetilde{R}R = 1$ and definition (6.19), we get

$$\begin{aligned} a \cdot \partial l(\gamma_\mu) &= (a \cdot \partial R)\gamma_\mu \widetilde{R} + R\gamma_\mu(a \cdot \partial\widetilde{R}) \\ &= (a \cdot \partial R)\widetilde{R}R\gamma_\mu\widetilde{R} + R\gamma_\mu\widetilde{R}R(a \cdot \partial\widetilde{R}) \\ &= -\frac{1}{2}\omega(a)l(\gamma_\mu) + l(\gamma_\mu)\frac{1}{2}\omega(a), \\ &= l(\gamma_\mu) \times \omega(a). \end{aligned}$$

In an analogous way, $a \cdot \partial l(\gamma^\mu) = l(\gamma^\mu) \times \omega(a)$. □

For the Darboux extensor field we have a remarkable identity.

Proposition 21. *For any smooth form fields a, b the following holds:*

$$a \cdot \partial\omega(b) - b \cdot \partial\omega(a) + \omega(a) \times \omega(b) = \omega([a,b]). \tag{6.21}$$

Proof. We have

$$a \cdot \partial\omega(b) = -2((a \cdot \partial b \cdot \partial R)\widetilde{R} + (b \cdot \partial R)(a \cdot \partial\widetilde{R})), \tag{6.22}$$

$$b \cdot \partial\omega(a) = -2((b \cdot \partial a \cdot \partial R)\widetilde{R} + (a \cdot \partial R)(b \cdot \partial\widetilde{R})). \tag{6.23}$$

Thus,

$$\begin{aligned}a \cdot \partial\omega(b) - b \cdot \partial\omega(a) &= -2(([a \cdot \partial b \cdot \partial]R)\widetilde{R} + (b \cdot \partial R)(a \cdot \partial\widetilde{R}) \\ &\quad - (a \cdot \partial R)(b \cdot \partial\widetilde{R})) \\ &= -2(([a, b] \cdot \partial R)\widetilde{R} + (b \cdot \partial R)(a \cdot \partial\widetilde{R}) \\ &\quad - (a \cdot \partial R)(b \cdot \partial\widetilde{R})) \\ &= \omega([a, b]) - 2((b \cdot \partial R)(a \cdot \partial\widetilde{R}) \\ &\quad - (a \cdot \partial R)(b \cdot \partial\widetilde{R}))\end{aligned} \tag{6.24}$$

where we used the identity $[a \cdot \partial, b \cdot \partial]X = [a, b] \cdot \partial X$, where a, b are smooth form fields and X is a smooth multiform field.

Also,

$$\omega(a)\omega(b) = 4(a \cdot \partial R)\widetilde{R}(b \cdot \partial R)\widetilde{R} = -4(a \cdot \partial R)(b \cdot \partial\widetilde{R}), \tag{6.25}$$

$$\omega(b)\omega(a) = -4(b \cdot \partial R)(a \cdot \partial\widetilde{R}), \tag{6.26}$$

which implies that

$$\omega(a) \times \omega(b) = -2((a \cdot \partial R)(b \cdot \partial\widetilde{R}) - (b \cdot \partial R)(a \cdot \partial\widetilde{R})). \tag{6.27}$$

Comparing (6.24) with (6.27), we finally obtain

$$a \cdot \partial\omega(b) - b \cdot \partial\omega(a) + \omega(a) \times \omega(b) = \omega([a, b]).$$

□

Definition 28. *The Hestenes directional covariant derivative (Hestenes derivative, for short)* d_a, *is defined by*

$$\hom[\Lambda^1(\mathcal{M}), \Lambda^1(\mathcal{M})] \ni b \mapsto d_a b \in \hom[\Lambda^1(\mathcal{M}), \Lambda^1(\mathcal{M})] \tag{6.28}$$

such that $d_a b \equiv l(a \cdot \partial l^{-1}(b)) = l(a \cdot \partial l^{\dagger}(b))$.

As the reader can easily prove it is indeed a well-defined directional covariant derivative since it satisfies the axiomatic in definition (2.1).

Proposition 22. *The associated derivatives* d_a^- *and* d_a^0 *coincide with* d_a, *i.e., for any smooth form field* b, *we have*

$$d_a^- b = d_a b = d_a^0 b. \tag{6.29}$$

Proof. Let a, b, c be arbitrary smooth form fields. Then

$$\begin{aligned}a \cdot \partial(b \cdot c) &= a \cdot \partial(l^{-1}(b) \cdot l^{\dagger}(c)) \\ &= a \cdot \partial l^{-1}(b) \cdot l^{\dagger}(c) + l^{-1}(b) \cdot a \cdot \partial l^{\dagger}(c) \\ &= l(a \cdot \partial l^{-1}(b)) \cdot c + b \cdot l^{\star}(a \cdot \partial l^{\dagger}(c)) \\ &= l(a \cdot \partial l^{-1}(b)) \cdot c + b \cdot l(a \cdot \partial l^{-1}(c)) \\ &= (d_a b) \cdot c + b \cdot (d_a c).\end{aligned} \tag{6.30}$$

Comparing the (6.30) with the identity (3.6) and employing definition (3.2), it follows that

$$d_a^- b = d_a b = d_a^0 b. \tag{6.31}$$

□

The result (6.31) means that the Hestenes derivative is i_d-compatible.

Finally, we will find the second connection extensor field associated to the Hestenes derivative.

Proposition 23. *The second connection extensor field associated to Hestenes derivative coincides with the Darboux extensor field*

$$\Omega(a) = \omega(a). \tag{6.32}$$

Proof.

$$\begin{aligned}\Omega(a) = -\frac{1}{2}\partial_n \wedge d_a n &= -\frac{1}{2}\partial_n \wedge l(a\cdot\partial l^{-1}(n)) \\ &= -\frac{1}{2}\gamma^\mu \wedge l(a\cdot\partial l^{-1}(\gamma_\mu)).\end{aligned} \tag{6.33}$$

Hence, using the remarkable identity

$$\gamma^\mu \wedge l(a\cdot\partial l^{-1}(\gamma_\mu)) + l^*(\gamma^\mu)\wedge a\cdot\partial l(\gamma_\mu) = 0$$

and (6.20), we have

$$\begin{aligned}\Omega(a) &= \frac{1}{2}l^*(\gamma^\mu)\wedge a\cdot\partial l(\gamma_\mu) \\ &= \frac{1}{2}l^*(\gamma^\mu)\wedge(l(\gamma_\mu)\times\omega(a)) \\ &= \frac{1}{2}\partial_n\wedge(n\times\omega(a)) \\ &= \omega(a).\end{aligned}$$

We have employed the invariant representation property for curl operator $\partial_n\wedge$ and the formula $\partial_n \wedge (n\times B) = 2B$. □

Remark 24. *Equations (6.29) and (6.20) imply that the curvature extensor field of the Hestenes derivative is null.*

Acknowledgments

The authors are grateful to Rafał Abłamowicz for the kind invitation to write this paper and to J. Emílio Maiorino for a careful reading of the manuscript.

References

[1] R. Adler, M. Bazin, and M. Schiffer, *Introduction to General Relativity,* McGraw-Hill, New York, 1965, Chapter 13.

[2] A. Challinor, A. Lasenby, C. Doran, et al., Massive, non-ghost solutions for the Dirac field coupled self-consistently to gravity, *Gen. Rel. Gravit.* **29** (1997), 1527–1544.

[3] Y. Dabrowski, M. P. Hobson, A. Lasenby, et al., Microwave background anisotropies and non-linear structures-II. Numerical computations, *Mon. Not. R. Astron Soc* **302** (1999), 757–770.

[4] C. Doran, A. Lasenby, A. Challinor, et al., Effects of spin-torsion in a gauge theory of gravity, *J. Math. Phys.* **39** (1998), 3303–3321.

[5] A. M. Moya, *Lagrangian Formalism for Multivector Fields on Minkowski Spacetime,* Ph.D. Thesis, IMECC-UNICAMP, 1999.

[6] A. M. Moya, V. V. Fernández, and W. A. Rodrigues, Jr., *Clifford and Extensor Fields. Mathematical Theory and Physical Applications,* book in preparation, 2000.

[7] V. V. Fernández, A. M. Moya, W. A. Rodrigues, Jr. Gravitational fields as distortion fields on Minkowski spacetime, submitted for publication, 2000a.

[8] D. Hestenes, *Space-time Algebra,* Gordon & Breach, New York, 1966.

[9] D. Hestenes G. and Sobczyk, *Clifford Algebra to Geometrical Calculus,* Reidel, Dordrecht, 1984.

[10] V. V. Fernández, A. M. Moya, and W. A. Rodrigues, Jr. The algebraic theory of connections, differential and Lie operators for Clifford and extensor fields, submitted for publication, 2000b.

[11] A. Lasenby, C. Doran, and S. Gull, Gravity, gauge theories and geometric algebra, *Phil. Trans. R. Soc* **356** (1998), 487–582.

[12] J. F. Pommaret, *Partial Differential Equations and Group Theory. New Perspectives and Applications,* Kluwer Acad. Publ., Dordrecht, 1994.

[13] P. Lounesto, *Clifford Algebras and Spinors,* Cambridge University Press, Cambridge, 1997.

[14] A. A. Logunov and Mestvirishvili, Relativistic theory of gravitation, *Found. Phys* **16** (1986), 1–26.

[15] W. A. Rodrigues, Jr., and M. A. F. Rosa, The meaning of time in relativity and Einstein later view of the twin paradox, *Found. Phys* **19** (1989), 705–724.

[16] W. A. Rodrigues, Jr., and Q. A. G. de Souza, The Clifford bundle and the nature of the gravitational field, *Found. Phys.* **23** (1993), 1456–1490.

[17] W. A. Rodrigues, Jr., Q. A. G. de Souza, and J. Vaz, Jr., The Dirac operator and the structure of on a Riemann-Cartan-Weyl spaces, in *Gravitation: The Spacetime Structure Proc. SILARG VIII,* Águas de Lindóia, SP, Brazil, 25–30, July, 1993, P. Letelier and W. A. Rodrigues, Jr., eds., World Scientific, Singapore, 1994, 522–531.

[18] W. A. Rodrigues, Jr., Q. A. G. de Souza, J. Vaz, Jr., and P. Lounesto, Dirac-Hestenes spinor fields on Riemann-Cartan manifolds, *Int. J. Theor. Phys.* **35** (1995), 1849–1900.

[19] W. A. Rodrigues, Jr., J. Vaz, Jr., and M. Pavsic, The Clifford bundle and the dynamics of the superparticle, *Banach Center Publ. Polish Acad. Sci.* **37** (1996), 295–314.

[20] H. Weyl, *Gravitation und Elektrizität,* Sitzber. Preuss. Akad. Wiss., Berlin, 1918, 465–480.

[21] H. Weyl, *Space, Time, Matter,* Methuen & Company, London, 1922, reprinted by Dover Publ., Inc, New York, 1952, 282–294.

Virginia V. Fernández
Institute of Physics Gleb Wataghin
IFGW-UNICAMP
13083-970 Campinas, SP, Brazil
E-mail: fernande@ime.unicamp.br

Antonio M. Moya
Institute of Mathematics, Statistics and Scientific Computation
IMECC-UNICAMP CP 6065
13081-970 Campinas, SP, Brazil
E-mail: moya@ime.unicamp.br

Waldyr A. Rodrigues, Jr.
Center for Research and Technology
CPTEC-UNISAL
Av. A. Garret 277
13087-290 Campinas, SP, Brazil
E-mail: walrod@ime.unicamp.br or walrod@cptec.br

Institute of Mathematics, Statistics and Scientific Computation
IMECC-UNICAMP CP 6065
13081-970 Campinas, SP, Brazil
E-mail: walrod@ime.unicamp.br or walrod@cptec.br

Received: October 4, 1999; Revised: January 3, 2000

An Introduction to Pseudotwistors: Spinor Solutions vs. Harmonic Forms and Cohomology Groups

Julian Ławrynowicz and Osamu Suzuki

ABSTRACT Penrose observed in 1976 [27] that the points of the Minkowski space-time can be represented by two-dimensional subspaces of a complex four-dimensional vector space on which an hermitian form of signature $(++--)$ is defined. He called this *flat twistor space*, and the method of investigating deformation of complex structures, yielded from there the *twistor program*. This initiated a series of papers and monographs by various authors. In the present paper, we deal with dynamical systems generated by the hermitian Hurwitz pairs of the signature (σ, s), $\sigma + s = 5 + 4\mu$, $|\sigma + 1 - s| = 2 + 4m$; $\mu, m = 0, 1, \ldots$ In particular for the signature $(3, 2)$ and its dual $(1, 4)$, the role of entropy was indicated as well as the relationship between Hurwitz and Penrose twistors, *Hurwitz twistors* being objects introduced by us. The signatures $(1, 8)$ and $(7, 6)$ give rise for introducing *pseudotwistors* and *bitwistors*, respectively. For pseudotwistors, we can prove a counterpart of the original fundamental Penrose theorem in the local version (on real analytic solutions of the spinor equations vs. harmonic forms, related to the original relativistic wave equations) and in the semi-global version (on holomorphic solutions of those equations vs. Dolbeault cohomology groups). This had to be preceded [24] by basic constructions, a study of the related pseudotwistors and spinor equations as well as complex structures on spinors. In particular, we proved a theorem (which we call the atomization theorem) saying that there exist complex structures on isometric embeddings for the hermitian Hurwitz pairs concerned so that the embeddings are real parts of holomorphic mappings. The atomization theorem enables us to introduce an analysis with respect to an embedding $\iota : G(2, 4) \hookrightarrow M$, M being a $\mathbb{C}^\infty$-manifold, which we call the quaternic analysis, and develop it from the point of view of quaternal spinors and quaternal harmonic forms, and finally to prove rigorously the above mentioned theorems on spinor solutions vs. harmonic forms and the cohomology group. Finally, a physical motivation is given for the double Cartan-like triality of the pairs concerned as well as an introduction to five-dimensional stochastical electrodynamics.

Research of the first author partially supported by the State Committee for Scientific Research (KBN) grant PB 2 P03A 010 17 (Sections 1-6 of the paper), and partially by the grants of the University of Łódź no. 505/626 and 252 (Sections 7-8).
AMS Subject Classification: 81R25, 32L25, 53A50, 15A66.

Keywords: Spinor, twistor, Clifford algebra, space-time, relativistic wave equation, harmonic form, cohomology group.

1 Introduction

In 1985, Ławrynowicz and Rembieliński [13] initiated a geometrization of the problem of A. Hurwitz, concerned with the composition of quadratic forms, and a geometrical study of the related differential operators of the Cauchy-Riemann-, Dirac- and Fueter-types by introducing the so-called Hurwitz pairs, also in the hyperbolic case [16] discussed by Penrose [27]. The results obtained quickly appeared in textbooks [30] and enabled [19] to formulate and prove *counterparts of two Penrose's fundamental theorems* within the theory of Hurwitz pairs.

Consider the Hurwitz pair consisting of the hermitian space $\mathbb{C}^{16}(\kappa) := (\mathbb{C}^{16}, \kappa)$, equipped with the metric

$$\kappa \equiv I_{8,8} := \begin{pmatrix} I_8 & 0 \\ 0 & -I_8 \end{pmatrix}$$

and the real space $\mathbb{R}^9(\eta) := (\mathbb{R}^9, \eta)$, equipped with the metric

$$\eta \equiv I_{8,1} := \begin{pmatrix} I_8 & 0 \\ 0 & -I_1 \end{pmatrix},$$

where I_n denotes the unit $n \times n$-matrix.

Let $(e_1, \dots, e_{16})$ be the canonical basis of $\mathbb{C}^{16}(\kappa)$. We consider a pair $H = (\mathbb{C}^{16}(\kappa), \mathbb{R}^9(\eta))$. If there exists a bilinear mapping $\circ : \mathbb{R}^9(\eta) \times \mathbb{C}^{16}(\kappa) \to \mathbb{C}^{16}(\kappa)$ called *multiplication* of the elements of $\mathbb{R}^9(\eta)$ by the elements of $\mathbb{C}^{16}(\kappa)$ such that, for $f \in \mathbb{C}^{16}(\kappa)$ and $\alpha \in \mathbb{R}^9(\eta)$, we have

$$\langle a, a\rangle_\eta \langle\langle f, f\rangle\rangle_\kappa = \langle\langle a \circ f, a \circ f\rangle\rangle_\kappa,$$

where

$$\langle\langle f, g\rangle\rangle_\kappa := f^* \kappa g, \ \ f, g \in \mathbb{C}^n; \ \ \langle a, b\rangle_\eta := a^T \eta b, \ \ a, b \in \mathbb{R}^p$$

and * denotes the hermitian conjugation; and, moreover, H is *irreducible*, i.e., there exists no subspace V of $\mathbb{C}^{16}$, $V \neq \{0\}$, $V \neq \mathbb{C}^{16}$, such that $\circ|\mathbb{R}^9(\eta) \times V : \mathbb{R}^5(\eta) \times V \to V$, then H is called an *hermitian Hurwitz pair* (cf. e.g. [19]). This is, of course, a particular case of the general definition. Further, let

$$\epsilon_\alpha \circ e_k = C^1_{\alpha k} e_1 + \dots + C^n_{\alpha k} e_n,$$

where $(\epsilon_1, \dots, \epsilon_9)$ is the canonical basis of $\mathbb{R}^9(\eta)$, and let $C_\alpha = (C^j_{\alpha k}), \alpha = 1, \dots, 9$. Since, traditionally, in the case of the five-dimensional space-time, the space-time element is defined as $ds^2 = c^2 dt^2 - dx^2 - dy^2 - dz^2 - d\tau^2$, where c is a positive constant interpreted as the light velocity, the five-dimensional space-time is $\mathbb{R}^5(I_{1,4})$. Therefore, when speaking about Hurwitz twistors, it seems convenient to associate them with $\mathbb{R}^5(I_{3,2})$ instead of $\mathbb{R}^5(I_{2,3})$; cf. [19, 21] and Figs. 1 and 2. The situation is somehow similar, because of the $(8,8)$-periodicity, if we shift 3 or/and 2 modulo 8, but *not* modulo 4. In fact, it seems to us worthwhile to study in detail the cases $(4+4,1)$, $(4,1+4)$, and $(3+4,2+4)$; cf. again [19] and Figs. 1 and 2. In the first two cases, we arrive at $\mathbb{R}^9(I_{8,1})$ and $\mathbb{R}^9(I_{4,5})$; in the third one, we arrive at $\mathbb{R}^{13}(I_{7,6})$.

We define the algebras $\mathcal{A}_{8,1}$ or, alternatively, $\mathcal{A}_{4,5}$, which are generated by the corresponding collections $\{C_\alpha^\# C_\beta : \alpha \leq \beta\}$, where $C_\alpha^\# = \kappa C_\alpha^* \kappa^{-1}$. An element $x \in \mathcal{A}_{8,1}$ is a *pseudotwistor*, or alternatively, an element $x \in \mathcal{A}_{4,5}$ is a *pseudobitwistor* [19] whenever x has the form

$$x = \sum_{\alpha<\beta} \xi_{\alpha\beta} C_\alpha^\# C_\beta, \quad \xi_{\alpha\beta} \in \mathbb{C}; \tag{1.1}$$

and $\operatorname{im} x^2 = 0$, where $\operatorname{im} x, x \in \mathcal{A}_{8,1}$ or $x \in \mathcal{A}_{4,5}$ is defined in the following manner: $x \in \mathcal{A}_{8,1}$. Alternatively, $x \in \mathcal{A}_{4,5}$ can be written uniquely as

$$x = \sum_{k=0}^{8} x_k, \quad \text{where} \quad x_k = \sum_{\alpha_1<\beta_1<\cdots<\alpha_k<\beta_k} \xi_{\alpha_1\beta_1\cdots\alpha_k\beta_k} C_{\alpha_1}^\# C_{\beta_1} \cdots C_{\alpha_k}^\# C_{\beta_k},$$

with $x_0 = \xi_0 I_4$. We define $\operatorname{im} x := x - x_0$ and denote the collection of pseudotwistors and pseudobitwistors by P_1 and P, respectively; they coincide and are given by

$$\mathcal{J}_H := \{x = \sum_{\alpha<\beta} \xi_{\alpha\beta} C_\alpha^\# C_\beta : \ \operatorname{im} x^2 = 0\};$$

cf. [20, 22] for details concerning identification of the space of pseudotwistors with the corresponding flag spaces.

Let $Z_{\mathcal{A}}^{(n)}(U)$ be the space of real analytic solutions of the corresponding spinor equation of the van der Waerden-type [19] of spin $\frac{1}{2}n$ on an open set $U \in \mathbb{C}^2$. Then the *first theorem* (Theorems 1, 2(i) and 3(i) below) says that the solutions in question can be written as harmonic forms, i.e., there exists a one-to-one correspondence between spinor solutions and harmonic forms with respect to the $(0,4)$-metric

$$ds^2 := -dz^1 d\bar{z}^1 - dz^2 d\bar{z}^2 - dz^3 d\bar{z}^3 - dz^4 d\bar{z}^4 \tag{1.2}$$

in the case of pseudotwistors and with respect to the $(2,2)$-metric

$$ds^2 := dz^1 d\bar{z}^1 + dz^2 d\bar{z}^2 - dz^3 d\bar{z}^3 - dz^4 d\bar{z}^4. \tag{1.3}$$

In the case of pseudobitwistors, it can be expressed as

$$Z_{\mathcal{A}}^{(n)}(U) \simeq \mathbb{H}^1(U, \mathbb{C}^{8n-8}),$$

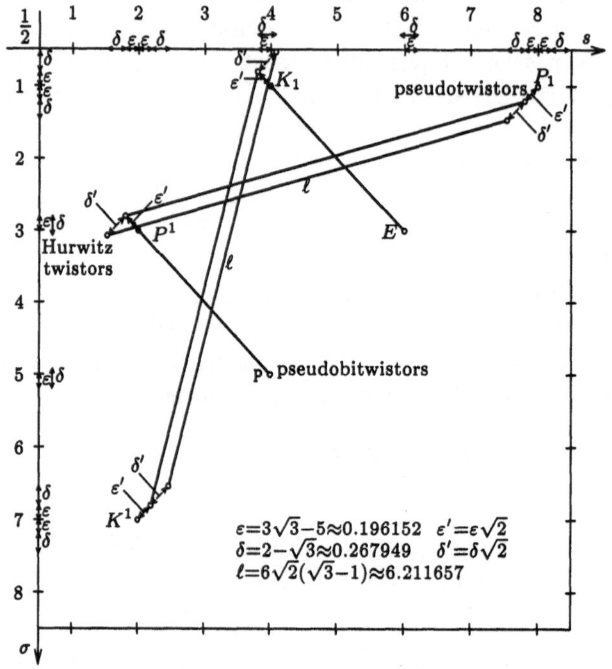

FIGURE 1. Passing from twistors to pseudotwistors

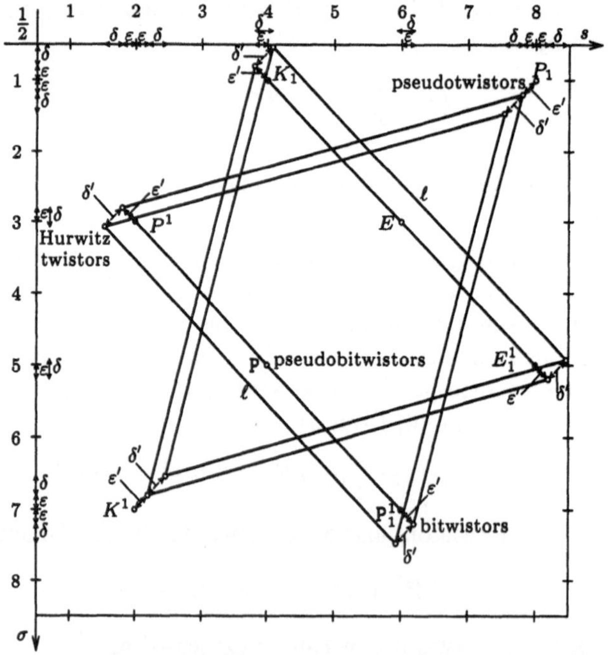

FIGURE 2. Double Cartan-like triality of hermitian Hurwitz pairs

where (cf. [20, 23])

$$\mathbb{H}^1(U,\mathbb{C}^{8n-8}) := \{\phi \in \Gamma^{1,0}(U,\mathbb{C}^{8n-8}) : \ \partial\phi = 0 \quad \text{and} \quad \vartheta\phi = 0\},$$

and ϑ is the formally adjoint operator of ∂ with respect to the indefinite fibre $(8,0)$-metric

$$d\rho^2 := d\zeta^1 d\bar{\zeta}^1 + d\zeta^2 d\bar{\zeta}^2 + \ldots + d\zeta^8 d\bar{\zeta}^8.$$

Similarly, let $Z^{(n)}_{\mathcal{H}}(U_2)$ be the space of holomorphic solutions of the corresponding spinor equations of the van der Waerden-type [19] of spin $\frac{1}{2}n$ on an open set U_2, whereas μ_2 and ν_2 are the related fibre bundles forming the Penrose-like diagram

$$\begin{array}{ccc} & P_1 & \\ {}^{\mu_2}\swarrow & & \searrow^{\nu_2} \\ \mathcal{P}^2_H & & \mathcal{U}^2_H \end{array} \quad \text{and} \quad \begin{array}{ccc} & \mathrm{P} & \\ {}^{\mu_2}\swarrow & & \searrow^{\nu_2} \\ \mathcal{P}^2_H & & \mathcal{U}^2_H \end{array}$$

$$\mathcal{J}_H = \left\{ (L^2_1,\ L^2_2) : \begin{array}{l} L^2_1, L^2_2 \quad \text{are linear subspaces of } \mathbb{C}^8 \\ \\ L^2_1 \subset L^2_2, \quad \dim L^2_1 = 1, \quad \dim L^2_2 = 2 \end{array} \right\},$$

$$\mathcal{P}^2_H = \{L^2_1 : L^2_1 \subset \mathbb{C}^8, \quad \text{linear subspace, } \dim L^2_1 = 1\} \ (\simeq \mathbb{P}^7(\mathbb{C})),$$

$$\mathcal{U}^2_H = \{L^2_2 : L^2_2 \subset \mathbb{C}^8, \quad \text{linear subspace,} \quad \dim L^2_2 = 2\} \quad (\simeq G(2,8)).$$

Further, let $U_2' = \nu_2^{-1}(U_2)$ and $U_2'' = \mu_2 \circ \nu_2^{-1}(U_2)$. Then the *second theorem* (Theorems 1, 2(ii) and 3(ii) below) says that if every fibre of μ_2 is connected, there then exists a one-to-one correspondence (cf. [20, 23]):

$$Z^{(n)}_{\mathcal{H}}(U_2) \simeq H^1(U_2'', \mathcal{O}(-\alpha n - \beta)),$$

where α and β, $\beta \geq 2$, are some positive integers, H^1 denotes the one-dimensional Dolbeault cohomology group, $\mathcal{O}(-n-2) = \mathcal{O}([e]^{-n-2})$, and $[e]$ is the canonical effective divisor of $\mathbb{P}^3(\mathbb{C})$.

2 Preliminaries: Atomization theorem on isometric embeddings

Let $(\mathbb{C}^{16}(I_{8,8}), \mathbb{R}^9(I_{\sigma,s}))$, $\sigma + s = 9$ be one of the Hurwitz pairs and let C_α, $\alpha = 1, 2, \ldots, 9$ be the corresponding Hurwitz matrices. Then we have the Hurwitz algebras $\mathcal{A} = \oplus^4_{k=0}\mathcal{A}_{2k}$, where

$$\mathcal{A}_{2k} = \{ \sum_{1 \leqq \alpha_1 < \beta_1 < \cdots < \alpha_k < \beta_k \leqq 9} \xi_{\alpha_1\beta_1\ldots\alpha_k\beta_k} C^{\#}_{\alpha_1} C_{\beta_1} \cdots C^{\#}_{\alpha_k} C_{\beta_k} \}.$$

Generally, under a *Hurwitz algebra*, we understand a central Clifford algebra $\mathcal{H}_{\sigma-1,s}$, $\sigma + s = p \in \mathbb{N}$, whose generators S_α satisfy the condition $S^{\#}_\alpha = S_\alpha$, $\alpha \neq t$, where t is fixed.

Definition 1. *An element $\xi \in \mathcal{A}$ is called a pseudotwistor of $\mathcal{A}$ if* $\operatorname{im}\xi^2 = 0$. *If $\xi \in \mathcal{A}_{2k}$, it is said to be of degree k. Here the non-scalar part of ξ^2 is denoted by* $\operatorname{im}\xi^2$.

In [23, 24], we have

Theorem A. *The pseudotwistor space of degree k*

$$\mathcal{J}^{(k)} = \{\xi \in \mathcal{A}_{2k} : \ \operatorname{im}\xi^2 = 0\}, \quad k = 1, 2, 3, 4,$$

has a decomposition

$$\mathcal{J}^{(k)} = \mathcal{J}_-^{(k)} + \mathcal{J}_+^{(k)}$$

and introduces a flag structure, i.e.,

(i) $\mathcal{J}_-^{(k)} \subset G(2k-1, 8)$ *and* $\mathcal{J}_+^{(k)} \subset G(2k, 8)$, *and*

(ii) $\xi = \xi_- + \xi_+$ *implies* $\xi_- \subset \xi_+$.

Theorem A yields the Penrose diagram

$$\begin{array}{ccc} & \mathcal{J}^{(k)} & \\ {}^{\nu_{(k)}}\swarrow & & \searrow^{\mu_{(k)}} \\ \mathcal{J}_-^{(k)} & \underset{\tau_k}{\rightsquigarrow} & \mathcal{J}_+^{(k)} \end{array}$$

so we can discuss the representations of pseudotwistors on Hurwitz algebras and look for the corresponding duality theorems.

As far as the Penrose diagram for a fixed representation is concerned, we choose a representation $\alpha_\rho : \mathcal{A} \to \mathcal{H}_\rho$; cf. [19, 21]. Let

$$\mathcal{J}_\rho = \{\alpha^\rho \xi \in \mathcal{H}_\rho : \ \xi \in \mathcal{J}\}.$$

Then we have $\mathcal{J}_\rho = \mathcal{J}_{\rho -} + \mathcal{J}_{\rho +}$ and the Penrose diagram

$$\begin{array}{ccc} & \mathcal{J}_\rho^{(k)} & \\ \swarrow & & \searrow \\ G(2k-1, 8) \supset \mathcal{J}_{\rho -}^{(k)} & & \mathcal{J}_{\rho -}^{(k)} \subset G(2k, 8). \end{array}$$

As a consequence,

$$\mathcal{J}_1^{(k)} = \mathcal{J}_{1-}^{(k)} + \mathcal{J}_{1+}^{(k)},$$

$$\mathcal{J}_{1-}^{(k)} = \{\sum \xi_{\bar{\beta}_1 \bar{\alpha}_2 \beta_2 \cdots \bar{\alpha}_k \beta_k} S_{\beta_1} S_{\alpha_2} \cdots S_{\bar{\alpha}_k} S_{\beta_k}\},$$

$$\mathcal{J}_{1+}^{(k)} = \{\sum \xi_{\bar{\alpha}_1 \beta_1 \bar{\alpha}_2 \beta_2 \cdots \bar{\alpha}_k \beta_k} S_{\bar{\alpha}_1} S_{\beta_1} \cdots S_{\bar{\alpha}_k} S_{\beta_k}\}.$$

Next, we proceed to the duality theorem for pseudotwistors. We choose two representations of $\mathcal{A}$:

$$\begin{array}{ccc} & \mathcal{A} & \\ {}^{\alpha^{(\rho)}}\swarrow & & \searrow^{\alpha^{(\rho')}} \\ \mathcal{H}^{(\rho)} & \underset{\gamma_{\rho\rho'}}{\longrightarrow} & \mathcal{H}^{(\rho')}. \end{array}$$

Then we have the duality mapping $\gamma_{\rho\rho'} : \mathcal{H}^{(\rho)} \to \mathcal{H}^{(\rho)'}$, and we arrive at

Theorem B (Duality Theorem for Pseudotwistors). *The duality mapping induces an isomorphism between twistors $\mathcal{J}_\rho^{(k)}$ and $\mathcal{J}_{\rho'}^{(k)}$ and the following diagram commutes:*

$$\begin{array}{ccccccc}
\mathcal{J}_{\rho-}^{(k)} & & & \overset{\gamma_{\rho\rho'}^{(-)}}{\longrightarrow} & & & \mathcal{J}_{\rho'-}^{(k)} \\
 & \nwarrow & & & & \nearrow & \\
 & & \mathcal{J}_{\rho}^{(k)} & \overset{\gamma_{\rho\rho'}}{\longrightarrow} & \mathcal{J}_{\rho'}^{(k)} & & \\
 & \swarrow & & & & \searrow & \\
\mathcal{J}_{\rho+}^{(k)} & & & \underset{\gamma_{\rho\rho'}^{(+)}}{\longrightarrow} & & & \mathcal{J}_{\rho'+}^{(k)}
\end{array}$$

Again, let $(\mathbb{C}^{16}(I_{8,8}), \mathbb{R}^9(I_{\sigma,s}))$, $\sigma + s = 9$, be a hermitian Hurwitz pair and let $\{C_\alpha\}$, $\alpha = 1, 2, \dots, 9$ be a collection of Hurwitz matrices. Then we can define the Hurwitz equation

$$\sum_{\alpha=1}^{9} C_\alpha \frac{\partial f}{\partial x^\alpha} = 0. \tag{2.1}$$

Definition 2. *A solution f of the equation (2.1) which is independent of one variable, say x_ρ, is called the spinor of class ρ, and the corresponding equation is called the spinor equation of class ρ.*

For the spaces of real analytic solutions of the spinor equation of class ρ in an open set $U_{(\rho)}$, we have [23]

Theorem C (Duality Theorem for Spinor Equations). *There exists an isomorphism*

$$\gamma_{\rho,\rho'}^* : S^{(\rho)}(U_{(\rho)}) \longrightarrow S^{(\rho')}(U_{(\rho')}).$$

Moreover, this isomorphism induces an isomorphism between the solutions of the corresponding Weyl equations

$$\gamma_{\rho,\rho'}^* : S_W^{(\rho)}(U_{(\rho)}) \longrightarrow S_W^{(\rho'')}(U_{(\rho')}),$$

which is realized using the Wick rotation of one variable of $\mathbb{R}^8$.

Finally, in [24] we considered complex structures of spinors; more exactly, we decided whether the isometric embedding

$$\mathbb{C}^4 \simeq \mathbb{R}^8 \overset{\iota}{\longrightarrow} G(8,16), \qquad \mathbb{R}^8 \ni \mathbf{x} \overset{\iota}{\longmapsto} \sum_{\alpha=1}^{7} x^\alpha S_\alpha + x^8 I_8$$

becomes the real part of a holomorphic mapping: $\iota = \iota_H \oplus \bar{\iota}_H$. Precisely, we have

Theorem D (Atomization Theorem). *There exist complex structures on isometric embeddings for hermitian Hurwitz pairs related to the algebras $\mathcal{H}_{0,8}$, $\mathcal{H}_{2,6}$, $\mathcal{H}_{4,4}$, $\mathcal{H}_{6,2}$, $\mathcal{H}_{8,0}$ so that the embeddings are real parts of holomorphic mappings.*

Explicitly, in addition to the cases with $\sigma + s = 9 - 1$, let us include, for the sake of completeness, the cases of $\mathcal{H}_{4,0}$ and $\mathcal{H}_{2,2}$ with $\sigma + s = 5 - 1$. Let

$$u = x^4 + ix^3, v = x^2 + ix^1 \text{ in } \mathcal{H}_{4,0},$$

$$u = x^4 + ix^3, v = x^1 + ix^2 \text{ in } \mathcal{H}_{2,2},$$

$$u = x^3 + ix^8, v = x^1 + ix^2, w = x^4 + ix^5, t = x^6 + ix^7 \text{ in } \mathcal{H}_{0,8} \text{ and } \mathcal{H}_{8,0},$$

$$u = x^4 + ix^3, v = x^2 + ix^1, w = x^5 + ix^6, t = x^7 + ix^8 \text{ in } \mathcal{H}_{6,2},$$

$$u = x^3 + ix^8, v = x^1 + ix^2, w = x^7 + ix^6, t = x^4 + ix^5 \text{ in } \mathcal{H}_{2,6},$$

$$u = x^3 + ix^8, v = x^1 + ix^2, w = x^6 + ix^7, t = x^4 + ix^5 \text{ in } \mathcal{H}_{4,4}.$$

Now, we introduce eleven 2×2 complex matrices which we call *atoms*:

$$A_1 = \begin{pmatrix} u & v \\ -\bar{v} & \bar{u} \end{pmatrix}, \quad A_2 = \begin{pmatrix} u & v \\ \bar{v} & \bar{u} \end{pmatrix}, \quad A_3 = \begin{pmatrix} \bar{u} & \bar{v} \\ v & -u \end{pmatrix},$$
$$A_4 = \begin{pmatrix} -u & -\bar{v} \\ -v & \bar{u} \end{pmatrix}, \quad A_5 = \begin{pmatrix} w & 0 \\ 0 & w \end{pmatrix}, \quad A_6 = \begin{pmatrix} \bar{w} & 0 \\ 0 & \bar{w} \end{pmatrix},$$
$$A_7 = \begin{pmatrix} t & 0 \\ 0 & t \end{pmatrix}, \quad A_8 = \begin{pmatrix} \bar{t} & 0 \\ 0 & \bar{t} \end{pmatrix}, \quad A_9 = \begin{pmatrix} t & 0 \\ 0 & -\bar{t} \end{pmatrix},$$
$$A_{10} = \begin{pmatrix} \bar{t} & 0 \\ 0 & -t \end{pmatrix}, \quad A_{11} = \begin{pmatrix} -\bar{u} & v \\ -\bar{v} & -u \end{pmatrix}.$$

The required embeddings, where $\zeta = (u, v) \in \mathbb{C}^2$ and $z = (u, v, w, t) \in \mathbb{C}^4$, are

$$\iota(\zeta) = A_1 \quad \text{for } \mathcal{H}_{0,4}, \qquad \iota(\zeta) = A_2 \quad \text{for } \mathcal{H}_{2,2},$$

$$\iota(z) = \begin{pmatrix} A_3 & A_5 & A_7 & 0 \\ A_6 & A_4 & 0 & A_7 \\ A_8 & 0 & A_4 & -A_5 \\ 0 & A_8 & -A_6 & A_3 \end{pmatrix} \quad \text{for } \mathcal{H}_{0,8} \text{ and } \mathcal{H}_{8,0}, \tag{2.2}$$

$$\iota(z) = \begin{pmatrix} A_9 & 0 & A_5 & A_1 \\ 0 & -A_{10} & A_{11} & A_6 \\ -A_6 & A_1 & -A_{10} & 0 \\ A_{11} & -A_5 & 0 & A_9 \end{pmatrix} \quad \text{for } \mathcal{H}_{6,2}, \tag{2.3}$$

$$\iota(z) = \begin{pmatrix} A_3 & A_7 & A_5 & 0 \\ A_8 & A_4 & 0 & A_5 \\ -A_6 & 0 & A_3 & -A_7 \\ 0 & -A_6 & -A_8 & A_4 \end{pmatrix} \quad \text{for } \mathcal{H}_{2,6}, \tag{2.4}$$

$$\iota(z) = \begin{pmatrix} A_3 & 0 & A_5 & A_7 \\ 0 & A_3 & -A_8 & A_6 \\ -A_6 & A_7 & A_4 & 0 \\ -A_8 & -A_5 & 0 & A_4 \end{pmatrix} \quad \text{for } \mathcal{H}_{4,4}. \tag{2.5}$$

3 Quaternal analysis of hermitian Hurwitz pairs: Quaternal spinors

In this section, we introduce a concept of quaternal analysis of HHPs (hermitian Hurwitz pairs), which will play a basic role in twistor theory. Here we treat quaternal spinors, whereas the next section will be devoted to harmonic forms.

Let M be a $\mathbb{C}^\infty$-manifold. Usually, we discuss complex analysis by use of the embedding $\iota : \mathbb{C} \to M$; for example, we define a concept of a holomorphic function φ on M, when $\iota : \mathbb{C} \to M$. Here we consider an embedding

$$\iota : G(2,4) \hookrightarrow M \tag{3.1}$$

and make an analysis with respect to this embedding, which will be called the *quaternal analysis*.

3.1 Generality

Definition 3.

(i) An embedding (3.1), where M is a C^∞-manifold, is called single quaternal embedding.

(ii) An embedding

$$\iota : G(2,4) \times G(2,4) \hookrightarrow M \tag{3.2}$$

is called double quaternal embedding.

In order to make the analysis definite, we have to consider a family of quaternal embeddings $\iota_{\mathbb{A}} : G(2,4) \hookrightarrow M$. The parameter $\mathbb{A}$ is called *quater*. In this section, we shall be concerned with a Grassmann manifold $M \sim G(8,16)$ and consider quaternal analysis of van der Waerden spinors. We make the following

Definition 4. *A section $\Phi \in \Gamma(U, S^{\odot n} \otimes \det S)$, S being the universal bundle of M and U an open set in M, is called a quaternal spinor if*

$$\iota_{\mathbb{A}}^{*}\Phi \in \Gamma(U_{\mathbb{A}}, \oplus S_{-}^{\prime \odot m} \otimes \det S_{-}'), \quad \nabla^{A,\dot{A}}(\iota_{\mathbb{A}}^{*}\Phi)_{\dot{A}\ldots\dot{D}} = 0$$

with a suitable m, where $\iota_{\mathbb{A}}(U_{\mathbb{A}}) \subset U$ and $\oplus$ implies a direct sum of several copies.

Throughout the paper, (z^{AB}) is a system of local coordinates of $G(2,4)$, e.g., $\{((z^{AB}), I_2)^t, z^{AB} \in \mathbb{C}\} \subset G(2,4)$ and the Einstein summation convention is used. Analogous definitions will be used for $G(8,16)$.

3.2 Quaters and quaternal embeddings of HHPs

Our program may now be summarized as follows. In the theory of complex variables, we describe the theory using the embedding $\iota : U \to M$, $U \in \mathbb{C}$. Here we wish to make the quaternal analysis in terms of a family of quaternal embeddings $\iota : G(2,4) \hookrightarrow G(8,16)$. We start with introducing the chart of quaternal embeddings and discuss solutions of van der Waerden spinors. Then we shall define the Hodge operators and quaternal harmonic forms.

In the following, we restrict our considerations to HHPs and discuss our quaternal analysis. We choose the Hurwitz pair $(\mathbb{C}^{16}(I_{8,8}), \mathbb{R}^9(I_{1,8}))$ and consider the isometric embedding $\iota(z)$ for $\mathcal{H}_{0,8}$, as given by (2.2). This depends on the choice of generators of $\mathcal{H}_{0,8}$. For the sake of simplicity we fix the generators of $\mathcal{H}_{0,8}$ and consider the corresponding embeddings. From the matrix in (2.2), we can find the following quaters, whose collection is called *the single quater* of $\mathcal{H}_{0,8}$:

$$(A_3, A_4, A_4^{c}, A_3^{c}), \quad \text{where} \quad A_3^{c} = \begin{pmatrix} w & t \\ \bar{t} & -\bar{w} \end{pmatrix}, \quad A_4^{c} = \begin{pmatrix} \bar{w} & t \\ \bar{t} & -w \end{pmatrix}.$$

The pairs (A_3, A_3^{c}) and (A_4, A_4^{c}) form the so-called *double quater* of $\mathcal{H}_{0,8}$.

For quaters, we have embeddings. For example, if we take $\mathbb{A} = A_3$, then

$$\iota_{\mathbb{A}} : G(2,4) \hookrightarrow G(8,16), \quad \iota_{\mathbb{A}} = \operatorname{diag}(\mathbb{A}, \mathbb{A}, \mathbb{A}, \mathbb{A}).$$

For a double quater $(\mathbb{A}, \mathbb{A}^{c})$, we have

$$\iota_{\mathbb{A}^*} : G(2,4) \times G(2,4) \hookrightarrow G(8,16).$$

$$\iota_{\mathbb{A}^*} = \begin{pmatrix} \mathbb{A} & A_5 & A_7 & 0 \\ 0 & 0 & 0 & 0 \\ 0 & 0 & 0 & 0 \\ 0 & A_8 & -A_6 & \mathbb{A} \end{pmatrix}, \quad \mathbb{A}^* = \mathbb{A} \cup \mathbb{A}^{c}.$$

These are called *single* (or *double*) *quaternal embeddings.*

3.3 Quaternal spinors

We make the following definition:

Definition 5.

(i) A van der Waerden spinor ϕ is called of degree 1 whenever $\iota^*_{\mathbb{A}}\phi$ is a spinor on $V_{\mathbb{A}}$ for any quater $\mathbb{A}$.

(ii) A van der Waerden spinor ϕ is called of degree 2 whenever $\iota^*\phi$ is a spinor of certain class; cf. equation (2.2) in [24].

We give several (typical) equations for spinors of degree 1 and 2.

Example 1. *Algebra $\mathcal{H}_{0,8}$, the single quater $\mathbb{A} = A_4$. The spinor equations of degree 1 read*

$$\begin{aligned} -\quad \phi_{1|u} - \phi_{2|\bar{v}} &= 0, \quad -\phi_{7|u} - \phi_{8|\bar{v}} = 0, \\ -\quad \phi_{1|v} + \phi_{2|\bar{u}} &= 0, \quad -\phi_{7|v} + \phi_{8|\bar{u}} = 0, \end{aligned} \tag{3.3}$$

where $\phi_{1|u} = \frac{\partial\phi_1}{\partial u}$, etc.

Example 2. $\mathcal{H}_{0,8}$, $\mathbb{A} = A_3$. *By analogy, we have*

$$\begin{aligned} \phi_{3|\bar{u}} + \phi_{4|\bar{v}} &= 0, \quad \phi_{5|\bar{u}} + \phi_{6|\bar{v}} = 0, \\ \phi_{3|v} - \phi_{4|u} &= 0, \quad \phi_{5|v} - \phi_{6|\bar{u}} = 0. \end{aligned} \tag{3.4}$$

Example 3. $\mathcal{H}_{0,8}$, $\mathbb{A} = A_4^c$. *We have*

$$\begin{aligned} -\quad \phi_{3|w} - \phi_{5|\bar{t}} &= 0, \quad -\phi_{3|t} + \phi_{5|\bar{w}} = 0, \\ -\quad \phi_{4|w} - \phi_{6|\bar{t}} &= 0, \quad -\phi_{4|t} + \phi_{6|\bar{w}} = 0. \end{aligned}$$

Example 4. $\mathcal{H}_{0,8}$, $\mathbb{A} = A_3^c$. *We have*

$$\begin{aligned} \phi_{7|w} - \phi_{1|t} &= 0, \quad -\phi_{1|\bar{w}} - \phi_{7\bar{t}} = 0, \\ \phi_{8|w} - \phi_{2|t} &= 0, \quad -\phi_{2|\bar{w}} - \phi_{8|\bar{t}} = 0. \end{aligned}$$

Example 5. $\mathcal{H}_{0,8}$. *The spinor equations of degree 2 read*

$$\begin{aligned} -\phi_{1|u} - \phi_{2|\bar{v}} - \phi_{3|w} - \phi_{5|\bar{t}} &= 0, \quad -\phi_{1|t} + \phi_{5|\bar{u}} + \phi_{6|\bar{v}} + \phi_{7|w} = 0, \\ -\phi_{1|v} + \phi_{2|\bar{u}} - \phi_{4|w} - \phi_{6|\bar{t}} &= 0, \quad -\phi_{2|t} + \phi_{5|v} - \phi_{6|u} + \phi_{8|w} = 0, \\ -\phi_{1|\bar{w}} + \phi_{3|\bar{u}} + \phi_{4|\bar{v}} - \phi_{7|\bar{t}} &= 0, \quad -\phi_{3|t} + \phi_{5|\bar{w}} - \phi_{7|u} - \phi_{8|\bar{v}} = 0, \\ -\phi_{2|\bar{w}} + \phi_{3|v} - \phi_{4|u} - \phi_{8|\bar{t}} &= 0, \quad -\phi_{4|t} + \phi_{6|\bar{w}} - \phi_{7|v} + \phi_{8|\bar{u}} = 0. \end{aligned} \tag{3.5}$$

Example 6. $\mathcal{H}_{6,2}$, $\mathbb{A} = \begin{pmatrix} t & w \\ -\bar{w} & -\bar{t} \end{pmatrix}$. *The spinor equations of degree 1 read*

$$\begin{aligned} -\quad \phi_{1|\bar{t}} + \phi_{5|\bar{w}} &= 0, \quad \phi_{3|t} + \phi_{7|w} = 0, \\ -\quad \phi_{1|w} + \phi_{5|t} &= 0, \quad -\phi_{3|\bar{w}} - \phi_{7|\bar{t}} = 0. \end{aligned}$$

Example 7. $\mathcal{H}_{6,2}$. *By analogy, we have*

$$\phi_{2|t} + \phi_{6|\bar{w}} = 0, \qquad -\phi_{4|\bar{t}} + \phi_{8|w} = 0,$$
$$-\phi_{2|w} - \phi_{6|\bar{t}} = 0, \qquad -\phi_{4|\bar{w}} + \phi_{8|t} = 0.$$

Example 8. $\mathcal{H}_{6,2}$, $\mathbb{A} = \begin{pmatrix} u & v \\ -\bar{v} & \bar{u} \end{pmatrix}$. *We have*

$$-\phi_{7|u} - \phi_{8|v} = 0, \qquad \phi_{5|\bar{u}} - \phi_{6|v} = 0,$$
$$\phi_{7|\bar{v}} - \phi_{8|\bar{u}} = 0, \qquad \phi_{5|\bar{v}} + \phi_{6|u} = 0.$$

Example 9. $\mathcal{H}_{6,2}$, $\mathbb{A} = \begin{pmatrix} -u & v \\ -\bar{v} & -u \end{pmatrix}$. *We have*

$$-\phi_{3|u} - \phi_{4|v} = 0, \qquad \phi_{1|\bar{u}} - \phi_{2|v} = 0,$$
$$\phi_{3|\bar{v}} - \phi_{4|\bar{u}} = 0, \qquad \phi_{1|\bar{v}} + \phi_{2|u} = 0.$$

Example 10. $\mathcal{H}_{6,2}$. *The spinor equations of degree* 2 *read*

$$-\phi_{1|\bar{t}} + \phi_{5|\bar{w}} - \phi_{7|u} - \phi_{5|v} = 0, \qquad -\phi_{1|w} - \phi_{3|u} - \phi_{4|v} + \phi_{5|t} = 0,$$
$$\phi_{2|t} + \phi_{6|\bar{w}} + \phi_{7|\bar{v}} - \phi_{5|\bar{u}} = 0, \qquad -\phi_{2|w} + \phi_{3|\bar{v}} - \phi_{4|\bar{u}} - \phi_{6|\bar{t}} = 0,$$
$$\phi_{3|t} + \phi_{5|\bar{u}} - \phi_{6|v} + \phi_{7|w} = 0, \qquad \phi_{1|\bar{u}} - \phi_{2|v} - \phi_{3|\bar{w}} - \phi_{7|\bar{t}} = 0,$$
$$-\phi_{4|\bar{t}} + \phi_{5|\bar{v}} + \phi_{6|u} + \phi_{8|w} = 0, \qquad \phi_{1|\bar{v}} + \phi_{2|u} - \phi_{4|\bar{w}} + \phi_{8|t} = 0.$$

Example 11. $\mathcal{H}_{2,6}$, $\mathbb{A} = \begin{pmatrix} \bar{u} & \bar{v} \\ v & -u \end{pmatrix}$. *The spinor equations of degree* 1 *read*

$$\phi_{1|\bar{u}} + \phi_{2|\bar{v}} = 0, \quad \phi_{5|\bar{u}} + \phi_{6|\bar{v}} = 0,$$
$$\phi_{1|v} - \phi_{2|u} = 0, \quad \phi_{5|v} + \phi_{6|u} = 0.$$

Example 12. $\mathcal{H}_{2,6}$, $\mathbb{A} = \begin{pmatrix} -u & -\bar{v} \\ -v & \bar{u} \end{pmatrix}$. *By analogy, we have*

$$-\phi_{3|u} - \phi_{4|\bar{v}} = 0, \quad -\phi_{7|u} - \phi_{8|\bar{v}} = 0,$$
$$-\phi_{3|v} - \phi_{4|\bar{u}} = 0, \quad -\phi_{7|v} + \phi_{8|\bar{u}} = 0.$$

Example 13. $\mathcal{H}_{2,6}$, $\mathbb{A} = \begin{pmatrix} t & w \\ -\bar{w} & -\bar{t} \end{pmatrix}$. *We have*

$$-\phi_{1|w} + \phi_{7|t} = 0, \quad -\phi_{2|\bar{t}} + \phi_{8|\bar{w}} = 0,$$
$$-\phi_{1|\bar{t}} + \phi_{7|\bar{w}} = 0, \quad -\phi_{2|w} + \phi_{8|t} = 0.$$

Example 14. $\mathcal{H}_{2,6}$, $\mathbb{A} = \begin{pmatrix} \bar{t} & w \\ -\bar{w} & -t \end{pmatrix}$. *We have*

$$-\phi_{4|w} + \phi_{6|\bar{t}} = 0, \qquad -\phi_{3|t} + \phi_{5|\bar{w}} = 0,$$
$$\phi_{4|t} + \phi_{6|\bar{w}} = 0, \qquad -\phi_{3|w} + \phi_{5|\bar{t}} = 0.$$

Example 15. $\mathcal{H}_{2,6}$. *The spinor equations of degree 2 read*

$$\begin{aligned}
&\phi_{1|\bar u} + \phi_{2|\bar v} - \phi_{3|t} + \phi_{5|\bar w} = 0, && -\phi_{1|w} + \phi_{5|\bar u} + \phi_{6|\bar v} + \phi_{7|t} = 0,\\
&\phi_{1|v} - \phi_{2|u} - \phi_{4|t} + \phi_{6|\bar w} = 0, && -\phi_{2|w} + \phi_{5|v} - \phi_{6|u} + \phi_{8|t} = 0,\\
&-\phi_{1|\bar t} - \phi_{3|u} - \phi_{4|\bar v} + \phi_{7|\bar w} = 0, && -\phi_{3|w} + \phi_{5|\bar t} - \phi_{7|u} - \phi_{8|\bar v} = 0,\\
&-\phi_{2|\bar t} - \phi_{3|v} + \phi_{4|\bar u} + \phi_{8|\bar w} = 0, && -\phi_{4|w} + \phi_{6|\bar t} - \phi_{7|v} + \phi_{8|\bar u} = 0.
\end{aligned}$$

Example 16. $\mathcal{H}_{4,4}$, $\mathbb{A} = \begin{pmatrix} -u & -\bar v \\ -v & \bar u \end{pmatrix}$. *The spinor equations of degree 1 read*

$$\begin{aligned}
&-\phi_{1|u} - \phi_{2|\bar v} = 0, && -\phi_{3|u} - \phi_{4|\bar v} = 0,\\
&-\phi_{1|v} + \phi_{2|\bar u} = 0, && -\phi_{3|v} + \phi_{4|\bar u} = 0.
\end{aligned}$$

Example 17. $\mathcal{H}_{4,4}$, $\mathbb{A} = \begin{pmatrix} \bar u & \bar v \\ v & -u \end{pmatrix}$. *By analogy, we have*

$$\begin{aligned}
&\phi_{5|\bar u} + \phi_{6|\bar v} = 0, && \phi_{7|\bar u} + \phi_{8|\bar v} = 0,\\
&\phi_{5|v} - \phi_{4|u} = 0, && \phi_{7|v} - \phi_{8|u} = 0.
\end{aligned}$$

Example 18. $\mathcal{H}_{4,4}$, $\mathbb{A} = \begin{pmatrix} w & t \\ -\bar t & \bar w \end{pmatrix}$. *We have*

$$\begin{aligned}
&\phi_{1|w} + \phi_{3|t} = 0, && \phi_{2|w} + \phi_{4|t} = 0,\\
&-\phi_{1|\bar t} + \phi_{3|\bar w} = 0, && -\phi_{2|\bar t} + \phi_{4|\bar w} = 0.
\end{aligned}$$

Example 19. $\mathcal{H}_{4,4}$, $\mathbb{A} = \begin{pmatrix} -\bar w & t \\ -\bar t & -w \end{pmatrix}$. *We have*

$$\begin{aligned}
&-\phi_{5|\bar w} + \phi_{7|t} = 0, && -\phi_{6|\bar w} + \phi_{8|t} = 0,\\
&-\phi_{5|\bar t} - \phi_{7|w} = 0, && -\phi_{6|\bar t} - \phi_{8|w} = 0.
\end{aligned}$$

Example 20. $\mathcal{H}_{2,6}$, $\mathbb{A} = \begin{pmatrix} \bar t & w \\ -\bar w & -t \end{pmatrix}$. *We have*

$$\begin{aligned}
&-\phi_{4|w} + \phi_{6|\bar t} = 0, && -\phi_{3|t} + \phi_{5|\bar w} = 0,\\
&\phi_{4|t} + \phi_{6|\bar w} = 0, && -\phi_{3|w} + \phi_{5|\bar t} = 0.
\end{aligned}$$

Example 21. $\mathcal{H}_{4,4}$. *The spinor equations of degree 2 read*

$$\begin{aligned}
&-\phi_{1|u} - \phi_{2|\bar v} + \phi_{5|\bar w} - \phi_{7|t} = 0, && -\phi_{1|w} - \phi_{3|t} + \phi_{5|\bar u} + \phi_{6|\bar v} = 0,\\
&-\phi_{1|v} + \phi_{2|\bar u} + \phi_{6|\bar w} - \phi_{8|t} = 0, && -\phi_{2|w} - \phi_{4|t} + \phi_{5|v} - \phi_{6|u} = 0,\\
&-\phi_{3|u} - \phi_{4|\bar v} + \phi_{5|\bar t} + \phi_{7|w} = 0, && \phi_{1|\bar t} - \phi_{3|\bar w} + \phi_{7|\bar u} + \phi_{8|\bar v} = 0,\\
&-\phi_{3|v} + \phi_{4|u} + \phi_{6|\bar t} + \phi_{8|w} = 0, && \phi_{2|\bar t} - \phi_{4|\bar w} + \phi_{7|v} - \phi_{8|u} = 0.
\end{aligned}$$

4 Quaternal analysis of hermitian Hurwitz pairs: Quaternal harmonic forms

4.1 A study of the Hodge operator and harmonic forms for $(\mathbb{C}^8(I_{2,2}), \mathbb{R}^5(I_{\sigma,s}))$, $\sigma + s = 5$

We start by giving basic definitions of harmonic forms for HHPs. Let $(\mathbb{C}^2, h)$ be the hermitian vector space with the metric

$$ds^2 = \lambda_1 du\, d\bar{u} + \lambda_2 dv\, d\bar{v}, \quad \lambda_1 = \pm 1, \; \lambda_2 = \pm 1.$$

We shall quote this metric as the (λ_1, λ_2)-metric. By this, we can define the *Hodge operator* $\hat{\star}$ and the ∂ operator in the usual manner:

$$\hat{\star} : \Gamma(\mathbb{C}^2, \Lambda^{*k}) \to \Gamma(\mathbb{C}^2, \Lambda^{*4-k}), \quad \partial : \Gamma(\mathbb{C}^2, \Lambda^{*k}) \to \Gamma(\mathbb{C}^2, \Lambda^{*k+1}),$$

and

$$\partial\phi = \frac{\partial\phi}{\partial u} du + \frac{\partial\phi}{\partial v} dv \qquad \text{for the 0-form,}$$

$$\partial\omega = \left(-\frac{\partial\phi_1}{\partial v} + \frac{\partial\phi_2}{\partial u} \right) du \wedge dv \qquad \text{for the 1-form } \omega = \phi_1 du + \phi_2 dv.$$

The formally adjoint operator of ∂ is defined by

$$\nu : \Gamma(\mathbb{C}^2, \Lambda^{*k+1}) \to \Gamma'(\mathbb{C}^2, \Lambda^{*k}), \qquad \nu = -\hat{\star}\bar{\partial}\hat{\star}.$$

We see that

$$\nu\omega = \frac{\partial\phi_1}{\partial\bar{u}} \pm \frac{\partial\phi_2}{\partial\bar{v}}, \quad \omega = \phi_1 du + \phi_2 dv,$$

where we have to choose + if the type is $(1,1)$ or $(-1,-1)$, and − if the type is $(1,-1)$. Hence, we conclude that $\omega = \phi_1 du + \phi_2 dv$ is harmonic if and only if

$$\frac{\partial\phi_2}{\partial u} - \frac{\partial\phi_1}{\partial v} = 0 \quad \text{and} \quad \frac{\partial\phi_1}{\partial\bar{u}} \pm \frac{\partial\phi_2}{\partial\bar{v}} = 0.$$

As we have shown in [23], this condition is equivalent to the solvability of spinor equations. The main purpose of this section is to generalize the above result to the HHPs in question.

4.2 A study of the basic forms

We are now ready to define the Hodge operators for forms $\iota^*_{\mathbf{A}}\Phi$ and $\iota^*_{\mathbf{A}^*}\Phi$, where Φ is a form on M and $\iota_{\mathbf{A}}$, $\iota_{\mathbf{A}^*}$ are quaternal embeddings which define harmonic forms. The main purpose of this section is to find a relationship between quaternal spinors and quaternal harmonic forms (the Key Lemma below). We shall be concerned with the HHP $(\mathbb{C}^{16}(I_{8,8}), \mathbb{R}^9(I_{1,8}))$ and the algebra $\mathcal{H}_{0,8}$. For the other HHPs we can conduct a discussion in a completely analogous manner. Hence, in Sect. 4.5, we include a short comment on how to treat them.

¿From the isometric embedding (2.2), we derive the forms

$$^t\Phi = [\phi_5, \phi_6, \phi_7, \phi_8, \phi_1, \phi_2, \phi_3, \phi_4],$$

$$\Theta = \begin{pmatrix} dA_4 & dA_7 & dA_5 & 0 \\ dA_7 & dA_3 & 0 & dA_6 \\ dA_5 & 0 & dA_3 & -dA_8 \\ 0 & dA_6 & -dA_8 & dA_4 \end{pmatrix},$$

where

$$dA_3 = \begin{pmatrix} -d\bar{v} & du \\ d\bar{u} & dv \end{pmatrix}, \quad dA_4 = \begin{pmatrix} d\bar{v} & -d\bar{u} \\ -du & -dv \end{pmatrix}, \quad dA_5 = \begin{pmatrix} dw & 0 \\ 0 & dw \end{pmatrix},$$

$$dA_6 = \begin{pmatrix} d\bar{w} & 0 \\ 0 & d\bar{w} \end{pmatrix}, \quad dA_7 = \begin{pmatrix} dt & 0 \\ 0 & dt \end{pmatrix}, \quad dA_8 = \begin{pmatrix} d\bar{t} & 0 \\ 0 & d\bar{t} \end{pmatrix};$$

$$\Pi_1 = \mathrm{diag}[d\bar{u} \wedge d\bar{v},\ du \wedge dv,\ du \wedge d\bar{v},\ d\bar{u} \wedge dv,\ du \wedge d\bar{v},\ d\bar{u} \wedge dv,\ d\bar{u} \wedge d\bar{v},\ du \wedge dv\,],$$

$$\Pi_2 = \mathrm{diag}[dw \wedge dt,\ dw \wedge dt,\ d\bar{w} \wedge dt,\ d\bar{w} \wedge dt,\ dw \wedge d\bar{t},\ dw \wedge d\bar{t},\ d\bar{w} \wedge d\bar{t},\ d\bar{w} \wedge d\bar{t}\,].$$

With the use of these forms, we define the basic forms:

Definition 6. *The following k-forms $\Omega^{(k)}$, $k = 1, 2, 3, 4$, are called basic k-forms:*

$$\Omega^{(1)} = \Theta\Phi_1^{(1)}, \qquad \Omega^{(3)} = \Theta\Pi_1\Phi_1^{(3)} + \Theta\Pi_2\Phi_2^{(2)},$$

$$\Omega^{(2)} = \Pi_1\Phi_1^{(2)} + \Pi_2\Phi_2^{(2)}, \qquad \Omega^{(4)} = \Pi_1\Pi_2\Phi_1^{(4)}.$$

4.3 A study of harmonic forms for a single quater

We are going to define the corresponding Hodge operators and harmonic forms. We define them for each $\mathbb{A}$ in question. For example, let us take

$$\mathbb{A} = \begin{pmatrix} \bar{u} & \bar{v} \\ u & v \end{pmatrix}. \tag{4.1}$$

Then we can find two copies of the quater $\mathbb{A}$:

$$\Theta = \begin{pmatrix} dA_4 & 0 \\ 0 & -{}^t(dA_3) \end{pmatrix}.$$

From this, we have the restriction to $V_{\mathbb{A}}$ by

$$^t\Phi_{\mathbb{A}} = [\phi_5, \phi_6, \phi_3, \phi_4], \qquad \Omega_{\mathbb{A}} = \Theta_{\mathbb{A}}\Phi_{\mathbb{A}}.$$

A study of the Hodge operator on Γ

We define the *Hodge operator*

$$\star : \Gamma(V_{\mathbb{A}}, \Lambda^{*1}) \to \Gamma(V_{\mathbb{A}}, \Lambda^{*3})$$

by

$$\star\Omega_{\mathbb{A}} = (\hat{\star}\Theta_{\mathbb{A}})\Gamma_{\mathbb{A}}\Pi_{\mathbb{A}}\Phi_{\mathbb{A}}, \tag{4.2}$$

where $\Gamma_{\mathbb{A}} = \mathrm{diag}^*(I_2, I_2)$. Explicit calculation shows that

$$\begin{aligned}
\star(\phi_5 d\bar{v} - \phi_6 d\bar{u}) &= (-\phi_3 du - \phi_4 dv)d\bar{u} \wedge d\bar{v},\\
\star(-\phi_5 du - \phi_6 dv) &= (\phi_3 d\bar{v} - \phi_4 d\bar{u})du \wedge dv,\\
\star(\phi_3 d\bar{v} - \phi_4 d\bar{u}) &= (\phi_5 du + \phi_6 dv)d\bar{u} \wedge d\bar{v},\\
\star(-\phi_3 du + \phi_4 dv) &= (-\phi_5 d\bar{v} + \phi_6 du)du \wedge dv.
\end{aligned}$$

We define

$$d_{\mathbb{A}} : \Gamma(V_{\mathbb{A}}, \Lambda^{*k}) \to \Gamma(V_{\mathbb{A}}, \Lambda^{*(k+1)})$$

by

$$d_{\mathbb{A}} = du\frac{\partial}{\partial u} + dv\frac{\partial}{\partial v} \quad \text{(note that} \quad d\bar{u}\frac{\partial}{\partial \bar{u}} + d\bar{v}\frac{\partial}{\partial \bar{v}} = 0)$$

and

$$\nu_{\mathbb{A}} : \Gamma(V_{\mathbb{A}}, \Lambda^{*(k+1)}) \to \Gamma(V_{\mathbb{A}}, \Lambda^{*k}), \quad \nu_{\mathbb{A}} = - \star d\mathbb{A}^*. \tag{4.3}$$

We shall give several examples of equations of a harmonic 1-form for a single quater $\mathbb{A}$ given by (4.1).

Example 22. *We take*

$$\omega_1 = \phi_5 d\bar{v} - \phi_6 d\bar{u}, \quad \omega_2 = -\phi_5 du - \phi_6 dv.$$

Hence,

$$\begin{aligned}
\star\omega_1 &= [-\phi_3 du - \phi_4 dv] \wedge d\bar{u} \wedge d\bar{v},\\
\star\omega_2 &= [\phi_3 d\bar{v} - \phi_4 d\bar{u}] \wedge du \wedge dv;
\end{aligned}$$

so (ω_1, ω_2) *is harmonic if and only if*

$$\begin{aligned}
\phi_{5|\bar{u}} + \phi_{6|\bar{v}} = 0, \quad \phi_{3|v} - \phi_{4|u} = 0,\\
\phi_{5|v} - \phi_{6|u} = 0, \quad \phi_{3|\bar{u}} + \phi_{4|\bar{v}} = 0.
\end{aligned} \tag{4.4}$$

A study of Hodge operators on 2-forms

Next, we define the Hodge operators on 2-forms. Also we start with a quater

$$\mathbb{A} = \begin{pmatrix} \bar{v} & u \\ v & \bar{u} \end{pmatrix}$$

and restrict the basic 2-forms on $V_{\mathbb{A}}$:

$$(\Pi_1)_{\mathbb{A}} = \mathrm{diag}(d\bar{u}\wedge d\bar{v}, du\wedge dv, d\bar{u}\wedge d\bar{v}, du\wedge dv),$$
$${}^t(\Phi_1)_{\mathbb{A}} = [\phi_5, \phi_6, \phi_3, \phi_4].$$

We set

$$\Omega^{(2)}_{\mathbb{A}} = (\Pi_1)_{\mathbb{A}}\Phi_{\mathbb{A}}$$

and define

$$\star\Omega^{(2)}_{\mathbb{A}} \pm (\Pi_1)_{\mathbb{A}}\Gamma_{\mathbb{A}}\Phi_{\mathbb{A}},$$

where

$$\hat{\star}(du\wedge dv) = d\bar{u}\wedge d\bar{v}, \quad \hat{\star}(d\bar{u}\wedge d\bar{v}) = du\wedge dv,$$
$$\hat{\star}(d\bar{u}\wedge dv) = du\wedge d\bar{v}, \quad \hat{\star}(du\wedge d\bar{v}) = d\bar{u}\wedge dv.$$

Explicit calculation shows that

$$\star(\phi_4 du\wedge dv) = \phi_5 d\bar{u}\wedge d\bar{v}, \quad \star(\phi_3 d\bar{u}\wedge d\bar{v}) = \phi_d u\wedge dv.$$

In the same manner, we can define the harmonic 2-forms. In this case, we can see that every 2-form becomes harmonic. We have no restrictions, and the details may be omitted.

A study of Hodge operators on 3-forms

For a single quater $\mathbb{A}$, we define the Hodge operator

$$\star_{\mathbb{A}} : \Gamma(U_{\mathbb{A}}, \Lambda^{*3}) \to \Gamma^1(U_{\mathbb{A}}, \Lambda^{*1})$$

so that $\star^2_{\mathbb{A}} = 1$. Explicit calculation shows that

$$\star[(\phi_5 d\bar{v} - \phi_6 d\bar{u})\wedge du\wedge dv] = -\phi_3 du - \phi_4 dv,$$
$$\star[(-\phi_5 du - \phi_6 dv)\wedge d\bar{u}\wedge d\bar{v}] = \phi_3 d\bar{v} - \phi_4 du,$$

etc., where $\mathbb{A} = A_3$. The harmonic forms are also defined, and the equations are completely the same as those for harmonic 1-forms; the details may be omitted.

A study of the Hodge operators on 4-forms

For a quater $\mathbb{A}$, we define

$$\star_{\mathbb{A}}\Omega^{(4)} = \Phi^{(4)}.$$

We see that $\star(du\wedge dv\wedge d\bar{u}\wedge d\bar{v}) = 1$. The harmonic 4-forms of $\star\Phi = (\phi_1, \phi_2)$ satisfy

$$\phi_{1|\bar{u}} = 0, \quad \phi_{1|\bar{v}} = 0, \quad \phi_{2|u} = 0, \quad \phi_{2|v} = 0.$$

We have defined the Hodge operators and harmonic forms for the quater $\mathbb{A}$. For other quaters, we can define the operations in a similar manner in four steps:

1. We choose $\mathbb{A}$;
2. We find two copies of $\mathbb{A}$;
3. We restrict Θ, Φ to $V_{\mathbb{A}}$;
4. We define $\star : \Gamma(V_{\mathbb{A}}, T^*) \to \Gamma(V_{\mathbb{A}}, T^*)$.

4.4 The Key Lemma

We are going to prove

Key Lemma.

(i) The fundamental 1*-form* $\Omega^{(1)}$ $(= \Theta\Phi^{(1)})$ *is a quaternal harmonic form with respect to any single quater if and only if* $\Phi^{(1)}$ *is a quaternal spinor with respect to any single quater.*

(ii) The fundamental $(3,1)$*-form* $\Omega^{(1,3)} = \Omega^{(1)} + \Omega^{(3)}$ *is a quaternal harmonic form with respect to any double quater if and only if* $\Phi^{(3)}$ *is a spinor of certain class.*

Proof. The proof is divided into two steps.

Step A. *A study of the Hodge operators and harmonic forms for double quaters.* Next, we define the Hodge operators for double quaters $\mathbb{A}$ and $\mathbb{A}^c$. In this case the corresponding forms are generated by the forms of $\mathbb{A}$ and $\mathbb{A}^c$. Hence, we can define the operator for double quaters. We demonstrate them by examples.

Example 23. *We choose*

$$\mathbb{A} = \begin{pmatrix} \bar{v} & u \\ \bar{v} & \bar{u} \end{pmatrix}, \quad \mathbb{A}^c = \begin{pmatrix} t & \bar{w} \\ \bar{t} & w \end{pmatrix}.$$

A 1*-form of* $(\mathbb{A}, \mathbb{A}^c)$ *is constructed by restricting* Θ *to* $\mathbb{A} \cup \mathbb{A}^c$ *:*

$$\begin{aligned} &\phi_5 dt - \phi_7 d\bar{v} + \phi_8 du + \phi_3 d\bar{w}, \quad \phi_5 dw - \phi_1 d\bar{v} + \phi_2 du - \phi_3 d\bar{t}, \\ &\phi_6 dt + \phi_7 d\bar{u} + \phi_8 dv + \phi_4 d\bar{w}, \quad \phi_6 dw + \phi_1 d\bar{u} + \phi_2 dv - \phi_4 d\bar{t}. \end{aligned}$$

Hence, we obtain

$$\begin{aligned} &\star\,(\phi_5 dt - \phi_7 d\bar{v} + \phi_8 du + \phi_3 d\bar{w}) \\ &\qquad\qquad = (\phi_6 dw + \phi_1 d\bar{u} + \phi_2 dv - \phi_4 d\bar{t})(\Pi^{(2)}_{\mathbb{A}} + \Pi^{(2)}_{\mathbb{A}^c}), \end{aligned}$$

$$\begin{aligned} &\star\,(\phi_6 dt + \phi_7 d\bar{u} + \phi_8 dv + \phi_4 d\bar{w}) \\ &\qquad\qquad = (\phi_5 dw - \phi_1 d\bar{v} + \phi_2 du - \phi_3 d\bar{t})(\Pi^{(2)}_{\mathbb{A}} + \Pi^{(2)}_{\mathbb{A}^c}). \end{aligned}$$

The Hodge operators on forms of higher degree can be defined in a similar manner. Explicit calculations may be omitted.

Example 24. *We shall construct the harmonic* 1*-form for a double quater*

$$\mathbb{A} = \begin{pmatrix} \bar{u} & \bar{v} \\ u & v \end{pmatrix}, \qquad \mathbb{A}^c = \begin{pmatrix} \bar{w} & \bar{t} \\ w & t \end{pmatrix}:$$

$$\begin{aligned} \omega_1 &= \phi_5 dt - \phi_7 d\bar{v} + \phi_8 du + \phi_3 d\bar{w}, \\ \omega_2 &= \phi_6 dt + \phi_7 d\bar{u} + \phi_2 dv + \phi_4 d\bar{w}. \end{aligned} \tag{4.5}$$

By (4.2), we have

$$\begin{aligned} \star\omega_1 &= (\phi_5 dw - \phi_1 d\bar{v} + \phi_2 du - \phi_3 d\bar{t})(d\bar{w} \wedge dt + d\bar{u} \wedge dv), \\ \star\omega_2 &= (\phi_6 dw + \phi_1 d\bar{u} + \phi_2 dv - \phi_4 d\bar{t})(d\bar{w} \wedge dt + du \wedge d\bar{v}). \end{aligned}$$

Hence, $\Omega = (\omega_1, \omega_2)$ *is harmonic if and only if*

$$\begin{gathered} \phi_{5|\bar{w}} - \phi_{3|t} = 0, \quad \phi_{6|\bar{w}} - \phi_{4|\bar{t}} = 0, \quad -\phi_{7|u} - \phi_{8|\bar{v}} = 0, \\ \phi_{7|v} + \phi_{8|\bar{u}} = 0, \\ -\phi_{5|\bar{t}} - \phi_{3|w} - \phi_{1|u} - \phi_{2|\bar{v}} = 0, -\phi_{6|\bar{t}} - \phi_{4|w} - \phi_{1|v} + \phi_{2|\bar{u}} = 0. \end{gathered} \tag{4.6}$$

Example 25. *Now we can construct harmonic* 2*-forms for a double quater. In this case, harmonic forms are direct sums of those of single quaters.*

Example 26. *We can proceed to constructing harmonic* 3*-forms for the double quater* $(\mathbb{A}, \mathbb{A}^c)$*. Namely, let us take*

$$\mu_1 = \omega_1[dw \wedge d\bar{t} + d\bar{u} \wedge dv], \quad \mu_2 = \omega_2[dw \wedge d\bar{t} + du \wedge d\bar{v}],$$

where ω_1, ω_2 *are given in (18). Hence,*

$$\star\mu_1 = \phi_5 dw - \phi_1 d\bar{v} + \phi_2 du - \phi_3 d\bar{t}, \quad \star\mu_2 = \phi_6 dw + \phi_1 d\bar{u} + \phi_2 d\bar{v} - \phi_4 d\bar{t};$$

so (μ_1, μ_2) *is harmonic if and only if*

$$\begin{gathered} -\phi_{5|\bar{t}} - \phi_{3|w} = 0, \quad -\phi_{6|\bar{t}} - \phi_{4|w} = 0, \quad -\phi_{1|u} - \phi_{2|\bar{v}} = 0, \\ -\phi_{1|v} + \phi_{2|u} = 0, \\ \phi_{5|\bar{w}} - \phi_{7|u} - \phi_{8|\bar{v}} - \phi_{3|t} = 0, \ \phi_{6|\bar{w}} - \phi_{7|v} + \phi_{8|\bar{u}} - \phi_{4|t} = 0. \end{gathered} \tag{4.7}$$

Here, we notice that the equations (4.5) and (24) are directly related to the equations of the spinor of degree 2 (cf. the definitions of a van der Waerden spinor and of that spinor of degree 1). In order to arrive at these equations, we consider the sum of basic forms of degree 1 and degree 3 :

$$\Omega^{(1,3)} = \Omega^{(1)} + \Omega^{(3)}.$$

Then, we can see that

$$\begin{aligned} d_{\mathbb{A}^*}\Omega^{(1,3)} &= (d_{\mathbb{A}^*}\Omega^{(1,3)})_2 + (d_{\mathbb{A}^*}\Omega^{(1,3)})_4, \\ \nu_{\mathbb{A}^*}\Omega^{(1,3)} &= (\nu_{\mathbb{A}^*}\Omega^{(1,3)})_2 + (\nu_{\mathbb{A}^*}\Omega^{(1,3)})_0, \end{aligned}$$

where $\mathbb{A}^* = A \cup A^c$, and the suffixes imply the degrees of forms. Here, we notice that

$$(d_{\mathbb{A}^*}\Omega^{(1,3)})_2 = 0 \quad \text{and} \quad (\nu_{\mathbb{A}^*}\Omega^{(1,3)})_2 = 0$$

yield

$$(d_{\mathbb{A}^*}\Omega^{(1,3)})_4 = 0 \quad \text{and} \quad (\nu_{\mathbb{A}^*}\Omega^{(1,3)})_0 = 0,$$

which implies that the equations of forms for a double quater $\mathbb{A}^*$ are reduced to a couple of equations of single quaters $\mathbb{A}$ and $\mathbb{A}^c$. In order to obtain non-reduced solutions, we have to consider the equations

$$(d_{\mathbb{A}^*}\Omega^{(1,3)})_{0,4} \simeq 0 \qquad \text{and} \qquad (\nu_{\mathbb{A}^*}\Omega^{(1,3)})_{0,4} \simeq 0, \tag{4.8}$$

where the subscript $0, 4$ signifies only the choice of the projected parts to the 0- and 4-forms, while $\simeq$ means that the equations hold modulo harmonic forms of single quaters. Hence, it is natural to call the solutions of equations (4.8) *harmonic forms for double quaters.*

Step B. *Quaternal spinors vs. quaternal harmonic forms.* Let us compare the spinor equations in Sect. 2 with the equations of harmonic forms (3.3) and (3.4) with (4.4), as well as (3.5) with (24) and (26). This suffices to conclude the proof. □

4.5 Quaternal harmonic forms for other Hurwitz algebras

Here we give a short description of how to treat the other cases. It is sufficient to give the basic 1-form for each Hurwitz algebra concerned.

The cases of $\mathcal{H}_{0,8}$ and $\mathcal{H}_{8,0}$

This case was already discussed in Sect. 4.1.

The case of $\mathcal{H}_{6,2}$

The isometric embedding is a counterpart of (2.2), listed on the same page as (2.3). The possible quaters are

$$\begin{pmatrix} -\bar{u} & v \\ -\bar{v} & -u \end{pmatrix}, \quad \begin{pmatrix} u & v \\ -\bar{v} & \bar{u} \end{pmatrix}, \quad \begin{pmatrix} t & w \\ -\bar{w} & -\bar{t} \end{pmatrix}, \quad \begin{pmatrix} -\bar{t} & \bar{w} \\ -w & t \end{pmatrix}.$$

The corresponding basic 1-form Θ reads

$$\Theta = \begin{pmatrix} dA_9 & 0 & -d^cA_{10} & dA_{11} \\ 0 & d^cA_9 & d^cA_{11} & dA_{10} \\ -dA_{10} & dA_{11} & d^cA_9 & 0 \\ d^cA_{11} & d^cA_{10} & 0 & dA_9 \end{pmatrix},$$

where

$$dA_9 = \begin{pmatrix} dw & 0 \\ 0 & -dw \end{pmatrix}, \quad dA_{10} = \begin{pmatrix} d\bar{t} & 0 \\ 0 & dt \end{pmatrix}, \qquad dA_{11} = \begin{pmatrix} dv & du \\ -d\bar{u} & d\bar{v} \end{pmatrix},$$

$$d^cA_9 = \begin{pmatrix} -d\bar{w} & 0 \\ 0 & d\bar{w} \end{pmatrix}, \quad d^cA_{10} = \begin{pmatrix} -dt & 0 \\ 0 & -d\bar{t} \end{pmatrix}, \quad d^cA_{11} = \begin{pmatrix} -dv & d\bar{u} \\ -du & -d\bar{v} \end{pmatrix}.$$

The case of $\mathcal{H}_{2,6}$

The isometric embedding is (2.4). The possible quaters are

$$\begin{pmatrix} \bar{u} & \bar{v} \\ v & -u \end{pmatrix}, \quad \begin{pmatrix} -u & -\bar{v} \\ -v & -\bar{u} \end{pmatrix}, \quad \begin{pmatrix} t & w \\ -\bar{w} & -\bar{t} \end{pmatrix}, \quad \begin{pmatrix} \bar{t} & w \\ -\bar{w} & t \end{pmatrix}.$$

The corresponding basic 1-form Θ reads

$$\Theta = \begin{pmatrix} dA_4 & dA_5 & dA_7 & 0 \\ dA_5 & dA_3 & 0 & dA_8 \\ -dA_7 & 0 & dA_4 & dA_6 \\ 0 & -dA_8 & -dA_6 & dA_3 \end{pmatrix}.$$

The case of $\mathcal{H}_{4,4}$

The isometric embedding is (2.5). The possible quaters are

$$(A_4, A_3, A_3^{c'}, A_4^{c'}), \quad \text{where} \quad A_3^{c'} = \begin{pmatrix} -\bar{w} & t \\ -\bar{t} & -w \end{pmatrix}, \quad A_4^{c'} = \begin{pmatrix} -w & t \\ -t & -\bar{w} \end{pmatrix}.$$

The corresponding basic 1-form Θ reads

$$\Theta = \begin{pmatrix} d^cA_3 & 0 & dA_7 & -dA_5 \\ 0 & d^cA_3 & d^cA_6 & dA_8 \\ -dA_7 & -dA_6 & d^cA_4 & 0 \\ dA_5 & -dA_8 & 0 & d^cA_4 \end{pmatrix},$$

where

$$d^cA_3 = \begin{pmatrix} -d\bar{v} & d\bar{u} \\ du & dv \end{pmatrix}, \quad d^cA_4 = \begin{pmatrix} d\bar{v} & -du \\ -d\bar{u} & -dv \end{pmatrix}.$$

5 Conclusions concerning the pseudotwistors

In this section, we give the Penrose-like pseudotwistor theory for the hermitian Hurwitz pairs concerned. At first, owing to the Key Lemma, we can consider the quaternal structures of pseudotwistor spaces. Then, we can give the Penrose-like correspondence for pseudotwistor spaces of degrees 1 and 3. Finally, we give an analogue of Penrose's twistor theory for our pairs.

5.1 Quaternal structure of pseudotwistors

Consider the family of quaternal embeddings

$$\iota_{\mathbb{A}} : G(2,4) \hookrightarrow G(8,16),$$

which were introduced in Sect. 3.2. We shall prove:

Theorem 1.

(i) Let

$$\begin{array}{ccc} & \mathcal{J}^{(1)} & \\ \swarrow & & \searrow \\ \mathcal{J}^{(1)}_{-} & & \mathcal{J}^{(1)}_{+} \end{array}$$

be the Penrose diagram for pseudotwistors of degree 1. *Then we have the following quaternal structure: For any quaternal embedding* $\iota_{\mathbb{A}} : V_{\mathbb{A}} \hookrightarrow G(8,16)$, *we set*

$$\iota^*_{\mathbb{A}}\mathcal{J}^{(1)} = \mathcal{J}^{(1)}_{\mathbb{A}}, \quad \iota^*_{\mathbb{A}}\mathcal{J}^{(1)}_{+} = \mathcal{J}^{(1)}_{+\mathbb{A}}, \quad \textit{and} \quad \iota^*_{\mathbb{A}}\mathcal{J}^{(1)}_{-} = \mathcal{J}^{(1)}_{-\mathbb{A}},$$

where $\mathcal{J}^{(1)}_{+\mathbb{A}} = \left(\mathcal{J}^{(1)}_{+}\right)_{\mathbb{A}}$. *Then we have the Penrose diagram*

$$\begin{array}{ccc} & \mathcal{J}^{(1)}_{\mathbb{A}} & \\ \swarrow & & \searrow \\ \mathcal{J}^{(1)}_{-\mathbb{A}} & & \mathcal{J}^{(1)}_{+\mathbb{A}} \end{array}$$

(ii) Let

$$\begin{array}{ccc} & \mathcal{J}^{(3)} & \\ \swarrow & & \searrow \\ \mathcal{J}^{(3)}_{-} & & \mathcal{J}^{(3)}_{+} \end{array}$$

be the Penrose correspondence for pseudotwistors of degree 3. *Then we have*

$$\mathcal{J}^{(3)}_{-} \subseteqq \bigoplus_{\mathbb{A}} \mathcal{J}^{(1)}_{\mathbb{A}_-} \cdot \mathcal{J}^{(1)}_{\mathbb{A}^c_+}, \qquad \mathcal{J}^{(3)}_{+} \subseteqq \bigoplus_{\mathbb{A}} \mathcal{J}^{(2)}_{\mathbb{A}_+} \cdot \mathcal{J}^{(2)}_{\mathbb{A}^c_+}. \tag{5.1}$$

Proof. In order to check (i), we choose $(a_1, b_1, a_2, b_2) \subset (2, 3, \ldots, 9)$, where $2 \leq a_1 < b_1 < a_2 < b_2 \leq 9$, and denote it by $\mathbb{A} = (a_1, b_1, a_2, b_2)$. The inclusion mapping is denoted by $\iota_{\mathbb{A}} : G(2,4) \hookrightarrow G(8,16)$ (see Sect. 3.2). Since the condition to be an element of $\mathcal{J}^{(1)}$ is stated in terms of $\mathbb{A}$, we have a one-to-one correspondence: $\xi \in \mathcal{J}^{(1)}$ if and only if $\iota^*_{\mathbb{A}} \in \mathcal{J}^{(1)}(\sigma, s)$, $\sigma + s = 9$ for any $\mathbb{A} \subset (2, 3, \ldots, 9)$, where $\mathcal{J}^{(1)}(\sigma, s)$ is the pseudotwistor space for $\mathcal{H}_{\sigma,s}$.

In order to demonstrate (ii), at first we notice that

$$(2, 3, \ldots, 9) = (\mathbb{A}, \mathbb{A}^c).$$

From this and Theorem A (i.e., Theorem 1 in [22], more precisely, by the relation

$$: \overline{x_2 x_2}^{(4)} := \sum_{\alpha_j,\alpha'_j;\beta_j,\beta'_j\geq 2} \xi_{\alpha_1\beta_1\alpha_2\beta_2\alpha_3\beta_3}\xi_{\alpha'_1\beta'_1\alpha'_2\beta'_2\alpha'_3\beta'_3} x_{\alpha_1}x_{\beta_1}x_{\alpha'_i}x_{\beta'_1} = 0$$

deduced in Step B of its proof), we have the second identity of (5.1). Here $\overline{}^{(l)}$ denotes the l-time contraction, and underlining indicates the way of contraction. From the relation

$$: \overline{x_1 x_2}^{(4)} := \sum_{\alpha_j,\alpha'_j;\beta_j,\beta'_j\geq 2} \xi_{1\beta_1\alpha_2\beta_2\alpha_3\beta_3}\xi_{\alpha'_1\beta'_1\alpha'_2\beta'_2\alpha_3\beta'_3} x_{\beta_1}x_{\alpha'_1}x_{\beta'_1} = 0,$$

that we deduced in Step B of the same proof, we arrive at the first identity of (5.1). This suffices to conclude the proof. □

5.2 A look at the Penrose-like Hurwitz twistor theory for HHPs, $\sigma + s = 5$

First recall the Penrose-like theory for HHPs, $\sigma + s = 5$. We then have three cases

$$(\alpha)\ (\mathbb{C}^4(I_{2,2}), \mathbb{R}^5(I_{2,3})),\quad (\beta)\ (\mathbb{C}^4(I_{2,2}), \mathbb{R}^5(I_{1,4})),\quad (\gamma)\ (\mathbb{C}^4(I_{4,0}), \mathbb{R}^5(I_{5,0}))$$

which reduce to two cases because cases (α) and (β) are essentially the same.

The case of $(\mathbb{C}^4(I_{2,2}), \mathbb{R}^5(I_{2,3}))$

The Hurwitz algebras are $\mathcal{H}_{1,3}$ and $\mathcal{H}_{2,2}$. We treat the representation for $\mathcal{H}_{2,2}$ and denote the collection of Hurwitz twistors of degree 1 by $\mathcal{J}^{(1)}$:

$$\mathcal{J}^{(1)} = \{\xi = \sum \xi_{1\beta} S_\beta + \sum \xi_{\alpha\beta} S_\alpha S_\beta : \operatorname{im} \xi^2 = 0\}.$$

Then, the Penrose-like correspondence becomes

$$\begin{array}{ccc} & \mathcal{J}^{(1)} & \\ {}^{\nu}\swarrow & & \searrow^{\mu} \\ G(1,4) \supset \mathcal{J}^{(1)}_{-} & & \mathcal{J}^{(1)}_{+} \subset G(2,4) \end{array}$$

that is

$$\begin{array}{ccc} & \begin{pmatrix} u & v \\ \bar{v} & \bar{u} \end{pmatrix} \times [s,t] & \\ {}^{\nu}\swarrow & & \searrow^{\mu} \\ s\begin{pmatrix} u \\ \bar{v} \\ 1 \\ 0 \end{pmatrix} + t\begin{pmatrix} v \\ \bar{u} \\ 0 \\ 1 \end{pmatrix} & \underset{\rho}{\rightsquigarrow} & \begin{pmatrix} u & v \\ \bar{v} & \bar{u} \\ 1 & 0 \\ 0 & 1 \end{pmatrix} \end{array}$$

and also

$$\begin{array}{ccc} & \omega \in H^1_{d_\nu}(U, \nu^*(S_-^{\odot n} \otimes \det S_-) \otimes [e]^{-1}) & \\ & {}^{\nu}\swarrow \qquad\qquad\qquad \searrow^{\mu} & \\ H^1(U_-, \mathcal{O}(-n-2)) \ni \omega & \rightsquigarrow & \{\phi_{\dot{A}\dot{B}\ldots\dot{D}}\} \in Z^{(n)}(U_+, S_-^{\odot n} \otimes \det S_-). \end{array}$$

Explicitly,

$$\{\phi_{\dot{A}\dot{B}\ldots\dot{D}}\} \iff \omega = \{\omega_{(0)}, \omega_{(\infty)}\}, \tag{5.2}$$

where

$$\begin{aligned} \omega_{(0)} &= \phi_{\dot{0}\dot{B}\ldots\dot{D}} dz^{0,0} + \phi_{\dot{1}\dot{B}\ldots\dot{D}} dz^{0,1}, \\ \omega_{(\infty)} &= \phi_{\dot{0}\dot{B}\ldots\dot{D}} dz^{1,0} + \phi_{\dot{1}\dot{B}\ldots\dot{D}} dz^{1,1}, \end{aligned}$$

with $z^{0,0} = u, z^{0,1} = v, z^{1,0} = \bar{u}, z^{1,1} = \bar{v}$. The following lemma is essential in proving the correspondence (5.2).

Lemma A. *There exists a one-to-one correspondence between*

$$\mathcal{H}^1(U, (\mathcal{O} \oplus \mathcal{O})^{n-1}) \quad \textit{and} \quad Z^{(n)}(U, S^{\odot n} \times \det S_-).$$

Here,

$$\mathcal{H}^1(U, (\mathcal{O} \oplus \mathcal{O})^{n-1}) = \{\phi \in \Gamma^{1,0}(U, (\mathcal{O} \oplus \mathcal{O})^{n-1}) : \partial\phi = 0 \textit{ and } \nu\phi = 0\},$$

where ν is the formally adjoint operator of ∂ with respect to the indefinite metric $ds^2 = dz_1^2 - dz_2^2$.

Remark 1. *The explicit form of the equations is*

$$\frac{\partial\phi_{0,1}}{\partial z^{0,0}} - \frac{\partial\phi_{0,1}}{\partial z^{0,1}} = 0, \qquad \frac{\partial\phi_{0,1}}{\partial \bar{z}^{0,0}} - \frac{\partial\phi_{0,0}}{\partial \bar{z}^{0,1}} = 0,$$

where $z^{0,0} = u$, $z^{0,1} = v$.

The case of $(\mathbb{C}^4(I_{2,2}), \mathbb{R}^5(I_{1,4}))$

In this case, the Hurwitz algebras are $\mathcal{H}_{0,4}$ and $\mathcal{H}_{1,3}$. We shall treat only the case of $\mathcal{H}_{0,4}$. The desired correspondence is completely similar to the previous one except for only one replacement:

$$\begin{pmatrix} u & v \\ \bar{v} & \bar{u} \end{pmatrix} \Longrightarrow \begin{pmatrix} u & v \\ \bar{v} & -\bar{u} \end{pmatrix}.$$

The equations read

$$\frac{\partial\phi_{0,0}}{\partial u} - \frac{\partial\phi_{0,1}}{\partial v} = 0, \qquad \frac{\partial\phi_{0,0}}{\partial \bar{v}} + \frac{\partial\phi_{0,1}}{\partial \bar{u}} = 0.$$

The case of $(\mathbb{C}^4(I_{4,0}), \mathbb{R}^5(I_{5,0}))$

Here, the Hurwitz algebra is $\mathcal{H}_{4,0}$. The required correspondence is

$$\begin{pmatrix} u & v \\ \bar{v} & \bar{u} \end{pmatrix} \Longrightarrow \begin{pmatrix} u & v \\ \bar{v} & \bar{u} \end{pmatrix}.$$

The equations read

$$\frac{\partial \phi_{0,0}}{\partial u} - \frac{\partial \phi_{0,1}}{\partial v} = 0, \qquad \frac{\partial \phi_{0,0}}{\partial \bar{v}} + \frac{\partial \phi_{0,1}}{\partial \bar{u}} = 0.$$

5.3 The Penrose-like pseudotwistor theory for HHPs, $\sigma + s = 9$

First we write down the Penrose correspondence in the case of pseudotwistors of degree k, $k = 1, 3$.

The case of pseudotwistors of degree 1

The generators of $\mathcal{H}_{0,8}$ are denoted by $S_2, \dots, S_9$. Then, the collection of twistors of degree 1 is given by

$$\begin{aligned} \mathcal{J}^{(1)} &= \{\xi = \sum_{\alpha_1 \geq 2} \xi_{1\alpha_1} S_{\alpha_1} + \sum_{2 \leq \alpha_2, \beta_2 \leq 9} \xi_{\alpha_2\beta_2} S_{\alpha_2} S_{\beta_2} : \operatorname{im} \xi^2 = 0\} \\ &= \mathcal{J}_-^{(1)} + \mathcal{J}_+^{(1)}, \end{aligned}$$

and we have

$$\begin{array}{ccc} & \mathcal{J}^{(1)} & \\ \swarrow & & \searrow \\ \mathcal{J}_-^{(1)} & & \mathcal{J}_+^{(1)} \end{array}$$

By Theorem 1, we have the following so-called *quaternal Penrose-like correspondence*: For an embedding $\iota_{\mathbf{A}} : V_{\mathbf{A}} \to G(8, 16)$, we have

$$\begin{array}{ccc} & \mathcal{J}^{(1)} & \\ {}^{\nu_{\mathbf{A}}}\swarrow & & \searrow^{\mu_{\mathbf{A}}} \\ \mathcal{J}_{\mathbf{A}-}^{(1)} & & \mathcal{J}_{\mathbf{A}+}^{(1)} \end{array}$$

where $\mu^{(1)} = \{\mu_{\mathbf{A}}\}$ and $\nu^{(1)} = \{\nu_{\mathbf{A}}\}$.

Theorem 2. *Suppose that*

$$\begin{array}{ccc} & U & \\ & \swarrow \qquad \searrow & \\ G(1,8) \supset U_- & & U_+ \subset G(2,8) \end{array} \tag{5.3}$$

is the Penrose diagram. Let $\iota_{\mathbf{A}} : G(2,4) \hookrightarrow G(8,16)$ *be an arbitrary single quaternal embedding, and let*

$$\begin{array}{ccc} & U_{\mathbf{A}} & \\ & \swarrow \qquad \searrow & \\ G(1,4) \supset (U_-)_{\mathbf{A}} & & (U_+)_{\mathbf{A}} \subset G(2,4) \end{array} \tag{5.4}$$

be the corresponding restriction. Then,

(i) $\iota_{\mathbb{A}}^*\Phi \in \Gamma(\mathcal{U}_{\mathbb{A}})_+ \oplus (S_-^{\odot n} \otimes \det S_-)$ *becomes a spinor on* $U_{\mathbb{A}}$ *for a van der Waerden spinor of degree* 1 *on* U_+.

(ii) *There exists a one-to-one correspondence between the following spaces*

$$\rho : H_Q^1(U_-, \mathcal{F} \oplus \mathcal{F}) \to \Gamma_Q(U_+, \iota^*(S_-^{\odot n} \otimes \det S_-)),$$

i.e., for any single quater $\mathbb{A}$, *we have*

$$\rho_{\mathbb{A}} : H^1(((U_{\mathbb{A}})_-, \mathcal{O}(-n-2) \oplus \mathcal{O}(-n-2)) \cong \Gamma((U_{\mathbb{A}})_+, \iota_{\mathbb{A}}^*(S_-^{\odot n} \otimes \det S_-))$$

where $\mathcal{F}$ *is a complex line bundle satisfying*

$$\mathcal{F}_{|V_{\mathbb{A}}} \cong \mathcal{O}(-n-2)), \tag{5.5}$$

and the complex structure is defined by the complex variables in the upper (or lower) line of the quaters.

Remark 2. $\iota_{\mathbb{A}}^* \det S_- = \det(S_-)_{\mathbb{A}}$.

Remark 3. *Relation (5.5) does not imply that* $\mathcal{F}$ *is a holomorphic line bundle over* U_+.

Proof of Theorem 2. We take a van der Waerden spinor Φ of degree 1 and get ω from Φ. Then $\omega \in \mathbb{H}_Q^1(U_+)$. We choose $\iota_{\mathbb{A}} : \mathbb{H} \hookrightarrow G(8,16)$. Then we have

$$\begin{aligned} \omega_{\mathbb{A}} &= \iota_{\mathbb{A}}^*\omega, & \omega_1 &= \phi^{(1)}_{0\dot{B}\dots\dot{D}}du + \phi^{(2)}_{\dot{1}\dot{B}\dots\dot{D}}d\bar{v}, \\ \omega_{\mathbb{A}} &= (\omega_1 + \omega_2), & \omega_2 &= \phi^{(7)}_{\dot{0}\dot{B}\dots\dot{D}}du + \phi^{(8)}_{\dot{1}\dot{B}\dots\dot{D}}d\bar{v}, \end{aligned} \tag{5.6}$$

where $\mathbb{A} = A_4$, for example. We can also see that $d_{\mathbb{A}}\omega_{\mathbb{A}} = 0$ and $\vartheta_{\mathbb{A}}\omega_{\mathbb{A}} = 0$.

Here we use the Penrose correspondence. We set $\tilde{\omega}_{\mathbb{A}} = \{\omega_{\mathbb{A}}^{(0)}, \omega_{\mathbb{A}}^{(\infty)}\}$ on $(U_{\mathbb{A}})_+ \times \mathbb{P}^1$ as follows: $\omega_{\mathbb{A}}^{(0)} = \omega_{\mathbb{A}}$, $\omega_{\mathbb{A}}^{(\infty)}$, having the same form of $\omega_{\mathbb{A}}$ except for changes of variables $(u, \bar{v}) \mapsto (\bar{u}, v)$ in (5.6) in analogy to (5.2).

Then we have an $\tilde{\omega}_{\mathbb{A}}$ on $(U_{\mathbb{A}})_+ \times \mathbb{P}^1$. From the diagram

$$\begin{array}{ccc} & (U_{\mathbb{A}})_+ \times \mathbb{P} & \\ {}^{\nu}\swarrow & & \searrow^{\mu} \\ (U_{\mathbb{A}})_- & & (U_{\mathbb{A}})_+, \end{array}$$

we have the related de Rham form $\tilde{\omega}$, i.e., $(d_{\mathbb{A}})_\nu \tilde{\omega}_{\mathbb{A}} = 0$, where d_ν is the exterior derivation along the fibre ϑ. Hence, we can see that there exists an $(\omega_-)_{\mathbb{A}} \in \Gamma((U_{\mathbb{A}})_-, \Lambda^{*1})$.

Here we notice that, from the construction of $(\omega_-)_{\mathbb{A}}$ yielded by the relation (5.2), $\{(\omega_-)_{\mathbb{A}}\}$ becomes a global relative one-form on U_-, with respect to the mapping ν. Therefore, finally, we obtain

$$\omega_- \in H_Q^1(U_-, \mathcal{F} \oplus \mathcal{F}),$$

as desired.

The converse direction of the procedure is also possible, thus completing the proof. □

The case of pseudotwistors of degree 3

By use of Theorem 1 we have the following Penrose-like correspondence:

$$\begin{array}{ccc} & \mathcal{J}^{(3)} & \\ {}^{\nu^{(3)}}\swarrow & & \searrow^{\mu^{(3)}} \\ \mathcal{J}_-^{(3)} & & \mathcal{J}_+^{(3)} \end{array}$$

where $\nu^{(3)} = \oplus_{\mathbb{A}} \nu_{\mathbb{A}}^{(1)} \mu_{\mathbb{A}^c}^{(2)}$ and $\mu^{(3)} = \oplus_{\mathbb{A}} \mu_{\mathbb{A}}^{(2)} \mu_{\mathbb{A}_c}^{(2)}$. Here we have followed the notation in (5.1). We can prove

Theorem 3. *Suppose that*

$$\begin{array}{ccc} & U & \\ & \swarrow \qquad \searrow & \\ G(1,4)\times G(2,4) \supset U_- & & U_+ \subset G(2,4)\times G(2,4) \end{array} \qquad (5.7)$$

is the Penrose diagram. Let $\iota_{\mathbb{A}^*} : G(2,4)\times G(2,4) \to G(8,16)$, *where* $\mathbb{A}^* = \mathbb{A}\cup\mathbb{A}^c$, *be an arbitrary double quaternal embedding. Then,*

(i) $\iota_{\mathbb{A}^*}^* \Phi \in \Gamma(\mathcal{U}_+, \iota_{\mathbb{A}^*}^*(S_-^{\odot n} \otimes \det S_-))$ *becomes a spinor on* $U_{\mathbb{A}^*}$ *of class* $(0,8)$ *for a van der Waerden spinor* Φ *of degree 2.*

(ii) There exists a one-to-one correspondence between the following spaces

$$\rho_{\mathbb{A}^*} : H_Q^{1,3}(U_-, \mathcal{F}\oplus\mathcal{F}) \cong \Gamma_Q(U_+, \iota_{\mathbb{A}^*}^*(S_-^{\odot n} \otimes \det S_-)),$$

where $\mathcal{F}$ *is given in (5.5).*

Proof. Verification of (i) is the direct consequence of (3.5) and related equalities and may be omitted.

In order to demonstrate (ii), we choose a van der Waerden spinor Φ and consider $\iota_{\mathbb{A}^*}^* \Phi$ $(= \Phi')$. Then, by Examples 12 and 15, we have a harmonic $(1,3)$-form $\Omega = \Omega^{(1)} + \Omega^{(3)}$. Here, we use the Penrose transform of (5.3). As in the case of degree 1, we can construct the corresponding Penrose-like transform:

$$\begin{array}{ccc} & \tilde{\Omega}^{(1)} & \\ {}^{\nu}\swarrow & & \searrow^{\mu} \\ \Omega_-^{(1)} & & \Omega_+^{(1)} \end{array} \qquad \Omega_+^{(1)} = \Omega^{(1)}.$$

Next, we consider $\Omega^{(2)}$. By construction, we have $\Omega^{(3)} = \Omega^{(1)'} \wedge \Pi^{(2)}$, where $\Omega^{(1)'}$ is an $\mathbb{A}\cup\mathbb{A}_c$-quaternal 1-form and $\Pi^{(2)}$ is the basic 2-form. We can construct the Penrose transform of $\Omega^{(1)'}$ as in (5.3):

$$\begin{array}{ccc} & \tilde{\Omega}^{(1)'} \wedge \Pi^{(2)} & \\ \swarrow & & \searrow \\ \Omega_-^{(1)'} \wedge \Pi^{(2)} & & \Omega_+^{(1)'} \wedge \Pi^{(2)}, \end{array}$$

and, hence, we get the desired correspondence. □

6 The double Cartan-like triality and an introduction to five-dimensional stochastical electrodynamics

The well-known E. Cartan's triality principle [2, 3, 28, 26] relates six distinct continuous families of linear substitutions, composing the group of the geometry of eight-dimensional Euclidean space about a point. These families correspond to six possible permutations of three classes of objects: vectors and semi-spinors of two types; the principle can be formulated in terms of rotation by the angles $\frac{2}{3}\pi$, $\frac{4}{3}\pi$, and 2π.

Lounesto [26] deduced from this that the triality automorphism of the group **Spin**(8) is a restriction of a polynomial mapping of second degree from the Clifford algebra of the Euclidean space $\mathbb{R}^8$ into itself, actually a product of two affine linear mappings trial $u = \text{trial}_1 u \ \text{trial}_2 u$, where

$$\text{trial}_1 u = \frac{1}{2}(1+e_{12\ldots 8})\langle\{[(e_8 u e_8^{-1}(1+e_{12\ldots 8}))\wedge e_8]e_8^{-1}\}(1+\mathbf{w})\rangle_{0,6},$$
$$\text{trial}_2 u = (3-\mathbf{w}[(u(1+e_{12\ldots 8}))\wedge e_8]e_8^{-1}(3-\mathbf{w})^{-1},$$

with $\mathbf{w} = \mathbf{v}e_{12\ldots 7}^{-1}$, $\mathbf{v} = e_{124}+e_{235}+e_{346}+e_{457}+e_{561}+e_{672}+e_{713}$ and e_8 as a real axis. This triality corresponds to the octonion product

$$\mathbf{x}\circ\mathbf{y} = (\mathbf{x}e_8\mathbf{y})(1+\mathbf{w})(1-e_{12\ldots 8}).$$

It seems that the mutual transformations of the physical systems corresponding to twistors, pseudotwistors, and bitwistors – as well as to their Kałuża-Klein-type duals in K_1, K^1 and K_1^1 [18, 19] – give a required physical motivation for the double Cartan-like triality of hermitian Hurwitz pairs; cf. Figs. 1 and 2.

By analogy with the quaternionic description of the classical electrodynamics [9, 10] – in the case of dynamical systems generated by the hermitian Hurwitz pairs of signatures $(3,2)$ and $(1,4)$ – we may construct the five-dimensional vector potential (A_j), $j=0,1,\ldots,4$, whose components allow us to consider the influence of forces on the particle in question. In order to describe their behaviour, we have to look for variational equations with respect to the particle trajectory as well as to the field components. At first, we introduce the field tensor

$$F_{jk} = (\partial/\partial x_j)A_k - (\partial/\partial x_k)A_j, \quad j,k=0,1,\ldots,4$$

and the invariant with respect to the field in the form $\Lambda = -\frac{1}{16\pi}F_{jk}^2$. In this case, the Lagrangian describing the motion of the particle embedded in the field is given by

$$L = -mc^2[1-(\frac{1}{c}v)^2-\eta(\frac{1}{c}v_t^\tau)^2]^{\frac{1}{2}} + \frac{e}{c}\sum_{j=0}^{4}A_j v^j - \Lambda,$$

where $\eta = 1$ (K_1-case) or $\eta = -1$ (P^1-case). The first term in the above corresponds to the free particle movement, and ν_t^T expresses the velocity of changes of the entropy in time. A variational procedure applied to the particle velocity leads to the system of equations

$$m\frac{d}{dt}\{v^j/[1-(\frac{1}{c}v)^2-\eta(\frac{1}{c}v_t^T)^2]^{\frac{1}{2}}\}-\frac{e}{c}\sum_{k=0}^{4}F_{jk}v^k=0. \tag{6.1}$$

An analogous procedure applied to the field components gives the system

$$\sum_{k=0}^{4}(\partial/\partial x_k)F_{jk}=4\pi\frac{e}{c}v_j,\text{ with}$$
$$(\partial/\partial x_\ell)F_{jk}+(\partial/\partial x_j)F_{k\ell}+(\partial/\partial x_k)F_{\ell j}=0. \tag{6.2}$$

In a parallel paper [17], the equations of motion (6.1) are considered in three particular cases: (M1) the velocities v^j, $j = 1, 2, 3$, correspond to one of three components in the three-dimensional configuration space; (M2) the velocity v^0 corresponds to its component considered as the velocity with respect to time, i.e., $v^0 = ic$ following from the usual interpretation of the four-dimensional Lorentz transformation; (M3) the velocity v^4 corresponds to a new parameter which formally can be considered as the velocity component, but in reality is connected with the dynamical behaviour in the stochastic dimension [7, 25], i.e., with the entropy: $v^4 = v_t^T$ in the K_1-case and $v^4 = iv_t^T$ in the P^1-case. Analogously, the field equations (6) can be considered in two cases:

(F1) the velocities v^j, $j = 0, 1, 2, 3$ are generators of the field components, including their stochastic counterparts,

(F2) the velocity v^4, whose physical nature should be the core subject of investigation, brings an additional source of the field; and, then, the equation (6) for $j = 4$ gives an additional condition for determination of the fifth component of the vector potential.

References

[1] V. M. Agranovich, *Teoriya èksitonov,* Izd. Nauka, Moscow, 1968.

[2] E. Cartan, Le principe de dualité et la théorie des groupes simples et semi-simples, *Bull. Sci. Math. France* **49** (1925), 367–374.

[3] —: *The Theory of Spinors, Second Edition,* Hermann, Paris, 1966, 120.

[4] E. I. Dinaburg, On the relations among various entropy characteristics of dynamical systems, in *Dynamical Systems,* Ya. G. Sina ĭ, ed., *Advanced Series in Nonlinear Dynamics 1,* Amer. Math. Soc., New York, 1972, 337–350.

[5] E. Fermi, *Thermodynamics, Second Edition,* Dover Publ., New York, 1956, 46–76.

[6] I. Furuoya, S. Kanemaki, J. Ławrynowicz, and O. Suzuki, Hermitian Hurwitz pairs, in *Deformations of Mathematical Structures II. Hurwitz-Type Structures and Applications to Surface Physics,* J. Ławrynowicz, Kluwer Academic, Dordrecht, 1994, 135–154.

[7] B. Gaveau, J. Ławrynowicz, and L. Wojtczak, Statistical mechanics of anharmonic crystals, *Physica Status Solidi (b)* **121** (1984), 47–58.

[8] H. Grauert, Discrete geometry, *Nachrichten Akad. Wiss. Göttingen* No. **6** (1996), 343–362.

[9] Sir W. R. Hamilton, *Lectures on Quaternions,* Dublin, 1853, and *The Mathematical Papers of Sir William Rowan Hamilton,* Vol. III, *Algebra,* Cambridge Univ. Press, London, 1967, 106–110.

[10] K. Imaeda, *Quaternionic Formulation of Classical Electrodynamics and Theory of Functions of a Biquaternion Variable,* Okayama Univ. of Science, Okayama, 1983.

[11] S. Kanemaki and O. Suzuki, Hermitian pre-Hurwitz pairs and the Minkowski space, in *Deformations of Mathematical Structures. Complex Analysis with Physical Applications,* J. Ławrynowicz, Kluwer Academic, Dordrecht, 1989, 225–232.

[12] J. Ławrynowicz, Type-changing transformations of pseudo-euclidean Hurwitz pairs, Clifford analysis, and particle lifetimes, in *Clifford Algebras and Their Applications in Mathematical Physics,* V. Dietrich, K. Habetha, and G. Jank, eds., Kluwer Academic, Dordrecht, 1998, 217–226.

[13] —, J. Rembieliński, Pseudo-euclidean Hurwitz pairs and generalized Fueter equations, (a) *Inst. of Math. Polish Acad. Sci.,* preprint No. **355** (1985), ii + 10, (b) in *Clifford Algebras and Their Applications in Mathematical Physics, Proceedings,* J. S. R. Chisholm and A. K. Common, eds., Canterbury, 1985, *NATO-ASI Series C: Mathematical and Physical Sciences 183, Reidel,* Dordrecht, 1986, 39–48.

[14] —, —: Complete classification for pseudoeuclidean Hurwitz pairs including the symmetry operations, (a) *Inst. of Phys. Univ. of Łódź,* preprint No. **86-5** (1986), 15, (b) *Bull. Soc. Sci. Lettres Łódź* **36** No. **29**, *Sér. Rech. Déform.* **4** No. **39** (1986), 15pp.

[15] —, —, Pseudo-euclidean Hurwitz pairs and the Kałuża-Klein theories, (a) *Inst. of Phys. Univ. of Łódź,* preprint, No. **86-8** (1986), 28 pp., (b) *J. Phys. A: Math. Gen.* **20** (1987), 5831–5848.

[16] —, —, On the composition of nondegenerate quadratic forms with an arbitrary index, (a) *Inst. of Math. Polish Acad. Sci.,* preprint No. **369** (1986), ii + 29 pp., (b) *Ann. Fac. Sci. Toulouse Math.* (2.2) **10** (1989), 141–168 [due to a printing error in Vol. **10**, the whole article was reprinted in Vol. **11** (1990), No. **1** of the same journal, 141–168].

[17] —, W. A. Rodrigues and L. Wojtczak, Stochastical electrodynamics in Clifford-analytical formulation related to entropy-depending structures, *Rep. Math. Phys.* **46** (2000), to appear.

[18] J. Ławrynowicz and O. Suzuki, (a) *Inst. of Math. Polish Acad. Sci.,* preprint, No. **569** (1997), ii + 30 pp., (b) *Advances Appl. Clifford Algebras* **8** (1998), 147–179.

[19] —, —, An approach to the 5-, 9-, and 13-dimensional complex dynamics I. Dynamical aspects, *Bull. Soc. Sci. Lettres Łódź* **48** *Sér. Rech. Déform.* **25** (1998), 7–39.

[20] —, —, An approach to the 5-, 9-, and 13-dimensional complex dynamics II. Twistor aspects, *ibid.* **48** *Sér. Rech. Déform.* **26** (1998), 23–48.

[21] —, —, Hurwitz-type and space-time-type duality theorems for hermitian Hurwitz pairs, in *Complex Analysis and Its Applications,* E. Ramírez de Arellano, M. Shapiro, L. M. Tovar, and N. Vasilevski, eds., Birkhäuser, Basel, 1999, 201–217.

[22] —, —, An introduction to pseudotwistors - basic constructions, in *Quaternionic Structures in Mathematical Physics,* S. Marchiafava and M. Pontecorvo, eds., Rome, 2000, to appear.

[23] —, —, An introduction to pseudotwistors – spinor equations, *Rev. Roumaine Math. Pures Appl.,* to appear.

[24] —, —, An introduction to pseudotwistors - spinor equations and atomization on isometric embeddings, *Internat. J. Theor. Phys.* **39** (2000), to appear.

[25] —, L. Wojtczak in cooperation with S. Koshi and O. Suzuki, Stochastical mechanics of particle systems in Clifford–analytical formulation related to Hurwitz pairs of bidimension (8, 5), in *Deformations of Mathematical Structures II. Hurwitz-Type Structures and Applications to Surface Physics,* J. Ławrynowicz, ed., Kluwer Academic, Dordrecht, 1994, 213–262.

[26] P. Lounesto, *Clifford Algebras and Spinors,* (London Math. Soc. Lecture Notes Series 239), Cambridge Univ. Press, Cambridge 1997, viii.

[27] R. Penrose, The twistor program , *Rep. Math. Phys.* **12** (1977), 65–76.

[28] I. R. Porteous, *Clifford Analysis and the Classical Groups,* (Cambridge Studies in Advanced Mathematics 50), Cambridge Univ. Press, Cambridge, 1995.

[29] A. Trautman, Fibre bundles associated with space-time, *Rep. Math. Phys.* **1** (1970), 29–62.

[30] R. O. Wells, Jr., *Complex Geometry in Mathematical Physics,* Séminaire de Mathématiques supérieures — Université de Montréal, Montréal, 1982.

Julian Ławrynowicz
Institute of Mathematics PAN
ul. Banacha 22
PL-90-238 Łódź, Poland

Chair of Solid State Physics
University of Łódź
ul. Pomorska 149/153
PL-90-236 Łódź, Poland
E-mail: jlawryno@krysia.uni.lodz.pl

Osamu Suzuki
Department of Mathematics
Nihon University
Sakurajosui 3-25-40, Setagaya-ku, Tokyo 156, Japan
E-mail: osuzuki@am.chs.nihon-u.ac.jp

Received: September 3, 1999; Revised: March 6, 2000

Ordinary Differential Equation: Symmetries and Last Multiplier

Zbigniew Oziewicz and José Ricardo R. Zeni

ABSTRACT This paper shows how to get a last multiplier for a differential n-form equivalent to an ordinary differential equation (ODE) in $(n+1)$-dimensions. The last multiplier makes it possible to find a closed differential n-form given by the ODE. Our construction is based on a set of n-symmetry vector fields admitted by an ODE. Our result is a generalization of Lie's theorem on integrating a factor for an ODE in two dimensions which admits a one-symmetry vector field.
Keywords: Exterior differential systems, Pffafian systems, Lie analysis, symmetry, last multiplier, integrating factor.

1 Introduction

The goal of our research is to set up a method to integrate a system of ordinary differential equations (ODE). This method is based on the symmetries admitted by the ODE. This paper presents a method to get a last multiplier for a differential n-form describing an $(n+1)$-dimensional ODE, that is, a method to obtain a closed n-form describing an ODE. This result generalizes a theorem by Lie who found a relation between a symmetry vector field and an integrating factor for an ODE in two dimensions.

For simplicity, we consider here only the scalar-valued Cartan exterior differential systems [Cartan 1946], i.e., the Cartan exterior differential systems on the manifolds without fibration. Matsyuk [1997, 1999] investigated symmetries of the vector-valued exterior differential systems on fibered manifolds and the related inverse problem. Note that the completely new inverse problem for the Newton law (second order) was posed by Gusiew-Czudżak [1999].

Zbigniew Oziewicz is supported by el Consejo Nacional de Ciencia y Tecnología (CONA-CyT) de México, Grant # 27670 E (1999–2000) and by UNAM, DGAPA, Programa de Apoyo a Proyectos de Investigación e Innovación Tecnológica, Proyecto IN-109599 (1999–2002). Zbigniew Oziewicz is a member of Sistema Nacional de Investigadores de México, Expediente # 15337.
AMS Subject Classification: 34A26, 22E60, 54H15, 53B25.

1.1 Notation and Cartan formalism

$\mathbb{R}$	the field of real numbers
$\mathcal{F}$	is an associative, unital and commutative $\mathbb{R}$-algebra
$\mathcal{X}$	is a finite dimensional $\mathcal{F}$-module of the differential one-forms
der $\mathcal{F}$	is a finite dimensional $\mathcal{F}$-module of the vector fields
$\mathcal{X}^*$	$\equiv \mathbf{mod}_{\mathcal{F}}(\mathcal{X}, \mathcal{F})$ is the $\mathcal{F}$-dual $\mathcal{F}$-module $\dim_{\mathcal{F}} \mathcal{X} = \dim_{\mathcal{F}} \mathcal{X}^* = n+1$
$\Lambda\mathcal{X}$ and $\Lambda(\mathcal{X}^*)$	the exterior $\mathcal{F}$-modules generated by $\mathcal{X}$ (and by $\mathcal{X}^*$)

The following definition uses Cartan formalism of exterior differential systems [Cartan 1946, Choquet-Bruhat 1968, Griffiths 1983].

Definition 1. *An ODE [in $(n+1)$-dimensions] is equivalent to any of the following data:*

i) A codimension one distribution $\mathcal{D} \subset \mathcal{X}$ of the differential one-forms,

ii) A one-dimensional distribution $\Gamma \subset \Lambda^n(\mathcal{X})$ of the differential forms of cograde $\Gamma = 1$,

iii) A one-dimensional distribution $\mathcal{D}^* \subset \mathcal{X}^* \approx der\mathcal{F}$ of the vector fields,

iv) A system of the first order differential equations, for $i = 1, \ldots, n$, $\dot{x}_i = v_i(t, x_i)$, $\dot{t} = 1$.

A distribution (i.e., an $\mathcal{F}$-module) $\mathcal{D}^*$ annihilates $\mathcal{D}$, $\mathcal{D}^*\mathcal{D} \equiv 0$. If $w_i \in \mathcal{D}$, than $\mathcal{D}^* = \cap_i \ker w_i$. The distributions Γ and $\mathcal{D}^*$ are related through the Poincaré isomorphism $\Lambda^{n+1-k}(\mathcal{X}) \approx \Lambda^k(\mathcal{X}^*)$, see Lemma 1 for more details. In what follows, a vector field $A \in \mathcal{D}^* \subset \text{der}\,\mathcal{F}$ generates $\mathcal{D}$, and $\{w_i \in \mathcal{X}, i = 1, 2, \cdots, n\}$ denotes a set of $\mathcal{F}$-linearly independent differential one-forms $0 \neq \Omega \equiv w_1 \wedge w_2 \wedge \cdots \wedge w_n \in \Gamma \subset \Lambda^n(\mathcal{D}) \subset \Lambda^n(\mathcal{X})$. A distribution $\mathcal{D}$ is generated by $\{w_1, w_2, \cdots, w_n\}$, and a distribution Γ is generated by Ω. Let $d \in \text{der}\,(\mathcal{F}, \mathcal{X})$ be a derivation. If an algebra $\mathcal{F}$ is commutative, then there exists only one contravariant additive endofunctor on $\mathcal{F}$-modules, $^* \equiv \mathbf{mod}_{\mathcal{F}}(-, \mathcal{F})$,

$$\begin{array}{ccc} \mathcal{F} & \xrightarrow{\;d\;} & \mathcal{X} \\ {\scriptstyle der}\big\downarrow & & \big\downarrow{\scriptstyle *} \\ \text{der}\,\mathcal{F} & \xleftarrow{\;d^*\;} & \mathcal{X}^* \end{array}$$

Here d^* gives an $\mathcal{F}$-linear isomorphism of $\mathcal{F}$-modules with $i \circ d^* = \text{id}_{\mathcal{X}^*}$ and $d^* \circ i = \text{id}_{\text{der}\,\mathcal{F}}$. A derivation $d \in \text{der}\,(\mathcal{F}, \mathcal{X})$ extends to the Cartan exterior graded (skew) derivation $d \in \text{der}\,(\Lambda\mathcal{X})$ with $d^2 = 0$ and grade 1. Then an $\mathcal{F}$-linear isomorphism

$$i = (d^*)^{-1} : \text{der}\,\mathcal{F} \ni S \longmapsto i_S \in \mathcal{X}^*$$

extends to graded derivation (also called a contraction)

$$\forall\, S, X, Y \in \operatorname{der} F, \qquad i_S \in \operatorname{der} \Lambda(\mathcal{X}), \qquad \operatorname{grade}(i_S) = -\operatorname{grade} S,$$
$$i_X \circ i_Y = -i_Y \circ i_X.$$

If $S \in \operatorname{der}\mathcal{F}$, then a Lie derivative along S is a derivation (of the entire tensor algebra of the tensor fields) of a zero grade

$$\pounds_S \in \operatorname{der} \Lambda\mathcal{X} \quad \text{and} \quad \pounds_S \in \operatorname{der} \Lambda\mathcal{X}^*, \quad \operatorname{grade} \pounds_S = 1 - \operatorname{grade} S,$$
$$\pounds_S|\Lambda(\mathcal{X}) = d \circ i_S + i_S \circ d, \qquad \forall\, X \in \operatorname{der}\mathcal{F}, \quad \pounds_S X \equiv [S, X]. \tag{1.1}$$

If $f \in \mathcal{F}$ and $X \in \operatorname{der}\mathcal{F}$, then

$$\pounds_X f \;=\; i_X\, d f = (i_X \circ d) f = (d^* \circ i)_X f = \; X f.$$

If $X, Y \in \operatorname{der}\mathcal{F}$ and $\operatorname{grade}\alpha = k$, then the following Leibniz rules hold:

$$\pounds_X \circ i_Y = i_{[X,Y]} + i_Y \circ \pounds_X,$$
$$i_X\,(\alpha \wedge \beta) = (i_X\, \alpha) \wedge \beta + (-1)^k\, \alpha \wedge i_X\, \beta, \quad \operatorname{grade}\alpha = k.$$

If an $\mathbb{R}$-algebra $\mathcal{F}$ is not necessarily commutative, then $\mathcal{X}$ must be a $\mathcal{F}$-bimodule of the differential one-forms, where for $f \in \mathcal{F}$ and $w \in \mathcal{X}$ $fw \neq wf$. This bimodule structure of $\mathcal{X}$ is reduced or commutative algebra $\mathcal{F}$ to $\mathcal{F}$-module. Moreover $\mathcal{F}$-bimodule $\mathcal{X}$ possesses two distinct $\mathcal{F}$-dual $\mathcal{F}$-bimodules, the right $\mathcal{F}$-dual ${}^*\mathcal{X}$ and the left $\mathcal{F}$-dual $\mathcal{X}^*$. Therefore, there are two kinds of partial "derivations," $\operatorname{im} d^*$ and $\operatorname{im} {}^*d$. An extension of the Cartan formalism for not necessarily commutative $\mathbb{R}$-algebra $\mathcal{F}$ is given by Borowiec [1996, 1997, 2000].

2 Last multiplier

A divergence of a vector field depends on a choice of a volume differential form.

Definition 2 (Divergence). *The divergence of a vector field $A \in \operatorname{der}\mathcal{F}$, with respect to a volume differential form $V \in \Lambda^{n+1}\mathcal{X}$, is a scalar field $\operatorname{div}_V A \in \mathcal{F}$ defined by*

$$\pounds_A V = (\operatorname{div}_V A)\, V, \quad \forall\, f \in \mathcal{F}, \quad df \wedge V \equiv 0.$$

Lemma 1. *Let $\dim_{\mathcal{F}} \mathcal{X} = n + 1$ and $V = w_0 \wedge \Omega \in \Lambda^{n+1}\mathcal{X}$, and let $A \in \operatorname{der}\mathcal{F}$ be such that $i_A\,\Omega = 0$ and $i_A\, w_0 = 1$. Then*

(i) $\quad i_A V = \Omega, \quad \pounds_A V = d\Omega \quad$ *and* $\quad \pounds_A \Omega = (\operatorname{div}_V A)\, \Omega$

(ii) $\quad \forall\, f \in \mathcal{F}, \quad (d f) \wedge \Omega = (A f)\, V.$

Proof. We will prove (ii). Since $df \wedge V = 0$, it follows that

$$0 = i_A (df \wedge V) = (i_A df)\, V - df \wedge i_A V. \quad \square$$

Recall the definition of a last multiplier [Flanders 1963].

Definition 3 (Last multiplier). *Let an ODE be given by a one-dimensional distribution $\Gamma \subset \Lambda^n \mathcal{X}$ generated by n-form $\Omega \in \Gamma$. An element $m \in \mathcal{F}$ is called a last multiplier for an ODE if*

$$d(m\,\Omega) = (dm) \wedge \Omega + m\, d\Omega = 0.$$

Proposition 1. *Let $A \in \operatorname{der} \mathcal{F}$ be a vector field and let V be a volume form such that $i_A V = \Omega$. Moreover let an ODE be given either by a one-dimensional distribution $\Gamma \subset \Lambda^n(\mathcal{X})$ generated by a n-form Ω or by a one-dimensional distribution $\mathcal{D} \subset \mathcal{X}^*$, this latter generated by a vector field $A \in \operatorname{der} \mathcal{F}$. If $m \in \mathcal{F}$ is such that*

$$A\,m + m \operatorname{div}_V A = 0,$$

then $m \in \mathcal{F}$ is a last multiplier for an ODE.

Proof. This is a consequence of Definitions 2 and 3. $\square$

Proposition 2. *Let $0 \neq h \in \mathcal{F}$ and $A \in \operatorname{der} \mathcal{F}$, which satisfies $A\,h = h \operatorname{div}_V A$. Then $m = h^{-1} \in \mathcal{F}$ is a last multiplier for the ODE given by a vector field $A \in \operatorname{der} \mathcal{F}$.*

3 Lie theorem

The notion of symmetry of a differential equation has different realizations according to the different descriptions that we can give to the differential equation [Barut, Zeni & Abdullah 1994, Barut & Zeni 2000].

Definition 4 (Symmetry I). *A vector field $S \in$ der $\mathcal{F}$ is called a symmetry of an ODE given by a vector field $A \in$ der $\mathcal{F}$ if*

$$\pounds_S A \equiv [S, A] = \lambda A, \quad \text{for some} \quad \lambda \in \mathcal{F}.$$

In the following definition an ODE is given by an $\mathcal{F}$-distribution generated by 1-forms, see Definition 1.

Definition 5 (Symmetry II). *A vector field S is called a symmetry of a k-dimensional distribution generated by 1-forms $\{w_1, w_2, \ldots, w_k\}$, and we write $S \in Sym\{w_1, w_2, \ldots, w_k\}$ if*

$$\pounds_S w_i = \sum_j \lambda_i^j w_j, \qquad \lambda_i^j \in \mathcal{F}.$$

Definitions 4 and 5 are equivalent if they refer to the same ODE, that is if $A \in \cap_i \ker w_i$.

Corollary 1. *If* $S \in \mathrm{Sym}\{w_1, w_2, \cdots, w_k\}$ *and* $\Omega \equiv w_1 \wedge w_2 \wedge \ldots \wedge w_k$, *then*

$$\pounds_S \Omega = \phi \Omega, \quad \textit{for some} \quad \phi \in \mathcal{F}.$$

The functions λ, λ_i^j and ϕ entering Definitions 4, 5 and Corollary 1 are not arbitrary but related to each other. It is instructive to review the Lie theorem on integrating factor for a two- dimensional ODE which admits a symmetry. In this case, the notion of last multiplier and integrating factor are identical since we are concerned with a differential one-form.

Theorem 1 (Lie). *Let* $\mathcal{X}$ *and* $\mathcal{X}^*$ *be two-dimensional* $\mathcal{F}$*-modules. Let* $\mathcal{D} \subset \mathcal{X}$ *and* $\mathcal{D}^* \subset \mathcal{X}^*$ *be one-dimensional distributions generated by a differential one-form* $w \in \mathcal{X}$ *and by a vector field* $A \in \mathrm{der}\,\mathcal{F}$, *respectively, such that* $\mathcal{D}^*$ *annihilates* $\mathcal{D}$. *If* $S \in \mathrm{der}\,\mathcal{F}$ *is a symmetry vector field admitted by* A *(Definition 4), such that* $S \wedge A \neq 0$, *then* $h^{-1} = (i_S w)^{-1} \in \mathcal{F}$ *is an integrating factor for a one-form* $w \in \mathcal{X}$.

The Lie Theorem says that $h^{-1} w = (i_S w)^{-1} w$ is a closed differential form $d(h^{-1} w) = 0$, $S \wedge A \neq 0 \Leftrightarrow i_S w \neq 0$.

Proof. According to Proposition 2, it is enough to prove that $\pounds_A h = h \,\mathrm{div}_V A$, where $V \in \Lambda^2(\mathcal{X}^*)$ is a volume two-form chosen in such a way that Lemma 1 is satisfied. We have

$$\begin{aligned} \pounds_A h = \pounds_A i_S w &= \left(i_{[A,S]} + i_S \pounds_A\right) w \\ &= \lambda i_A w + (\mathrm{div}_w A) i_S w, \end{aligned}$$

where to get the second line we make use of the fact that S is a symmetry vector admitted by A, see Definition 4. Also, we have used Definition 2 that $\pounds_A w = (\mathrm{div}_w A) w$. Since by the hypothesis $i_A w = 0$, the proof is complete. □

4 The main theorem

In this section, the Lie theorem is generalized to any dimension. Hereafter, for $S_j \in \mathrm{der}\,\mathcal{F}$ we denote $i_j \equiv i_{S_j}$.

Main Theorem. *Let* $\dim_{\mathcal{F}} \mathcal{X} = \dim_{\mathcal{F}} \mathcal{X}^* = n+1$. *Consider one-dimensional distributions* $\mathcal{D}^* \subset \mathcal{X}^*$ *and* $\Gamma \subset \Lambda^n(\mathcal{X})$, *generated by a vector field* A *and a differential* n*-form* Ω, *such that for some volume form* $V \in \Lambda^{n+1}(\mathcal{X})$ *(cf. Lemma 1),*

$$i_A \Omega = 0, \qquad \textit{and} \qquad i_A V = \Omega.$$

Let $\{S_1, S_2, \dots, S_n\}$ be a set of the symmetry vector fields admitted by the vector field $A \in \operatorname{der}\mathcal{F}$, such that

$$A \wedge S_1 \wedge S_2 \wedge \cdots \wedge S_n \neq 0, \quad h \equiv i_n\, i_{n-1} \cdots i_2\, i_1\, \Omega \quad \in \mathcal{F}.$$

Then $m = h^{-1} \in \mathcal{F}$ is a last multiplier for Ω, that is $d\,(m\,\Omega) = 0$.

Proof. According to Proposition 2, it is enough to prove that

$$\pounds_A\, h = h\, \mathrm{div}_V A. \tag{4.1}$$

For the symmetry conditions, given in Definition 4, set

$$[S_j, A] = \lambda_j\, A, \qquad (\lambda_j \in \mathcal{F}, \quad j = 1, 2, \cdots, n). \tag{4.2}$$

Now, in the expression for $\pounds_A\, h$, we are going to commute the Lie derivative along vector A with the contraction with each vector S_j as it follows that

$$\begin{aligned} \pounds_A\, h &= \pounds_A\, i_n\, i_{n-1} \cdots i_2\, i_1\, \Omega \\ &= (i_{[A,S_n]} + i_n\, \pounds_A)\, i_{n-1} \cdots i_2\, i_1\, \Omega. \end{aligned} \tag{4.3}$$

The first term in the last row is zero since from symmetry we have

$$i_{[A,S_n]}\, i_{n-1} \cdots i_2\, i_1\, \Omega = \lambda_n\, i_A\, i_{n-1} \cdots i_2\, i_1\, \Omega = 0 \tag{4.4}$$

and by definition $i_A\, \Omega = 0$. Therefore, $\pounds_A\, h$ is given only by the second term in the last row of (4.3). Repeat the same reasoning for this term to get

$$\begin{aligned} \pounds_A\, h &= i_n\, \pounds_A\, i_{n-1}\, i_{n-2} \cdots i_2\, i_1\, \Omega \\ &= i_n\, (i_{[A,S_{n-1}]} + i_{n-1}\, \pounds_A)\, i_{n-2} \cdots i_2\, i_1\, \Omega \\ &= i_n\, i_{n-1} \pounds_A\, i_{n-2} \cdots i_2\, i_1\, \Omega. \end{aligned} \tag{4.5}$$

Continuing this procedure, after commuting the Lie derivative with all contraction, we will obtain

$$\pounds_A\, h = i_n\, i_{n-1} \cdots i_2\, i_1\, \pounds_A, \Omega. \tag{4.6}$$

From Lemma 1, $\pounds_A\, \Omega = (\mathrm{div}_V A)\, \Omega$, we can complete the proof that $\pounds_A\, h = h\, (\mathrm{div}_V A)$, which is equivalent to the theorem above. □

4.1 Example: Three-dimensional ODE

Let $\dim_{\mathcal{F}} \mathcal{X} = \dim_{\mathcal{F}} \mathcal{X}^* = 3$.

Theorem 2. *Let $\mathcal{D}^* \subset \mathcal{X}^*$ and let $\Gamma \subset \Lambda^2(\mathcal{X})$ be a one-dimensional distribution generated by a vector field $A \in \operatorname{der}\mathcal{F}$ and by a differential two-form Ω, respectively, such that they are related to each other through a volume form $V \in \Lambda^3(\mathcal{X}^*)$, as in Lemma 1,*

$$i_A \Omega = 0, \qquad \text{and} \qquad i_A V = \Omega.$$

If $S_1, S_2 \in \mathcal{X}$ are both symmetry vector fields admitted by A, such that $S_1 \wedge S_2 \wedge A \neq 0$, and let $h \equiv i_{S_2} i_{S_1} \Omega \in \mathcal{F}$, then $m = h^{-1} \in \mathcal{F}$ is a last multiplier for the 2-form Ω.

Proof. We need to show that $\pounds_A h = h \operatorname{div}_V A$. Let

$$\text{for } i = 1,2, \quad [S_i, A] = \lambda_i A, \qquad \text{for some } \lambda_i \in \mathcal{F}.$$

Now, we have

$$\begin{aligned}
\pounds_A h &= \pounds_A i_{S_2} i_{S_1} \Omega \\
&= \left(i_{[A,S_2]} + i_{S_2} \pounds_A\right) i_{S_1} \Omega \\
&= \lambda_2 i_A i_{S_1} \Omega + i_{S_2} \left(i_{[A,S_1]} + i_{S_1} \pounds_A\right) \Omega \\
&= -\lambda_2 i_{S_1} i_A \Omega + \lambda_1 i_{S_2} i_A \Omega + i_{S_2} i_{S_1} \pounds_A \Omega \\
&= (\operatorname{div}_V A)\, i_{S_2} i_{S_1} \Omega
\end{aligned}$$

since by the hypothesis $i_A \Omega = 0$ and also $\pounds_A \Omega = (\operatorname{div}_V A)\, \Omega$, (Lemma 1). □

5 Conclusion

In this paper, we obtain a last multiplier for an ordinary differential equation (ODE). A last multiplier is a useful object to completely integrate an ODE [Flanders 1963]. Our construction is intrinsic, independent of local coordinates as well as independent of choice of the differential one-forms representing the ODE. We make use of the Cartan formalism of exterior differential systems, which is the most powerful method to investigate the subject [Cartan 1946, Choquet-Bruhat 1968, Griffiths 1983]. We assume no additional structure (like a Riemannian or a symplectic structure) on the ODE. Our discussion applies to every ODE admitting a set of symmetry.

References

[1] A. O. Barut, José R. Zeni, and I. A. Abdallah, On the Lie group analysis of linear systems of O.D.E., *Lie Groups and Their Applications* **1 2** (1994), 1–17.

[2] A. O. Barut and José R. Zeni, *Group Theory and Differential Equations,* World Scientific Publications, Singapore, to appear, 2000.

[3] A. Borowiec, Cartan pairs, *Czechoslovak Journal of Physics* **46 (12)**, q-alg/9609011, MR#98a:17015 (1996), 1197–1202.

[4] A. Borowiec, Vector fields and differential operators: noncommutative case, *Czechoslovak Journal of Physics* **47 (12)**, q-alg/9710006 (1997), 1093–1202.

[5] A. Borowiec and Guillermo Arnulfo Vázquez Coutiño, Some topics in coalgebra calculus, *Czechoslovak Journal of Physics,* **50 (1)**, mathQA/9910018 (2000), 23–28.

[6] R. L. Briant, S. Chern, R. Gardner, H. Goldschmidt, and P. Griffiths, *Exterior Differential Systems,* Springer Verlag, 1991.

[7] E. Cartan, *Les Systèmes Différentiels Extérieurs et Leurs Applications Géométriques,* Hermann, 1946.

[8] Y. Choquet-Bruhat, *Géométrie Différentielle et Systèmes Extérieurs,* Dunod, 1968.

[9] H. Flanders, *Differential Forms with Applications to the Physical Sciences,* Dover, 1963, 1989.

[10] P. A. Griffiths, *Exterior Differential Systems and the Calculus of Variations,* Birkhäuser, 1983.

[11] M. Gusiew-Czudżak, The new inverse problem of the Newton law, in *Theoretical Physics, Fin de Siècle,* in honour of Professor Jan Łopuszański, Lectures Notes in Physics, Karwowski Witold, Bernard Jancewicz, Andrzej Borowiec, Wojciech Cegła, and Arkadiusz Jadczyk, eds., Springer–Verlag, Berlin, Heidelberg, 1999.

[12] R. Ya. Matsyuk, Symmetries of vector exterior differential systems and the inverse problem in second order Ostrohrads'kyj mechanics, *Journal of Nonlinear Mathematical Physics* **4 (1/2)** (1997), 89–97.

[13] R. Ya. Matsyuk, Integration by parts and vector differential forms in higher order variational calculus on fibered manifolds, *Matematychni Studii* **11 (1)** (1999), 85–107.

[14] J. R. Zeni, Lie analysis of differential equations: the role of canonical variables, PUC, Departamento de Matematica, preprint 05/1996, 1996.

[15] J. R. Zeni, Teaching Lie analysis of differential equations to undergraduate students, in *Modern Group Analysis, Proceedings of VII MOGRAN,* N. Ibragimov, K. R. Naqvi, and E. Straumer, eds., MARS Publishers, 1999, 339–344.

Zbigniew Oziewicz
Universidad Nacional Autónoma de México
Facultad de Estudios Superiores
Apartado Postal # 25,
C.P. 54700 Cuautitlán Izcalli, Estado de México
E-mail: oziewicz@servidor.unam.mx

Uniwersytet Wrocławski
Instytut Fizyki Teoretycznej
Plac Maksa Borna 9
PL 50-204 Wrocław, Poland
E-mail: oziewicz@ift.uni.wroc.pl

José Ricardo R. Zeni
Universidade do Oeste Paulista, UNOESTE
Pro-Reitoria de Pós-Graduação e Pesquisa
Rodovia Raposo Tavares, km 572, Bairro Limoeiro
Presidente Prudente, SP, Brazil, 19.067-175
E-mail: josez@posgrad.unoeste.br, josez@apecnt.unoeste.br,
lfzeni@stetnet.com.br

Received: September 30, 1999; Revised: March 9, 2000

Universal Similarity Factorization Equalities Over Complex Clifford Algebras

Yongge Tian

ABSTRACT A set of valuable universal similarity factorization equalities is established over complex Clifford algebras $\mathcal{C}_n$. Through them matrix representations of complex Clifford algebras $\mathcal{C}_n$ can directly be derived, and their properties can easily be determined.
Keywords: Clifford algebras, matrix representations, universal similarity factorizations.

1 Introduction

Let $\mathcal{C}_n$ be the complex Clifford algebra, with the identity 1, defined on n generators $e_1, e_2, \dots, e_n$ subject to the multiplication laws:

$$e_i^2 = -1, \quad i = 1, 2, \dots, n, \tag{1.1}$$
$$e_i e_j + e_j e_i = 0, \quad i \neq j,\ i, j = 1, 2, \dots, n, \tag{1.2}$$

and $e_1 e_2 \cdots e_n \neq \pm 1$. In that case $\mathcal{C}_n$ is spanned as a 2^n-dimensional vector space with 2^n basis $\{e_A\}$, where the multi-index A ranges all naturally ordered subsets of the first positive integer set $\{1, 2, \dots, n\}$; the basis element e_A, where $A = (i_1, i_2, \dots, i_k)$ with $1 \leq i_1 < i_2 < \cdots < i_k \leq n$, is defined as the product

$$e_A \equiv e_{(i_1, i_2, \dots, i_k)} = e_{i_1} e_{i_2} \cdots e_{i_k}, \quad e_{A=\emptyset} = 1.$$

For simplicity, the volume element $e_{12\dots n} = e_1 e_2 \cdots e_n$ of $\mathcal{C}_n$ will be denoted by $e_{[n]}$ in the sequel. The square of the volume element is

$$e_{[n]}^2 = (-1)^{\frac{1}{2}n(n-1)} e_1^2 e_2^2 \cdots e_n^2.$$

AMS Subject Classification: 15A23, 15A66.

In that case, all $a \in \mathcal{C}_n$ can be expressed as

$$a = \sum_A a_A e_A, \quad a_A \in \mathcal{C},$$

where A ranges all naturally ordered subsets of $\{1, 2, \ldots, n\}$. We shall adopt the following notation from now on: $\mathcal{C}_n := \mathcal{C}\{e_1, \ldots, e_n\}$.

Clifford algebras (real or complex) have been studied for many years, and their algebraic properties are well-known. In particular, all Clifford algebras are classified as matrix algebras, or as direct sums of matrix algebras over the fields of real or complex numbers, or the quaternion ring (see, e.g., [2, 3, 5, 7, 8, 9, 10, 11]). For complex Clifford algebras $\mathcal{C}_n$, it is well-known that they can faithfully be realized as certain matrix algebras over $\mathcal{C}$, and the general algebra isomorphism is

$$\mathcal{C}_n \simeq \begin{cases} \mathcal{C}(2^{\frac{n}{2}}) & \text{if } n \text{ is even,} \\ {}^2\mathcal{C}(2^{\frac{n-1}{2}}) & \text{if } n \text{ is odd,} \end{cases} \tag{1.3}$$

where $\mathcal{C}(s)$ stands for the $s \times s$ total complex matrix algebra, and ${}^2\mathcal{C}(s)$ stands for the complex matrix algebra

$${}^2\mathcal{C}(s) = \left\{ \begin{bmatrix} A & 0 \\ 0 & B \end{bmatrix} \,\middle|\, A, B \in \mathcal{C}(s) \right\}.$$

In this article we improve this relationship to a new level by establishing a set of valuable universal similarity factorization equalities between elements of $\mathcal{C}_n$ and complex matrices over $\mathcal{C}(2^{\frac{n}{2}})$ or $\mathcal{C}(2^{\frac{n-1}{2}})$ for all n. Through these universal factorization equalities the complex matrix representations of elements in $\mathcal{C}_n$ can explicitly be established. Based on them various results in the complex matrix theory can directly be extended to complex Clifford numbers. Moreover, one can also easily develop matrix analysis over complex Clifford algebras.

We first establish some basic universal similarity factorization equalities for elements in Clifford algebras with dimensions 2 and 4, which will be directly applied to establish some more general results for $\mathcal{C}_n$.

Lemma 1. *Let $\mathcal{F}$ be an algebraically closed field and $\mathcal{F}_1 = \mathcal{F}\{e \mid e^2 = u\}$ be a Clifford algebra defined on a generator e with $e^2 = u \in \mathcal{F}$ and $u \neq 0$. Then all $a \in \mathcal{F}_1$ can be written as $a = a_0 + a_1 e$, where $a_0, a_1 \in \mathcal{F}$. Moreover, define $\overline{a} = a_0 - a_1 e$. In that case, a and $\overline{a}$ satisfy the following universal similarity factorization equality*

$$P \begin{bmatrix} a & 0 \\ 0 & \overline{a} \end{bmatrix} P^{-1} = \begin{bmatrix} a_0 + \sqrt{u}a_1 & 0 \\ 0 & a_0 - \sqrt{u}a_1 \end{bmatrix}, \tag{1.4}$$

where P and P^{-1} have the following universal forms (no relation with a) :

$$P = \frac{1}{2}\begin{bmatrix} 1+\frac{1}{\sqrt{u}}e & -(\sqrt{u}-e) \\ \frac{1}{u}(\sqrt{u}-e) & 1+\frac{1}{\sqrt{u}}e \end{bmatrix},$$

$$P^{-1} = \frac{1}{2}\begin{bmatrix} 1+\frac{1}{\sqrt{u}}e & \sqrt{u}-e \\ -\frac{1}{u}(\sqrt{u}-e) & 1+\frac{1}{\sqrt{u}}e \end{bmatrix}. \tag{1.5}$$

Proof. Note that $e^2 = u$. It is easy to verify that

$$\begin{bmatrix} 1 & e \\ e^{-1} & -1 \end{bmatrix}\begin{bmatrix} a & 0 \\ 0 & \overline{a} \end{bmatrix}\begin{bmatrix} 1 & e \\ e^{-1} & -1 \end{bmatrix} = \begin{bmatrix} a_0 & ua_1 \\ a_1 & a_0\overline{a} \end{bmatrix}.$$

On the other hand, it is also easy to verify that

$$\begin{bmatrix} 1 & \sqrt{u} \\ \frac{1}{\sqrt{u}} & -1 \end{bmatrix}\begin{bmatrix} a_0+ua_1 & 0 \\ 0 & a_0-ua_1 \end{bmatrix}\begin{bmatrix} 1 & \sqrt{u} \\ \frac{1}{\sqrt{u}} & -1 \end{bmatrix} = \begin{bmatrix} a_0 & ua_1 \\ a_1 & a_0\overline{a} \end{bmatrix}.$$

Combining the above two equalities yields (1.4) and (1.5). □

Lemma 2. *Let $M_2(\mathcal{F})$ be the 2×2 total matrix algebra over an arbitrary field $\mathcal{F}$ with its basis satisfying the multiplication rules*

$$\tau_{pq}\tau_{st} = \begin{cases} \tau_{pt}, & q = s, \\ 0, & q \neq s, \end{cases} \tag{1.6}$$

for $p, q, s, t = 1, 2$. Then all $a = a_{11}\tau_{11} + a_{12}\tau_{12} + a_{21}\tau_{21} + a_{22}\tau_{22} \in M_2(\mathcal{F})$, where $a_{pq} \in \mathcal{F}$, satisfy the following universal similarity factorization equality

$$Q\begin{bmatrix} a & 0 \\ 0 & a \end{bmatrix}Q^{-1} = \begin{bmatrix} a_{11} & a_{12} \\ a_{21} & a_{22} \end{bmatrix}, \tag{1.7}$$

where Q has the following universal form

$$Q = Q^{-1} = \begin{bmatrix} \tau_{11} & \tau_{21} \\ \tau_{12} & \tau_{22} \end{bmatrix}. \tag{1.8}$$

Proof. Follows directly from a verification. □

Lemma 3. *Let $\mathcal{F}$ be an algebraically closed field of characteristic not two, and $\mathcal{F}_2 = \mathcal{F}\{e_1, e_2\}$ be a Clifford algebra defined on two generators e_1, e_2 with $e_1^2 = u \in \mathcal{F}, e_2^2 = v \in \mathcal{F}$ and $u \neq 0, v \neq 0$. Then all $a \in \mathcal{F}_2$ can be written as*

$$a = a_0 + a_1e_1 + a_2e_2 + a_3e_{12}, \tag{1.9}$$

where $a_0, \ldots, a_3 \in \mathcal{F}$. *In that case,* aI_2 *satisfies the following universal similarity factorization equalities*

$$R\begin{bmatrix} a & 0 \\ 0 & a \end{bmatrix}R^{-1} = \begin{bmatrix} a_0 + \sqrt{u}a_1 & v(a_2 + \sqrt{u}a_3) \\ a_2 - \sqrt{u}a_3 & a_0 - \sqrt{u}a_1 \end{bmatrix}, \tag{1.10}$$

and

$$T\begin{bmatrix} a & 0 \\ 0 & a \end{bmatrix}T^{-1} = \begin{bmatrix} a_0 + \sqrt{v}a_1 & u(a_1 - \sqrt{v}a_3) \\ a_1 + \sqrt{v}a_3 & a_0 - \sqrt{v}a_1 \end{bmatrix}, \tag{1.11}$$

where

$$R = R^{-1} = \frac{1}{2}\begin{bmatrix} 1 + \frac{1}{\sqrt{u}}e_1 & e_2 - \frac{1}{\sqrt{u}}e_{12} \\ \frac{1}{v}(e_2 + \frac{1}{\sqrt{u}}e_{12}) & 1 - \frac{1}{\sqrt{u}}e_1 \end{bmatrix}, \tag{1.12}$$

and

$$T = T^{-1} = \frac{1}{2}\begin{bmatrix} 1 + \frac{1}{\sqrt{v}}e_2 & e_1 + \frac{1}{\sqrt{v}}e_{12} \\ \frac{1}{u}(e_1 - \frac{1}{\sqrt{v}}e_{12}) & 1 - \frac{1}{\sqrt{v}}e_2 \end{bmatrix}. \tag{1.13}$$

Proof. By Lemma 2, we take the change of basis of $\mathcal{F}_2$ as follows

$$\tau_{11} = \frac{1}{2}\left(1 + \frac{1}{\sqrt{u}}e_1\right), \qquad \tau_{12} = \frac{1}{2v}\left(e_2 + \frac{1}{\sqrt{u}}e_{12}\right), \tag{1.14a}$$

$$\tau_{21} = \frac{1}{2}\left(e_2 - \frac{1}{\sqrt{u}}e_{12}\right), \qquad \tau_{22} = \frac{1}{2}\left(1 - \frac{1}{\sqrt{u}}e_1\right). \tag{1.14b}$$

Then it is not hard to verify that this new basis satisfies the multiplication rules in (1.6). In this new basis, any $a = a_0 + a_1e_1 + a_2e_2 + a_3e_{12} \in \mathcal{F}_2$ can be rewritten as

$$a = (a_0 + \sqrt{u}a_1)\tau_{11} + (va_2 + \sqrt{u}a_3)\tau_{12} + (a_2 - \sqrt{u}a_3)\tau_{21} + (a_0 - \sqrt{u}a_1)\tau_{22}.$$

Substituting this and (1.14) into (1.7) and (1.8), we can obtain (1.10) and (1.12). By the similar approach, we can get (1.11) and (1.13). □

2 Main results

Notice that $\mathcal{C}$ is an algebraically closed field. We then can establish a set of universal similarity factorization equalities for elements in $\mathcal{C}_n$ on the basis of Lemmas 1 and 3.

Theorem 1. *Let $a \in \mathcal{C}_1 = \mathcal{C}\{e\}$ be given. Then a can be written as $a = a_0 + a_1 e$, where $a_0, a_1 \in \mathcal{C} = \{x + iy \,|\, x, y \in \mathcal{R}\}$. Moreover denote by $\overline{a} = a_0 - a_1 e$, called the conjugate of a. Then a and $\overline{a}$ satisfy the following universal similarity factorization equality*

$$P_1 \begin{bmatrix} a & 0 \\ 0 & \overline{a} \end{bmatrix} P_1^{-1} = \begin{bmatrix} a_0 + a_1 i & 0 \\ 0 & a_0 - a_1 i \end{bmatrix}, \tag{2.1}$$

where P_1 and P_1^{-1} have the following universal forms (no relation with a) :

$$P_1 = \frac{1}{2}\begin{bmatrix} 1 - ie & -(i-e) \\ -(i-e) & 1 - ie \end{bmatrix}, \quad P_1^{-1} = \frac{1}{2}\begin{bmatrix} 1 - ie & i - e \\ i - e & 1 - ie \end{bmatrix}. \tag{2.2}$$

Proof. Let $\mathcal{F} = \mathcal{C}$ and $u = -1$ as in Lemma 1. Then (2.1) and (2.2) follow directly from (1.4) and (1.5). □

It is easy to verify that

$$\overline{\overline{a}} = a, \quad \overline{a+b} = \overline{a} + \overline{b}, \quad \overline{ab} = \overline{a}\overline{b}, \quad \overline{\lambda a} = \overline{a\lambda} = \lambda\overline{a}. \tag{2.3}$$

hold for all $a, b \in \mathcal{C}_1$ and $\lambda \in \mathcal{C}$. According to (2.1), we define a map from $\mathcal{C}_1$ to the double field $\mathcal{C} \oplus \mathcal{C}$ by

$$\phi_1 : a = a_0 + a_1 e \in \mathcal{C}_1 \longrightarrow \begin{bmatrix} a_0 + a_1 i & 0 \\ 0 & a_0 - a_1 i \end{bmatrix} \in \mathcal{C} \oplus \mathcal{C}. \tag{2.4}$$

Then it is easy to derive from (2.1) and (2.3) the following properties.

Corollary 1. *Let $a = a_0 + a_1 e, b = b_0 + b_1 e \in \mathcal{C}_1, \lambda \in \mathcal{C}$ be given, and ϕ_1 be defined by (2.4). Then*

(a) $a = b \Longleftrightarrow \phi_1(a) = \phi_1(b)$.

(b) $\phi_1(a+b) = \phi_1(a) + \phi_1(b)$, $\phi_1(ab) = \phi_1(a)\phi_1(b)$, $\phi_1(\lambda a) = \lambda\phi_1(a)$, $\phi_1(1) = I_2$.

(c) $\phi_1(\overline{a}) = \begin{bmatrix} 0 & 1 \\ 1 & 0 \end{bmatrix} \phi_1(a) \begin{bmatrix} 0 & 1 \\ 1 & 0 \end{bmatrix}$.

(d) Denote $a^{\#} = \overline{a_0} - \overline{a_1} e$, *then* $\phi_1(a^{\#}) = \phi_1^*(a)$, *the conjugate transpose of the complex matrix* $\phi_1(a)$.

(e) $a = \frac{1}{4}[1 - ie, i - e]\phi_1(a)[1 - ie, e - i]^T$.

(f) $\det[\phi_1(a)] = a_0^2 + a_1^2$.

(g) a is invertible $\Longleftrightarrow$ *$\phi_1(a)$ is invertible, in which case* $\phi_1(a^{-1}) = \phi_1^{-1}(a)$.

The properties in Corollary 1 (a) and (b) show that through the bijective map (2.4), the Clifford algebra $\mathcal{C}_1$ is algebraically isomorphic to the double field $\mathcal{C} \oplus \mathcal{C}$, and $\phi_1(a)$ is a faithful matrix representation of a in $\mathcal{C} \oplus \mathcal{C}$.

Notice that P_1 and P_1^{-1} in (2.1) have no relation with a. Thus the equality in Theorem 1 can also be extended to all matrices over the complex Clifford algebra $\mathcal{C}_1$.

Theorem 2. *Let $A = A_0 + A_1 e \in \mathcal{C}_1^{m\times n}$ be given, where $A_0, A_1 \in \mathcal{C}^{m\times n}$. Then A and its conjugate $\overline{A} = A_0 - A_1 e$ satisfy the following universal factorization equality*

$$J_{2m}\begin{bmatrix} A & 0 \\ 0 & \overline{A} \end{bmatrix} J_{2n}^{-1} = \begin{bmatrix} A_0 + A_1 i & 0 \\ 0 & A_0 - A_1 i \end{bmatrix}, \tag{2.5}$$

where

$$J_{2m} = \frac{1}{2}\begin{bmatrix} (1-ie)I_m & -(i-e)I_m \\ -(i-e)I_m & (1-ie_1)I_m \end{bmatrix},$$

$$J_{2n}^{-1} = \frac{1}{2}\begin{bmatrix} (1-ie_1)I_n & (i-e)I_n \\ (i-e)I_n & (1-ie)I_n \end{bmatrix}. \tag{2.6}$$

In particular, if $m = n$, then (2.5) becomes a universal similarity factorization equality over $\mathcal{C}_1$.

It is easy to verify that for any matrices $A, B \in \mathcal{C}_1^{m\times n}, C \in \mathcal{C}_1^{n\times p}$, and $\lambda \in \mathcal{C}$,

$$\overline{\overline{A}} = A, \quad \overline{A+B} = \overline{A} + \overline{B}, \quad \overline{AC} = \overline{A}\,\overline{C}, \quad \overline{\lambda A} = \overline{A\lambda} = \lambda\overline{A}. \tag{2.7}$$

Now according to (2.5), we define the complex matrix representation of a matrix $A = A_0 + A_1 e \in \mathcal{C}_1^{m\times n}$ by $\Phi_1(A) = \begin{bmatrix} A_0 + A_1 i & 0 \\ 0 & A_0 - A_1 i \end{bmatrix}$. Then the following properties can be easily derived from (2.5) and (2.7).

Corollary 2. *Let $A, B \in \mathcal{C}_1^{m\times n}, C \in \mathcal{C}_1^{n\times p}$ and $\lambda \in \mathcal{C}$ be given. Then*

(a) $A = B \Longleftrightarrow \Phi_1(A) = \Phi_1(B)$.

(b) $\Phi_1(A+B) = \Phi_1(A) + \Phi_1(B)$.

(c) $\Phi_1(AC) = \Phi_1(A)\Phi_1(C)$, $\Phi_1(\lambda A) = \lambda\Phi_1(A)$, $\Phi_1(I_m) = I_{2m}$.

(d) $\Phi_1(\overline{A}) = \begin{bmatrix} 0 & I_m \\ I_m & 0 \end{bmatrix} \Phi_1(A) \begin{bmatrix} 0 & I_n \\ I_n & 0 \end{bmatrix}$.

(e) *Let* $A = A_0 + A_1 e$ *and denote* $A^{\#} = A_0^* - A_1^* e$, *where* A_0^* *and* A_1^* *are the conjugate transposes of the complex matrices* A_0 *and* A_1. *Then* $\Phi_1(A^{\#}) = \Phi_1^*(A)$, *the conjugate transpose of the complex matrix* $\Phi_1(A)$.

(f) $A = \frac{1}{4}[(1-ie)I_m, (i-e)I_m]\Phi_1(A)[(1-ie)I_n, (e-i)I_n]^T$.

(g) A *is invertible* $\Longleftrightarrow$ $\Phi_1(A)$ *is invertible, in which case* $\Phi_1(A^{-1}) = \Phi_1^{-1}(A)$.

(h) $p_A(A) = 0$, *where* $p_A(x) = \det[xI_{2m} - \Phi_1(A)]$.

Theorem 3. *Let* $a \in \mathcal{C}_2 = \mathcal{C}\{e_1, e_2\}$, *the complex quaternion algebra. Then* a *can be written as*

$$a = a_0 + a_1 e_1 + a_2 e_2 + a_3 e_{12},$$

where $a_0, \dots, a_3 \in \mathcal{C}$. *In that case,* aI_2 *satisfies the following universal similarity factorization equality*

$$P_2 \begin{bmatrix} a & 0 \\ 0 & a \end{bmatrix} P_2^{-1} = \begin{bmatrix} a_0 + a_1 i & -(a_2 + a_3 i) \\ a_2 - a_3 i & a_0 - a_1 i \end{bmatrix}, \tag{2.8}$$

where P_2 *has the universal form*

$$P_2 = P_2^{-1} = \frac{1}{2} \begin{bmatrix} 1 - ie_1 & e_2 + ie_{12} \\ -e_2 + ie_{12} & 1 + ie_1 \end{bmatrix}. \tag{2.9}$$

Proof. Let $\mathcal{F} = \mathcal{C}$ and $u = v = -1$ in Lemma 3. Then (2.8) and (2.9) follow from (1.10) and (1.12). □

According to (2.8), define a map from $\mathcal{C}_2$ to the 2×2 complex matrix algebra $\mathcal{C}^{2\times 2}$ by

$$\phi_2 : a = a_0 + a_1 e_1 + a_2 e_2 + a_3 e_{12} \in \mathcal{C}_2 \longrightarrow \begin{bmatrix} a_0 + a_1 i & -(a_2 + a_3 i) \\ a_2 - a_3 i & a_0 - a_1 i \end{bmatrix} \in \mathcal{C}^{2\times 2}. \tag{2.10}$$

Then we can easily derive from (2.8) the following properties.

Corollary 3. *Let* $a, b \in \mathcal{C}_2 = \mathcal{C}\{e_1, e_2\}$ *and* $\lambda \in \mathcal{C}$ *be given. Then*

(a) $a = b \Longleftrightarrow \phi_2(a) = \phi_2(b)$.

(b) $\phi_2(a + b) = \phi_2(a) + \phi_2(b)$, $\phi_2(ab) = \phi_2(a)\phi_2(b)$, $\phi_2(\lambda a) = \lambda\phi_2(a)$, $\phi_2(1) = I_2$.

(c) *Let* $a = a_0 + a_1e_1 + a_2e_2 + a_3e_{12} \in C_2$, *and denote* $a^{\#} = \overline{a_0} - \overline{a_1}e_1 - \overline{a_2}e_2 - \overline{a_3}e_{12}$. *Then* $\phi_2(a^{\#}) = \phi_2^*(a)$, *the conjugate transpose of the complex matrix* $\phi_2(a)$.

(d) $a = \frac{1}{4}[1 - ie, e_2 + ie_{12}]\phi_2(a)[1 - ie, -e_2 + ie_{12}]^T$.

(e) $\det[\phi_2(a)] = a_0^2 + a_1^2 + a_2^2 + a_3^2$.

(f) a *is invertible* $\Longleftrightarrow$ $\phi_2(a)$ *is invertible, in which case* $\phi_2(a^{-1}) = \phi_2^{-1}(a)$.

(g) *The two elements* a *and* b *in* C_2 *are similar, i.e., there is an invertible* $x \in C_2$ *such that* $ax = xb$ *if and only if the two complex matrices* $\phi_2(a)$ *and* $\phi_2(b)$ *are similar over* C.

The properties in Corollary 3 (a) and (b) clearly show that through the bijective map (2.10) the Clifford algebra C_2, i.e., the complex quaternion algebra, is algebraically isomorphic to the complex matrix algebra $C^{2\times 2}$, and $\phi_2(a)$ is a faithful matrix representation of a in $C^{2\times 2}$.

Notice that P_2 and P_2^{-1} in (2.9) have no relation to a. Thus the equality in Theorem 3 can also be extended to all matrices over C_2.

Theorem 4. *Let* $A = A_0 + A_1e_1 + A_2e_2 + A_3e_{12} \in C_2^{m\times n} = C^{m\times n}\{e_1, e_2\}$ *be given where* $A_0, \dots, A_3 \in C^{m\times n}$. *Then* A *satisfies the following universal factorization equality*

$$K_{2m}\begin{bmatrix} A & 0 \\ 0 & A \end{bmatrix}K_{2n}^{-1} = \begin{bmatrix} A_0 + A_1i & -(A_2 + A_3i) \\ A_2 - A_3i & A_0 - A_1i \end{bmatrix}, \tag{2.11}$$

where

$$K_{2t} = K_{2t}^{-1} = \frac{1}{2}\begin{bmatrix} (1 - ie_1)I_t & (e_2 + ie_{12})I_t \\ (-e_2 + ie_{12})I_t & (1 + ie_1)I_t \end{bmatrix}. \tag{2.12}$$

In particular, if $m = n$, *then (2.11) becomes a universal similarity factorization equality over* C_2.

According to (2.11), we define the complex representation of a matrix $A = A_0 + A_1e \in C_2^{m\times n}$ by $\Phi_2(A) = \begin{bmatrix} A_0 + A_1i & -(A_2 + A_3i) \\ A_2 - A_3i & A_0 - A_1i \end{bmatrix}$. Then the following properties are easy to verify by (2.11).

Corollary 4. *Let* $A, B \in C_2^{m\times n}, C \in C_2^{n\times p}$, *and* $\lambda \in C$ *be given. Then*

(a) $A = B \Longleftrightarrow \Phi_2(A) = \Phi_2(B)$.

(b) $\Phi_2(A + B) = \Phi_2(A) + \Phi_2(B)$.

(c) $\Phi_2(AC) = \Phi_2(A)\Phi_2(C)$, $\Phi_2(\lambda A) = \lambda\Phi_2(A)$, $\Phi_2(I_m) = I_{2m}$.

(d) Let $A = A_0 + A_1e_1 + A_2e_2 + A_3e_{12}$ *and denote* $A^{\#} = A_0^* - A_1^*e_1 - A_2^*e_2 - A_3^*e_{12}$*; then* $\Phi_2(A^{\#}) = \Phi_2^*(A)$*, the conjugate transpose of the complex matrix* $\Phi_2(A)$.

(e) $A = \frac{1}{4}[(1-ie_1)I_m, (e_2+ie_{12})I_m]\Phi_2(A)[(1-ie_1)I_n, (-e_2+ie_{12})I_n]^T$.

(f) A *is invertible* $\Longleftrightarrow$ $\Phi_2(A)$ *is invertible, in which case* $\Phi_2(A^{-1}) = \Phi_2^{-1}(A)$.

(g) $p_A(A) = 0$*, where* $p_A(\lambda) = \det[\lambda I_{2m} - \Phi_2(A)]$*, the characteristic polynomial of* $\Phi_2(A)$.

(h) Two square matrices A *and* B *are similar over* $\mathcal{C}_2$*, i.e., there is an invertible matrix* X *over* $\mathcal{C}_2$ *such that* $AX = XB$ *if and only if* $\Phi_2(A)$ *and* $\Phi_2(B)$ *are similar over* $\mathcal{C}$.

In the next several results we only present the basic universal similarity factorization equalities without listing their operation properties and their extensions to matrices over $\mathcal{C}_n$.

Theorem 5. *Let* $a \in \mathcal{C}_3 = \mathcal{C}\{e_1, e_2, e_3\}$ *be given. Then* a *can factor as*

$$a = a_0 + a_1 e_{[3]}, \tag{2.13}$$

where

$$a_0,\ a_1 \in \mathcal{C}_2 = \mathcal{C}\{e_1, e_2\}, \quad e_{[3]}^2 = 1.$$

Moreover, define $\overline{a} = a_0 - a_1e_{[3]}$*. In that case, the diagonal matrix* $D_a = \mathrm{diag}(aI_2, \overline{a}I_2)$ *satisfies the following universal similarity factorization equality*

$$P_3D_aP_3^{-1} = \begin{bmatrix} \phi_2(a_0) + \phi_2(a_1) & 0 \\ 0 & \phi_2(a_0) - \phi_2(a_1) \end{bmatrix}$$

$$:= \phi_3(a) \in {}^2\mathcal{C}^{2\times 2}, \tag{2.14}$$

where $\phi_2(a_t)$, $t = 0, 1$, *is the matrix representation of* a_t *in* $\mathcal{C}^{2\times 2}$ *defined in (2.10) and*

$$P_3 = \frac{1}{2}\begin{bmatrix} (1+e_{[3]})P_2 & -(1-e_{[3]})P_2 \\ (1-e_{[3]})P_2 & (1+e_{[3]})P_2 \end{bmatrix}, \tag{2.15}$$

$$P_3^{-1} = \frac{1}{2}\begin{bmatrix} P_2^{-1}(1+e_{[3]}) & P_2^{-1}(1-e_{[3]}) \\ -P_2^{-1}(1-e_{[3]}) & P_2^{-1}(1+e_{[3]}) \end{bmatrix}, \tag{2.16}$$

where P_2 *and* P_2^{-1} *are given by (2.9).*

Proof. Notice that $be_{[3]} = e_{[3]}b$ holds for all $b \in \mathcal{C}_2 = \mathcal{C}\{e_1, e_2\}$. We have by applying (2.8) to a in (2.13) that

$$\begin{aligned} P_2(aI_2)P_2^{-1} &= P_2(a_0I_2)P_2^{-1} + P_2(a_1I_2)P_2^{-1}e_{[3]} \\ &= \phi_2(a_0) + \phi_2(a_1)e_{[3]} := \psi(a), \end{aligned}$$

and

$$P_2(\overline{a}I_2)P_2^{-1} = \phi_2(a_0) - \phi_2(a_1)e_{[3]} := \psi(\overline{a}).$$

Next we build, according to Lemma 1, a matrix and its inverse as follows

$$V = \frac{1}{2}\begin{bmatrix} (1+e_{[3]})I_2 & -(1-e_{[3]})I_2 \\ (1-e_{[3]})I_2 & (1+e_{[3]})I_2 \end{bmatrix},$$

$$V^{-1} = \frac{1}{2}\begin{bmatrix} (1+e_{[3]})I_2 & (1-e_{[3]})I_2 \\ -(1-e_{[3]})I_2 & (1+e_{[3]})I_2 \end{bmatrix}.$$

and then calculate to get

$$V\begin{bmatrix} \psi(a) & 0 \\ 0 & \psi(\overline{a}) \end{bmatrix}V^{-1} = \begin{bmatrix} \phi_2(a_0)+\phi_2(a_1) & 0 \\ 0 & \phi_2(a_0)-\phi_2(a_1) \end{bmatrix}.$$

Finally substituting $\psi(a) = P_2(aI_2)P_2^{-1}$ and $\psi(\overline{a}) = P_2(\overline{a}I_2)P_2^{-1}$ into the left-hand side of the above equality yields (2.14), (2.15), and (2.16). □

Theorem 6. *Let $a \in \mathcal{C}_4 = \mathcal{C}\{e_1, e_2, e_3, e_4\}$ be given. Then a can factor as*

$$a = a_0 + a_1e_{123} + a_2e_{124} + a_3e_{43} = a_0 + e_{123}a_1 + e_{124}a_2 + e_{43}a_3, \quad (2.17)$$

where

$$a_0, a_1, a_2, a_3 \in \mathcal{C}_2 = \mathcal{C}\{e_1, e_2\},$$
$$e_{123}^2 = 1, \quad e_{124}^2 = 1, \quad e_{43} = e_{123}e_{124} = -e_{124}e_{123}.$$

In that case, aI_4 satisfies the following universal similarity factorization equality

$$\begin{aligned} P_4(aI_4)P_4^{-1} &= \begin{bmatrix} \phi_2(a_0)+\phi_2(a_1) & \phi_2(a_2)+\phi_2(a_3) \\ \phi_2(a_2)-\phi_2(a_3) & \phi_2(a_0)-\phi_2(a_1) \end{bmatrix} \\ &:= \phi_4(a) \in \mathcal{C}^{4\times 4}, \end{aligned} \quad (2.18)$$

where $\phi_2(a_t)$, $t = 0, \ldots, 3$, is the matrix representation of a_t in $\mathcal{C}^{2\times 2}$ defined in (2.10) and

$$P_4 = P_4^{-1} = \frac{1}{2}\begin{bmatrix} (1+e_{[3]})P_2 & (e_{124}-e_{43})P_2 \\ (e_{124}+e_{43})P_2 & (1-e_{[3]})P_2 \end{bmatrix}, \quad (2.19)$$

where P_2 is given in (2.9).

Proof. Note that the commutative rules

$$be_{123} = e_{123}b, \quad be_{124} = e_{124}b, \quad be_{43} = e_{43}b$$

hold for all $b \in \mathcal{C}_2 = \mathcal{C}\{e_1, e_2\}$. Thus, it follows from (2.8) that

$$\begin{aligned} & P_2(aI_2)P_2^{-1} \\ &= P_2(a_0I_2)P_2^{-1} + P_2(a_1I_2)P_2^{-1}e_{123} + P_2(a_2I_2)P_2^{-1}e_{124} + P_2(a_3I_2)P_2^{-1}e_{43} \\ &= \phi_2(a_0) + \phi_2(a_1)e_{123} + \phi_2(a_2)e_{124} + \phi_2(a_3)e_{43} \\ &:= \psi(a). \end{aligned}$$

Next building, according to Lemma 3, a matrix and its inverse as follows

$$V = V^{-1} = \frac{1}{2}\begin{bmatrix} (1+e_{123})I_2 & (e_{124}-e_{43})I_2 \\ (e_{124}+e_{43})I_2 & (1-e_{123})I_2 \end{bmatrix},$$

and applying them to $\psi(a)$ given above, we obtain

$$V\begin{bmatrix} \psi(a) & 0 \\ 0 & \psi(a) \end{bmatrix}V^{-1} = \begin{bmatrix} \phi_2(a_0)+\phi_2(a_1) & \phi_2(a_2)+\phi_2(a_3) \\ \phi_2(a_2)-\phi_2(a_3) & \phi_2(a_0)-\phi_2(a_1) \end{bmatrix}.$$

Finally substituting $\psi(a) = P_2(aI_2)P_2^{-1}$ into its left-hand side yields (2.18) and (2.19). □

By induction, we have the following two general results.

Theorem 7. *Suppose that there is an independent invertible matrix P_n over $\mathcal{C}_n = \mathcal{C}\{e_1, \ldots, e_n\}$ with n even such that*

$$P_n(aI_{2^{\frac{n}{2}}})P_n^{-1} := \phi_n(a) \in \mathcal{C}^{2^{\frac{n}{2}} \times 2^{\frac{n}{2}}} = \mathcal{C}(2^{\frac{n}{2}}) \tag{2.20}$$

holds for all $a \in \mathcal{C}_n$. Now let $a \in \mathcal{C}_{n+1} = \mathcal{C}\{e_1, \ldots, e_{n+1}\}$. Then a can factor as

$$a = a_0 + a_1e_{[n+1]} = a_0 + e_{[n+1]}a_1, \tag{2.21}$$

where

$$a_0, a_1 \in \mathcal{C}_n = \mathcal{C}\{e_1, \ldots, e_n\}, \quad e_{[n+1]}^2 = (-1)^{\frac{1}{2}(n+1)(n+2)} := r.$$

Moreover define $\overline{a} = a_0 - a_1e_{[n+1]}$. In that case, $D_a = \mathrm{diag}(aI_{2^{\frac{n}{2}}}, \overline{a}I_{2^{\frac{n}{2}}})$ satisfies the following universal similarity factorization equality

$$\begin{aligned} P_{n+1}D_aP_{n+1}^{-1} &= \begin{bmatrix} \phi_n(a_0)+\sqrt{r}\phi_n(a_1) & 0 \\ 0 & \phi_n(a_0)-\sqrt{r}\phi_n(a_1) \end{bmatrix} \\ &:= \phi_{n+1}(a) \in {}^2\mathcal{C}(2^{\frac{n}{2}}), \end{aligned} \tag{2.22}$$

where

$$P_{n+1} = \frac{1}{2}\begin{bmatrix} (1+\frac{1}{\sqrt{r}}e_{[n+1]})P_n & -(\sqrt{r}-e_{[n+1]})P_n \\ \frac{1}{r}(\sqrt{r}-e_{[n+1]})P_n & (1+\frac{1}{\sqrt{r}}e_{[n+1]})P_n \end{bmatrix}, \tag{2.23}$$

and

$$P_{n+1}^{-1} = \frac{1}{2}\begin{bmatrix} P_n^{-1}(1+\frac{1}{\sqrt{r}}e_{[n+1]}) & P_n^{-1}(\sqrt{r}-e_{[n+1]}) \\ -P_n^{-1}\frac{1}{r}(\sqrt{r}-e_{[n+1]}) & P_n^{-1}(1+\frac{1}{\sqrt{r}}e_{[n+1]}) \end{bmatrix}. \tag{2.24}$$

Proof. Note that the commutative rule $be_{[n+1]} = e_{[n+1]}b$ holds for all $b \in \mathcal{C}_n = \mathcal{C}\{e_1, \ldots, e_n\}$. By applying (2.20) to (2.21), we obtain

$$\begin{aligned} P_n(aI_{2^{\frac{n}{2}}})P_n^{-1} &= P_n(aI_{2^{\frac{n}{2}}})P_n^{-1} + P_n(a_1I_{2^{\frac{n}{2}}})P_n^{-1}e_{[n+1]} \\ &= \phi_n(a_0) + \phi_n(a_1)e_{[n+1]} := \psi(a). \end{aligned}$$

and

$$P_n(\overline{a}I_{2^{\frac{n}{2}}})P_n^{-1} = \phi_n(a_0) - \phi_n(a_1)e_{[n+1]} := \psi(\overline{a}).$$

Next, setting

$$V = \frac{1}{2}\begin{bmatrix} (1+\frac{1}{\sqrt{r}}e_{[n+1]})I_{2^{\frac{n}{2}}} & -(\sqrt{r}-e_{[n+1]})I_{2^{\frac{n}{2}}} \\ \frac{1}{\sqrt{r}}(\sqrt{r}-e_{[n+1]})I_{2^{\frac{n}{2}}} & (1+\frac{1}{\sqrt{r}}e_{[n+1]})I_{2^{\frac{n}{2}}} \end{bmatrix},$$

$$V^{-1} = \frac{1}{2}\begin{bmatrix} (1+\frac{1}{\sqrt{r}}e_{[n+1]})I_{2^{\frac{n}{2}}} & (\sqrt{r}-e_{[n+1]})I_{2^{\frac{n}{2}}} \\ -\frac{1}{r}(\sqrt{r}-e_{[n+1]})I_{2^{\frac{n}{2}}} & (1+\frac{1}{\sqrt{r}}e_{[n+1]})I_{2^{\frac{n}{2}}} \end{bmatrix},$$

and applying them to $D_a = \mathrm{diag}(\psi(a), \psi(\overline{a}))$ we get

$$V\begin{bmatrix} \psi(a) & 0 \\ 0 & \psi(\overline{a}) \end{bmatrix} = \begin{bmatrix} \phi_n(a_0)+\sqrt{r}\phi_n(a_1) & 0 \\ 0 & \phi_n(a_0)-\sqrt{r}\phi_n(a_1) \end{bmatrix}.$$

Finally, substituting $\psi(a) = P_n(aI_{2^{\frac{n}{2}}})P_n^{-1}$ and $\psi(\overline{a}) = P_n(\overline{a}I_{2^{\frac{n}{2}}})P_n^{-1}$ into its left-hand side yields the desired results. □

Theorem 8. *Suppose that there is an independent invertible matrix P_n over $\mathcal{C}_n = \mathcal{C}\{e_1, \ldots, e_n\}$ with n even such that*

$$P_n(aI_{2^{\frac{n}{2}}})P_n^{-1} := \phi_n(a) \in \mathcal{C}(2^{\frac{n}{2}}) \tag{2.25}$$

holds for all $a \in \mathcal{C}_n$. Now, let $a \in \mathcal{C}_{n+2} = \mathcal{C}\{e_1, \ldots, e_{n+2}\}$. Then it can factor as

$$a = a_0 + a_1e_{[n]}e_{n+1} + a_2e_{[n]}e_{n+2} + a_3\mu_{n+2}, \tag{2.26}$$

where

$$a_0, a_1, a_2, a_3 \in \mathcal{C}_n = \mathcal{C}\{e_1, \ldots, e_n\},$$
$$(e_{[n]}e_{n+1})^2 = (e_{[n]}e_{n+2})^2 = (-1)^{\frac{1}{2}(n+1)(n+2)} = r,$$
$$\mu_{n+2} = (e_{[n]}e_{n+1})(e_{[n]}e_{n+2}) = -(e_{[n]}e_{n+2})(e_{[n]}e_{n+1}).$$

In that case, $aI_{2^{\frac{n+1}{2}}}$ *satisfies the following universal similarity factorization equality*

$$P_{n+2}(aI_{2^{\frac{n+1}{2}}})P_{n+2}^{-1} = \begin{bmatrix} \phi_n(a_0) + \sqrt{r}\phi_n(a_1) & r[\phi_n(a_2) + \sqrt{r}\phi_n(a_3)] \\ \phi_n(a_2) - \sqrt{r}\phi_n(a_3) & \phi_n(a_0) - \sqrt{r}\phi_n(a_1) \end{bmatrix}$$
$$:= \phi_{n+2}(a) \in \mathcal{C}(2^{\frac{n+1}{2}}),$$

where

$$P_{n+2} = \frac{1}{2}\begin{bmatrix} (1 + \frac{1}{\sqrt{r}}e_{[n+1]})P_n & (e_{[n]}e_{n+2} - \frac{1}{\sqrt{r}}\mu_{n+2})P_n \\ \frac{1}{r}(e_{[n]}e_{n+2} + \frac{1}{\sqrt{r}}\mu_{n+2})P_n & (1 - \frac{1}{\sqrt{r}}e_{[n+1]})P_n \end{bmatrix},$$

$$P_{n+2}^{-1} = \frac{1}{2}\begin{bmatrix} P_n^{-1}(1 + \frac{1}{\sqrt{r}}e_{[n+1]}) & P_n^{-1}(e_{[n]}e_{n+2} - \frac{1}{\sqrt{r}}\mu_{n+2}) \\ P_n^{-1}\frac{1}{r}(e_{[n]}e_{n+2} + \frac{1}{\sqrt{r}}\mu_{n+2}) & P_n^{-1}(1 - \frac{1}{\sqrt{r}}e_{[n+1]}) \end{bmatrix}.$$

The proof of this result is analogous to that of Theorem 6, and is therefore omitted here.

3 Conclusions

In this article we have established a set of universal similarity factorization equalities for elements over the complex Clifford algebra $\mathcal{C}_n$. These equalities reveal two basic facts about $\mathcal{C}_n$:

(i) Each element a in $\mathcal{C}_n$ has a complex matrix representation $\phi_n(a)$. Moreover, a diagonal matrix constructed by the element a is uniformly similar to its complex matrix representation $\phi_n(a)$.

(ii) Conversely, each element a in $\mathcal{C}_n$ could be regarded as an eigenvalue of its complex representation matrix $\phi_n(a)$. In other words, all complex matrices with the form $\phi_n(a)$ can uniformly be diagonalized over $\mathcal{C}_n$.

Based on the above two facts, one easily see that almost all known results in complex matrix theory can be extended to complex Clifford algebras. On the other hand, some problems related to complex matrices can also be transformed to the problems related to Clifford numbers. One such problem (see [1]) is concerning the exponential e^A of a complex matrix A. In fact, we can see from Theorems 14 and 15 that for any $a \in \mathcal{C}_n$, there is

$$P_n(e^a I)P_n^{-1} = e^{\phi_n(a)},$$

or

$$P_n \mathrm{diag}(e^a I, e^{\overline{a}} I)P_n^{-1} = e^{\phi_n(a)},$$

which implies that the exponential of a complex matrix $\phi_n(a)$ can be determined by the exponential of its corresponding Clifford numbers.

References

[1] R. Abłamowicz, Matrix exponential via Clifford algebras, *J. Nonlinear Math. Physics* **5** (1998), 294–313.

[2] M. F. Atiyah, R. Bott, and A. Shapiro, Clifford modules, *Topology* **3** (1961), 3–38.

[3] G. N. Hile and P. Lounesto, Matrix representations of Clifford algebras, *Linear Algebra Appl.* **128** (1990), 51–63.

[4] T. Y. Lam, *The Algebraic Theory of Quadratic Forms,* W. A. Benjamin, Reading, Mass., 1973.

[5] M. Karoubi, *K-Theory,* Springer-Verlag, Berlin, 1978.

[6] M. Marcus, *Finite Dimensional Multilinear Algebra (Part II),* Marcel Dekker Inc., New York, 1975.

[7] S. Okubo, Real representations of finite Clifford algebras I, classification, *J. Math. Phys.* **32** (1991), 1657–1668.

[8] S. Okubo, Real representations of finite Clifford algebras II, explicit construction and pseudo-octonion, *J. Math. Phys.* **32** (1991), 1669–1673.

[9] S. Okubo, Representations of Clifford algebras and its applications, *Math. Japn.* **41** (1995), 59–79.

[10] I. R. Porteous, *Topological Geometry,* Cambridge U. P., Cambridge, 1981.

[11] I. R. Porteous, *Clifford Algebras and the Classical Groups,* Cambridge U. P., Cambridge, 1995.

[12] Y. Tian, Universal similarity factorization equalities over real Clifford algebras, *Adv. Appl. Clifford Algebras* **8** (1998), 365–402.

[13] V. V. Varlamov, Modulo 2 periodicity of complex Clifford algebras and electromagnetic field, preprint, 1997.

Yongge Tian
Department of Mathematics and Statistics
Queen's University
Kingston, Ontario, Canada K7L 3N6
E-mail: ytian@mast.queensu.ca

Received: August 6, 1999; Revised: October 16, 1999

Index

B.

C.

D.

E.

H.

I.

J.

N.

O.

P.

Q.

T.

U.

V.

W.

Y.

Z.